T0405495

Semiconductor Power Devices

Josef Lutz · Heinrich Schlangenotto ·
Uwe Scheuermann · Rik De Doncker

Semiconductor Power Devices

Physics, Characteristics, Reliability

Springer

Prof. Dr. -Ing. Prof. h.c. Josef Lutz
Chemnitz University of Technology
Faculty of ET/IT
Chair Power Electronics and
Electromagnetic Compatibility
Reichenhainer Str. 70
D-09126 Chemnitz
Germany
josef.lutz@etit.tu-chemnitz.de

Prof. Dr. Heinrich Schlangenotto
Stoltzestr. 86
D-63262 Neu-Isenburg
Germany
hschlangenotto@arcor.de

Dr. Uwe Scheuermann
Semikron Elektronik GmbH & Co. KG
Sigmundstr. 200
D-90431 Nürnberg
Germany
uwe.scheuermann@semikron.com

Prof. Dr. ir. Rik De Doncker
RWTH Aachen University
Faculty of ET&IT
Chair Power Generation and Storage
Systems (PGS) at E.ON ERC
Mathieustrasse 6
D-52074 Aachen
Germany
dedoncker@rwth-aachen.de

ISBN 978-3-642-11124-2 e-ISBN 978-3-642-11125-9
DOI 10.1007/978-3-642-11125-9
Springer Heidelberg Dordrecht London New York

Library of Congress Control Number: 2010934110

Cover design: eStudio Calamar S.L., Heidelberg

Printed on acid-free paper

Springer is part of Springer Science+Business Media (www.springer.com)

Preface

Power electronics is gaining more and more importance in industry and society. It has the potential to substantially increase the efficiency of power systems, a task of great significance. To exploit this potential, not only engineers working in the development of improved and new devices but also application engineers in the field of power electronics need to understand the basic principles of semiconductor power devices. Furthermore, since a semiconductor device can only fulfil its function in a suitable environment, interconnection and packaging technologies with the related material properties have to be considered as well as the problem of cooling, which has to be solved for reliable applications.

This book was written for students and for engineers working in the field of power device design and power electronics application. The focus was set on modern semiconductor switches such as power MOSFETs and IGBTs together with the essential freewheeling diodes. The engineer in practice may start his work with the book with the specific power device. Each chapter presents first the device structure and the generic characteristics and then a more thorough discussion is added with the focus on the physical function principles. The in-depth discussions require the principles of semiconductor physics, the functioning of pn-junctions, and the basics of technology. These topics are treated in depth such that the book will also be of value for the semiconductor device specialist.

Some subjects are treated in particular detail and presented here for the first time in an English textbook on power devices. In device physics, this is especially the emitter recombination which is used in modern power devices to control forward and switching properties. A detailed discussion of its influence is given using parameters characterizing the emitter recombination properties. Furthermore, because of the growing awareness of the importance of packaging technique for reliable applications, chapters on packaging and reliability are included. During the development of power electronic systems, engineers often are confronted with failures and unexpected effects with the consequence of time-consuming efforts to isolate the root cause of these effects. Therefore, chapters on failure mechanisms and oscillation effects in power circuits are included in this textbook to supply guidance based on long-time experience.

The book has emerged from lectures on "Power devices" held by J. Lutz at Chemnitz University of Technology and from earlier lecture notes on "Power devices" from H. Schlangenotto held at Darmstadt Technical University in 1991–2001. Using these lectures and adding considerable material on new devices, packaging, reliability, and failure mechanisms, Lutz published the German book *Halbleiter-Leistungsbauelemente – Physik, Eigenschaften, Zuverlässigkeit* in 2006. The English textbook presented here is far more than a translation; it was considerably extended with new material.

The basic chapters on semiconductor properties and pn-junctions and a part of the chapter on pin-diodes were revised and enhanced widely by H. Schlangenotto. J. Lutz extended the chapters on thyristors, MOSFETs, IGBTs, and failure mechanisms. U. Scheuermann contributed the chapter on packaging technology, reliability, and system integration. R. De Doncker supplied the introduction on power devices as the key components. All the authors have contributed, however, also to other chapters not written mainly by themselves.

Several researchers in power devices have supported this work with helpful discussions, support in translations, suggestions, and comments. These are especially Arnost Kopta, Stefan Linder and Munaf Rahimo from ABB Semiconductors, Dieter Polenov from BMW, Thomas Laska, Anton Mauder, Franz-Josef Niedernostheide, Ralf Siemieniec, and Gerald Soelkner from Infineon, Martin Domeij and Anders Hallén from KTH Stockholm, Stephane Lefebvre from SATIE, Michael Reschke from Secos, Reinhard Herzer and Werner Tursky from Semikron, Wolfgang Bartsch from SiCED, Dieter Silber from University of Bremen, Hans Günter Eckel from the University of Rostock. Several diploma and Ph.D. students at Chemnitz University of Technology have supported part of the work, especially Hans-Peter Felsl, Birk Heinze, Roman Baburske, Marco Bohlländer, Tilo Pollera Matthias Baumann, and Thomas Basler. Thomas Blum and Florian Mura from RWTH Aachen have translated the chapter on MOSFETS, and Mary-Joan Blümich has given support with improvements of the English text. Finally, the authors thank the many other researchers and students in power electronics, who supported this work with critical comments and discussions.

Chemnitz, Germany
Neu-Isenburg, Germany
Nürnberg, Germany
Aachen, Germany
March 2010

Josef Lutz
Heinrich Schlangenotto
Uwe Scheuermann
Rik De Doncker

Contents

Chapter 1
Power Semiconductor Devices – Key Components for Efficient Electrical Energy Conversion Systems

1.1 Systems, Power Converters, and Power Semiconductor Devices

In a competitive market, technical systems rely on automation and process control to improve their productivity. Initially, these productivity gains were focused on attaining higher production volumes or less (human) labor-intensive processes to save costs. Today, attention is paid toward energy efficiency because of a global awareness of climate change and, above all, questions related to increasing energy prices, as well as security of energy and increasing urbanization. Consequently, it is expected that the trend toward more electrical systems will continue and accelerate over the next decades. As a result, the need to efficiently process electrical energy will dramatically increase.

Devices that are capable of converting electrical energy from one form into another, i.e. transforming electrical energy, have been a major breakthrough technology since the beginning of electrical power systems and are considered key enabling technologies. For example, without transformers, large-scale power generation, transmission, and distribution of electrical power would not have been possible. Interestingly, very few people today are aware that without this invention, initially called secondary generator [Jon04], we would not have been able to create such an efficient, safe, and (locally) environmentally clean power source. Of course, as transformers, or generally speaking electromagnetic devices, can only transform voltage or control reactances, their use in automation systems remained limited. At the beginning of electrification, frequency and phase control could only be realized using electromechanical conversion devices (i.e. motors, generators). However, these machines were bulky, required maintenance, had high losses, and remained expensive. Furthermore, these electromechanical devices had rather low control bandwidth. Therefore, they operated mostly at fixed set points. Today, most automation and process control systems require more flexible energy conversion means to vary dynamically voltage or to regulate current, frequency, phase angle, etc.

At present, power electronics is the most advanced electrical energy conversion technology that attains both high flexibility and efficiency. As an engineering field, power electronics came into existence about 50 years ago, with the development and

J. Lutz et al., *Semiconductor Power Devices*, DOI 10.1007/978-3-642-11125-9_1,

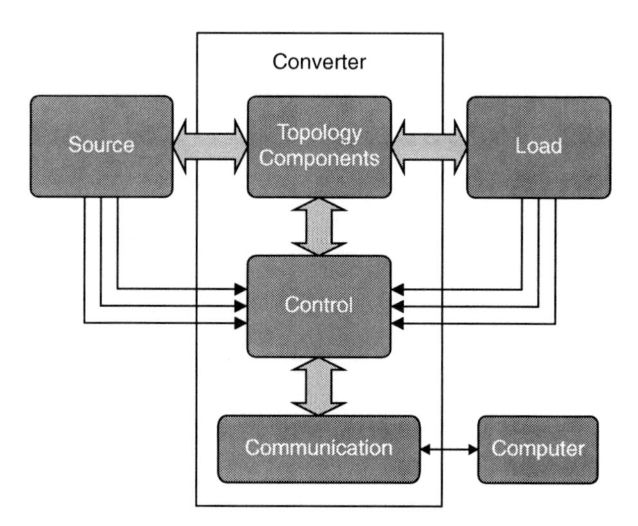

Fig. 1.1 Power electronic systems convert and control electrical energy in an efficient manner between a source and a load. Sensor interfaces to the source and load, as well as information and communication links, are often integrated

the market introduction of the so-called silicon controlled rectifier, known today as the thyristor [Owe07, Hol01]. Clearly, power electronics and power semiconductor devices are closely intertwined fields. Indeed, in its *Operations Handbook*, the IEEE Power Electronics Society defines the field of power electronics as *"This technology encompasses the effective use of electronic components, the application of circuit theory and design techniques, and the development of analytical tools toward efficient electronic conversion, control, and conditioning of electric power"* [PEL05].

Simply stated, *a power electronics system is an efficient energy conversion means using power semiconductor devices.* A power electronics system can be illustrated with the block diagram shown in Fig. 1.1.

A special class of power electronic systems are electrical drives. A block diagram of an electrical drive is illustrated in Fig. 1.2. Electrical drives are used in propulsion systems, power generation (wind turbines), industrial and commercial drives, for example, in heating ventilation and air conditioning systems, and in motion control. In an electrical drive, the control of the electromechanical energy converter, the latter being a highly sophisticated load from a control perspective, is integrated in the power electronics converter control. Most research institutions that deal with power electronic converter technology also work on electrical drive technology, because this field still represents one of the largest application areas (expressed in installed apparent power) of power electronic converters [Ded06]. In the near future, despite the increased use of power converters in photovoltaic systems and computer power supplies, it is expected that this dominance of drives will remain. Most experts predict that the existing industrial markets for drives will continue to grow and will be complemented by newly developed markets, such as wind turbines, more electric ships and aircrafts, and electric mobility, i.e. trains, trams, trolley busses, automobiles, scooters, and bikes.

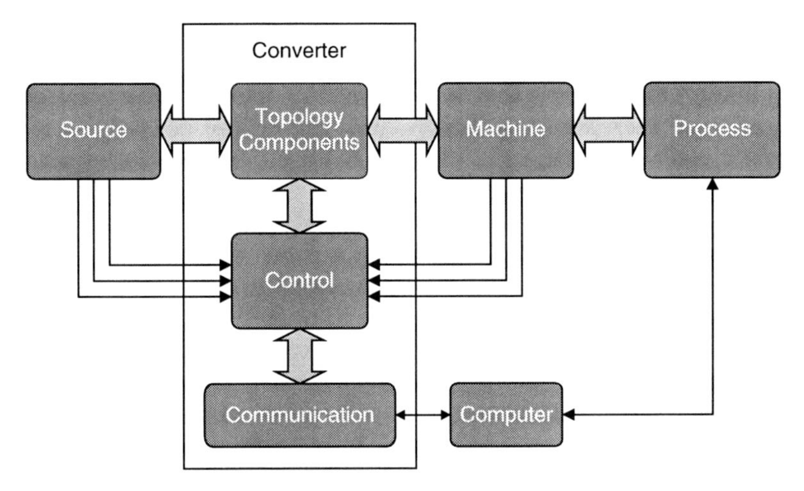

Fig. 1.2 Highly dynamic electrical drives systems comprise power electronic converters and electrical machines or actuators with dedicated control to convert electrical energy into mechanical motion

1.1.1 Basic Principles of Power Converters

Looking into the generic power electronic converter block diagram of Fig. 1.1, more details can be revealed when considering the operating principles and the topology of a modern power electronic converter. Basically, to make power electronic converters work, three types of components are needed:

- Active components, i.e. the power semiconductor components that turn on and off the power flow within the converter. The devices are either in the off-state (forward or reverse blocking) or in the on-state (conducting).
- Passive components, i.e. transformers, inductors, and capacitors, which temporarily store energy within the converter system. Based on the operating frequency, voltage, cooling method, and level of integration, different magnetic, dielectric and insulation materials are used. For a given power rating of the converter, higher operating (switching) frequencies enable smaller passive components.
- Control unit, i.e. analog and digital electronics, signal converters, processors, and sensors, to control the energy flow within the converter such that the internal variables (voltage, current) follow computed reference signals that guarantee proper behavior of the converter according to the external commands (that are obtained via a digital communication link). Today, most control units also provide status and system level diagnostics.

As power electronic converters ought to convert electrical energy efficiently (efficiencies above 95%), linear operation of power devices is no option. Rather, the devices are operated in a switching mode. Hence, in the power supply area, to make this distinction, power converters are called "switched-mode power supplies." The basic idea behind all power converters to control and convert the electrical energy flowing through the converter is to break down this continuous flow of energy into small packets of energy, process these packets, and deliver the energy in another, but again a continuous, format at the output. Hence, power converters are true power processors! In doing so, all converter topologies must respect fundamental circuit theory principles. Most importantly, the principle that electrical energy can only be exchanged efficiently via a switching network when energy is exchanged between dual components, i.e. energy stored in capacitors or voltage sources, should be transferred to inductors or current sources.

As described in guidelines and standards, for example, IEEE 519-1992 [IEE92] and IEC 61000-3-6 [IEC08], and to protect sources and loads, the energy flow at the input and at the output of the converter has to be continuous and substantially free from harmonics and electromagnetic noise. To make the energy flow continuous, filter components are necessary. Note that in many applications these filter components can be part of the source or the load. To minimize cost of filter components, to comply with international standards, and to improve efficiency, the control units of inverters, DC-to-DC converters, and rectifiers tend to switch the power devices at constant switching frequency, using pulse width modulation (PWM) techniques, sometimes called duty cycle control. Basic circuit theory and component design proves that higher switching frequencies will lead to smaller passive elements and filter components. Hence, all converter designs strive to increase switching frequencies to minimize overall converter costs. However, as will be discussed in the next sections, higher switching frequencies impact converter efficiency. As a result, a balance has to be found between investment material and production costs and efficiency. Note that efficiency also determines the energy costs of the conversion process over the entire life span of the converter.

1.1.2 Types of Power Converters and Selection of Power Devices

Power electronic converters can be categorized in various ways. Today, with power electronics, it is possible to convert electrical energy from AC to DC (rectifier), from DC to DC (DC-to-DC converter), and from DC back to AC (inverter).

Although some converters can convert AC directly to AC (matrix converter and cycloconverters), most AC-to-AC conversion is done using a series connection of a rectifier and an inverter. Hence, as shown in Fig. 1.3, most converters possess at least one DC link, where the energy is temporarily stored between the different conversion stages. Based on the type of DC link used, the converters can be divided into current source and voltage source converters. Current source converters use an inductor to store the energy magnetically and operate with near-constant current in the DC link. Their dual, i.e. the voltage source converter, uses a capacitor to keep the DC voltage constant.

In case of AC supplies and loads, the converter could take advantage of the fact that the fundamental component of the line current or the load current crosses zero.

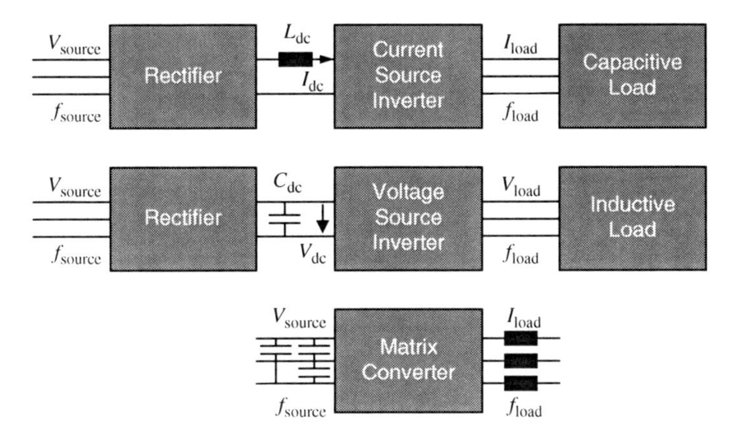

Fig. 1.3 DC-Link converters and matrix converters can convert electrical power between (three-phase) AC supplies and loads. Most converters use a combination of rectifier and inverter

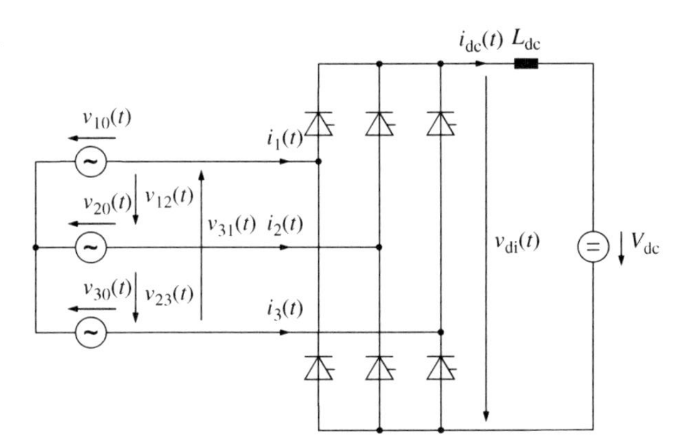

Fig. 1.4 Elementary diagram of line-commutated rectifier circuit based on thyristors

These converters are called line-commutated or load-commutated converters and are still common in controlled rectifiers and high-power resonant converters as well as synchronous machine drives, using thyristors. A three-phase bridge rectifier is shown in Fig. 1.4. Detailed analysis shows that these converters produce line side harmonics and cause considerable lagging reactive power [Moh02]. Consequently, large filters and reactive power (so-called VAR-) compensation circuits are needed to maintain high power quality. As these filters cause losses and represent a substantial investment cost, line-commutated (rectifier) circuits are slowly phased out in favor of forced commutated circuits that use active turn-off power semiconductor devices, i.e. power transistors (MOSFET, IGBT) or turn-off thyristors (GTO, IGCT). Active rectifier circuits (actually inverters operating in rectifying mode) can eliminate the need for VAR compensators and reduce or eliminate harmonic filter components.

However, not only the type of converter (rectifier, inverter, or DC-to-DC converter), but also the type of topology selected (voltage source or current source)

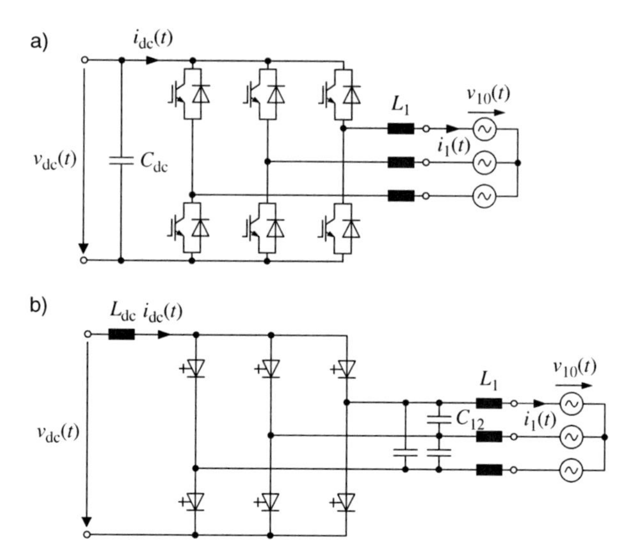

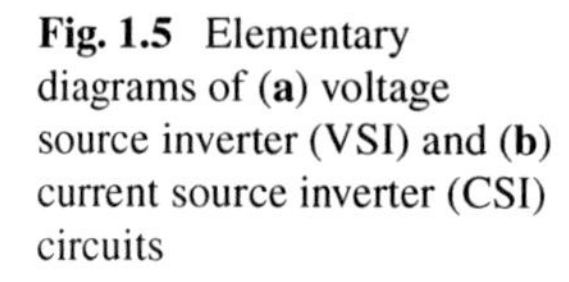

Fig. 1.5 Elementary diagrams of (**a**) voltage source inverter (VSI) and (**b**) current source inverter (CSI) circuits

has a profound impact on the characteristics and the type of semiconductor devices that are required. A three-phase current source inverter and a three-phase voltage source inverter are illustrated in Fig. 1.5 to point out the operating differences of the devices.

In current source converters, the devices need to have forward and reverse blocking capability. These devices are called symmetrical voltage blocking devices. Although symmetrical blocking turn-off devices do exist, in practice, the reverse blocking capability is often realized by connecting or integrating a diode in series with the active turn-off semiconductor switch (transistor or turn-off thyristor). Hence, in this case, higher conduction losses must be tolerated as compared to asymmetric blocking devices. As will be shown in this book, the physics of power semiconductor switches leads to the fact that the design of symmetrical blocking turn-off devices (with integrated reverse blocking pn-junction) somehow relates to thyristor-based structures (see Chap. 8 and Sect. 10.7). As these devices are more suitable for high-power applications (voltages above 2.5 kV), some high-power converter manufacturers still use symmetrical (GCT) devices in high-power (above 10 MVA) current source converters [Zar01]. The main advantage of such converters is the fact that a current source converter is fault tolerant against internal and external short circuits.

Voltage source converters require a reverse conducting device because they inevitably drive inductive loads at their AC terminals. Hence, to avoid voltage spikes, when a device turns off current, a freewheeling path is needed. This reverse conduction or freewheeling capability of semiconductor switches can be realized by connecting or integrating a diode anti-parallel to the turn-off device. As this additional junction is not in series with the main turn-off device, no additional voltage drop occurs in the current path of the converter. Hence, with the present state of device technology, voltage source converters tend to be more efficient than current source converters, especially at partial load conditions [Wun03]. Indeed, at partial load the current source converter still has a high circulating current in the DC link of the converter, while the voltage source operates at reduced current, even when the DC-link capacitor carries full voltage.

In practice, due to the lower losses in the DC-link capacitor as compared to the DC-link inductor, the size of a voltage source converter can become considerably smaller than that of a current source converter. In addition, most loads and sources behave inductively (at the switching frequency). Hence, voltage source converters may not require additional impedances or filters, while a current source converter requires capacitors at its output terminals. Taking all these engineering considerations into account, one can understand the growing importance of voltage source converters as compared to current source converters. The device manufacturers have responded to this growing market by optimizing far better asymmetric transistors and thyristors with respect to conduction and switching losses, which led to considerable efficiency improvements and less cooling costs. Furthermore, most voltage source converter topologies use the two-level phase leg configuration that was shown in Fig. 1.5a. This phase leg topology has become so universal that device manufacturers offer complete phase legs integrated in single modules as elementary building blocks, called *power electronic building blocks* (PEBBs), thereby reducing manufacturing cost and improving reliability (see Chap. 14). As power electronics is becoming a mature technology, one can state that in the near future most new converter designs (with ratings from few millivolt amperes to several gigavolt amperes) will be voltage source-type converters.

1.2 Operating and Selecting Power Semiconductors

When designing a power converter, many details need to be considered to achieve the design goals. Typical design specifications are low cost, high efficiency, or high power density (low weight, small size). Ultimately, thermal considerations, i.e. device losses, cooling, and maximum operating temperature, determine the physical limits of a converter design. When devices are operated within their (electrical) safe operation area (SOA), conduction and switching losses dominate device losses. The underlying physics of these losses are described and analyzed in this book. However, it should be noted that the converter designer can substantially minimize these losses by making proper design decisions. In general, the outcome of the design depends greatly on the selection of the following:

- device type (unipolar, bipolar, transistor, thyristor) and rating (voltage and current margins, frequency range)
- switching frequency
- converter layout (minimizing parasitic stray inductances, capacitances, and skin effects)
- topology (two level, multi-level, hard switching, or soft switching)

- gate control (switching slew rate)
- control (switching functions, minimizing filters, EMI)

Furthermore, as losses cannot be avoided, the design of the cooling system (liquid cooling, air cooling) has a strong impact on the selection of the type of packaging. Several types of packaging technologies are currently available in the market: discrete, module, and press-type packages. Whereas the discrete and module packages can be electrically insulated, which allows all devices of a converter to be mounted on one heat sink, the press-type packages can be cooled on both sides. Typically, discrete devices are used in switched mode power supplies (up to 10 kW). Higher power levels, up to 1 MW, require parallel connection of multiple semiconductor chips and make use of the module-type package, while double-sided cooled packages (single wafer disc-type designs or multiple-chip presspacks) are used at the highest power levels up to several gigawatts. Details of system architecture are discussed in Chap. 11.

As already stated, the device switching power (product of maximum blocking voltage and repetitive turn-off current) and its maximum switching frequency are important criteria for a first selection of a power semiconductor in many applications. Next to this theoretical application limit, the practical application range of silicon devices depends also on cooling limits and economical factors.

At present several device structures have been developed, each offering specific advantages. The structures of today's most important power semiconductor devices are shown in Fig. 1.6. However, it is worth mentioning that in modern applications classical bipolar transistors have been superseded by IGBTs (insulated gate bipolar transistors), being basically a MOS-controlled bipolar device. Details on each of the devices in Fig. 1.6 will be given in Chaps. 5, 6, 7, 8, 9, and 10.

As production of silicon devices has made great progress over the past 50 years, the application range of silicon devices has expanded and became better understood. Figure 1.7 illustrates the practical application range of each type of silicon device in classical power converters (rectifiers and hard-switching power converters).

Note that for these applications the operation ranges are within a hyperboloid. In other words, the product of switching power (product of maximum voltage and current) and switching frequency that can be attained per device in practical conversion systems using silicon devices, assuming classical hard-switching converter configurations and similar type of cooling, appears to be fairly constant:

$$P_{\mathrm{sw}} - \mathrm{hard} \cdot f_{\mathrm{sw}} = V_{\max} - \mathrm{hard} \cdot I_{\max} - \mathrm{hard} \cdot f_{\mathrm{sw}} \approx 10^9 \ \mathrm{V\,A/s}$$

This frequency–power product is a good performance indicator for how well the designer was able to maximize utilization of the power semiconductors and to improve the power density of the converter. Indeed, as pointed out earlier, increasing switching frequency also reduces size of transformers, machines, and filter components (at constant apparent power). Actually, if passive components of the same

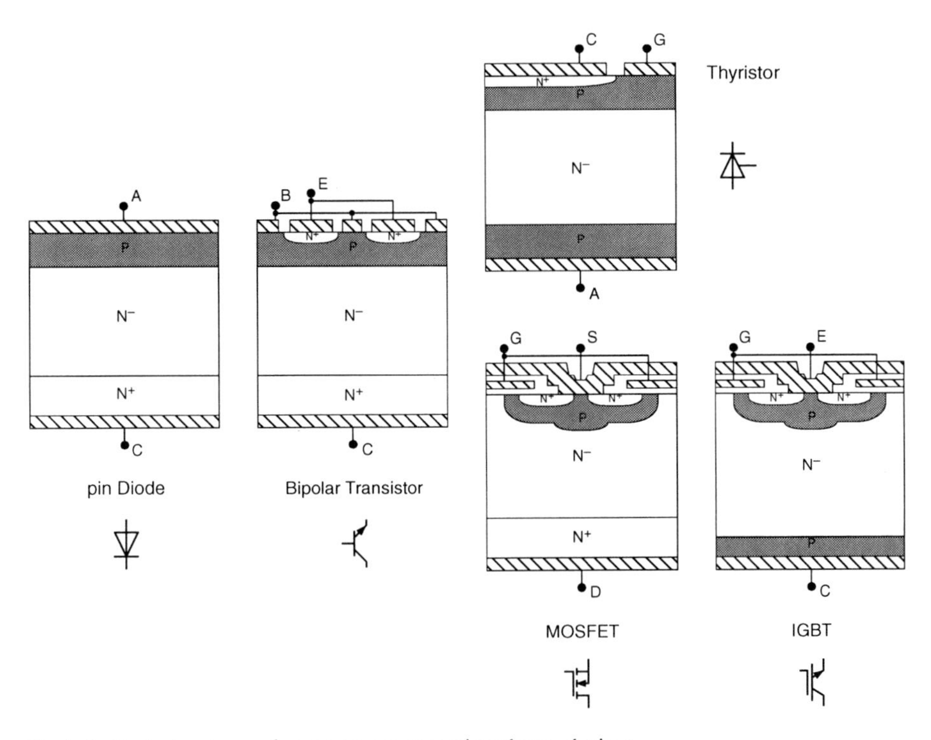

Fig. 1.6 Basic structures of common power semiconductor devices

type (electromagnetic or electrostatic) are being considered, for example, inductors, transformers, and machines, they also experience a similar frequency–power product barrier as they use the same materials (copper, silicon steel, and insulation materials), operating at same maximum temperatures.

To reduce switching losses, soft-switching converters or wide bandgap materials, such as silicon carbide (SiC) devices, can break this technology barrier. For example, soft-switching resonant converters are being applied successfully in switched-mode power supplies and DC-to-DC converters. In soft-switching converters, not only the switching losses of the turn-off devices are being reduced, but also the reverse recovery losses of the power diodes are mostly eliminated. As will be shown in this book, reverse recovery effects in power diodes not only increase switching losses but also are a root cause for high HF noise (EMI) in converters. To limit these EMI effects, designers are forced to slow down switching transients, which leads to higher switching losses in hard-switching converters. As soft-switching converters utilize resonant snubber techniques, these losses do not occur and the switching frequency can be increased or, alternatively, the output power of the converter may be augmented. Soft-switching (resonant or transition-resonant) converters typically improve the frequency–power product by a factor of up to 5:

$$P_{\mathrm{sw}} - \mathrm{soft} \cdot f_{\mathrm{sw}} \approx 5 \times 10^{9}\ \mathrm{V\,A/s}$$

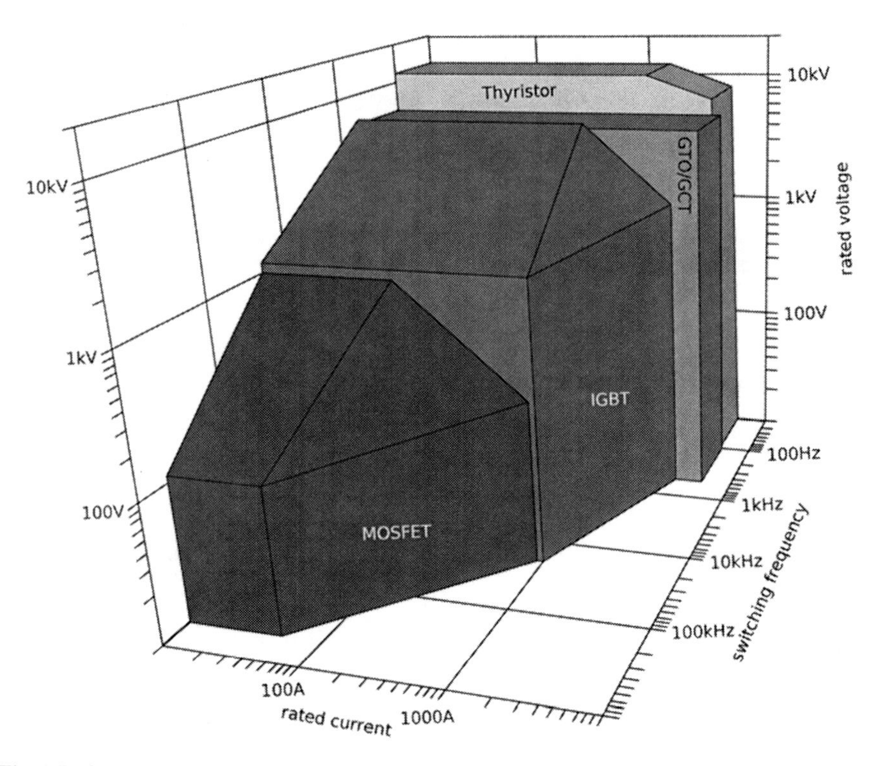

Fig. 1.7 Operating range of silicon power semiconductor devices

As was stated, yet another approach to increase power density of converters is the use of SiC diodes. SiC diodes have near-zero reverse recovery current. Hence, the silicon turn-off devices can be operated with higher turn-on and turn-off slew rates. These hybrid silicon–SiC designs are currently under investigation as SiC diodes are becoming available at higher power levels. Combining this hybrid concept with high-frequency soft-switching principles, i.e. using the parasitic elements of the devices (capacitances) and packages (inductances) as resonant components, the highest power densities can be attained. These concepts already find their implementation in ultra compact power supplies. Also, high-power DC-to-DC converters start to make use of these principles. One can estimate that the frequency–power product of the silicon switches in these hybrid converters can become as high as 10^{10} V A/s.

1.3 Applications of Power Semiconductors

One can conclude that the field of power electronics and power semiconductors is still evolving at a rapid pace. Soon, all electric power will pass not only through copper, dielectric, or magnetic materials but also through semiconductors, because most applications require energy conversions or because increased efficiency is required in these energy conversion processes.

As was mentioned above, converters are being used over a wide power range, with ratings from milliwatts or mVA (technically speaking, it is more correct to use apparent power) up to gigawatts. Depending on the required voltage and current ratings of the power semiconductors, different types of power semiconductors are being used. At the low power end (1 VA up to 1 kVA), switched-mode power supplies for battery chargers, mostly for portable communication devices and power tools, as well as for electronic systems (audio, video, and controllers) and personal computer systems form a major global market. Pushed by legislation, these power supplies have steadily augmented efficiency by improving control and developing better power devices and passive components. Modern power supplies also have reduced standby losses. The trend is toward higher switching frequencies because less material is needed for filter components. Hence, most power supplies in this power range are using power MOSFET devices to convert electrical energy.

Another major market for power electronic systems is electronic ballasts in lighting systems. New energy-efficient light sources (fluorescent, gas discharge lamps, LED, OLED) require control and conversion of the electrical power to operate. The main challenge is to develop power electronic circuits that are cheap and that can be mass produced. Moreover, the overall life cycle assessment (to assess impact on environment) of light sources seems to favor more efficient lighting systems [Ste02]. New legislation in the EU will phase out incandescent light bulbs.

Drive applications span a power range from few 10 VA up to 100 MVA. In automotive applications, many small drives (100 VA up to 1 kVA) are fed from the onboard power source, nominally 12 or 24 V. Hence, MOSFET devices are most common in these applications. On the other hand, grid-connected drives have to cope with the different grid standard voltage levels. For example, single-phase systems for households in North America and power systems in the aircraft industry have 115 V (rms) phase voltage at 60 or 400 Hz, respectively. Higher power single-phase systems offer 230 V line to line. In Europe, single-phase systems are 230 V, while three-phase line-to-line voltages equal 400 V. Canada and the United States also have 460 V three-phase power systems. Typically, the highest low-voltage power systems have 660 V line voltage (IEC 60038 defines low-voltage systems up to 1000 V). To cope with all standards and to lower production costs, device manufacturers have settled on few voltage levels that cover most grid-connected applications (rectifiers and inverters). Consequently, power devices with breakdown voltages of 600, 1200, and 1700 V have been developed. As transistor-type devices offer short-circuit protection at low cost, IGBTs are predominantly being used in drives fed from power grids. Medium-voltage drives (grid voltage from 1000 V up to 36 kV) use, depending on drive rating, transistor (IGBTs), and turn-off thyristor (GTO or GCT)-type devices. Above 3 kV, i.e. at higher voltage and power ratings (above 5 MW), three-level converters [Nab81] based on GCTs seem to dominate the market. However, at very high power levels above 15 MW, load-commutated inverters (LCIs) using thyristors are still produced by some manufacturers, for example, in rolling mills and compressor drives [Wu08].

Drives in traction applications such as locomotives, trains, and trams also face many different voltage standards. In Europe, several DC (600, 1500, and 3000 V) and AC (16.7 and 50 Hz) systems are used. Older converter designs used thyristors to control torque of various types of machines (DC, synchronous, and asynchronous machines). Typically, one converter would drive multiple motors (multi-axle design). More and more, IGBT-based converters are being used and single-axle designs are preferred. Hence, the required rating of the converters in traction systems has gone down, which favors designs based on transistor-type devices. Most importantly, the load cycle capacity of the converter is essential for the required reliability, especially in traction applications. In this area, research is ongoing to improve device package and cooling system reliability to improve converter life cycle costs (for details see Chap. 14).

Yet another modern drive application at the lower power spectrum (10 W) is the electronic toothbrush. This household appliance is a true power electronics marvel. A switched-mode power supply transforms the power of the AC line (115 or 230 V phase voltage) to medium frequency (50 k Hz) AC power to allow a contactless energy transfer (via a split transformer core) to the handheld battery fed toothbrush. A rectifier converts the medium frequency to DC. A step-down converter regulates the charging current to the battery and the electronics. An electronic commutated brushless PM machine drives the mechanical gears that move the brush in a rocking motion. Note that the complexity of this toothbrush approaches that of an electric vehicle. At these power levels, control and power devices are highly integrated to make mass production possible at reasonable cost. However, often these low-power applications are precursors of what can be achieved with high-level integration at higher power levels in the future.

Power electronics is used in generator systems whenever constant speed operation of a turbine or an engine cannot be guaranteed. A typical application is maximum power point tracking of generators driven by combustion engines (10–1000 kW range). More recently, power generation with wind turbines is inverter driven. Power levels of wind turbines have grown from 50 kW in 1985 to 5.0 MW in 2004 [Ack05]. Wind turbine manufacturers expect off-shore wind turbines to reach 10 MW per unit in the future. These large units will be "full converter" units in contrast to the doubly fed generators systems that are currently mostly used in on-shore applications. Doubly fed generators (also called rotating transformers) use AC-to-AC converters that are rated typically lower than 60% of the turbine power. This solution tends to be economical advantageous when using low-voltage (400 V or 690 V) generators, up to 5.0 MW. Note that worldwide approximately 120 GVA of inverter apparent peak power has been installed in the last decade to satisfy the demand for wind power [WEA09].

Another high-power application is transport of electrical energy over long distances using high-voltage DC (HVDC) transmission. Classical HVDC systems use three-phase bridge-type rectifiers based on thyristors. Some variants use direct light-triggered thyristors, although the requirement of diagnostic status feedback (via a glass fiber, due to the high-voltage BIL (basic insulation level) requirements) often favors separate light-triggered thyristors or thyristors triggered using a classical gate driver (both methods use energy stored in the snubber capacitor to trigger the thyristor via a glass fiber). The first HVDC systems date from 1977 and are still in use. However, increasing power demand over long distances (mostly hydropower), for example, in the so-called BRIC countries (Brazil, Russia, India, and China) has given HVDC a new boost. HVDC technology is now operating with ±500 kV, delivering 3 GW of power, while new systems will operate at ±800 kV, transmitting 6 GW [Ast05]. These transmission systems are current source-type converters and are designed to deliver power from point to point. Voltage source-type transmission systems are being implemented in those areas where more decentralized power generation takes place. These systems (called HVDC Light or HVDC Plus) currently use presspack IGBTs or IGBT modules. The functional advantages of voltage source systems, i.e. independent active and reactive power control, PWM voltage control, lower harmonics, and smaller filter requirements, have enabled voltage source converter technology to compete economically against classical HVDC at power levels up to 1 GW [Asp97]. Currently, off-shore wind power plants are under construction using voltage source systems to transmit power via undersea cables.

Electrolysers for electrowinning and electroplating are yet another high-power application in power electronics. Contrary to HVDC, very high DC currents at modest voltage (200–500 V) have to be controlled [Wie00]. Units delivering more than 100 kA have been constructed based on thyristor rectifiers. In the future, electrolysers may play a growing role when energy from renewable power sources is converted and stored in hydrogen [Bir06].

A growing market for power electronics are converters for photovoltaic (PV) systems, especially grid-connected PV systems. High efficiency, also at partial load, drives the design of PV converters. Units from 150 W (module converters), 5 kW (string converters) up to 1 MW (central converters) are being produced [Qin02]. Most designs use IGBT devices. Depending on geographical latitude, most road maps of PV cell manufacturers foresee PV at parity with electrical energy cost by 2015 (southern Europe) and 2020 (central Europe). Large-scale PV systems as well as solar thermal systems are envisaged in the near future around the equator. To transport the electrical energy, HVDC transmission systems will be needed that span entire continents. These super-grids are under study and can be realized with today's state-of-the-art power electronics [Zha08].

The more the energy demand of the world will rely on renewable power sources, the more electrical storage capacity will be needed. High-power battery storage systems are being demonstrated for over a decade in Japan using high-temperature sodium–sulfur batteries [Bit05]. Lithium-ion-based materials will further increase power density and energy density [Sau08]. Furthermore, if electric vehicles, all driven by power electronic converters, are used on a massive scale, it is anticipated that these vehicles can provide sufficient storage capacity to substantially load-level renewable power sources.

While this first chapter discussed power conversion systems from the application view, we will, after discussing in depth the physics and technology of power electronic devices and components, return to the system design at the end of this book from a bottom-up perspective.

One can conclude that with power electronics, vast amounts of energy can be saved (due to efficient control of processes). In addition, power electronics is a key enabling technology to make the electrical energy supply more robust and flexible, so that a more sustainable energy supply can be realized. By definition, at the heart of power electronics are power semiconductor devices that enable this efficient energy conversion. Consequently, a deep understanding of power semiconductors is a must for any electrical engineer who wishes to contribute toward a more sustainable world.

References

[Ack05] Ackermann T: Wind Power in Power Systems. Wiley, Chichester (2005)

[Asp97] Asplund G, Eriksson K, Svensson K: "DC transmission Based on Voltage Source Converters", CIGRE SC14 Colloquium, South Africa (see also library.abb.com) (1997)

[Ast05] Astrom U, Westman B, Lescale V, Asplund G: "Power Transmission with HVDC at Voltages Above 600 kV", IEEE Power Engineering Society Inaugural Conference and Exposition in Africa, pp. 44–50 (2005)

[Bir06] Birnbaum U, Hake JF, Linssen J, Walbeck M: "The hydrogen economy: Technology, logistics and economics", Energy Mater: Mater. Sci. Eng. Energy Syst., vol. 1, pp.152–157 (2006)

[Bit05] Bito A: "Overview of the sodium-sulfur battery for the IEEE stationary battery committee". Power Eng. Soc. General Meeting, vol. 2, pp. 1232–1235 (2005)

[Ded06] De Doncker RW: "Modern Electrical Drives: Design and Future Trends", CES/IEEE 5th International Power Electronics and Motion Control Conference (IPEMC 2006), 1, pp. 1–8 (2006)

[Hol01] Holonyak N: "The silicon p-n-p-n switch and controlled rectifier (thyristor)", IEEE Trans. Power Electronics, vol. 16, pp. 8–16 (2001)

[IEE92] IEEE 519-1992: "IEEE Recommended Practices and Requirements for Harmonic Control in Electrical Power Systems", Institute of Electrical and Electronics Engineers (1993)

[IEC08] IEC 61000-3-6: "Electromagnetic Compatibility (EMC) - Part 3-6: Limits - Assessment of Emission Limits for the Connection of Distorting Installations to MV, HV and EHV Power Systems", International Electrotechnical Commission (2008)

[Jon04] Jonnes J: Empires of Light: Edison, Tesla, Westinghouse, and the Race to Electrify the World. Random House, NewYork, NY (2004)

[Moh02] Mohan N, Undeland TM, Robbins WP: Power Electronics: Converters, Applications, and Design, Wiley 3rd edn,. New York, NY (2002)

[Nab81] Nabae A, Takahashi I, Akagi H: "A new neutral-point-clamped PWM inverter". IEEE Trans. Ind. Appl., vol. 17, pp. 518–523 (1981)

[Owe07] Owen EL: "SCR is 50 years", IEEE Ind. Appl. Mag., vol. 13, pp. 6–10 (2007)

[PEL05] PELS Operations Handbook: IEEE PELS Webpages http://ewh.ieee.org/soc/pels/pdf/PELSOperationsHandbook.pdf) (2005)

[Qin02] Qin YC, Mohan N, West R, Bonn RH: "Status and Needs of Power Electronics for Photovoltaic Inverters", Sandia National Laboratories Report, SAND2002-1535 (2002)

[Sau08] Sauer DU: Storage systems for reliable future power supply networks. In: Droege P (ed) Urban Energy Transition – From Fossil Fuels to Renewable Power. Elsevier, pp. 239–266 (2008)

[Ste02] Steigerwald DA, Bhat JC, Collins D, Fletcher RM, Holcomb MO, Ludowise MJ, Martin PS, Rudaz SL: Illumination with solid state lighting technology. IEEE J. Selected Topics Quantum Electronics, vol. 8, pp. 310–320 (2002)

[WEA09] World Wind Energy Association: "World Wind Energy Report 2008", http://www.wwindea.org (2009)

[Wie00] Wiechmann EP, Burgos RP, Holtz J: Sequential connection and phase control of a high-current rectifier optimized for copper electrowinning applications", IEEE Trans. Ind. Electronics, vol. 47, pp. 734–743 (2000)

[Wu08] Wu B, Pontt J, Rodríguez J, Bernet S, Kouro S: "Current-source converter and cycloconverter topologies for industrial medium-voltage drives", IEEE Trans. Ind. Electronics, vol. 55, pp. 2786–2797 (2008)

[Wun03] Wundrack B, Braun M: "Losses and Performance of a 100 kVA dc Current Link Inverter", European Power Electronics Association Conference EPE 2003, Toulouse, topic 3b, pp. 1–10 (2003)

[Zar01] Zargari NR, Rizzo SC, Xiao Y, Iwamoto H, Satoh K, Donlon J F: "A new current-source converter using a symmetric gate-commutated Thyristor (SGCT)". IEEE Trans. Ind. Appl., vol. 37, pp. 896–903 (2001)

[Zha08] Zhang XP, Yao L: "A Vision Of Electricity Network Congestion Management with FACTS and HVDC", IEEE Conference on Electric Utility Deregulation and Restructuring and Power Technologies, 3rd Conference, pp. 116–121 (2008)

Chapter 2
Semiconductor Properties

2.1 Introduction

Research on semiconductors has a long history [Smi59, Lar54]. Phenomenologically, they are defined as substances whose electrical resistivity covers a wide range, about $10^{-4}-10^{9}\ \Omega$ cm, between that of metals and insulators and which at high temperatures decreases with increasing temperature. Other characteristics are light sensitivity, rectifying effects, and an extreme dependency of the properties even on minute impurities. After reaching a basic understanding of their physical nature in the 1930s and 1940s, semiconductors are defined now often by the band model and impurity levels leading to the observed phenomena: Semiconductors are solids whose conduction band is separated from the valence band by an energy gap E_g and at sufficiently low temperatures is completely empty, whereas all states of the valence band are occupied. Most important for application in devices, however, is that the conductivity can be controlled over a wide temperature range by impurities and that there are two types of impurities, donors which release electrons causing n-type conductivity and acceptors which provide positive carriers, the holes, leading to p-type conductivity. This allows the fabrication of pn-junctions.

Semiconductors for power devices must have a sufficiently large energy gap or "bandgap" to insure that the intrinsic carrier concentration, present also without doping, stays below the doping concentration of the weakest doped region up to a sufficient operation temperature, for example, 450 K. This is the precondition that the doping structure remains effective. A sufficient bandgap is also advantageous because the critical field strength, up to which the material can withstand electric fields without breakdown, increases with bandgap. Unnecessary large energy gaps, on the other hand, can prove disadvantageous because the ionization energy of the impurities becomes larger with increasing bandgap and hence they release only an unfavorably low number of carriers at room temperature. Also built-in and threshold voltages get larger with wider bandgap.

Another essential demand on the semiconductor is that the mobilities of the free carriers are sufficiently high. We will come back to this point in Sect. 2.6. Furthermore, besides physical also chemical properties can be significant. An example is the stable native oxide of silicon which is a prerequisite for

J. Lutz et al., *Semiconductor Power Devices*, DOI 10.1007/978-3-642-11125-9_2,

almost the whole modern semiconductor technology. The impact of semiconductor properties on device characteristics and technology will come up at many points in the book.

Semiconductors for power devices such as silicon are always monocrystalline, of single crystal form, because, first, only single crystals guarantee a homogeneous space charge under blocking bias and as few as possible energy levels in the bandgap. High blocking voltages and low leakage currents are possible only under these conditions. Second, carrier mobilities in single crystal semiconductors are much higher than in polycrystalline. This allows correspondingly higher current densities and the use of smaller devices.

The advantages obtained passing from polycrystalline to monocrystalline material can be seen looking at the first commercial semiconductor power rectifiers made of polycrystalline selenium. These rectifiers were produced from about 1940 until 1970 [Spe58, Pog64]. With current densities of at best 1 A/cm^2 and blocking voltages up to 40 V per cell, packages with relatively large area were necessary to handle the given currents and voltages. The properties were to a good part due to the polycrystalline state, although partly due to the semiconductor Se itself. Silicon diodes replacing Se rectifiers (and Ge diodes) from the end of the 1950s allow current densities and blocking voltages more than two orders of magnitude higher.

The era of modern semiconductors started at the end of the 1940s with the advent of hyperpure single crystal germanium together with a breakthrough in understanding pn-junctions and the invention of the transistor. Ge was replaced by silicon in the 1950s, and nowadays it has no importance for power devices. Its main drawback is the relative small bandgap of only 0.67 eV (at 300 K) which results in a low allowed operating temperature ($\approx$ 70°C). Si is by far the most commonly used semiconductor also for power devices, and one can read more about it in this book. The bandgap of 1.1 eV and other properties of Si are very suitable for most applications. Si belongs to group IV of the periodic system, following carbon, C, and preceding Ge. A compound semiconductor consisting of the group III element gallium and the group V element arsenic is gallium arsenide, GaAs. This semiconductor has gained significance for microwave devices, since it allows high switching frequencies. It has a bandgap of 1.4 eV and a very high electron mobility (five times that of Si). In the field of power devices, the high electron mobility renders it suitable for high-voltage Schottky diodes. Schottky diodes of GaAs for 300 V and higher are available on the market.

Another compound semiconductor, which has attracted much research and development efforts in the last two decades and was often called the ideal material for power devices, is silicon carbide, SiC [Fri06]. It has a large bandgap ranging from 2.3 to 3.3 eV depending on the crystal modification (polytype, see the next section). Consequently, the maximum operating temperature up to which the intrinsic carrier concentration is small enough and the critical field strength at which breakdown occurs are much higher than for Si. Whereas the higher possible temperature could not be utilized much till now because metallurgical properties of contacts and packaging limit the operating temperature, extensive use was made of the benefits of the high critical field. These are a much higher allowed doping concentration and smaller width of the base region required for a given blocking voltage. This means that fast Schottky diodes and other unipolar devices for high voltages can be fabricated, and pn-devices with extremely high blocking voltage (possibly up to 20 kV) and low power losses are possible. Although SiC like Si possesses a native oxide, the lower quality of the SiO_2–SiC interface still hinders the fabrication of competitive MOS (*m*etal *o*xide *s*emiconductor) devices. A significant difference to Si technology is that diffusion of impurities cannot be used as a doping method because the diffusion constants in SiC are too small. Although from these and other reasons the technology of SiC is less versatile and developed and more expensive than that of Si, it has apparently good chances to win some application areas of power electronics. Schottky diodes in SiC for voltages up to 1600 V are on the market; junction field effect and bipolar transistors are available as prototypes. We will come back to SiC devices at several points in this book.

In recent years very interesting investigations have been performed on microwave and power devices on the base of GaN using a heterojunction with AlGaN [Ued05]. The base material GaN is with regard to the bandgap of 3.4 eV and a correspondingly high critical field similar to 4H-SiC. The operation principle of the AlGaN/GaN devices, however, is based specifically on a highly conducting two-dimensional electron gas induced by the heterojunction with the AlGaN which is deposited as a thin layer. Hence the devices have a lateral structure. Because the fabrication of large single crystals and wafers of GaN still offers problems, the GaN layer is grown epitaxially on substrates of 4H-SiC, saphire (Al_2O_3), or silicon. Field effect transistors with a blocking voltage up to 1.8 kV and a current of more than 100 A have been demonstrated [Ike08]. Owing to the high critical field the lateral and vertical dimensions of the GaN layer can be chosen small, so an on-resistance not much larger than for vertical devices of 4H-SiC was obtained. There is also the potential of combining high-density lateral GaN power devices with conventional control circuits on a silicon substrate [Bri09]. How this new technology will develop further-on and if power devices on the base of GaN will find a place in future power electronics cannot be predicted at present. Silicon based devices are always strong rivals.

A further subject of investigations in view of a possible future material for power devices has been the diamond modification of carbon. Diamond has an energy gap of 5.5 eV and in pure form is a semi-insulator. Its critical field is extremely high (10–20 MV/cm). If it contains impurities like boron or phosphorus, it behaves as a semiconductor whose carriers have higher mobilities than in silicon. A handicap of diamond is the high ionization energy of the dopants, especially of donors. Hence the impurities are ionized only to a small extent at room temperature. Concepts to partially overcome this drawback have been proposed and samples of Schottky diodes have been demonstrated [Ras05]. The employment of diamond devices in power electronics is expected at best in the longer term.

2.2 Crystal Structure

The atoms in a crystal have their resting center at lattice points around which they can vibrate. The lattice is to a great extent determined by the type of bonding between the atoms. In all the named semiconductors (apart from Se) the bonds in all directions are covalent, i.e. formed by an electron pair with opposite spins. This is the rule for group IV elements. Because they have four valence electrons in the outer shell, an atom can undergo covalent bonds with four nearest neighbors which, like the central atom, contribute an electron to the respective bond. In this way, a closed shell of eight shared electrons is gained for each atom establishing a very stable configuration. Similar considerations hold for group III–V compounds such as GaAs which *on average* have four valence electrons per atom. Symmetrically arranged, the four nearest neighbors are located at the corners of a tetrahedron, the midpoint of which is occupied by their common bonding partner. For element semiconductors, this tetrahedral bonding can be realized only by one lattice structure, the diamond lattice. This is hence the lattice of Ge, Si, and the diamond modification of carbon.

Figure 2.1 (a) shows the cubic unit cell of the diamond lattice. It can be described as superposition of two face-centered cubic (fcc) sub-lattices, which by definition have lattice points at the corners and midpoints of the side faces of the cube. The two sub-lattices are shifted by a quarter of the space diagonal. In the lower left part of the figure, the covalent bonding of an atom on the space diagonal with its four nearest neighbors is marked. This part of the figure is redrawn in Fig. 2.1(b) to show the tetrahedral bonding geometry more clearly.

GaAs has the same lattice structure, except that now one of the face-centered cubic sub-lattices is occupied by Ga and the other by As atoms. If in Fig 2.1(b) an As atom is in the center, the four neighbors at the corners are Ga and vice versa. This so-called zincblende lattice is the structure of most group III–V semiconductors [Mad64].

An exception is GaN which possesses the wurtzite structure, a hexagonal lattice called after a second polytype of ZnS appearing besides zincblende. In the wurtzite

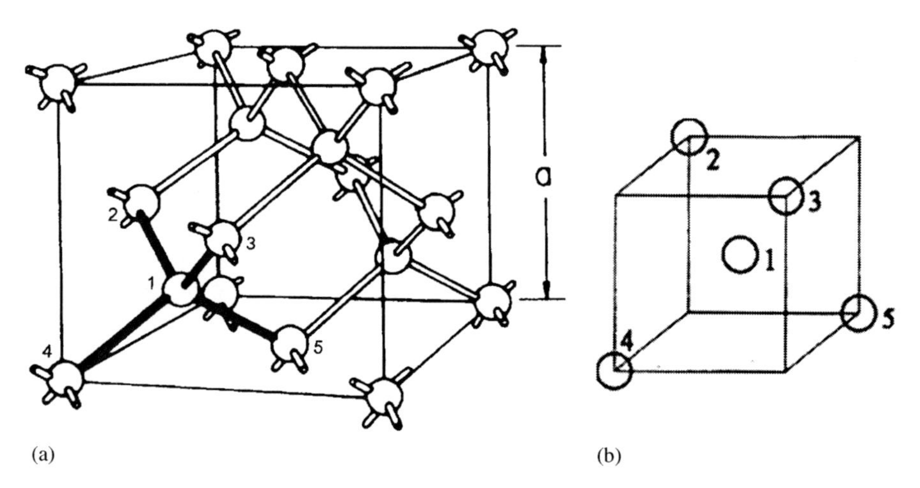

Fig. 2.1 **(a)** Cubic unit cell of the diamond lattice. The tetrahedral bonding between an atom 1 on the space diagonal and four nearest neighbors 2–5 is marked. **(b)** Central atom with four neighbor atoms at the corners of a tetrahedron. The lattice constant for silicon is $a = 5.43$ Å, the distance between the center of Si atoms $d = \sqrt{3}/4\,a = 2.35$ Å (adapted from [Sze81] John Wiley & Sons Inc., reproduced with permission and [Hag93] Aula-Verlag GmbH, reproduced with permission)

lattice the tetrahedral bonding between the neighbors of both types is realized as well, only the orientation of neighboring tetrahedra toward each other is different from the zincblende structure. Actually, the tetrahedral bonding system of *XY*-compounds is compatible with many other lattices. This is shown in fact by SiC which crystallizes in the zincblende structure and likewise in many other polytypes, most of which are hexagonal. In all cases, the arrangement of nearest neighbors is identical, each silicon atom being surrounded by four carbon atoms at the corners of a tetrahedron with the Si in the center and vice versa. Only the arrangement of more distant atoms varies from polytype to polytype [Mue93]. The atomic distance between nearest neighbors is always 0.189 nm = 1.89 Å, nearly the mean value between the atomic distances in diamond with 1.542 Å and Si with 2.35 Å. For power devices, the hexagonal polytype called 4H-SiC is preferred. The prefix indicates that the structure repeats itself after stacking of four Si–C double planes. In this nomenclature GaN has a 2H structure, since the wurtzite lattice repeats itself after two double planes of GaN.

The relevant crystal structures are compiled in the following list:

Ge, Si, diamond C	Diamond lattice, cubic
GaAs	Zincblende lattice, cubic
GaN	Wurtzite lattice, 2H hexagonal
4H-SiC	4H hexagonal

The crystal orientation influences to some extent processing parameters and physical properties. In silicon, the thermal oxidation rate, the epitaxial deposition rate, the density of surface states, and the elastic constants depend on orientation, whereas mobilities and diffusion constants due to the cubic lattice are (isotropic) scalars. To describe crystal planes and directions one uses the Miller indices. These are defined via the reciprocals of the intercepts of the considered plane with the crystallographic axes, expressing their ratios by a triple of smallest integers. For example, the six faces of a cube intersect two of the axes in the point ∞, hence two of the Miller indices are 0, and the planes are denoted as (1 0 0), (0 1 0), (0 −1 0), and so on. The family of these planes is denoted as {100} planes. The crystal planes which have equal intercepts on the axes are the {111} planes. In silicon technology crystal wafers (slices) with a {100} or a {111} surface are used.

2.3 Energy Gap and Intrinsic Concentration

The energy gap results because all the valence electrons of the semiconductor atoms are required for the complete covalent bonds and because a certain amount of energy is necessary for breaking electrons out of the bonds so that they can move freely and hence conduct current. This is expressed by the band diagram shown in Fig. 2.2, where the energy of the electrons is plotted versus a space coordinate x. The valence band contains the states for the electrons in the bonds, and the conduction band represents the states of the electrons which are free for conduction. Between the top

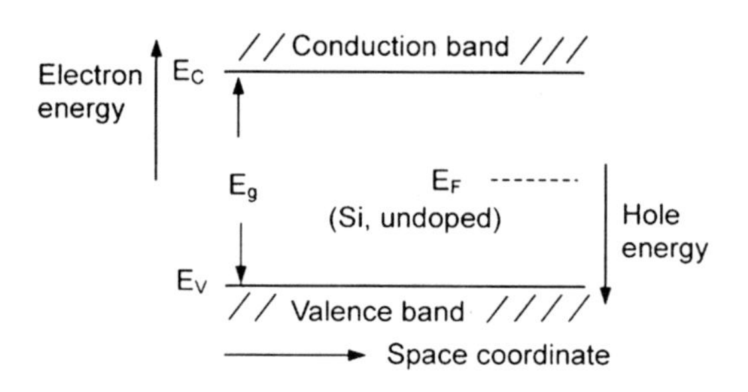

Fig. 2.2 Energy band model

of the valence band, E_V, and the bottom of the conduction band, E_C, no energy levels are present, $E_g = E_C - E_V$ is the energy gap. Here we refer to an "intrinsic" semiconductor whose properties are not influenced by impurities but are due to the semiconductor itself. At least the energy E_g is necessary to break an electron out of the bonds and create a conduction electron. At zero absolute temperature, no valence electron has this energy (external activation excluded), and since no carriers are present, the semiconductor behaves like an insulator. At $T > 0$ a certain number of valence electrons is elevated into the conduction band. By this process, however, not only conduction electrons are created but also an equal number of empty electron places remain in the valence band and these voids, called "holes," can conduct current likewise. As a missing valence electron the hole has the opposite, i.e. a positive charge. Because in an electric field the empty place is filled repeatedly by a neighboring valence electron, the hole travels in opposite direction to the electrons. This is similar to a bubble in water moving against gravity. It follows that one can ascribe also an energy to the holes which is directed downward in Fig. 2.2 (see the calculation below).

So far, the picture of a (classical) vacancy at a point x in the otherwise full band of valence electrons is successful. With this understanding, the hole is an auxiliary quantity for a simpler description of the motion of all involved valence electrons which are the actual carriers. However, there are other decisive experiments, particularly measurements of the Hall effect, which contradict this classical picture and show the holes as positive charge carriers on their own. This is confirmed by quantum theory according to which the holes are independent stable (quasi-) particles on the same grounds as conduction electrons. A detailed discussion of these features will be given in the next paragraph.

We calculate now the intrinsic concentration of carriers in a pure semiconductor and introduce simultaneously some more general relationships for electrons and holes which are applicable also when impurities are present. Since the thermal generation of carriers is counteracted by recombination annihilating them, this leads to a thermal equilibrium described by statistical physics. The occupation probability of states with energy E, defined as the number of electrons n_E per number of states with that energy, N_E, is given by the Fermi distribution:

$$\frac{n_E}{N_E} = \frac{1}{1 + e^{\frac{E - E_F}{kT}}} \tag{2.1}$$

In this equation k denotes the Boltzmann constant, T the absolute temperature, and E_F the Fermi energy, which in statistical thermodynamics is called also "electrochemical potential." E_F is a constant of the system determined so that the sum over n_E returns the total electron concentration. As shown by Eq. (2.1), the states with energies smaller than E_F are mostly occupied by an electron, while the states with $E > E_F$ are mostly empty. The number of occupied divided by the number of unoccupied states of energy E, called the occupation degree, is

$$\frac{n_E}{N_E - n_E} = e^{-(E-E_F)/kT} \tag{2.2}$$

In the intrinsic case and often also in doped semiconductors (see Sect. 2.5), the occupation probability is small, $n_E << N_E$. In this so-called non-degenerate case, Eq. (2.2) turns into the classical Boltzmann or Maxwell–Boltzmann distribution:

$$\frac{n_E}{N_E} = e^{-(E-E_F)/kT} \tag{2.3}$$

Since primarily only states near the bottom of the conduction band are occupied and primarily only states near the top of the valence band are empty, we assume at the moment that the density of states in the bands (number of states per infinitesimal energy interval and per volume) is concentrated at the edges and in integrated form is given by effective densities of states (number per volume) N_C, N_V. Taking Eq. (2.3) at the energy E_C one obtains for the concentration of conduction electrons in thermal equilibrium:

$$n = N_C \cdot e^{-E_C - E_F/kT} \tag{2.4}$$

Using Eq. (2.2) at$E = E_V$ and considering that the density of unoccupied states in the valence band is identical with the hole concentration p, whereas the density of occupied states is $N_{V-p} \approx N_V$, the hole concentration in thermal equilibrium is obtained as follows:

$$p = N_V \cdot e^{-E_F - E_V/kT} \tag{2.5}$$

This shows that the holes behave statistically as particles with energy scale inverted compared with electrons. Multiplication of Eqs. (2.4) and (2.5) using $E_C - E_V = E_g$ yields

$$n\,p = n_i^2 = N_C N_V \cdot e^{-E_g/kT} \tag{2.6}$$

where $n_i = n = p$ is the intrinsic concentration. Equation (2.6) is the mass law equation of the reaction (0) $\leftrightarrow$ $n + p$, describing generation and, inversely, the recombination of an electron–hole pair.

Since the condition $n = p = n_i$ has not been used in the derivation, the intrinsic conduction represents only a special case of Eqs. (2.4), (2.5), and (2.6), actually they are applicable also to doped semiconductors where $n \neq p$. Doped semiconductors will be treated in detail in Sect. 2.5. Only the Boltzmann distribution (2.3) is assumed for thermal equilibrium, meaning that the doping is not too high (case of non-degeneracy). As shown by Eq. (2.6), the np product is a constant independent of the Fermi level, but dependent on the bandgap.

The Fermi level for the intrinsic case is obtained from Eqs. (2.4) and (2.5) setting $n = p$:

$$E_\mathrm{i} = \frac{E_\mathrm{V} + E_\mathrm{C}}{2} - \frac{kT}{2} \ln \frac{N_\mathrm{C}}{N_\mathrm{V}} \tag{2.7}$$

Because of similar values of the densities of states N_C, N_V, the Fermi level in intrinsic semiconductors lies close to the middle of the bandgap.

In spite of the simplifying assumption on the distribution of the density of states, Eqs. (2.4), (2.5) and (2.6) are applicable to the actual situation. To take into account that with increasing T more states above, respectively below the band edges are occupied, this results only in a temperature dependency of the effective densities of states N_C and N_V. The density of states N_E increases with distance ΔE from the edges as $\sqrt{\Delta E}$. Multiplying this with the Boltzmann factor of Eq. (2.3) and integrating, one obtains again Eqs. (2.4) and (2.5), where N_C, N_V now are proportional to $T^{3/2}$ [Sze02]. Considering that also the band parameters themselves vary a little with T, one obtains for Si [Gre90]:

$$\begin{aligned} N_\mathrm{C} &= 2.86 \times 10^{19} \left(\frac{T}{300}\right)^{1.58} /\mathrm{cm}^3 \\ N_\mathrm{V} &= 3.10 \times 10^{19} \left(\frac{T}{300}\right)^{1.85} /\mathrm{cm}^3 \end{aligned} \tag{2.8}$$

These numbers are large compared with the doping concentrations in most cases, as will be seen later. Compared with the number of Si atoms per cm^3, 5.0×10^{22}, they are small.

The bandgap is approximately a constant, more precisely considered, however, it decreases slightly with temperature. This can be expressed for Si and other semiconductors in the form [Var67]

$$E_\mathrm{g}(T) = E_\mathrm{g}(0) - \frac{\alpha T^2}{(T + \beta)} \tag{2.9}$$

The bandgap parameters of this equation together with the effective densities of states are compiled for Si, GaAs, 4H-SiC, and GaN in Table 2.1 [Thr75, Gre90, Lev01, Mon74]

The intrinsic carrier densities calculated with these data are shown in Fig. 2.3 as functions of temperature. From Si to SiC the intrinsic concentration decreases extremely for a given temperature due to the exponential dependence on energy gap. Considering a given value of n_i, the absolute temperature at which this value is adopted is to a rough approximation proportional to E_g, neglecting the temperature dependence of the pre-exponential factor in Eq. (2.6).

The temperature at which n_i reaches a value comparable with the impurity concentration of the lowest doped region in a device, defines a limit, above which the

Table 2.1 Bandgap parameters and effective density of states of some semiconductors

	Si	GaAs	4H-SiC	GaN
$E_g(0)$ (eV)	1.170	1.519	3.263	3.47
$\alpha \times 10^4$ (eV/K)	4.73	5.405	6.5	7.7
β (K)	636	204	1300	600
$E_g(300)$ (eV)	1.124	1.422	3.23	3.39
$N_C(300)$ (cm^{-3})	2.86×10^{19}	4.7×10^{17}	1.69×10^{19}	2.2×10^{18}
$N_V(300)$ (cm^{-3})	3.10×10^{19}	7.0×10^{18}	2.49×10^{19}	4.6×10^{19}

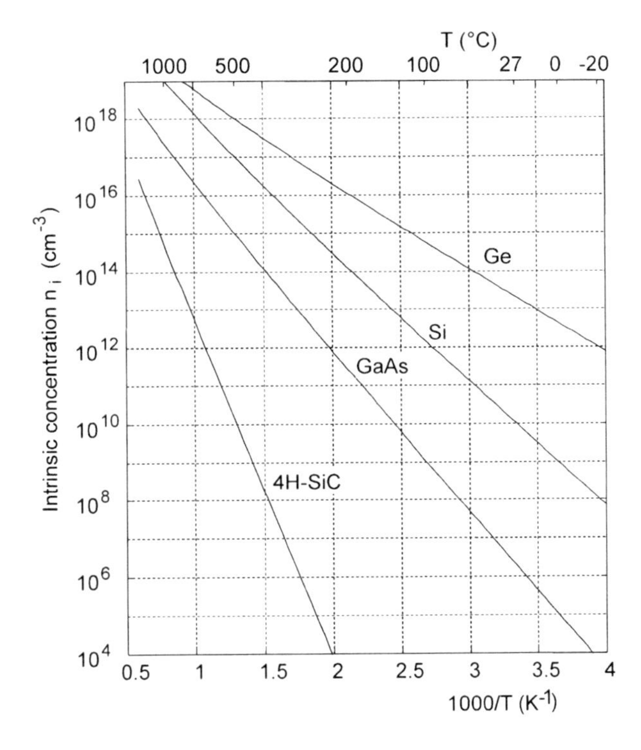

Fig. 2.3 Intrinsic carrier density for Ge, Si, GaAs, and 4H-SiC as a function of temperature

pn-structure begins to be leveled out and loses its normal function. If n_i is dominating, its exponential increase with temperature and the corresponding decrease of resistance, together with thermal feedback, can lead to current constriction and destruction of the device. In a Si device with 1000 V blocking voltage, a doping in the range of 10^{14} cm^{-3} is necessary for the base region to sustain the voltage. To meet the condition $n_i < 10^{14}$ cm^{-3}, the temperature must remain below about 190°C, as is shown in Fig. 2.3. With 4H-SiC, temperatures of more than 800°C would be allowed for 1000 V devices from this requirement. In practice, interconnect and packaging materials, however, set a much smaller temperature limit.

Equation (2.6) shows why semiconductors with a wide bandgap, if nearly intrinsic, behave like insulators. In 4H-SiC, the intrinsic concentration as given by above data is even at 400 K only 0.3 cm^{-3}, corresponding to a resistivity of

$\approx 2 \times 10^{16}\,\Omega$ cm. Although the practically attainable resistivity is several orders of magnitude smaller, SiC can be used for insulating layers with high thermal conductance.

For later use in the case of doped semiconductors we note an alternative form of Eqs. (2.4) and (2.5) which is obtained using the intrinsic concentration and intrinsic Fermi level to eliminate the effective densities of states and energies of band edges. Dividing Eq. (2.4) by the specialized form $n_\mathrm{i} = N_\mathrm{C} \exp(-(E_\mathrm{C} - E_\mathrm{i})/kT)$ it converts to

$$n = n_\mathrm{i} \cdot \mathrm{e}^{(E_\mathrm{F}-E_\mathrm{i})/kT} \tag{2.10}$$

The hole concentration in thermal equilibrium can be written as follows:

$$p = n_i \cdot \mathrm{e}^{(E_i-E_\mathrm{F})/kT} \tag{2.11}$$

2.4 Energy Band Structure and Particle Properties of Carriers

Besides the $E(x)$ band diagram of Fig. 2.2, there is the more detailed energy band representation in k-space, $E(\vec{k})$, which allows further insight into fundamental semiconductor properties. Here the electron energy is plotted versus the wave vector $\vec{k}$ of a wave packet, which solves the quantum mechanical Schrödinger equation for an electron in the crystal. Of main interest are the valence $E(\vec{k})$ band with highest maximum and the conduction $E(\vec{k})$ band with lowest minimum. In Fig. 2.4 these bands are shown for Si and GaAs for specified directions in $\vec{k}$-space. The energy difference between the absolute minimum of the conduction band and the maximum of the valence band is the bandgap E_g shown in Fig. 2.2. The maximum of the valence band is nearly always at $\vec{k} = 0$. In GaAs, also the minimum of the conduction band is located at this position. A semiconductor of this kind is called a direct semiconductor. In Si, the conduction band has a minimum far away from $\vec{k} = 0$. Semiconductors of this type are called indirect. Since in Si a minimum lies in each of the $\{100\}$ directions, there are six minima in a unit cell.

Whether a semiconductor has a direct or indirect bandgap is decisive for the probability of transitions between the bands. This determines the suitability for optical devices, but has some significance also for power devices. The influence on transition probability follows because the crystal momentum $\vec{p} = \hbar\,\vec{k}$ ($\hbar = h/2\pi$, h is Planck's constant) has the property of a momentum of the electron regarding its reaction on external forces. Internal forces arising from the periodic potential in the crystal are taken into account in another way (see below). An external force $\vec{F}$, owing to an electric or magnetic field, causes an acceleration which, as in Newton's second law of mechanics, is given by $\mathrm{d}\vec{p}/\mathrm{d}t = \vec{F}$. Furthermore, in transitions between bands the change in crystal momentum $\vec{p}$ has to be counted in the law of conservation of momentum. The recombination of an electron at the bottom of the conduction band with a hole at $\vec{k} = 0$ occurs under emission of a photon, which receives nearly the whole energy released, but has negligible momentum. In

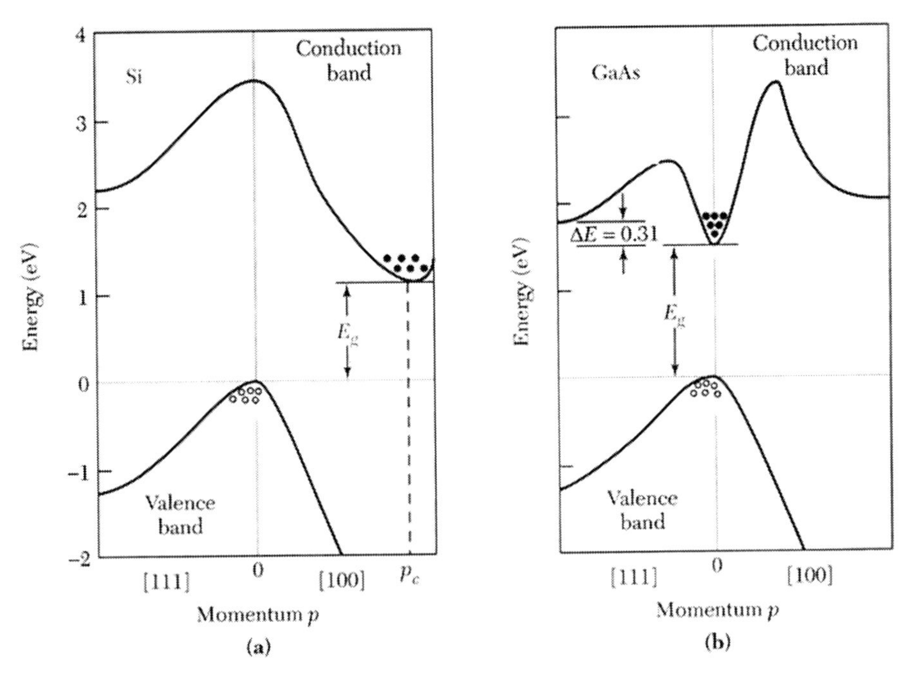

Fig. 2.4 Energy band $E(k)$ for Si (indirect semiconductor) and GaAs (direct semiconductor). Figure from Sze [Sze02] John Wiley & Sons Inc., reproduced with permission

indirect semiconductors the recombination can occur only if the crystal momentum of the conduction electron can be transferred to a quantum of lattice vibrations, a phonon. Hence, compared with a direct semiconductor such as GaAs, where the recombination occurs without participation of a phonon, the radiative band-to-band recombination in indirect semiconductors has a much lower probability. Therefore, only direct semiconductors are used for LEDs and lasers. The lifetime associated with the radiative band-to-band recombination represents an upper limit of the lifetime of minority carriers. For GaAs, a radiative minority carrier lifetime $\tau = 1/(BN)$ (B is the recombination constant) of 6 μs is estimated for a doping concentration of $N = 1 \times 10^{15}$ cm^{-3}; for 10^{17} cm^{-3} this means $\tau = 60$ ns [Atk85]. This is sufficient to allow satisfactory operation of pin diodes for a medium voltage range, but it is detrimental for bipolar transistors and thyristors with commonly used doping structures. In Si the radiative recombination constant B is four orders of magnitude smaller [Sco74]. Hence very high lifetime values are possible in Si. The recombination radiation in Si, having a wavelength $\lambda \simeq hc/E_g = 1.1$ μm, is often used as a tool to investigate and test the internal operation of devices. Besides Si also Ge and all polytypes of SiC are indirect semiconductors. Some of the group III–V compound semiconductors are of the direct type like GaAs, while others are of the indirect type like GaN.

The $E(\vec{k})$ bands are also basic for the behavior of electrons and holes as charge carriers. We will outline this here shortly, referring for a more complete discussion to the books of Moll [Mol64] and Spenke [Spe58]. If an external force is applied, the electrons near the minimum of the conduction band are accelerated. However, the increase in kinetic energy is only relatively small, because the acceleration is stopped after a short relaxation time owing to the non-ideality of the crystal, i.e. by scattering by phonons and impurities. Thus the wave vector and kinetic energy are reduced nearly to their initial values (on the statistical average) and remain in the band not far away from the minimum. Hence the kinetic energy, $E_{\mathrm{n,kin}} = E - E_{\mathrm{C}}$, can be expressed using Taylor's expansion as

$$E_{\mathrm{n,kin}} = \frac{1}{2}\frac{d^2E}{\mathrm{d}k^2} \cdot (\vec{k} - \vec{k}_{\mathrm{m}})^2 \tag{2.12}$$

where $\vec{k}_{\mathrm{m}}$ denotes the $\vec{k}$-vector at the band minimum. We assume for simplicity that the $E(\vec{k})$ function depends only on the absolute value k of $\vec{k}$ and not on the orientation. Defining a "particle momentum" as $\vec{p}_n = \hbar(\vec{k} - \vec{k}_{\mathrm{m}})$, Eq. (2.12) turns into

$$E_{\mathrm{n,kin}} = \frac{1}{2\hbar}\frac{d^2E}{\mathrm{d}k^2}\,\vec{p}_{\mathrm{n}}^{\,2} = \frac{\vec{p}_{\mathrm{n}}^{\,2}}{2m_{\mathrm{n}}} \tag{2.13}$$

where we have defined, furthermore,

$$m_{\mathrm{n}} \equiv 2\hbar^2 \Big/ \frac{d^2E}{\mathrm{d}k^2} \tag{2.14}$$

m_{n} has the dimension of a mass and is called the effective mass of the electrons. It has the order of magnitude of electron mass in vacuum, but is not equal to it. The velocity of the electron is $\vec{v}_{\mathrm{n}} = \vec{p}_{\mathrm{n}}/m_{\mathrm{n}}$. Because the momentum vector $\vec{p}_{\mathrm{m}}$ of the conduction band minimum in $\vec{p}_{\mathrm{n}} = \vec{p} - \vec{p}_{\mathrm{m}}$ is constant, the relation between force and acceleration can be written also as $\vec{F} = \mathrm{d}\vec{p}_{\mathrm{n}}/\mathrm{d}t$. Hence, quantum mechanics leads to the result that conduction electrons in the lattice-periodic potential obey relationships of mechanics; they react on external forces like mass points with a positive effective mass m_{n} and a negative charge $-q$. The interaction with the internal field of the periodic potential is taken into account by the effective mass. The effective mass is not a scalar but a tensor, but this is not important for an essential understanding of the model.

Holes behave quite analogously. This is obtained representing a hole by the full valence band minus an electron. The full band does not conduct current because the parts from $+\vec{k}$ and $-\vec{k}$ compensate each other. The contribution of an electron to the current density for a unit of volume is $-q\,\vec{v}_n$, and that of the missing electron, the hole, $+q\vec{v}_n$. Since the velocity of the hole is identical to that of the missing electron, the equation of motion yields $\dot{\vec{v}}_{\mathrm{n}} = \vec{F}_{\mathrm{n}}/m_{\mathrm{nv}} = \dot{\vec{v}}_{\mathrm{p}} = \vec{F}_{\mathrm{p}}/m_{\mathrm{p}}$. Here $\vec{F}_{\mathrm{n}}$ denotes the force on the electron, $\vec{F}_{\mathrm{p}}$ the force on the hole, which due to the opposite charge has an opposite sign for electric and magnetic fields, $\vec{F}_{\mathrm{p}} = -\vec{F}_{\mathrm{n}}$. Hence we have to define $m_{\mathrm{p}} = -m_{\mathrm{nv}}$ to obtain the classical relation between force and acceleration. Because at the top of the valence band $\mathrm{d}^2E/\mathrm{d}k^2 < 0$, the effective mass m_{nv} of the valence electron is negative and that of the hole positive:

$$m_{\mathrm{p}} = -2\hbar^2 \bigg/ \frac{\mathrm{d}^2E}{\mathrm{d}k^2}(E = E_{\mathrm{V}}) \tag{2.15}$$

For $E_{\mathrm{kin,p}} \equiv E_{\mathrm{V}} - E$ one obtains similarly as for the electrons: $E_{\mathrm{kin,p}} = \vec{p}^2/2m_{\mathrm{p}} = m_{\mathrm{p}}\vec{v}_{\mathrm{p}}^2/2$. Hence we have obtained that also holes obey the relationships of mechanics, they behave like mass points with a positive charge q and positive effective mass. Quantum mechanics leads to a particle picture, called quasi-particle model, which is justified on the same basis for holes as for conduction electrons. It has a wide range of applicability and is used throughout device physics.

Quantitatively, the situation is more complicated because the effective mass of electrons in Si and other semiconductors is a tensor and depends strongly on orientation. However, the effective mass entering the mobility and conductivity, the "conductivity effective mass" $m_{\mathrm{n,c}}$, is an average over the equivalent minima in k-space, and this averaging yields a scalar in cubic crystals [Smi59]. The conductivity effective mass in Si is $m_{\mathrm{n,c}} = 0.27\, m_0$ (m_0 free electron mass) [Gre90]. Also for holes in Si a scalar value is used for the conductivity effective mass, it amounts to $m_{\mathrm{p,c}} = 0.4\, m_0$ at 300 K . Both masses depend only very weakly on temperature.

Regarding the quasi-particle model for holes, the question is whether it implicates essentially different results compared with the previous classical picture of a void or bubble in the sea of bonding electrons. The answer is yes. Most evidently this is shown by the Hall effect, measurements of which are the main experimental tool to investigate the basic processes underlying conductivity. In these experiments, a magnetic field $\vec{B}$ is applied perpendicular to a semiconductor strip in which a longitudinal current is flowing, the lateral voltage V_{H} at the strip is measured, see Fig. 2.5. Since the carriers are forced to flow along the strip, the Lorentz force $Q\,\vec{v} \times \vec{B}$ is balanced by the formation of a lateral electric field $\vec{E}_H = -\vec{v} \times \vec{B}$ where $\vec{v}$ is the vector of (drift) velocity of the carriers and Q their charge. If x and z are

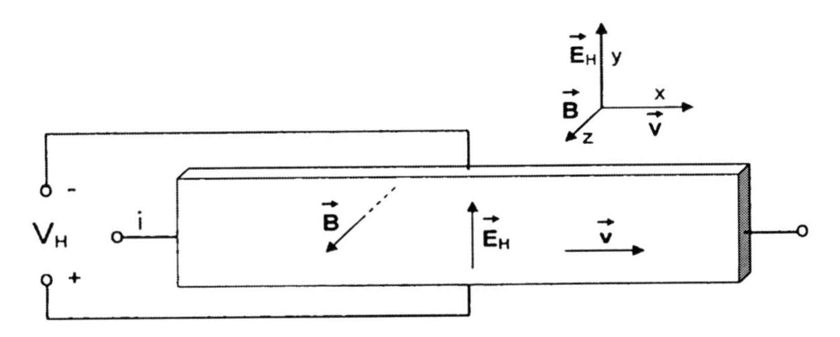

Fig. 2.5 Basic arrangement for Hall measurements

the coordinates in the directions of $\vec{v}$ and $\vec{B}$, respectively, then the "Hall field" $\vec{E}_H$ has the direction of the positive y-coordinate. With the scalar components in this coordinate system denoted by ordinary letters (so that $\vec{v} = (v, 0, 0)$, $\vec{B} = (0, 0, B)$, $\vec{E}_H = (0, E_H, 0)$), one obtains

$$E_H = v \cdot B = \frac{j}{Q \cdot C} B = R_H \cdot B \cdot j \text{ with } R_H = \frac{1}{Q \cdot C} \tag{2.16}$$

Here the velocity was substituted by the current density according to $j = Q \cdot C \cdot v$ where C denotes the carrier concentration. R_H is called "Hall constant." If in a p-type semiconductor the hole current is in principle achieved by a motion of valence electrons as in the classical bubble model, the Hall constant would have the same negative sign as for an n-type specimen since then also $Q = -q$ would hold. Furthermore, because the concentration of valence electrons, C, is much higher than that of the empty states and also much higher than the concentration of electrons in an n-type semiconductor of equal conductivity, the absolute value of R_H would be found to be very small if the classical bubble model would apply. Actually, however, the Hall constant measured for specimens doped with acceptors is positive in contrast to that of n-type samples and in magnitude the R_H is comparable for both doping types. Hence the holes manifest themselves as independent entities which in a magnetic field experience a Lorentz force like positively charged mass points. It is for this reason that p-type conductivity is a conductivity type of its own which is equivalent to n-type conduction. Before the quantum theory of solids was developed, the positive Hall constant found in many specimens was a very irritating phenomenon.

2.5 The Doped Semiconductor

If in a silicon crystal some atoms in the lattice are replaced by atoms of an element of group V in the periodic table, e.g. phosphorus, each impurity atom has one electron more in the outer shell than necessary for the four covalent bonds. Therefore, one electron is only loosely bonded and needs only a small amount of energy – available probably as thermal energy – to be removed from the impurity atom and freed for conduction. These elements donating their fifth electron to the conduction band are called donors. The result is an extrinsic, n-type semiconductor with "n" pointing to the negative charge of the carriers. On the other hand, if the silicon atoms are replaced at some lattice points by atoms of an element of group III, e.g. boron, each impurity atom has one electron less than necessary for the four covalent bonds. Since the bonds between the impurity and the neighboring silicon atoms are nearly as tight as between the silicon atoms themselves, there is only little energy necessary for moving an electron out of a Si–Si bond in the neighborhood to the impurity to complete its bonds with the four silicon neighbors. Accepting an electron of the valence band and thus generating a mobile hole, these impurities leading to p-type conductivity are called acceptors. In the energy band picture, the donors have energy

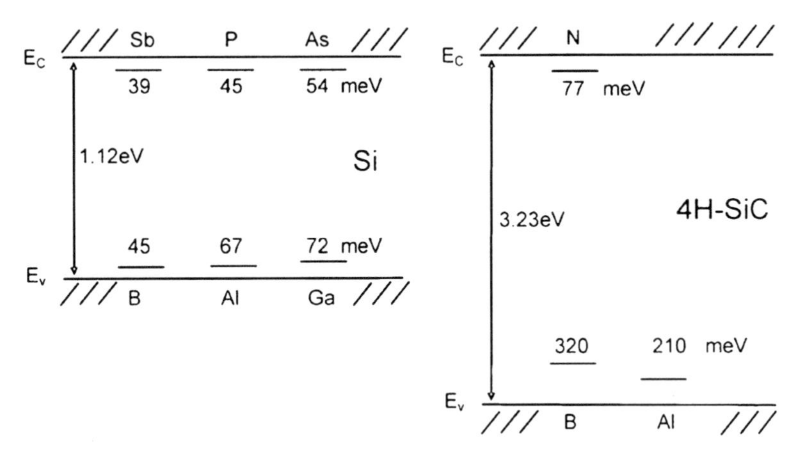

Fig. 2.6 Energy levels of doping impurities in Si and 4H-SiC. The differences from the respective band are given, i.e. the ionization or activation energies for elevating an electron or hole from the neutral impurity atom into the conduction or valence band, respectively

levels close to the conduction band, the acceptors close to the valence band edge. The energy levels for important dopants in Si and 4H-SiC are shown in Fig. 2.6.

Because of the small ionization energies $\Delta E_\mathrm{D} = E_\mathrm{C} - E_\mathrm{D}, \Delta E_\mathrm{A} = E_\mathrm{A} - E_\mathrm{V}$, and the many states in the conduction and valence bands available for electrons and holes, most donors and acceptors in Si are ionized at room temperature. This follows from Eq. (2.1) or (2.2) if the Fermi level lies below the level of the donor or above the acceptor level in the respective case of doping. Considering the degree of ionization in detail we use the notation in the case of donor doping denoting the total impurity concentration with N_D and the concentrations of neutral and ionized impurities with N_D^0, N_D^+, respectively. Then the number of occupied donor levels in Eq. (2.2) is $n_\mathrm{E} = N_D^0$ and the number of unoccupied levels $N_\mathrm{E} - n_\mathrm{E} = N_\mathrm{D} - N_\mathrm{D}^0 = N_\mathrm{D}^+$, hence the ratio of both, the occupation degree, is

$$\frac{N_\mathrm{D}^0}{N_\mathrm{D}^+} = g \cdot \mathrm{e}^{-(E_D - E_F)/kT} \tag{2.17}$$

where we have added a "degeneracy factor" g. This factor, which for donors is $g = 2$, is necessary because the neutral donors D^0exist in two states depending on the spin orientation of the trapped electron. Nevertheless only one electron can be trapped because for a second the Coulomb field is no longer present. Since the state D^- does not appear, the Fermi energy does not include the degeneracy of the D^0 state and hence one has to take into account it separately [Spe58, Sho59]. The Fermi energy can be eliminated dividing Eq. (2.17) by Eq. (2.4), which yields

$$\frac{N_\mathrm{D}^0/N_\mathrm{D}^+}{n/N_\mathrm{C}} = g \cdot \mathrm{e}^{-(E_C - E_D)/kT} \tag{2.18}$$

As long as the intrinsic concentration and hence the hole density $p = n_i^2/n$ is small, the neutrality condition is $N_D^+ = n$ so that $N_D^0 = N_D - n$. Inserting this and solving for n one obtains after simple conversion

$$n = \frac{N_D}{\frac{1}{2} + \sqrt{\frac{1}{4} + \frac{N_D}{N_C} \cdot g \cdot e^{\Delta E_D/kT}}} \tag{2.19}$$

For small ΔE_D and $N_D << N_C$, this yields $n \approx N_D$ as expected. For acceptor doping the hole concentration is obtained exchanging N_D, N_C, and ΔE_D by the respective acceptor quantities N_A, N_V, and ΔE_A (see the more general equation (2.23) at $N_D = 0$). However, the degeneracy factor of acceptor levels in Si, Ge, and SiC is $g = 4$ since there are two degenerate valence bands at $k = 0$, from which the levels are split off, and this results in a further degeneracy in addition to spin degeneracy [Bla62].[1] For both n- and p-type conductivity, the degeneracy factor reduces the ionization ratio n/N_D or p/N_A, respectively.

As is confirmed by inserting the ionization energies of Fig. 2.6 and the effective density of states from Eq. (2.8), dopants in Si are nearly completely ionized at room temperature up to doping concentrations of 10^{17} cm^{-3}. Considerable deionization is obtained at higher concentrations. In this doping range one has to take into account, however, that the ionization energy decreases with increasing concentration starting from the values given in Fig. 2.6 for small densities. For a Ga impurity concentration of 5×10^{17} cm^{-3}, which is about the maximum doping of the p-base in a thyristor, an activation energy of 65 meV was determined [Wfs60]. With this ΔE_A an ionization ratio of 66% of the total Ga concentration is obtained at 300 K. As function of temperature, the carrier concentration is shown in Fig. 2.7 for this example and additionally for the case of phosphorus doping with a concentration of 10^{14}cm^{-3}. Whereas for the Ga concentration 5×10^{17}cm^{-3} the deionization is noticeable up to 400 K, for a doping concentration of phosphorous of 10^{14} cm^{-3} the ionization remains complete down to a temperature of 80 K. Above $T = 430$ K the intrinsic concentration becomes comparable with $N_D = 10^{14}$ cm^{-3} and causes an increase of the carrier concentration. To include this effect, the neutrality condition inserted into Eq. (2.18) has to take into account the minority carrier concentration. Writing $n = N_D^+ + n_i^2/n$ where in this range $N_D^+ = N_D$, one obtains

$$n = \frac{N_D}{2} + \sqrt{\left(\frac{N_D}{2}\right)^2 + n_i(T)^2} \tag{2.20}$$

In spite of the explicit dependence on N_D in Eq. (2.19) the ionization ratio does not decrease with increasing doping concentration over the *whole* available doping

[1] Actually the occupation of excited states causes an additional degeneracy and enhancement of g [Bla62, p. 140 ff]. In Si this effect seems to be relatively small owing to the high energy difference between the excited states and the ground state. Since a calculation under real conditions is not available, the effect is not taken into account.

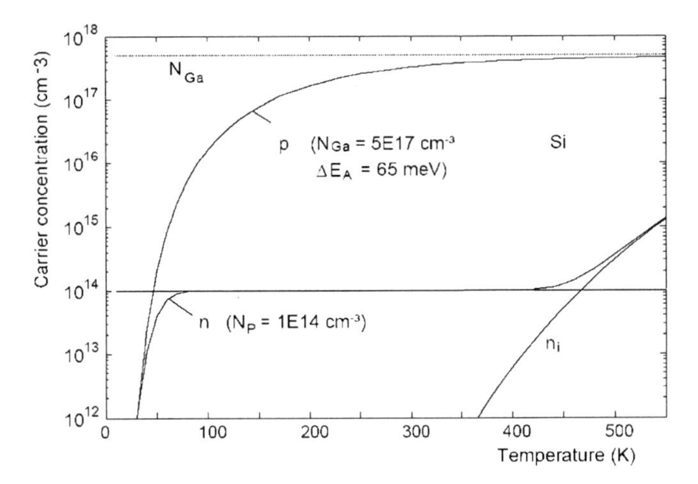

Fig. 2.7 Carrier concentration as a function of temperature for a Ga doping of 5×10^{17} cm^{-3} and the case of a phosphorus doping of 10^{14} cm^{-3}

range. Above 5×10^{17} cm^{-3}, the explicit dependence is strongly counteracted and later overcompensated by the mentioned decrease of the ionization energy. This is caused by screening of the impurity charge by free carriers, e.g. D^+ by electrons, as well as by the formation of a tail of states at the neighbored band because the periodicity of the lattice potential is disturbed by the Coulomb field of the large number of statistically distributed impurities. Additionally, the levels of the impurity atoms spread and form an impurity band because the wave functions of the bound states of the impurities overlap at high concentrations and hence the levels split up. These effects begin more than an order of magnitude below the concentration $N_{C,V}$, where degeneracy sets in. A calculation of Kuzmicz [Kuz86], considering these effects yielded the dependence on impurity concentration of the ionization ratio as shown in Fig. 2.8. The figure refers to phosphorus and boron in Si, the most commonly used dopants. The curves were recalculated using analytical expressions developed in [Kuz86] on the base of numerical calculations. The deionization has its maximum near 2×10^{18} cm^{-3} and amounts to 9% for phosphorus and 20% for boron at 300 K. Above 10^{19} cm^{-3}, the ionization is again complete. The deionization in the intermediate range is noticeable in the temperature dependency of the resistivity.

In SiC the energy levels of dopants lie deeper in the bandgap, and hence a larger part of the doping atoms remains neutral, particularly for acceptors. For Al, the preferred acceptor dopant in SiC devices, the acceptor version of Eq. (2.19) with $N_V = 2.5 \times 10^{19}$cm^{-3} [Gol01], yields for a doping concentration $N_A = 1 \times 10^{16}$cm^{-3} an ionization ratio of only 35% at 300 K. This can strongly influence the device characteristics. Boron in SiC shows even a smaller degree of ionization.

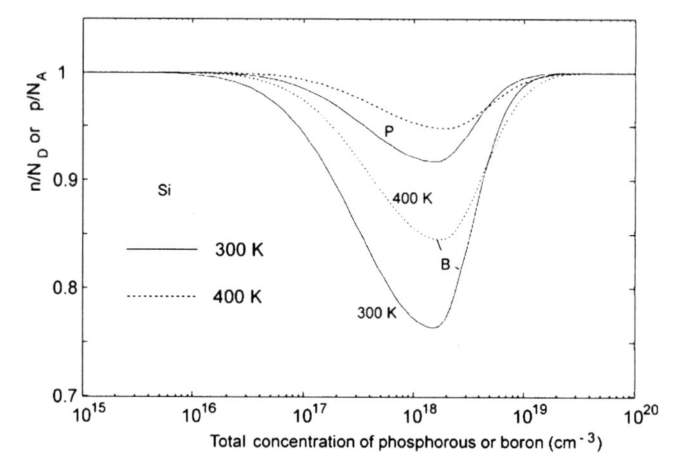

Fig. 2.8 Ionization ratio of phosphorus and boron in Si at 300 and 400 K as a function of the total concentration of the respective impurity. After Kuzmicz [Kuz86]

What if both donors and acceptors are present simultaneously? This is always the case in devices doped by impurity diffusion and particularly in the vicinity of diffused pn-junctions, which are defined by the condition $N_A = N_D$, i.e. that both types of impurities compensate each other. To analyze this compensation effect, we consider the case that the acceptor doping predominates and that the difference $N_A - N_D$ is large against the intrinsic concentration n_i. The donors will then transfer all their surplus electrons to the deeper lying acceptor levels so that $N_D^+ = N_D$. Hence only the portion $N_A - N_D$ of the acceptors is free to accept electrons from the valence band and generate holes. If also this part is completely ionized, the hole density is $p = N_A - N_D = N_{A,net}$, hence equals the net impurity concentration. Usually this is assumed analyzing the functioning of devices similarly as for uncompensated doping. However, the compensation leads to an appreciable reduction of the ionization of the net impurity concentration. To consider this we start as above with Eq. (2.2) to obtain the ionization degree N_A^-/N_A^0. Multiplying this with Eq. (2.5) results in

$$\frac{N_A^-}{N_A^0} p = \frac{N_V}{g} e^{-(E_A - E_V)/kT} \equiv N_{va} \tag{2.21}$$

where again a degeneracy factor g is included for the occupation number of the neutral impurity state. The concentration N_{va} containing the thermal activation term is introduced for compact writing. In addition to Eq. (2.21) one has for the charged and neutral acceptor densities the neutrality condition $N_A^- = p + N_D$ and the equation $N_A^0 = N_A - N_A^- = N_A - p - N_D$ following from it. Inserting this one obtains as equation for the hole density

$$\frac{(p + N_D) \cdot p}{N_A - N_D - p} = N_{va} \tag{2.22}$$

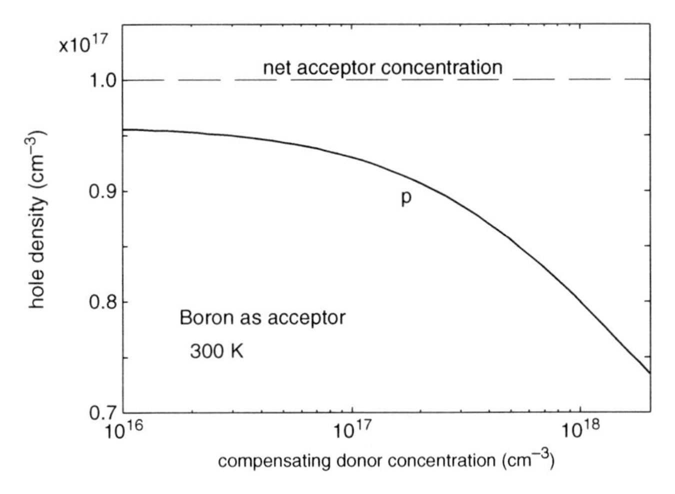

Fig. 2.9 Hole concentration for a constant net acceptor doping $N_A - N_D = 1 \times 10^{17}$ cm^{-3} versus compensating donor density, assuming boron as acceptor

This has the solution

$$p = \frac{2\,(N_\mathrm{A} - N_\mathrm{D})}{1 + \dfrac{N_\mathrm{D}}{N_\mathrm{va}} + \sqrt{\left(1 + \dfrac{N_\mathrm{D}}{N_\mathrm{va}}\right)^2 + 4\left(\dfrac{N_\mathrm{A} - N_\mathrm{D}}{N_\mathrm{va}}\right)}} \tag{2.23}$$

This equation together with Eq. (2.21) is a generalization of the acceptor version of Eq. (2.19) into which it passes with $N_\mathrm{D} = 0$. The equation shows that the fraction of the ionized net acceptor doping does not only depend on $N_\mathrm{A} - N_\mathrm{D}$ itself but also on the density N_D of the compensating doping. *For a given net impurity concentration the ionized fraction thereof decreases with compensation.* This is a significant effect, although the explicit dependence of Eq. (2.23) is weakened at high concentrations on account of the decrease of ionization energies. According to Eq. (2.21) this results in an increase of N_va with doping concentration and hence compensation. We consider first the case of a net p-type doping of 5×10^{17} cm^{-3} achieved by a Ga concentration $N_\mathrm{A} = 1 \times 10^{18}$cm^{-3} and a compensating donor density N_D of 5×10^{17} cm^{-3}. Using an ionization energy of 57 meV for the estimated total ion concentration of $\approx 8 \times 10^{17}$cm^{-3} [Wfs60], Eq. (2.23) results in an ionization ratio $p/(N_\mathrm{A} - N_\mathrm{D})$ of 53%, compared with 66% in the uncompensated case discussed above. As another example, Fig. 2.9 shows the ionization in the case of boron doping as function of the compensating donor density, where the net p-type doping density is kept constant at 1×10^{17} cm^{-3}. The decrease of the ionization energy of boron was expressed as $\Delta E_\mathrm{A} = 46 - 3 \times 10^{-5} \cdot (N_\mathrm{A}^+ + N_\mathrm{D})^{1/3}$ meV [Li78]. The figure shows that the decrease of ionization with increasing compensation is quite significant. Whereas, however, the ionized net impurity portion $p = N_\mathrm{A}^- - N_\mathrm{D}^+$ is considerably reduced by the compensation, not only the donors in our case are completely ionized, also the ionized acceptor density $N_A^- = p + N_D$ is enhanced: For $N_D = 5 \times 10^{17}\text{cm}^{-3}$, $N_A = 6.0 \times 10^{17}\,\text{cm}^{-3}$ in Fig. 2.9, the hole concentration is $p = 0.855(N_A - N_D) = 8.55 \times 10^{16}\,\text{cm}^{-3}$ and $N_A^- = 5.86 \times 10^{17}\,\text{cm}^{-3} = 0.976\,N_A$. The high ionization degree of N_A follows because the donors give their electrons to the acceptors enhancing the number N_A^-. Although the calculation above was simplified considering high doping effects only via ΔE_A, it will provide a good first approximation. In literature, the ionization in cases of doping with more than one impurity species has been discussed early and extensively using equations like Eq. (2.23) [Bla62, p. 132 ff].

Till now we have assumed that the semiconductor is neutral and in thermal equilibrium. In devices, however, there are regions with space charge where carriers are depleted and, on the other hand, regions with injected carriers adding to those supplied by dopants. Departing from above calculations, the ionization in space charge regions is nearly always complete because the depletion of carriers ($p << N_A^-$) shifts the reaction $A \Leftrightarrow A^- + \oplus$ to higher N_A^-. Under conditions of high injection, on the other hand, the deionization is enhanced, and especially in this case it can essentially affect device operation. In Si the equilibrium equations (2.18) and (2.21) can be used in most cases to calculate these effects.

In spite of the limits made obvious by this discussion, it is often a useful approximation for Si devices to assume complete ionization throughout the device. This is actually done in most analytical calculations and programs for device simulation. We can now estimate where a more precise treatment may be advisable.

For complete ionization and temperatures where $n_i << N_D, N_A$, the Fermi level is related in a simple manner to the doping concentration. Inserting $n = N_D$ into Eq. (2.4) one obtains

$$E_C - E_F = k \cdot T \cdot \ln\left(\frac{N_C}{N_D}\right) \tag{2.24}$$

This and the analogous equation for acceptor doping are plotted in Fig. 2.10 for Si at 300 K. The Fermi energy is a linear function of the logarithm of the doping density.

Furthermore, we note the terms *majority carriers* and *minority carriers* which can be used independently of the type of doping. The former are the carriers of the type supplied by the doping atoms and the latter are carriers of the second kind, which are in minority and related with the first by the mass law equation $np = n_i^2$ in thermal equilibrium. In this case, the density of minority carriers is usually many orders of magnitude smaller than the majority concentration. For a donor density $N_D = 1 \times 10^{14}\text{cm}^{-3}$ one obtains at room temperature with $n_i = 10^{10}\,\text{cm}^{-3}$ a majority carrier concentration $n = 1 \times 10^{14}\,\text{cm}^{-3}$ and a minority concentration $p = n_i^2/n = 1 \times 10^6\text{cm}^{-3}$. At higher temperatures, the minority carrier concentration can be calculated inserting Eq. (2.20) into $p = n_i^2/n$, see also Fig. 2.7. Despite the fact that at usual temperatures the minority carriers play virtually no role in the behavior of bulk material in thermal equilibrium, they play a mayor role in the

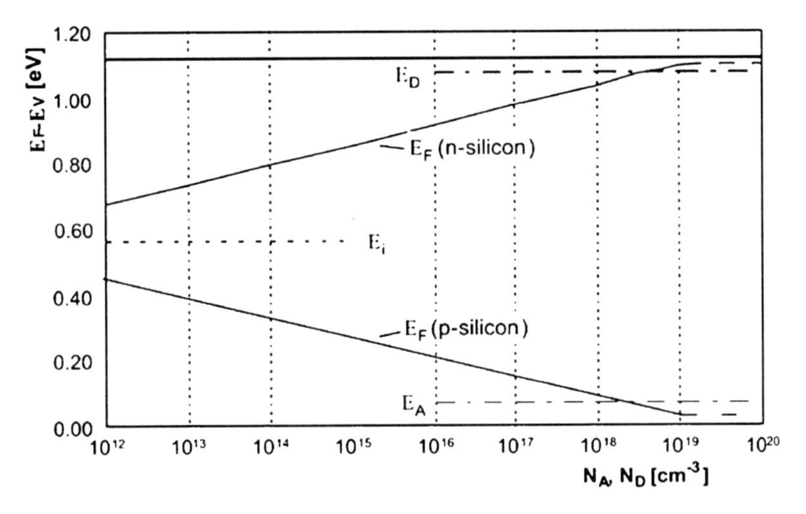

Fig. 2.10 Fermi level in n- and p-type Si versus doping concentration at $T = 300$ K

functioning of pn-devices which is essentially based on injection and extraction of minority carriers. This will be further investigated later.

The high doping effects mentioned in connection with the ionization degree are responsible for another phenomenon which influences the characteristics of power devices: the reduction of bandgap and the corresponding increase of the intrinsic concentration n_i at high doping densities. From measurements on the p-base of bipolar transistors Slotboom and De Graaff [Slo76] obtained the following empirical relationship for the bandgap narrowing, ΔE_g, in dependence of doping concentration N:

$$\Delta E_g = 9 \times 10^{-3}\text{eV} \cdot \left(\ln \frac{N}{10^{17}\text{cm}^{-3}} + \sqrt{\left(\ln \frac{N}{10^{17}\text{cm}^{-3}} \right)^2 + 0.5} \right) \tag{2.25}$$

$$\simeq 18\,\text{meV} \times \ln \frac{N}{10^{17}\text{cm}^{-3}} \quad \text{for } N > 5 \times 10^{17}\,\text{cm}^{-3}$$

Calculations considering band tails and impurity band were found to be in rough accordance with Eq. (2.25) [Slo77]. Another effect which has been shown by Lanyon and Tuft [Lan79] to result in an effective reduction of bandgap is the electrostatic field energy required when an electron–hole pair is created within the surrounding of majority carriers. This stored energy included in the bandgap becomes smaller with increasing (majority) carrier density due to screening. Based on this mechanism, the following theoretical formula has been derived in the range where Maxwell–Boltzmann statistics apply [Lan79]:

$$\Delta E_g = \frac{3q^2}{16\pi\varepsilon} \left(\frac{q^2 n}{\varepsilon kT} \right)^{1/2} = 22.6 \left(\frac{n}{10^{18}} \frac{300}{T} \right)^{1/2} \text{meV} \tag{2.26}$$

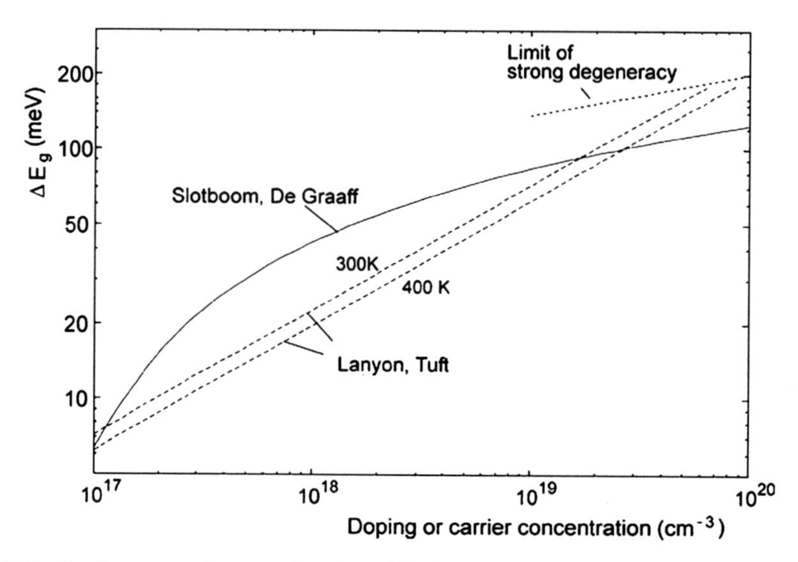

Fig. 2.11 Bandgap narrowing as a function of doping or carrier density according to [Slo76], Eq. (2.25) (*solid line*), and [Lan79], Eq. (2.26) (*dashed lines*), respectively. The *dotted line* represents a limiting expression for strong degeneracy [Lan79]

where ΔE_g depends here on the carrier concentration n (in case of n doping) rather than the doping concentration, which can be of significance especially in space charge regions. The numerical expression on the right-hand side of Eq. (2.26) is obtained for Si using the dielectric constant $\varepsilon = 11.7\varepsilon_0$. Although band tails and impurity band are not considered, Eq. (2.26) is also claimed to describe largely the measurements which were partly reinterpreted [Lan79]. Differing from the results of Slotboom and De Graaff, Eq. (2.25), the bandgap reduction according to Eq. (2.26) depends on temperature. A plot of the two relationships is shown in Fig. 2.11. Around 10^{18} cm^{-3}, the equation of Lanyon and Tuft yields a considerably smaller bandgap narrowing than that of Slotboom and De Graaff, above 2×10^{19} cm^{-3} it is the other way round. Since both theoretical and experimental methods are not free of assumptions, further work seems to be necessary to understand the contribution of the various effects more clearly.

The bandgap narrowing is equivalent to an enhancement of the intrinsic concentration. Inserting the reduced bandgap $E_g = E_{g0} - \Delta E_g$ into Eq. (2.6) one obtains

$$n \cdot p = n_i^2 = N_C N_V e^{-E_g(N)/kT} = n_{i0}^2 e^{\,\Delta E_g/kT} \tag{2.27}$$

where n_{i0} signifies the intrinsic concentration at low doping (or carrier) concentrations corresponding to the bandgap E_{g0}. At $N, n = 2 \times 10^{-18}$ cm^{-3} and 400 K, n_i^2 is enhanced by a factor 4.88 if calculated with Eqs. (2.25), (2.27), and by a factor 2.23 using Eq. (2.26).

In devices with layers of different doping levels, a superscript index "–" or "+" is added to the symbols for the conductivity type to indicate the doping level. So n^-, p^- signify n- and p-layers with a much lower doping than neighboring regions; likewise n^+, p^+ mean layers with a much higher doping level. This notation is used also often to indicate absolute doping ranges where the meaning is typically as follows:

$$
\begin{aligned}
n^-, p^- &: 10^{12} - 10^{14}\ \text{cm}^{-3}\\
n, p &: 10^{15} - 10^{18}\text{cm}^{-3}\\
n^+, p^+ &: 10^{19} - 10^{21}\ \text{cm}^{-3}
\end{aligned}
$$

Normally a power device contains in the interior a weakly doped n^--region which essentially determines the characteristics. Heavily doped n^+-and p^+-layers border on the metalized surfaces. Further topics on doping, such as the solubility of the impurities and the methods of introducing them into the semiconductor, for example diffusion, are discussed in Chap. 4 on technology.

2.6 Current Transport

2.6.1 Carrier Mobilities and Field Currents

As used already in Sect. 2.3, electrons and holes in a semiconductor behave essentially like free particles which, however, are scattered by vibrating lattice atoms, impurity ions and other scattering centers. Like molecules in a gas they have a kinetic thermal energy which on statistical average is

$$E_{kin} = \frac{m}{2} v_{th}^2 = \frac{3}{2} kT \tag{2.28}$$

where v_{th} is the mean thermal velocity and m represents the effective mass of the respective carrier. Already at room temperature the mean thermal velocity is very high: With the electron effective mass $m = m_n = 0.27m_0$, one obtains from Eq. (2.28) $v_{th} = \sqrt{(3kT/m_n)} = 2.2 \times 10^7$ cm/s $= 220\,\mu$m/ns at 300 K. The mean free path between two scattering events is of the order 10 nm in lightly doped Si and still smaller if many impurities are present. Since the thermal motion is statistically distributed over all directions, the current resulting from the large number of carriers is zero between terminals on equal potential. The thermal velocity used in Eq. (2.28) is defined as the root of the mean value of the squared velocity, a rough measure for the mean *absolute* velocity. Contrary to it, the normal *linear* mean value of the thermal velocities taking account also of their directions is zero for both electrons and holes, if no field is present.

When an electric field E is applied, each carrier experiences a force $\pm\, qE$ and is accelerated between two collisions. Hence the thermal velocity is superimposed by an additional velocity which for holes has the direction of the electric field and for electrons the opposite. Averaging linearly over time and the carriers of each type, now *non-zero* mean velocities v_n, v_p of the electrons and holes result. These velocities caused by the field are called *drift* velocities. For low fields, defined by the condition that the drift velocities are small compared with the thermal velocity v_{th}, the mean free time τ_c between two collisions which depend on the total velocity is independent of the field. Therefore, in this range, the drift velocities are proportional to the field strength

$$v_{n,p} = \mp\mu_{n,p} \cdot E \tag{2.29}$$

The proportionality factors μ_n for electrons and μ_p for holes are called *mobilities*. Due to the minus sign used in the case of electrons both mobilities are positive constants. Inserting Eq. (2.29) into the condition $v << v_{th}$, and omitting now the indices n or p, one obtains as condition which the field must satisfy for constant mobilities

$$E << \frac{v_{th}}{\mu} \tag{2.30}$$

The significance of the mobilities follows from their connection with the macroscopic electric current densities. These are obtained from Eq. (2.29) as

$$\begin{aligned} j_n &= -q \cdot n \cdot v_n = q \cdot \mu_n \cdot n \cdot E \\ j_p &= q \cdot p \cdot v_p = q \cdot \mu_p \cdot p \cdot E \end{aligned} \tag{2.31}$$

with the concentrations n and p of electrons and holes, respectively. The sum of both is the total current density j:

$$\begin{aligned} j = j_n + j_p &= q \cdot (\mu_n \cdot n + \mu_p \cdot p) \cdot E \\ &= \sigma \cdot E = E/\rho \end{aligned} \tag{2.32}$$

where

$$\sigma \equiv q \cdot (n \cdot \mu_n + p \cdot \mu_p) = 1/\rho \tag{2.33}$$

is the electrical conductivity and ρ the resistivity. According to these equations, the mobilities are material parameters which determine the ohmic voltage drop $V = E\Delta x = \rho j \Delta x$ for a given current density and hence the power loss density jV and heat generation. Hence, the mobilities determine the maximum allowed current density of devices. Together with other characteristics they decide on the suitability of a semiconductor for power devices.

An overview of mobilities in semiconductors is given in Table 2.2. It shows at first that the mobility of holes is in all cases significantly smaller than that of electrons. Hence, n- and p-regions in devices are not equivalent. Especially in unipolar devices which conduct current only by majority carriers, the weakly doped region required for the blocking voltage is chosen preferentially of n-type doping. As mentioned in Sect. 2.1, the very high electron mobility of GaAs offers the possibility to make Schottky diodes with a low on-state voltage drop and simultaneously a relative large thickness necessary for a high blocking voltage. For the hexagonal

Table 2.2 Mobilities of various semiconductors at room temperature and for light doping (1×10^{14} cm^{-3}). The table contains also the saturation drift velocity of electrons, a property discussed at the end of this chapter

	μ_n(cm^2/Vs)	μ_p(cm^2/Vs)	$v_{sat(n)}$(cm/s)
Ge	3900	1900	6×10^6
Si	1420	470	1.05×10^7
GaAs	8000	400	1×10^7
4H-SiC	1000	115	2×10^7
GaN	<1000	<200	2.5×10^7
Diamond	2200	1800	2.7×10^7

semiconductors SiC and GaN the mobilities are anisotropic, i.e. they are different in directions parallel ($\mu_{||}$) and perpendicular ($\mu_{\perp}$) to the hexagonal axis. The values in the table are the mobilities parallel to the hexagonal axis, which in vertical devices is the direction of main current flow. In 4H-Si the anisotropy is small, $\mu_{n||} \simeq 1.2\,\mu_{n\perp}$ [Scr94] in contrast to the polytype 6H-SiC whose electron mobility $\mu_{n||}$ is about a factor 5 smaller than $\mu_{n\perp}$ and also than $\mu_{n||}$ in 4H-SiC. This is the main reason why 4H-SiC is preferred now against 6H-SiC, which polytype attracted much research efforts in the 1990s. To judge different semiconductors regarding conduction losses in devices, one has to use the mobilities together with other relevant properties such as the necessary thickness and doping density allowed for a given blocking voltage. Although the mobilities of 4H-SiC are lower than those of Si, in combination with the much smaller thickness and higher allowed doping of the base region, a much smaller on-state resistance of unipolar devices can be achieved. This holds also in comparison with GaAs. GaN is with respect to the mobilities comparable to SiC. High mobilities are measured in diamond, but because of large ionization energies of the dopants the carrier concentration is small.

The mobilities are constant with regard to the field (if small enough), but depend on doping concentration and temperature. An accurate knowledge of their dependence on doping concentration is very important particularly because the mobility determines the relation between resistivity and doping density which is used daily to conclude from the simply measurable resistivity to doping density N. Assuming complete ionization one has for an n-type wafer

$$\rho(N) = \frac{1}{q \cdot \mu_n \cdot n} = \frac{1}{q \cdot \mu_n(N)} \cdot \frac{1}{N} \tag{2.34}$$

In an undoped and weakly doped semiconductor, the mobilities are determined by scattering on phonons, i.e. vibrating lattice atoms. Above a doping concentration of 10^{15}cm^{-3}, the mobilities are noticeably and at higher concentrations strongly reduced by collisions with doping ions. At still higher concentrations the scattering by ions is limited by the carriers themselves by screening of the impurity charge. The experimental dependence of mobilities in Si on carrier concentration (= concentration of donor or acceptor ions, respectively) is depicted in Fig. 2.12 which

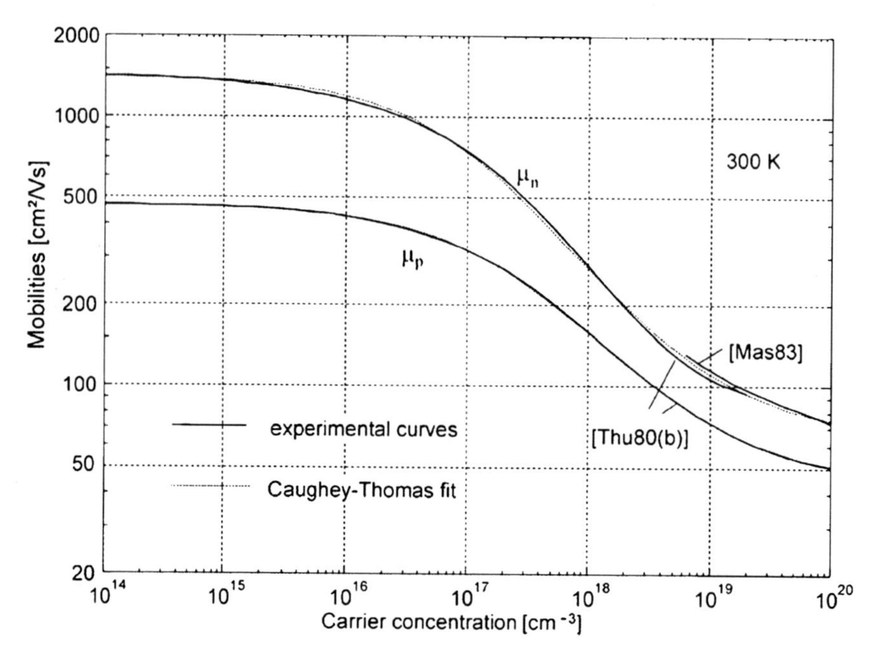

Fig. 2.12 Mobilities in Si as function of carrier density. For μ_p the Caughey–Thomas fit of Eq. (2.35) falls within the line width of the experimental curve

also shows fitting curves to the measurements. The figure is valid at room temperature. The experimental curves represent measurements of Thurber and coworkers on phosphorus-doped silicon for μ_n [Thu80] and boron-doped silicon for μ_p [Thu80b] using their analytical representation of the measurements. For $n > 8 \times 10^{18}\text{cm}^{-3}$ an empirical dependence for the electron mobility of Masetti et al. [Mas83] is shown, who determined the mobilities at very high doping levels. The dashed lines are fits to the experiments by the often used formula of Caughey and Thomas [Cau67]:

$$\mu = \mu_\infty + \frac{\mu_0 - \mu_\infty}{1 + (N/N_{\text{ref}})^\gamma} \tag{2.35}$$

The limiting values μ_0 and μ_∞ for low and high concentrations and the concentration N_{ref} at which the mobility adopts the mean value between them, are matched to the experiments. N denotes the carrier concentration n or p. μ_0 is the mobility due to lattice scattering. As is seen, this simple approach is well suited to describe the experimental dependence up to electron and hole densities of 1×10^{20} cm^{-3}. The parameter values used for the fit are given in Appendix A, where also their variation with temperature is described. Around 2×10^{18} cm^{-3} the carrier density deviates noticeably from the total impurity density as discussed in the previous paragraph. To obtain the mobility for a given total impurity concentration in this range, first the ionized impurity density has to be determined using Fig. 2.8. Although Eq. (2.35) can be used also with N defined as total impurity density, the plot versus carrier density is preferred, because the carrier and ionized impurity concentration are the quantities determining the mobilities, while scattering by impurities in the neutral state may be neglected. Hence the plot is applicable with good accuracy also for doping with As, Ga, and Al, which are ionized to a different extent. Often complete ionization is assumed when discussing mobilities in the range of operation temperatures of devices. We will also do this in what follows.

Besides impurities, carriers are scattered significantly by carriers of the opposite type, i.e. electrons by holes and holes by electrons (electron–hole scattering). For the mobilities in the base region of power devices with high injected carrier densities this results in a significant increase of the on-state voltage drop. Measurements of the mobility sum $\mu_n + \mu_p$ at high injection levels, where $n = p >> N_D, N_A$, have been performed up to a carrier concentration of 1×10^{18} cm^{-3} [Dan72, Kra72]. The experimental sum $\mu_n + \mu_p$ as a function of the concentration $c = n = p$ is within the experimental accuracy equal to the mobility sum which is calculated from Eq. (2.35) (with the respective parameter sets) inserting the concentration $N = c$. Hence, both for doped silicon in thermal equilibrium and in the case of high injection level, the experimental results can be described uniquely defining N in Eq. (2.35) as the total concentration of scatterers of the respective carriers , i.e. $N = N_D + p$ in the $\mu_n(N)$-function and $N = N_A + n$ in $\mu_p(N)$. Obviously this represents a plausible model also for the general case, where impurity scattering *and* electron hole scattering have to be considered. Donor ions and holes can be put together because they are (approximately) equally effective in scattering of electrons as shown by the measurements, and likewise this holds for the scatterers of holes. Scattering of electrons by electrons and holes by holes is a second-order effect and neglected.

In regions with compensation, the mobility is influenced by repulsive ions in addition to attractive. Although of considerable practical interest, experimental investigations of mobilities in compensated semiconductors are very scarce. Theoretically, attractive and repulsive Coulomb fields have equal scattering cross section and hence the same effect on mobilities, if the concentrations are not too high (smaller than about 10^{18} cm^{-3} at room temperature), so that an approximate scattering theory (Born approximation) applies [Mol64, Smi59, Kla92] and, additionally, screening can be neglected. Hence, the sum of donor and acceptor concentrations, $N = N_A + N_D$, has to be inserted into Eq. (2.35) in this case. Generally, if also injected carriers are present, the mobilities are given by

$$\begin{aligned} &\text{a)} \qquad \mu_n = f_n(N_D + N_A + p) \\ &\text{b)} \qquad \mu_p = f_p(N_D + N_A + n) \end{aligned} \tag{2.36}$$

where $f_n(N)$ and $f_p(N)$ are functions (2.35) with the μ_n and μ_p parameter set, respectively. This model includes also the case of mobility of minority carriers which is determined by repulsive impurities. For example, the electron mobility in a p-type region is obtained with $N_A \geq N_D$ in Eq. (2.36a). Therefore, this model can be broadly used. An important restriction, however, concerns the concentration of repulsive scatterers: It must be smaller than about 10^{18} cm^{-3} in order that the results are in sufficient agreement with measurements and the theory.

At higher concentrations, the mobilities in compensated material and the minority carrier mobilities are influenced by screening, which depends on the carrier density and hence is given (approximately) by the *difference* of the doping concentrations in thermal equilibrium. In the empirical equation (2.35), screening is included for the majority mobility in uncompensated silicon in thermal equilibrium. Additionally, a rigorous quantum mechanical scattering theory shows that scattering on repulsive Coulomb centers is weaker than on attractive ions, and this becomes significant at high concentrations of the scatterers [Ben83]. A model including these effects has been developed by Klaassen [Kla92] using Eq. (2.35) as an empirical base. To get beyond the special case of this formula, a screening term selected from Eq. (2.35) has been generalized considering the relevant theory, and weighting factors $G < 1$ were calculated with which the concentrations of repulsive scatterers are multiplied. According to this generally applicable model the minority mobilities at high doping densities are considerably higher than majority mobilities, which is in agreement with measurements [Swi87, Kla92, Dzi79]. At concentrations of repulsive scatterers below 10^{18} cm^{-3}, the model reduces gradually to the simplified model described above (minor differences ignored). For details we refer to [Kla92, Kla92b].

The *temperature* dependence of mobilities is determined by the same effects as the dependence on doping density. At *low* doping, scattering on thermal lattice vibrations predominates and yields a decrease of mobilities with temperature approximately proportional to T^{-2}. Impurity scattering is shown theoretically to yield a mobility μ_i *increasing* with temperature nearly as $T^{3/2}$, if screening is neglected. This comes from the increase of the thermal velocity causing a decrease of scattering by the Coulomb centers. The increase of thermal velocity, however, implicates also a decrease of screening which at very high doping density inverts the temperature dependence of μ_i to a slow *decrease* with *T*. These effects of the partial mobilities μ_l and μ_i find themselves also in the total mobility μ which is formed according to the approximate rule

$$\frac{1}{\mu} = \frac{1}{\mu_\mathrm{l}} + \frac{1}{\mu_\mathrm{i}} \tag{2.37}$$

Considering the temperature range 250–450 K, the result is that at low doping the (total) mobilities in Si show a strong decrease with increasing *T*, an approximate independency on *T* in the doping range 10^{18} to 10^{19}cm^{-3} and at very high doping a minor decrease. The experimental temperature dependence can be well described – together with doping dependence – using temperature-dependent parameters μ_0, μ_∞, N_ref and γ in formula (2.35) as given in Appendix A.

In Fig. 2.13 the temperature dependency of μ_n and μ_p is shown for two doping densities, one which is typical for the n-base region of power devices (left-hand side) and the other for the p-base in thyristors, IGBTs, and bipolar transistors (right-hand side). The temperature dependence at 3×10^{17} cm^{-3} is considerably weaker than

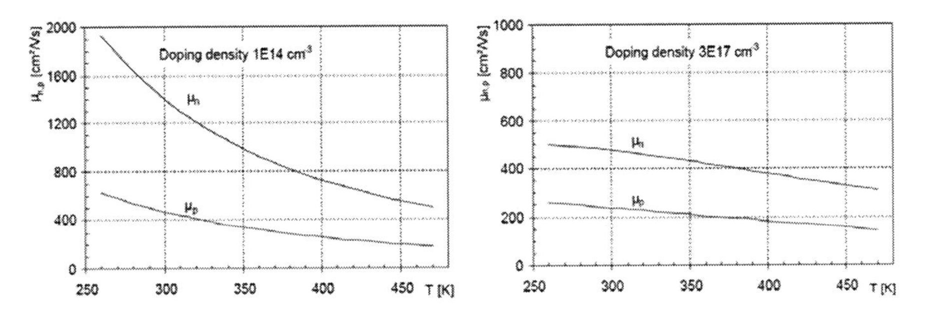

Fig. 2.13 Temperature dependence of mobilities in Si at two doping densities (see text)

at 1×10^{14} cm^{-3}. That the resistivity $\rho = 1/(q\mu_p N_A)$ of the p-base of mentioned devices increases relatively weakly with T is very desirable for the *lateral* resistance of the p-base (see Chaps. 8 and 10). Mainly the mobilities and their temperature dependency are of great significance for the on-state voltage drop of power devices, including the on-resistance of MOSFETs. The dependence on injected carriers and compensating doping can be taken into account also in the temperature-dependent model using Eqs. (2.36) and (2.35) and Appendix A.

2.6.2 High-Field Drift Velocities

At high electric fields, where condition (2.30) is not satisfied, the drift velocity is no longer proportional to the field, but increases weaker. At very high-field strength it approaches a limiting value, the saturation drift velocity v_{sat}. The dependency of the drift velocities of electrons and holes on the electric field is often expressed in the form [Cau67, Tho80]:

$$v_{n,p} = \frac{\mu_{n,p}^{(0)} \cdot E}{\left(1 + \left(\frac{\mu_{n,p}^{(0)} \cdot E}{v_{sat(n,p)}}\right)^{\beta}\right)^{1/\beta}} \tag{2.38}$$

where $\mu_{n,p}^{(0)}$ are the low-field mobilities discussed above. The additional marking is introduced, because Eq. (2.29) is used also for the non-linear range of fields where the mobilities $\mu_{n,p} \equiv v_{n,p}/E$ are field dependent. For small E, Eq. (2.38) turns into Eq. (2.29) with $\mu_{n,p} = \mu_{n,p}^{(0)}$. The exponent β was proposed by Caughy and Thomas [Cau67] to be $\beta_n = 2$ for electrons and $\beta_p = 1$ for holes. Jacobini and coworkers [Jac77] have determined the β's renewed and as functions of temperature. They obtained $\beta_n = 2.57 \times 10^{-2} \cdot T^{0.66}$, $\beta_p = 0.46 \times T^{0.17}$, which yields a value near 1 at 300 K in *both* cases. Besides $\mu_{n,p}$ and $\beta_{n,p}$, also the saturation velocities are functions of temperature. According to Jacobini et al. [Jac77] they decrease with T as $v_{sat(n)} = 1.53 \times 10^9/T^{0.87}$; $v_{sat(p)} = 1.62 \times 10^8/T^{0.52}$ in cm/s.[2]

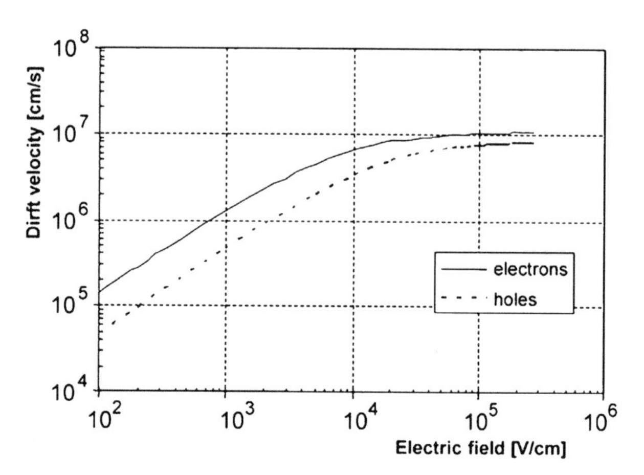

Fig. 2.14 Drift velocity of electrons and holes as function of the electric field. Temperature 300 K

Using these results in Eq. (2.38), one obtains at 300 K the field dependences shown in Fig. 2.14. Below 10^3 V/cm, fields which occur in devices during the conducting state, one obtains the linear dependency described by constant mobilities. Soon above this value, however, the increase of v_n and somewhat later that of v_p becomes sub-linear, as is expected also from condition (2.30). At 3×10^4 V/cm the drift velocities approach the respective saturation velocity $v_{sat(n)}$ or $v_{sat(p)}$. These range near 1×10^7 cm/s and thus already reach the order of magnitude of the mean thermal velocity. Fields in the range up to typically 2×10^5 V/cm appear in space charge regions during blocking of devices. Hence for a wide range of blocking voltages, carriers in the space charge region move with saturation velocity.

Relationship (2.38) has been verified experimentally in pure silicon [Jac77], for doped silicon experimental data of the high-field drift velocity are not available. Using the doping-dependent mobility $\mu_{n,p}^{(0)}(N)$ in Eq. (2.38), however, a physically reasonable dependence on E and doping density N is obtained which satisfies general scaling requirements [Tho80]. The saturation velocities $v_{sat(n,p)}$ are assumed to be independent of N. An alternative approach for $v_{n,p}$ as functions of E and N has been given by Scharfetter and Gummel [Scf69], see Appendix A1.

2.6.3 Diffusion of Carriers and Current Transport Equations

Unlike the situation in metals, the current in semiconductor devices is caused often not only by an electric field but also by diffusion of carriers. Generally, if some mobile particles have a spatially variable concentration C, they diffuse from a region of high to a region of low concentration. The particle current density J thus arising

[2] In [Jac77] a second expression for the saturation velocity of electrons has been given, which reads:

$$v_{sat(n)} = \frac{2.4 \times 10^7 \text{cm/s}}{1 + 0.8 \cdot \exp(T/600)}$$

is proportional to the negative concentration gradient (Fick's first law):

$$J = -D \cdot \nabla C \tag{2.39}$$

where the proportionality factor D is the diffusion constant. This holds also for electrons and holes, whose concentrations can vary because of a variation of the doping concentration or as result of injection of carriers. Due to the charge of the carriers the particle currents are connected with electrical currents. Multiplying with $\pm\, q$ and assuming the concentration gradient to appear in x-direction, the electrical diffusion current densities are

$$j_{\mathrm{n,diff}} = q \cdot D_{\mathrm{n}} \cdot \frac{\mathrm{d}n}{\mathrm{d}x} \tag{2.40}$$

$$j_{\mathrm{p,diff}} = -q \cdot D_{\mathrm{p}} \cdot \frac{\mathrm{d}p}{\mathrm{d}x} \tag{2.41}$$

Together with the field currents, Eq. (2.31), the total current densities are given by the transport equations:

$$j_{\mathrm{n}} = q \cdot \left(\mu_{\mathrm{n}} \cdot n \cdot E + D_{\mathrm{n}} \cdot \frac{\mathrm{d}n}{\mathrm{d}x} \right) \tag{2.42}$$

$$j_{\mathrm{p}} = q \cdot \left(\mu_{\mathrm{p}} \cdot p \cdot E - D_{\mathrm{p}} \cdot \frac{\mathrm{d}p}{\mathrm{d}x} \right) \tag{2.43}$$

The diffusion constants D_{n}, D_{p} depend on the same scattering mechanisms as the mobilities. In fact, they are related to the mobilities by the following simple relationship

$$D_{\mathrm{n,p}} = \frac{kT}{q} \cdot \mu_{\mathrm{n,p}} \tag{2.44}$$

which is called *Einstein relation*. This can be derived from the case of thermal equilibrium [Sho59]: As follows from Eq. (2.43) with the thermal equilibrium condition $j_{\mathrm{p}} = 0$, a concentration gradient $\mathrm{d}p/\mathrm{d}x$ is connected with a field E and hence with an electrical potential $V(x) = -\int E dx$. Now the holes obey the Boltzmann distribution:

$$p(x) \propto \exp\left(-\frac{qV(x)}{kT} \right) \tag{2.45}$$

as can be concluded from Eq. (2.2). The energy states are located here at varying points in space. On the other hand, one obtains from Eq. (2.43) with $j_{\mathrm{p}} = 0$

$$\begin{gathered} \frac{\mathrm{d}\ln p}{\mathrm{d}x} = \frac{\mu_{\mathrm{p}}}{D_{\mathrm{p}}} E = -\frac{\mu_{\mathrm{p}}}{D_{\mathrm{p}}} \times \frac{\mathrm{d}V}{\mathrm{d}x} \\ p \propto \exp\left(\frac{\mu_{\mathrm{p}}}{D_{\mathrm{p}}} V(x) \right) \end{gathered} \tag{2.46}$$

This is compatible with Eq. (2.45) only if $D_p/\mu_p = k\ T/q$. Similarly one can proceed for electrons. Hence the Einstein relation (2.44) is an immediate consequence of the Boltzmann distribution. It is valid independently of the doping concentration and the type of doping. Although derived from thermal equilibrium, it is applicable also to non-equilibrium as long as $\mu_{n,p}$ and $D_{n,p}$ stay constant or change by the same factor in the non-equilibrium state. Therefore, the Einstein relation is applicable generally if the field (see condition (2.30)) and the excess carrier concentration are relatively small. Actually, it is used also under conditions of high injection where the applicability is not guaranteed. The quantity kT/q has the dimension of a voltage and is named thermal voltage. Its value at 300 K is 25.85 mV. Hence the diffusion constants measured in cm^2/s amount to only about 1/40th of the respective mobility in $cm^2/(Vs)$.

2.7 Recombination-Generation and Lifetime of Non-equilibrium Carriers

In thermal equilibrium, charge carriers are continuously generated thermally and they disappear with the same rate by recombination. In devices during operation, however, carrier densities in the active region are not in thermal equilibrium, they are higher or lower than the densities according to the equilibrium Eqs. (2.4), (2.5), and (2.6). A non-equilibrium state tends to restore itself to equilibrium. The time within which the system strives to achieve this, if injection/extraction and external generation are turned off, is determined by the *lifetime* τ of the non-equilibrium carriers. This is a technologically adjustable quantity which is decisive for the dynamic as well as static behavior of power devices.

The definition of the lifetime uses the net recombination rates R_n and R_p of electrons and holes which are defined as the difference between the thermal recombination rates $r_{n,p}$ and thermal generation rates $g_{n,p}$:

$$R_n \equiv r_n - g_n,\ R_p \equiv r_p - g_p$$

These thermodynamic quantities, which are zero in thermal equilibrium, describe the decrease of n and p with time due to the net thermal recombination. Hence one has

$$\begin{aligned} R_n &\equiv r_n - g_n = -\left(\frac{\partial n}{\partial t}\right)_{rg} \\ R_p &\equiv r_p - g_p = -\left(\frac{\partial p}{\partial t}\right)_{rg} \end{aligned} \tag{2.47}$$

where the index "rg" marks the part of the time derivatives owing solely to recombination and generation. An in/outflow of carriers into/out of the considered volume element as well as external generation, for example by light, is excluded in Eq. (2.47). R_n, R_p depend on doping and carrier densities and increase with increasing deviation of the respective carrier density from the equilibrium density n_0

or p_0. The relationship between R_n, R_p, and the excess concentration $\Delta n \equiv n - n_0$, respectively $\Delta p \equiv p - p_0$, is usually not very far from a linear dependency. Therefore it is useful to define lifetimes τ_n, τ_p by the equations:

$$R_n \equiv \frac{\Delta n}{\tau_n}, \quad R_p \equiv \frac{\Delta p}{\tau_p} \tag{2.48}$$

Compared with R_n, R_p, the lifetimes depend not very strongly on the respective excess concentration, and in some significant cases they are actually constant. This holds for the lifetime of injected minority carriers, the *minority carrier lifetime*, if the injection level is low (density of minority carriers is small compared to the majority concentration).

For the decay of a homogeneous excess concentration, e.g. of $\Delta p(t)$, Eqs. (2.47) and (2.48) yield

$$\frac{\mathrm{d}\Delta p}{\mathrm{d}t} = -\frac{\Delta p}{\tau_p} \quad , \; \Delta p = p(0) \cdot e^{-t/\tau_p} \tag{2.49}$$

where in the latter equation τ_p was assumed to be constant. From this the meaning of τ as a *lifetime* of the excess carriers becomes obvious. In the *stationary* case the disappearance of carriers due to recombination (for $R_n > 0$) is compensated by a net inflow of carriers or by external generation. In this case the net recombination rates of electrons and of holes are equal:

$$R_n = R_p = R \tag{2.50}$$

since the number of electrons leaving the conduction band must equal the number of electrons entering the valence band, the charge on possible intermediate levels being constant in the stationary case. If energy levels in the bandgap are not involved, Eq. (2.50) is valid also for time-dependent processes.

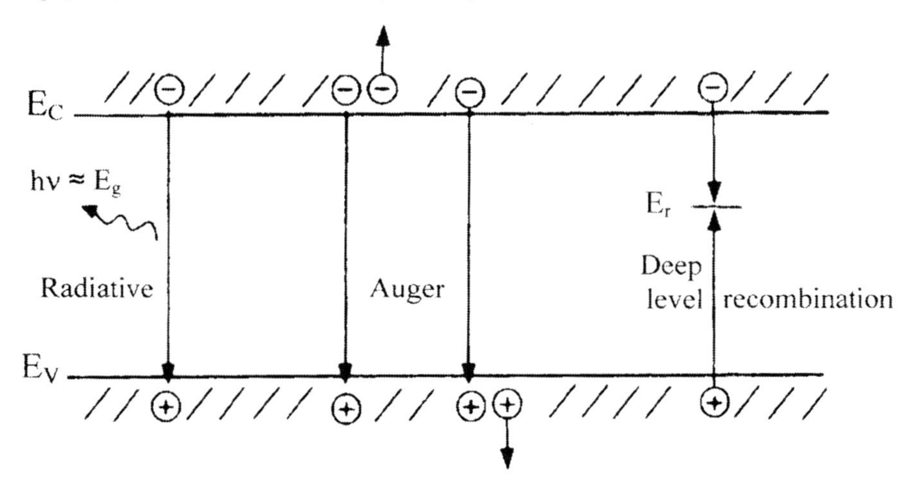

Fig. 2.15 Mechanisms of recombination: Radiative band-to-band recombination, Auger recombination with energy transfer to a third carrier, and recombination via a deep energy level

There are mainly three physical *mechanisms of recombination*, via which the processes of recombination and generation take place in devices: (i) recombination at recombination centers formed by "deep impurities" or traps which have energy levels deep in the bandgap, (ii) band-to-band Auger recombination, and (iii) radiative band-to-band recombination. The latter two mechanisms occur in the semiconductor lattice itself and depend only on carrier concentrations, not directly on the density of normal and deep impurities. The three mechanisms are illustrated in Fig. 2.15. The total recombination rate is made up additive of the single parts, hence according to Eq. (2.48) the total *lifetime* is calculated by adding the *reciprocal* values of the single lifetimes. The intrinsic mechanisms of radiative and Auger recombination are described first.

2.7.1 Intrinsic Recombination Mechanisms

a. **Radiative band-to-band recombination:** As shown in Sect. 2.4, this direct recombination of an electron and a hole under transfer of the released energy to a light quant has a high probability only in direct semiconductors. According to simple statistics the net recombination rate is

$$R = B\,(np - n_i^2) \tag{2.51}$$

with the radiative recombination probability B.

The radiative lifetime, for example the hole lifetime in an n-type semiconductor, is obtained with Eq. (2.48) as $\tau_{\mathrm{p,rad}} = \Delta p/R = 1/(Bn)$ since $np - n_\mathrm{i}^2 = n_0\Delta p + p_0\Delta n + \Delta n\Delta p = n\Delta p$ assuming $p_0 << n_0$. Hence, the radiative minority carrier lifetime is inversely proportional to the majority carrier density. In most cases $np >> n_\mathrm{i}^2$, so that $R \simeq Bnp$. In GaAs the radiative lifetime is estimated to be 6 μs at a doping density $N = 1 \times 10^{15}\ \mathrm{cm}^{-3}$ or 60 ns at $1\times10^{17}\ \mathrm{cm}^{-3}$ [Atk85]. This small lifetime limits the applicability of GaAs for bipolar devices. In silicon, the recombination constant at 300 K is $B \simeq 1 \times 10^{-14}\ \mathrm{cm}^3/\mathrm{s}$ [Sco75] yielding $\tau_{\mathrm{rad}} = 1$ ms at a majority carrier density $n = 1 \times 10^{17}\ \mathrm{cm}^{-3}$. In practice such a high lifetime is not measured in silicon devices, because other recombination mechanisms are more effective (see Sect. 2.7.2). Using the connection between the injected carrier concentrations and the intensity of the recombination radiation, the radiation is used to investigate the inner operation of devices.

b. **Band-to-band Auger recombination:** In Auger recombination, the energy released during the recombination event is transferred not to a light quant but to a third electron or hole, where participation of a phonon can be required for conservation of momentum. Therefore, the recombination probability B in Eq. (2.51) has to be replaced by a factor which is proportional to the carrier concentrations. Hence the Auger recombination rate is

$$R_\mathrm{A} = (c_{\mathrm{A,n}} \cdot n + c_{\mathrm{A,p}} \cdot p)\,(n \cdot p - n_\mathrm{i}^2) \tag{2.52}$$

The coefficients $c_{A,n}$, $c_{A,p}$ determine the recombination rate for the cases that the third carrier taking the energy away is an electron, respectively a hole. Since the concentrations appear with power 3, the probability of this mechanism increases strongly with carrier concentration, and the lifetime *decreases* strongly. Therefore, the Auger recombination is important mainly in highly doped regions. In an n^+-region with a small concentration of injected holes, with $p << n$ and $np >> n_i^2$ Eq. (2.52) turns into $R_A = c_{A,n}n^2p$, and for the hole lifetime one obtains from Eq. (2.48)

$$\tau_{A,p} = \frac{p}{R_A} = \frac{1}{c_{A,n} \cdot n^2} \tag{2.53}$$

where the extremely small equilibrium concentration p_0 has been neglected. The formula for the electron lifetime in a p^+-region is formed analogously. The Auger coefficients in silicon are in the range of $10^{-31}\text{cm}^6/\text{s}$, according to [Dzi77] their values are

$$c_{A,n} = 2.8 \times 10^{-31}\text{cm}^6/\text{s}, \quad c_{A,p} = 1 \times 10^{-31}\text{cm}^6/\text{s} \tag{2.54}$$

They are approximately independent of temperature. For a doping density of 1×10^{19} cm^{-3}, the Auger electron lifetime in a p^+-region is $\tau_{A,n} = 1/(c_{A,p}p^2) = 0.1\ \mu\text{s}$, and the hole lifetime in an n^+-region is 0.036 μs. The small lifetime in highly doped regions is a constituent part of the *h* parameters via which the properties of these regions influence the characteristics of devices. This will be shown in Sect. 3.4.

Another case where the Auger recombination is of significance in devices is that of high concentrations of injected carriers in a weakly doped base region. With neglect of the doping density the neutrality requires $p \simeq n$ which inserted into Eq. (2.52) yields

$$R_{A,HL} = (c_{A,n} + c_{A,p}) \cdot p^3 \tag{2.55}$$

Hence, the high-level Auger lifetime is

$$\tau_{A,HL} = \frac{1}{(c_{A,n} + c_{A,p}) \cdot p^2} \tag{2.56}$$

At $n = p = 3 \times 10^{17}\ \text{cm}^{-3}$, this relation together with Eq. (2.54) results in an Auger lifetime of 29 μs. At high current densities, the Auger recombination in the base region of high-voltage devices becomes significant.

2.7.2 Recombination and Generation at Recombination Centers

The recombination via deep energy levels in the bandgap, associated with appropriate impurities or lattice imperfections, is the dominant mechanism of recombination in lowly and intermediately doped regions in silicon devices. By means of this

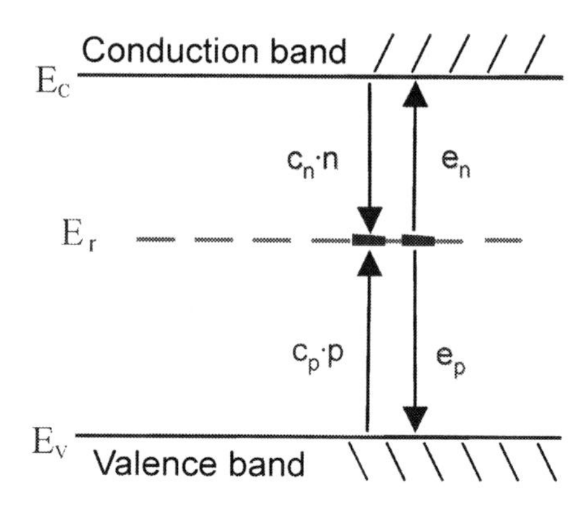

Fig. 2.16 Capture and emission of carriers at a recombination center

mechanism the lifetime can be controlled over a wide range by the density of the deep impurities or traps. This is commonly used to adjust the trade-off between dynamical and stationary device properties. The doping with deep impurities is carried out after the normal doping which determines the conductivity. In the history of device technology, at first gold was used as deep impurity for lifetime control in silicon. Meanwhile, many power devices are diffused with platinum, and most important is now the generation of lattice defects with deep levels by electron, proton, or α-particle irradiation.

The recombination at a deep impurity proceeds in two steps: the capture of a conduction electron which then occupies the deep energy level and thereafter a falling down of the electron into an empty place of the valence band, meaning capture of a hole by the impurity (see Fig. 2.16). Vice versa, an electron–hole pair is generated by thermal emission of a valence electron first to the impurity level, i.e. emission of a hole from the impurity to the valence band, and then emission of the electron from the impurity level to the conduction band. The energy released during capture of a carrier is transferred to lattice vibrations, and the energy required conversely for generation is taken up from the lattice. Because of the large band-to-level distance a series of phonons are emitted respectively absorbed during capture and emission. These multi-phonon processes of capture and emission, however, are considered as a whole and described by overall capture and emission probabilities.

We consider now in detail the recombination at a center R which can appear in a neutral and a negative charge state R^0, R^-. The impurity level is called in this case an acceptor level independent of its position in the bandgap. Similarly, if the charge state of the impurity atom changes from positive to neutral when the level is occupied by an electron, the level is called a donor level. The capture of electrons by centers R^0 defines an electron recombination rate $r_n = c_n n N_r^0$ where N_r^0 is the concentration of the neutral centers and c_n a constant called capture probability capture or capture rate. The electron *generation* rate, given by emission of electrons from centers R^- to the conduction band, is proportional to the concentration of negatively charged centers N_r^-, $g_n = e_n N_r^-$, where the constant e_n is the emission probability also called emission rate of electrons. Hence the net recombination rate R_n is

$$R_n = c_n n N_r^0 - e_n N_r^- \tag{2.57}$$

From thermal equilibrium with $R_n = 0$, one obtains, indicating the concentrations by a "0" and using Eqs. (2.21) and (2.6), the following relationship between emission and capture probability:

$$e_n = c_n \frac{n_0 N_{r0}^0}{N_{r0}^-} = c_n n_r \text{ with } n_r = N_c g \exp\left(-\frac{E_c - E_r}{kT}\right) \tag{2.58}$$

Similarly, the capture of holes at R^- and emission (generation) of holes from R^0 result in the net recombination rate of holes:

$$R_p = c_p p N_r^- - e_p N_r^0 \tag{2.59}$$

where c_p is the hole capture probability and e_p the emission probability of holes. A relationship between these quantities is obtained again from thermal equilibrium with $R_p = 0$ using Eq. (2.21)

$$\begin{aligned} e_p &= c_p \frac{p_0 N_{r0}^-}{N_{r0}^0} = c_p p_r \text{ with} \\ p_r &= \frac{N_v}{g} \exp\left(-\frac{E_r - E_v}{kT}\right) = n_i^2 / n_r \end{aligned} \tag{2.60}$$

The relation $n_r p_r = n_i^2$ is easily verified. The concentrations n_r, p_r which relate the emission probabilities to the corresponding capture probabilities are apart from the degeneracy factor g identical with the electron or hole density, respectively, which are present if the Fermi level coincides with the recombination level. As mentioned earlier, in the stationary case and generally if the time variation of charge on the deep levels is negligible, one has $R_n = R_p$. Equating hence the right-hand sides of Eqs. (2.57) and (2.59) and using the total concentration $N_r = N_r^0 + N_r^-$, one can solve for N_r^- and N_r^0 to obtain

$$N_r^- = \frac{c_n n + c_p p_r}{c_n(n + n_r) + c_p(p + p_r)} N_r \quad N_r^0 = N_r - N_r^- \tag{2.61}$$

Inserting this into Eq. (2.57) or Eq. (2.59) the net recombination rate is obtained as the following function of n and p:

$$\begin{aligned} R_n = R_p = R &= c_n c_p N_r \frac{n \cdot p - n_i^2}{c_n(n + n_r) + c_p(p + p_r)} \\ &= \frac{n \cdot p - n_i^2}{\tau_{p0} \cdot n + \tau_{n0} \cdot p + \tau_g \cdot n_i} \end{aligned} \tag{2.62}$$

τ_{n0}, τ_{p0} and τ_g are lifetime quantities defined as

$$\tau_{p0} = \frac{1}{N_r c_p} \quad \tau_{n0} = \frac{1}{N_r c_n} \tag{2.63}$$

$$\tau_{\mathrm{g}} = \frac{n_{\mathrm{i}}}{N_{\mathrm{r}}}\left[\frac{1}{e_{\mathrm{n}}} + \frac{1}{e_{\mathrm{p}}}\right] \tag{2.64}$$

These equations with Eq. (2.62) in the center are called the model of Shockley, Read, and Hall [Sho52, Hal52]. They are widely used to describe the effect of deep impurities on device properties. The equations are valid also for a donor level except that in Eqs. (2.57), (2.58), (2.59), (2.60), and (2.61) N_{r}^{0} has to be replaced by N_{r}^{+} and N_{r}^{-} by N_{r}^{0} and the degeneracy factor g by 1/g. The capture coefficients have typically values in the range $10^{-9} - 2 \times 10^{-7}\ \mathrm{cm^{-3}s^{-1}}$ at 300 K, they decrease mostly slightly with temperature. So, a concentration N_{r} of only $1 \times 10^{13}\ \mathrm{cm^{-3}}$ results in lifetimes values τ_{n0}, τ_{p0} in the range 0.5–100 μs.

We consider now some consequences of the SRH model. Together with Eq. (2.48), Eq. (2.62) yields the lifetime as function of n and p. In an n-type region with $n_0 >> p_0$ and $np - n_{\mathrm{i}}^2 = n_0\Delta p + p_0\Delta n + \Delta n\ \Delta p = n\Delta p$, the hole lifetime $\tau_{\mathrm{p}} = \Delta p/R$ is obtained as follows:

$$\tau_{\mathrm{p}} = \tau_{\mathrm{p0}} + \tau_{\mathrm{n0}}\frac{p}{n} + \tau_{\mathrm{g}}\frac{n_{\mathrm{i}}}{n} = \tau_{\mathrm{p0}}\left(1 + \frac{n_{\mathrm{r}}}{n}\right) + \tau_{\mathrm{n0}}\left(\frac{p + p_{\mathrm{r}}}{n}\right) \tag{2.65}$$

In the case of neutrality, n is given according to $n = n_0 + p$ by the hole concentration, if charge on the traps can be neglected. According to Eqs. (2.58) and (2.60), n_{r}, p_{r} depend exponentially on the position of the recombination level and upon temperature. At higher temperatures, at least the higher of the concentration of n_{r}, p_{r} has a considerable effect on τ, assuming the concentration n not much higher than about $10^{14}\ \mathrm{cm^{-3}}$. The lifetime at low injection level ($n \simeq n_0 >> p$) is

$$\tau_{\mathrm{p,LL}} = \tau_{\mathrm{p0}}\left(1 + \frac{n_{\mathrm{r}}}{n_0}\right) + \tau_{\mathrm{n0}}\ \frac{p_{\mathrm{r}}}{n_0} \tag{2.66}$$

and this becomes equal to τ_{p0} if the recombination level is located near the middle of the gap or the temperature is low ($n_0 >> n_{\mathrm{r}},\ p_{\mathrm{r}}$) . Similar holds for the electron lifetime τ_{n} in a p-region. At high injection levels defined by the condition $n = p >> n_0,\ p_0$, both lifetimes turn into the high level lifetime

$$\tau_{\mathrm{HL}} = \tau_{\mathrm{no}} + \tau_{\mathrm{p0}} + (\tau_{\mathrm{p0}}n_{\mathrm{r}} + \tau_{\mathrm{n0}}p_{\mathrm{r}})/n \tag{2.67}$$

which in a range where also $n = p >> n_{\mathrm{r}}$, $\mathrm{p_r}$ approaches the limiting value

$$\tau_{\mathrm{HL,lim}} = \tau_{\mathrm{n0}} + \tau_{\mathrm{p0}} \tag{2.68}$$

For recombination levels near one of the band edges and particularly at elevated temperatures, form (2.67) of the high-level lifetime applies to a considerable range of concentrations. The lifetime $\tau_{\mathrm{HL,lim}}$ is smaller than the low-level lifetime if $\tau_{\mathrm{p0}}n_{\mathrm{r}} + \tau_{\mathrm{n0}}p_{\mathrm{r}} > \tau_{\mathrm{n0}}n_0$. Between the limiting values the lifetime varies monotonously,

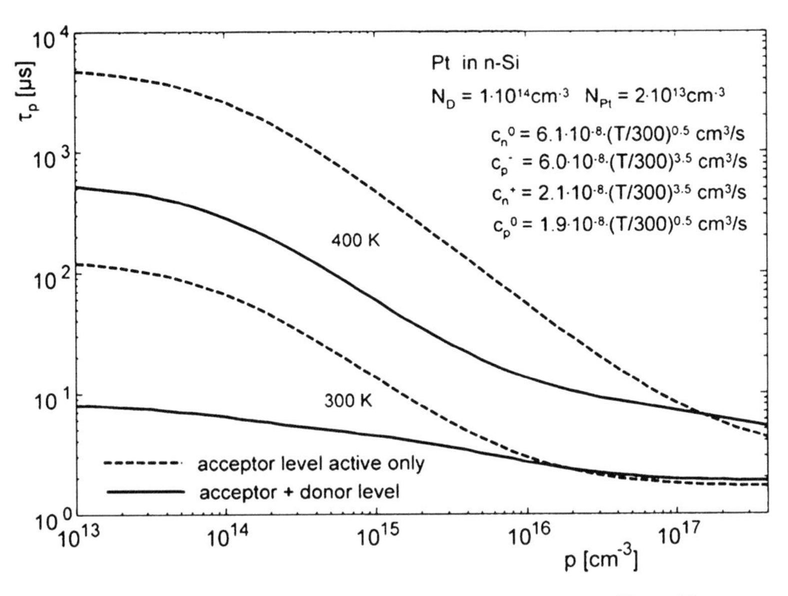

Fig. 2.17 Hole lifetime for the platinum recombination center in n-type silicon. The capture rates were taken essentially from [Con71] considering also other results. As degeneracy factor, $g = 2$ has been used for the donor level and $g = 4$ for the acceptor level

as is obtained from Eq. (2.65) assuming $p = p_0 + n - n_0$ as approximate neutrality condition.

In Fig. 2.17 the variation of the hole lifetime in n-Si is shown for an acceptor level with $E_r = 0.23$ eV below the conduction band (*dashed* lines). This level is identical with the acceptor level of platinum in silicon (see Fig. 2.18). The capture rates are called here for distinction c_n^0, c_p^- (see the text below). As shown by the figure, the lifetime decreases from a very high value at small hole concentration to a much lower lifetime at high injection, at 400 K the factor between the two extremes is 10^3. Owing to the extremely small e_p in Eq. (2.59), the lifetime can be written as $\tau_p = 1/(c_p N_r^-)$. The very high lifetime at low injection levels now results from the fact that the electron concentration is small compared with the n_r value (which in the present case is 1.6×10^{16} cm^{-3}) and hence the concentration N_r^- according to Eq. (2.61) is also very small. In the range 10^{16}–10^{17} cm^{-3} which is important for the on-state voltage drop of bipolar devices, the lifetime has not reached its limiting value $\tau_{\mathrm{HL,lim}}$. For switching and recovery times the strong increase of the lifetime with decreasing p is unfavorable. In the calculation for the figure the charge on the traps was taken into account in the neutrality condition, but it has only a small effect. The *solid* curves in the figure will be discussed below where impurities with *two* energy levels are considered.

In a depletion or space charge region as found in a reverse-biased pn-junction, the np product is *smaller* than n_i^2 and usually negligible. With $n \simeq p \simeq 0$, Eq. (2.62) yields the generation rate

$$-R = G = \frac{n_i}{\tau_g} = \frac{N_r}{1/e_n + 1/e_p} \tag{2.69}$$

Determining the generation, τ_g is called a generation lifetime. Since it does not obey Eq. (2.48), however, it is strictly speaking not a carrier lifetime in the usual sense. τ_g and hence G depend strongly on the position of the level in the bandgap. Owing to the inverse interrelation between e_n and e_p (Eq. (2.60)), τ_g reaches its minimum and G its maximum at the energy $E_r \equiv E_{rm}$ where $e_n = e_p = \sqrt{c_n c_p}\, n_i$. Inserting this into Eq. (2.64) the generation lifetime at its minimum is obtained as $\tau_{g\,min} = 2/(N_r\sqrt{c_n c_p}) = 2\sqrt{\tau_{n0}\tau_{p0}}$: the minimal value of τ_g is a little larger than the mean value of τ_{n0} and τ_{p0}. The energy of maximal G is given by $E_{rm} = (E_c + E_v)/2 + kT/2 \times \ln(c_p N_v/(c_n N_c g^2))$ which is not far from the bandgap middle. The generation rate can be written as $G = G_{max}/\cosh((E_r - E_{rm})/kT)$. A few kT away from E_{rm}, G is proportional to the smaller one of the emission rates and hence decreases strongly with decreasing distance of the level from the more distant band. If short switching times of a device are required, but on the other side also a small generation in the space charge region and hence low blocking current, this can be reached by choosing a deep impurity whose recombination level is well distant from the middle of the bandgap, but not *too* close to one of the band edges.

Although the density of recombination centers is often considerably lower than the doping of the weakly doped base region, this may be not so in very fast devices. Here the charge on the traps can have undesirable effects, such as a compensation of the normal doping (reduction of the conductivity), a reduction of the breakdown voltage or a premature punch through of the space charge region. The influence of the traps on the carrier concentration in thermal equilibrium depends on the position of their level relative to the Fermi energy and likewise on the type of the level. For example, an acceptor-like impurity in a neutral n base with a level a few kT above the Fermi energy will be neutral to a large part and hence will not affect the free electron concentration essentially. A donor level at the same position on the other hand will *increase n*. In a space charge region of a reverse-biased pn-junction, the charge is given by the condition that the generation rate of electrons must equal to that of holes at steady state. For an acceptor level one has $e_n N_r^- = e_p N_r^0$, hence $N_r^-/N_r^0 = e_p/e_n$. To obtain a low concentration of charged deep acceptors (e.g., to avoid punch-through), a higher emission rate e_n than e_p is required. Generally, the density of charged traps in a stationary non-equilibrium state is given by Eq. (2.61). During switching processes, the densities N_r^-, N_r^0 follow the variation of n and p not instantaneously. Their time dependence is described by the equation

$$\frac{dN_r^-}{dt} = c_n n N_r^0 - c_p p N_r^- - e_n N_r^- + e_p N_r^0 = -\frac{dN_r^0}{dt} \tag{2.70}$$

which follows immediately from the capture and emission events recharging the impurity. Together with Eqs. (2.57), (2.59), and (2.47) for n and p, the charged trap density as a function of time can be numerically calculated. Some effects of the temporary charge on deep impurities will come up in Sect. 13.3.

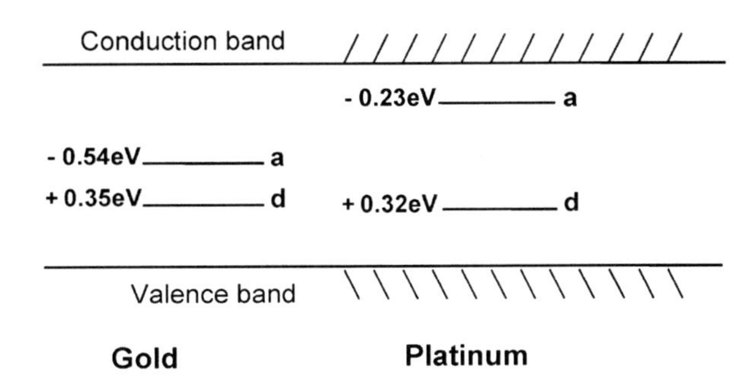

Fig. 2.18 Energy levels of gold and platinum in silicon. Numbers with a plus sign indicate the distance of the level from the valence band edge, numbers with a minus sign the distance from the conduction band. Donor levels are labeled with "*d*" and acceptor levels with "*a*"

Deep impurities often possess not only one, but two or even more energy levels. This is the case also for gold and platinum whose levels in silicon are shown in Fig. 2.18. Both possess a donor level in the lower half of the bandgap and an acceptor level at higher energy. Whereas the acceptor level of gold is located very near to the middle of the gap (0.54 eV below the conduction band), both levels of Pt are far away from the middle (for gold see [Bol66, Fai65, Wuf82], for platinum [Mil76, Con71, Soe94, Sie01]).

The levels are coupled to one another, since they describe transitions between charged states of the same impurity, whose total density N_{r} is distributed among the three charged states involved. In the present case these states are positively charged (R^+), neutral (R^0) and negatively charged centers (R^-). According to the definition given above, the donor level belongs to the transition between R^+ and R^0 and is occupied, when R^+ captures an electron or emits a hole, and is emptied if R^0 captures a hole or emits an electron. The acceptor level describes the transitions between a neutral and a negatively charged center, R^0 and R^-, caused similarly by capture and emission of a carrier. Indicating the capture rates for definiteness with the charge state of the involved center before capture, c_{n}^+, c_{p}^0 are the capture rates belonging to the donor level and c_{n}^0, c_{p}^- those of the acceptor level. Because of the electrostatic energy gained during capture of an electron by R^+, but not by R^0, the donor level is (generally) below the acceptor level. This differs from the familiar positions of donor and acceptor levels of normal doping atoms (see Fig. 2.5). The net recombination rate now is given by the sum of two Shockley–Read–Hall equations (2.62), one for each level. However, N_{r} is replaced for the donor level by $N_{\mathrm{d}} = N_{\mathrm{r}}^+ + N_{\mathrm{r}}^0$ and for the acceptor level by $N_{\mathrm{a}} = N_{\mathrm{r}}^0 + N_{\mathrm{r}}^-$, and these concentrations are not constant, but depend on n and p. This can be calculated putting $R_{\mathrm{n}} = R_{\mathrm{p}}$ for each level and using the constant total trap density $N_{\mathrm{r}} = N_{\mathrm{r}}^+ + N_{\mathrm{r}}^0 + N_{\mathrm{r}}^-$ by which the two levels are coupled. Calculations of the lifetime for two coupled levels and partly for gold and radiation-induced centers were carried out in [Sez66, Sco82, Abb84]. For platinum, the course of the hole lifetime in n-type silicon calculated by considering both levels is shown by the solid curves in Fig. 2.17. The dashed curves take into account only the acceptor level, as has been discussed already above. A very interesting result is that recombination via the donor level yields a strong reduction of the lifetime over the entire concentration range up to nearly $1 \times 10^{17}\ \mathrm{cm}^{-3}$. In similar cases it is often assumed that the recombination in n-type silicon at small minority concentrations takes place only via the acceptor level and similarly in p-Si only via the donor level. This follows if the traps in n-Si appear nearly all as negatively charged centers R^- and in p-Si as positively charged centers R^+. As is shown by the figure, however, this is not fulfilled in our case. One reason is that already a minority density of 10^{13} cm^{-3} leads to a strong deviation of N_r^+, N_r^0, N_r^- from thermal equilibrium values. Furthermore, if the system is in fact nearly in thermal equilibrium ($p \simeq p_0 << n_0$), then the centers R^0, not R^-, predominate in n-type silicon up to a doping concentration of about 1×10^{16} cm^{-3} in the case of Pt. Since the R^0 centers can capture a hole as well as an electron, both levels participate in the recombination also in this case. As is seen also, the donor level reduces the variation of τ_p with injection and with temperature.

The strong increase of the lifetime with temperature, which still remains, is unwanted because it results in a strong decrease of the forward voltage of diodes with increasing temperature [Lut94], a detrimental property for paralleling devices in modules. An example will be given later in Fig. 5.11. The strong enhancement of lifetime with decreasing injection leads to a somewhat higher recovery time of platinum-diffused devices than in the case of gold, particularly for thyristors. As follows from the above discussion for one level, the generation of carriers in a space charge region is determined only by the level which is nearest to the middle of the gap. In the case of gold this is the acceptor level. The generation by gold atoms in a space charge region is more than a decade higher than with Pt even at 150°C. Since this results in a high leakage current of pn-junctions, platinum diffusion or particle irradiation is now preferred for adjustment of carrier lifetime in fast diodes, destined e.g. for IGBT applications. This holds in spite of above-mentioned drawbacks of platinum. Moreover, an advantage of platinum for some applications is that it does not considerably reduce the conductivity of the weakly doped n-base region, because its acceptor level is located typically above the Fermi level and hence the Pt atoms in contrast to Au are mostly neutral.

Often irradiation with high-energy electrons, protons, or He ions is used nowadays to generate recombination centers and control the carrier lifetime. The energy of irradiation is usually in the range 1–15 MeV. While electron irradiation produces

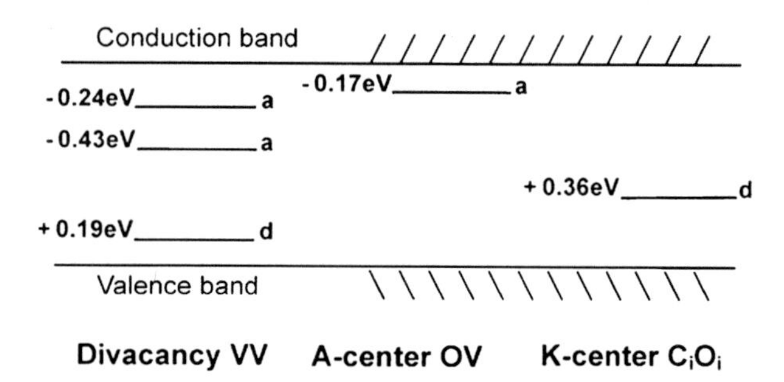

Fig. 2.19 Energy levels of important radiation-induced centers

a homogeneous density of recombination centers, a narrow region of high trap concentration can be created with H and He ion irradiation. The radiation methods show a good reproducibility. As depicted in Fig. 2.19, mainly three independent centers with various levels are generated: the divacancy (VV), the A-center which is an oxygen–vacancy complex (OV), and the K-center, probably an association of an interstitial carbon atom and an interstitial oxygen atom (C_iO_i) [Niw08]. The type of the levels and distances from band edges are indicated as in Fig. 2.18. The divacancy has a donor level and two acceptor levels, the upper of which refers to transitions between R^- and R^{2-}. In contrast, the A-center has a single acceptor level near the conduction band. The relative concentrations of the centers depend on the radiation energy as well as on tempering processes following the irradiation. For lifetime control in neutral regions, especially at high injection level, the A-center is considered to be most efficient because of high capture rates. The carrier generation in space charge regions is determined by the divacancy whose level at 0.43 eV below the conduction band is located closest to the middle of the bandgap. Owing to this energy level, the lifetime control by radiation leads to a significantly lower blocking current than obtained using gold, but it is higher than generated by platinum. The possibility to use simplified models is limited in the case of radiation-induced centers. More details on the radiation technique for lifetime control can be found in Sect. 4.9, including consequences for devices.

As has to be noted finally, caution is advisable regarding the use of Eqs. (2.58) and (2.60) for calculation of emission from capture rates and vice versa. Using the accepted experimental position of the levels in the bandgap and reasonable degeneracy factors, the calculated results differ often from measurements even qualitatively [Tac70, Ral78, Lag80]. For the gold acceptor level, the ratio c_n/c_p has been determined to be about 1/85 [Wuf82, Fai65]. Using this in Eqs. (2.58) and (2.60) together with $E_c - E_r = 0.54$ eV and $g = 4$, one obtains at 300 K for the emission rates $e_n/e_p = 0.82$, whereas experimentally $e_n/e_p \approx 10$ was found [Sah69, Eng75]. To get agreement with the experimental e_n/e_p a degeneracy g of 13 would be necessary here. Such high values cannot be understood as electronic degeneracy [Ral78]. They are interpreted now generally as due to an entropy change ΔS occurring during capture or emission of a carrier by the impurity. This is reasonable because the occupation probability of impurity states is determined by the *free* energy $F = E - TS$ to be used in the Boltzmann factor if entropy changes can occur. In contrast the position of the energy level as determined from thermal activation measurements is given by an *energy* difference ΔE from the respective band edge [Vec76, Eng78]. Therefore, the degeneracy g in the above formulae has to be reinterpreted as an entropy factor $\exp(\Delta S/kT)$, which besides the electronic degeneracy contains an entropy change caused by a change of vibrational frequency of the impurity [Eng78]. Since the entropy factors are not known today from independent determinations, the calculation of emission or capture data from Eqs. (2.58), (2.60) contains an uncertainty. Definite results are obtained if measured data of both capture and emission rates are available and these are used to define $n_r \equiv e_n/c_n$ and $p_r \equiv e_p/c_p$ in the above equations, without using Eqs. (2.58) and (2.60).

2.8 Impact Ionization

The electric field which a semiconductor junction can sustain is limited by impact ionization which leads to avalanche multiplication. This effect determines the breakdown voltage in the whole voltage range, including very small voltages (below 12 V in Si) where the blocking *current* is dominated by quantum mechanical tunneling of carriers. Avalanche multiplication and the "critical" field strength resulting from it are most important for dimensioning of power devices. The critical field strength is a property of the semiconductor. Via the width of the base region required for a given blocking voltage, the critical field determines static and dynamic limits of devices. In this section, we introduce impact ionization as a further generation mechanism. The detailed consequences for devices will be treated in later sections.

Impact ionization occurs if the electric field in a junction is high enough, such that a noticeable number of electrons or holes in the statistical distribution gain sufficient kinetic energy that they can lift a valence electron by impact into the conduction band. Each ionizing carrier generates a pair of a free electron and hole, which again can generate further electron–hole pairs, thus giving rise to an avalanche event. Impact and avalanche generation are therefore used as synonymous expressions. Because the produced secondary particles have a kinetic energy following from conservation of momentum, the ionization energy which the primary carrier at least must have is about 3/2 E_g (see [Moe69] where an extensive review on the physics of impact ionization is given). A few carriers in the thermal distribution have this energy already at zero field strength. However, the generation at zero field has been taken into account already as Auger generation, whose microscopic mechanism is identical with that of impact ionization. Impact ionization is defined as the generation arising from the *enhancement* of the carrier velocities by the field. It is represented by impact ionization rates α_n, α_p defined as the number of electron–hole pairs generated per electron or hole per length of the path which the assembly travels with *drift* velocity v_n or v_p, respectively. The number of electron–hole pairs generated per unit of time, i.e. the avalanche generation rate G_{av}, is then given by

$$G_{av} = \alpha_n \cdot n \cdot v_n + \alpha_p \cdot p \cdot v_p = \frac{1}{q}(\alpha_n j_n + \alpha_p j_p) \tag{2.71}$$

On the right-hand side, the densities of the field currents are replaced by the total current densities, the diffusion currents are neglected because of the high fields.

Much experimental and theoretical work has been carried out to determine the ionization rates and their field dependency, which is very strong. Reviews have been given by Mönch [Moe69] and Maes et al. [Mae90]. Theoretically, Wolff predicted very early the relationship $\alpha \propto \exp(-b/E^2)$ [Wof54]. Experimental ionization rates, on the other hand, were found by Chynoweth [Chy58] and later most other authors, to follow the relationship

Table 2.3 Coefficients of ionization rates in Eq. (2.72) for Silicon as obtained from different publications

Source	Electrons		Holes		Field for $\alpha_{\text{eff}} = 100$/cm
Field range (E in 10^5 V/cm)	a (10^6/cm)	b (10^6 V/cm)	a (10^6/cm)	b (10^6 V/cm)	(10^5 V/cm)
Lee et al. [Lee64] $1.8 < E < 4$	3.80	1.77	9.90	2.98	1.99
Ogawa [Oga65] $1.1 < E < 2.5$	0.75[a]	1.39[a]	0.0188[a]	1.54[a]	1.88[a]
			4.65[b]	2.30[b]	1.80[b]
Overstraaten, De Man [Ove70] $1.75 < E < 4$	0.703	1.231	1.582	2.036	1.66
Our choice [Sco91] $1.5 < E < 4$	1.10	1.46	2.10	2.20	1.81

$T = 300$ K
[a] Directly measured
[b] From α_{eff}, α_n

$$\alpha = a\, e^{-b/E}, \tag{2.72}$$

both for electrons and holes. With E we denote here the *positive* field strength in direction of the field. In a logarithmic plot versus $1/E$, the values of α_n and α_p follow a straight line. Shockley derived this relationship [Sho61] using a simplified model for the manner the carriers reach the high ionization energy. As shown then by Baraff [Baf62], the field dependences of Wolff and of Chynoweth are limiting cases of a more general theory. The experimental determination is done usually by measuring carrier multiplication factors M_n and M_p, consisting of double integrals over the ionization rates (see Appendix B), which then have to be extracted. The measured field dependence obeys the relationship (2.72) down to low fields, as is not expected to this extent from theory [Oga65, Mae90]. Table 2.3 contains the constants a and b for silicon at room temperature as gathered from some often cited papers. From Ogawa's work [Oga65], two sets for α_p are given, the first one determined from multiplication factors like other values in the table. The second set describes the hole ionization rate determined in [Oga65] from measurements of the electron ionization rate α_n and the *effective* ionization rate

$$\alpha_{\text{eff}} = \frac{\alpha_n - \alpha_p}{\ln(\alpha_n/\alpha_p)} \tag{2.73}$$

which was determined from the breakdown voltage of pin-diodes (see below). The extremely varying values of $a(= \alpha(E = \infty))$ in the table are mainly due to the extrapolation with different slopes in the plot versus $1/E$ owing to the different b values. Nevertheless, also in the experimental range the diversity of the obtained ionization rates is large. For example, the α_{eff} values obtained from [Lee64] and [Ove70] at a field strength of 2×10^5 V/cm differ by more than a factor 4. Although the differences in the *field* for a given α_{eff} are smaller because of the strong field dependence, they too are considerable. In the last column of Table 2.3, the field strengths belonging to $\alpha_{\text{eff}} = 100/\text{cm}$ are given. This field can be interpreted as the critical field E_c of a pin-diode whose width of the intrinsic i-region is $w = 1/\alpha_{\text{eff}} = 100\,\mu\text{m}$, since the breakdown condition of a pin diode is $\alpha_{\text{eff}}(E_c)\,w = 1$ (see Eq. (2.75) below and Chaps. 3 and 5 on pn-junctions and pin-diodes). The breakdown voltage $V_B = w\,E_c$ calculated with the data of [Lee64] is 20% higher than obtained from [Ove70]. For p^+n-junctions the difference is still larger (about 35% for $V_B > 1000$ V).

Calculations for a tight dimensioning of devices need ionization rates whose results are in close agreement with the measured blocking behavior. We have compared calculations using different α-sets with measurements on thyristors and diodes, whose weakly doped n-base region was very homogeneously doped by neutron transmutation (see Sect. 4.2), a method not used before 1975. Mainly devices in the blocking range 1200–6000 V were used for comparison, although data in the lower blocking range have also been considered. The ionization rates of Ogawa with the second set of α_p is found to be well suited for *diodes*. The blocking behavior of pn^-p-structures in thyristors is better described according to our tests by a course of the hole ionization rate *between* the two strongly differing α_p curves of [Oga65]. From these comparisons and by considering also measurements with low blocking devices [Lee64, Ove70], the α-parameters given in the last line of Table 2.3 are proposed for the simulation of power devices [Sco91]. They agree also well with a renewed determination of the ionization rates reported by Valdinoci et al. [Val99], although these authors use a more complex fitting expression.

In Fig. 2.20 these ionization rates for silicon at 300 K (Table 2.3, last line) are plotted versus the inverse field strength. Within a relative small field range the α's vary by several orders of magnitude. The ionization rate of electrons is much larger than that for holes, and the effective ionization rate defined by Eq. (2.73) runs between $\alpha_n(E)$ and $\alpha_p(E)$. As is seen, α_{eff} too can be described with excellent approximation by a straight line in this plot and hence can be described by Eq. (2.72). Best agreement with Eq. (2.73) together with the parameter set in the last line of Table 2.3 is obtained with

$$\alpha_{\text{eff}} = 1.06 \times 10^6 \cdot \mathrm{e}^{-1.68\cdot 10^6/E} = a_{\text{eff}} \cdot e^{-b_{\text{eff}}/E} \tag{2.74}$$

(E in V/cm, α_{eff} in cm^{-1}) which up to a few percent agrees with the exact α_{eff} in the range 1.5×10^5 to 4×10^5 V/cm. Equation (2.74) is nearly equal with the result of Ogawa [Oga65].

The effective ionization rate is defined with Eq. (2.73) in a manner that solely this α_{eff} determines the breakdown voltage of pn-junctions [Wol60, Oga65]. The breakdown condition which the field $E(x)$ and hence voltage at breakdown must satisfy is

$$\int \alpha_{\text{eff}}(E(x))dx = 1 \tag{2.75}$$

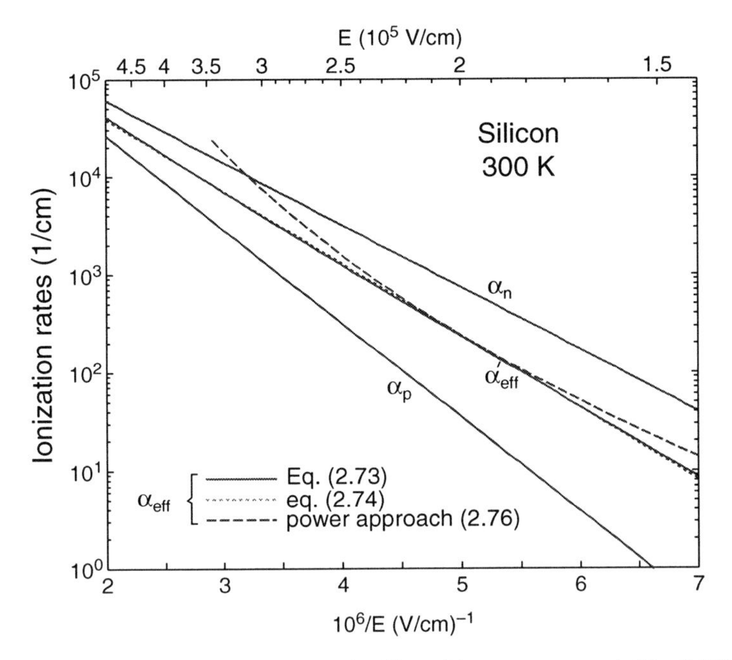

Fig. 2.20 Field dependence of ionization rates in silicon. The power approach (Eq. (2.76)) [Shi59, Ful67] is matched at the point $E_0 = 2 \times 10^5$ using Eq. (2.77)

where the integration extends over the space charge region. The derivation is given in Appendix B. Although this holds exactly only if the ratio α_n/α_p is independent of E, it is a very good approximation also in most other cases. To enable easy analytical integration for simple field shapes, Shields [Shi59] and Fulop [Ful67] approximated the field dependency by a power law, which normalized to a field E_0 reads

$$\alpha_{\text{eff}}(E) \simeq C \left(\frac{E}{E_0} \right)^n \tag{2.76}$$

Based on this approach, practicable and often used relationships for the breakdown voltage can be derived (See Sects. 3.3.2 and 5.3). If Eq. (2.76) together with its derivative is fitted to Eq. (2.74) at the point E_0, the constants are obtained as

$$n = b_{\text{eff}}/E_0, \quad C = a_{\text{eff}} \exp(-n) \tag{2.77}$$

where according to Eq. (2.74) $a_{\text{eff}} = 1.06 \times 10^6\ \text{cm}^{-1}$ and $b_{\text{eff}} = 1.68 \times 10^6$ V/cm. The value $n = 7$ used by Shields and Fulop is obtained for $E_0 = b_{\text{eff}}/7 = 2.40 \times 10^5$ V/cm. Matching at the field $E_0 = 2 \times 10^5$ V/cm, Eq. (2.77) yields $n = 8.40$ and $C = 238\ \text{cm}^{-1}$. Using these values, approximation (2.76) is plotted additionally in Fig. 2.20. Although the approximation is not very good over a wide range of E, very satisfying results can be obtained choosing the matching point E_0 near the maximum of the considered field distribution.

Whereas for diodes only α_{eff} is decisive for the breakdown voltage, for transistor structures such as the pn$^-$p-structure in thyristors and IGBTs both ionization rates separately have an effect on the breakdown behavior. Of course, also the single ionization rates α_n and α_p can be approximated by a power law according to Eqs. (2.76) and (2.77). In this book, the Shields–Fulop approach will often be used to describe the blocking capability in an analytical way. If the *field* distribution $E(x)$ in Eq. (2.75) does not allow an analytical integration, the power approximation loses its meaning.

The ionization rates decrease with increasing temperature because the mean free path between collisions with phonons decreases. As shown by Grant [Gra73] and Maes et al. [Mae90], the temperature dependence can be expressed by an increasing coefficient b in Eq. (2.72), while the pre-exponential factor a may be left constant. For electrons, the temperature coefficient db/dT was determined in [Gra73] to be 1300 V/(cm K), whereas from [Mae90] a mean value of 710 V/(cm K) is obtained; for holes Grant found $db/dT = 1100$ V/(cm K). Using a value of 1100 V/(cm K) in both cases, the temperature dependence of blocking behavior in the range −20 to 150°C is well described. Hence the following field and temperature dependence is obtained:

$$\begin{aligned}\alpha_n &= 1.1 \times 10^6 \cdot \exp\left(-\frac{1.46 \times 10^6 + 1100(T - 300\,K)}{E}\right) \text{cm}^{-1} \\ \alpha_p &= 2.1 \times 10^6 \cdot \exp\left(-\frac{2.2 \times 10^6 + 1100(T - 300\,K)}{E}\right) \text{cm}^{-1}\end{aligned} \tag{2.78}$$

where the field is scaled in volts per centimeter. Inserting this into Eq. (2.73) and using Eq. (2.74) at 300 K, the same T-dependence is obtained for α_{eff}:

$$\alpha_{eff} \cong 1.06 \times 10^6 \exp\left(-\frac{1.68 \times 10^6 + 1100(T - 300\,\text{K})}{E}\right) \text{cm}^{-1} = a_{eff} \exp\left(-\frac{b_{eff}(T)}{E}\right) \tag{2.79}$$

The temperature dependence of the constants in the Shields approximation follows inserting the coefficient $b_{eff}(T)$ of the last equation into Eq. (2.77):

$$\begin{aligned}n(T) &= \frac{b_{eff}(T)}{E_0} = \frac{1.68 \times 10^6 + 1100(T - 300\,\text{K})}{E_0} \\ C(T) &= \frac{a_{eff}}{e^{n(T)}} = \frac{1.68 \times 10^6/\text{cm}}{\exp(n(T))}\end{aligned} \tag{2.80}$$

In the mentioned paper of Valdinoci et al. [Val99], the temperature dependency of the ionization rates in the range 300–670 K is described. The results [Val99] agree well with Eq. (2.78). Singh and Baliga [Sin93] have determined the temperature dependence of the effective ionization rate in the range 77–300 K using the approach of Shields and Fulop. Although the temperature dependences of their constants differ partly strongly from Eq. (2.80), the ionization rate α_{eff} itself varies quite

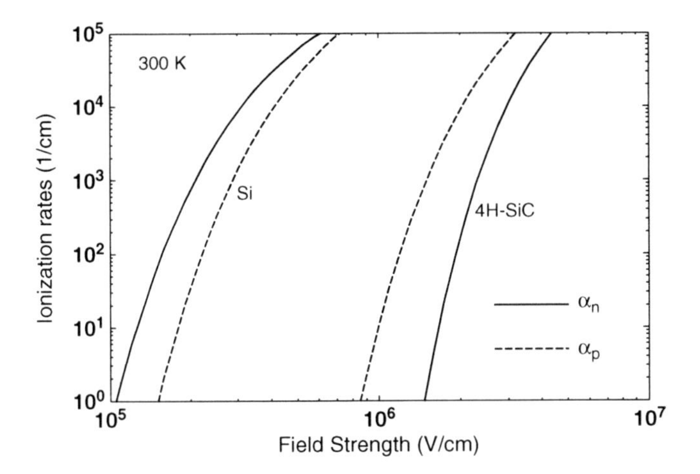

Fig. 2.21 Ionization rates of 4H-SiC and Si at 300 K

similarly. According to [Sin93], α_{eff} increases by a factor 2.4, if the temperature is lowered from 300 to 100 K. The same factor follows from above equations for a field strength $E = E_0 = 2.5 \times 10^5$ V/cm.

As mentioned in Sect. 2.1, SiC has the advantage of an exceptionally high critical field, that is, the field at which impact ionization becomes important is very high. Measurements of the ionization rates in 4H- SiC, the preferred polytype, have been performed by Konstantinov et al. [Kon98], Ng et al. [Ng03], and Loh et al. [Loh08]. The results were fitted by the following equations [Loh08], a modified form of Eq. (2.72):

$$\begin{aligned} \alpha_n &= 2.78 \times 10^6 \exp\left[-\left(\frac{1.05\times 10^7}{E}\right)^{1.37}\right] \text{cm}^{-1} \\ \alpha_p &= 3.51 \times 10^6 \exp\left[-\left(\frac{1.03\times 10^7}{E}\right)^{1.09}\right] \text{cm}^{-1} \end{aligned} \qquad (2.81)$$

where E has to be used in volts per centimeter. These ionization rates together with those of silicon (Eq. (2.78)) are plotted in Fig. 2.21 versus E. As is seen, the field required for a given value of ionization rates is in 4H-SiC nearly an order of magnitude higher than in silicon. Contrary to silicon, the hole ionization rate α_p is larger than α_n in 4H-SiC.

The effective ionization rate defined according to Eq. (2.73) can be described in the interesting range, $50/\text{cm} < \alpha_{\text{eff}} < 10^4/\text{cm}$, approximately by the power law

$$\alpha_{\text{eff}} \simeq 4 \times 10^{-54} \cdot E^9 \text{cm}^{-1} \qquad (2.82)$$

where E is scaled in volts per centimeter. Recently Bartsch et al. [Bar09] have published measurements of the breakdown voltage of 4H-SiC p^+nn^+-diodes for different doping concentrations of the base region. From their measurements which include the temperature dependence they conclude that

$$\alpha_{\text{eff}} \cong 2.185 \times 10^{-48} E^{8.03} \text{cm}^{-1} \quad (E \text{ in V/cm}) \tag{2.82b}$$

at 300 K. According to this result α_{eff} is somewhat lower than stated in Eq. (2.82). The field for an effective ionization rate of 1×10^3/cm is 2.04 MV/cm according to Eq. (2.82b) compared with 1.85 MV/cm calculated from Eq. (2.82). Correspondingly higher are the measured breakdown voltages in [Bar09]. Improvements in material quality of the SiC wafers and in SiC device manufacturing processes are possibly the cause of this.

2.9 Basic Equations of Semiconductor Devices

The device operation depends on the processes with which the carriers and the electric field in the interior react on terminal currents and voltages. This is described by some basic equations which we will discuss now. A central part of these equations are the continuity equations of electrons and holes

$$-\frac{\partial n}{\partial t} = \text{div}\,\vec{J}_{\text{n}} + R_{\text{n}} \tag{2.83}$$

$$-\frac{\partial p}{\partial t} = \text{div}\,\vec{J}_{\text{p}} + R_{\text{p}} \tag{2.84}$$

Here $\vec{J}_{\text{n}}$, $\vec{J}_{\text{p}}$ are the vectors of particle current densities. The equations represent the time decrease of a carrier concentration (e.g. $-\partial n/\partial t$) as a flow of carriers out of the considered volume element ($div\,\vec{J}_{n,p}$) plus a disappearance of carriers with a rate R_{n}, respectively R_{p}. In these exact mathematical equations one has to insert the previously derived models for the current densities and excess recombination rates R_{n}, R_{p}. The latter have to include generally, besides the thermal recombination generation rates treated in Sect. 2.7 and called now $R_{\text{n,p,th}}$, the impact generation rate G_{av}:$R_{\text{n,p}} = R_{\text{n,p,th}} - G_{\text{av}}$. The electrical current densities have been given in one-dimensional form in Eqs. (2.42) and (2.43). In two or three dimensions the particle current densities are

$$\vec{J}_{\text{n}} = -\mu_{\text{n}} n \vec{E} - D_{\text{n}}\,\text{grad}\,n \tag{2.85}$$

$$\vec{J}_{\text{p}} = \mu_{\text{p}} p \vec{E} - D_{\text{p}}\text{grad}\,p \tag{2.86}$$

The continuity equations include the law of conservation of charge

$$\text{div}\vec{j} + \frac{\partial \rho}{\partial t} = 0 \tag{2.87}$$

where $\vec{j}$ is the total electrical current density transported by the carriers (conduction current), $\vec{j} = q(\vec{J}_{\text{p}} - \vec{J}_{\text{n}})$, and ρ the charge density or "space charge." Equation (2.87) is obtained by subtracting Eq. (2.83) from Eq. (2.84) and considering that the difference of the excess recombination rates $R_{\text{n}} - R_{\text{p}}$, if it is non-zero, signifies a recharging of impurities, usually of the recombination centers. Hence $R_{\text{n}} - R_{\text{p}}$

contributes to a change of charge density, which includes the charged flat and deep impurities in the form

$$\rho = q(p - n + N_D^+ - N_A^- + N_r^+ - N_r^-) \tag{2.88}$$

A third differential equation on the level of the continuity equations results from the fundamental law

$$\text{div}\,\vec{D} = \rho \tag{2.89}$$

which states that an electric charge is a source of a displacement field $\vec{D}$. The latter is proportional to the electric field strength, $\vec{D} = \varepsilon \vec{E}$, where the permittivity constant ϵ splits up into the absolute permittivity (permittivity of vacuum) and the relative permittivity as a material constant, $\varepsilon = \varepsilon_r \cdot \varepsilon_0$. The ε_r-values of some materials are compiled in Appendix D. With Eq. (2.88), Eq. (2.89) can be written as follows:

$$\text{div}\,\varepsilon\vec{E} = \rho = q\left(p - n + N_D^+ - N_A^- + N_r^+ - N_r^-\right) \tag{2.90}$$

The further procedure now is to express the electric field in Eqs. (2.85), (2.86) and (2.90) as the negative gradient of a potential V:

$$\vec{E} = -\text{grad}\,V \tag{2.91}$$

Equation (2.90) then turns into the Poisson equation:

$$\text{div grad}\,V = -\frac{q}{\varepsilon}\left(p - n + N_D^+ - N_A^- + N_r^+ - N_r^-\right) \tag{2.92}$$

where the semiconductor is assumed homogeneous and isotropic with regard to ε. For non-cubic crystals ε_r is a tensor, but in current simulation programs it is used as a scalar. The differential operator on the left-hand side of Eq. (2.92) is the Laplace operator, div grad $= \partial^2/\partial x^2 + \partial^2/\partial y^2 + \partial^2/\partial z^2$.

The Poisson equation and the continuity equations (2.83), (2.84) with (2.91) substituted for $\vec{E}$ in the current densities form a system of three partial differential equations with the unknown variables V, n, and p. The doping structure and the density of recombination centers as well as boundary conditions at the surface and the characteristics of the external circuit are required to be known. The normal donors and acceptors are assumed usually to be completely ionized and the concentration of deep impurities to be small. The concentration of charged impurities is then given by the net doping concentration $N_D - N_A$, and the recombination rates R_n, R_p are approximately equal (see Sect. 2.7). The three differential equations are then sufficient to calculate the potential V and the carrier concentrations n and p as functions of space and time. Equations (2.83), (2.84), (2.85), (2.86), (2.91) and (2.92) with $N_D^+ = N_D$, $N_A^- = N_A$, $N_r^+ = N_r^- = 0$ and $R_n = R_p = R$ are called therefore the basic semiconductor equations. The behavior of this system is investigated in the book of Selberherr [Sel84]. Simulation programs based on these equations in one to three dimensions are on the market. The two- and three-dimensional programs are rather complex with all implications. In special one-dimensional cases the equations are used widely for analytical calculations.

In fast switching devices, the charge of deep traps is often considerable, and additionally the incomplete ionization of normal dopants can be a subject of interest, particularly in semiconductors like SiC. In these cases, Eqs. (2.90), (2.91), and (2.92) have to be used in their general form, and for each impurity level, whose variable partial occupation has to be considered, one has to add an equation of the form (2.70) to the system. A program including these effects is Sentaurus^{TCAD} [Syn07].

So far we have assumed that the device has the same temperature at all points. Often this is a good approximation because the thickness of the semiconductor chips is relatively small and the thermal conductivity of silicon is high. On the other hand, phenomena of current constriction, thermal runaway, and other processes responsible for destruction at heavy loads are connected often with a strongly inhomogeneous temperature distribution. For power devices, heat conduction is the primary criterion to be considered for the area and package of a device. To supplement the above system for inclusion of variable temperature and to provide a basis for thermal estimates, the basic relationships of heat conduction will be summarized now. The heat current density, the heat energy transported through a surface element per unit area and time, is proportional to the negative gradient of temperature

$$J_{\text{heat}} = -\lambda \operatorname{grad} T \tag{2.93}$$

where the proportionality factor λ is the thermal conductivity. The thermal energy per volume belonging to a temperature rise ΔT is $Q = \rho_{\text{m}} c \Delta T$, where ρ_{m} denotes the specific (mass) density and c the specific heat of the material. Using this, the continuity equation describing the conservation of thermal energy is obtained in the form

$$\rho_{\text{m}} c \frac{\partial T}{\partial t} = \operatorname{div}(\lambda \operatorname{grad} T) + H \tag{2.94}$$

where H signifies the heat generation rate per unit volume. This consists of the ohmic energy dissipation $\vec{E} \cdot \vec{j}$ and the heat produced by the net recombination of carriers, $R \cdot E_{\text{g}}$ [Sel84]:

$$H = \vec{E} \cdot \vec{j} + R \cdot E_{\text{g}} \tag{2.95}$$

where the non-degenerate case is assumed. R is defined here again as the total net recombination rate including the impact generation: $R = R_{\text{th}} - G_{\text{av}}$. By adding the heat flow equation (2.94) to the above system of partial differential equations, the temperature distribution can be calculated in addition to V, n, and p. All quantities in these equations, such as the mobilities and the carrier lifetime, have to be used in their temperature-dependent form. The equations of current densities may be unchanged for a first approximation only. A more detailed discussion can be found in [Sel84]. We note that in silicon $\rho_{\text{m}} c$ is practically independent of temperature and has the value $\rho_{\text{m}} c = 2.0$ Ws/(cm^{3}K), whereas the heat conductivity λ decreases considerably with T. In the range 220–600 K, the temperature dependence of λ is described by the formula [Kos74]

$$\lambda = \frac{320}{T - 82} \frac{W}{\text{cmK}} \tag{2.96}$$

where T is scaled in Kelvin. The one-dimensional increase of temperature from the lower surface of a chip toward the interior can be simply estimated for a given load using these equations.

Till now the ***Maxwell equations*** as a fundamental description of all electromagnetic phenomena have not been mentioned. The question arises, how far they are satisfied by the above calculation method. Law (2.89) is one of the Maxwell equations. Furthermore, the equation of conservation of charge (2.87) can be deduced from the first Maxwell equation

$$\text{rot } \vec{H} = \vec{j} + \frac{\partial \vec{D}}{\partial t} \tag{2.97}$$

by applying the div operator and using Eq. (2.89). Beyond this, however, the Maxwell equations are not taken into account. According to Eq. (2.97) the current is accompanied by a magnetic field $\vec{H}$ and a magnetic induction $\vec{B} = \mu\mu_0\vec{H}$. Fast changes of the current density $\vec{j}$ and the displacement current density $\partial\vec{D}/\partial t$ cause a corresponding variation of $\vec{H}$ and $\vec{B}$, and according to the second Maxwell equation

$$\text{rot } \vec{E} = -\frac{\partial \vec{B}}{\partial t} \tag{2.98}$$

this leads to an induced electric field. Since rot $\vec{E} \neq 0$, this field cannot be represented by the gradient of a potential as in Eq. (2.91); hence it is not contained in the described differential equations. These effects of self-induction, which play an important part in *circuits* of power electronics, become noticeable (e.g. as skin effect) in semiconductors, however, only at extremely fast current changes. Another neglect is the omission of the Lorentz force $\pm q\vec{v} \times \vec{B}$ on holes and electrons in the current equations. It is the cause of the Hall effect discussed in Sect. 2.4. Ignoring this, large external magnetic fields are excluded. The *current-induced* magnetic field, on the other hand, is nearly always too small as to lead to a noticeable influence of the Lorentz force on the current. Hence the incomplete fulfillment of Maxwell's equations and omission of the Lorentz force do not much restrict the applicability of the basic device equations for the majority of applications.

2.10 Simple Conclusions

We consider now a few simple, but important consequences of the above equations. A homogeneous n-type semiconductor is assumed, and the decay of minority carrier density or charge density is studied either as a function of time after external generation or, on the other hand, as a function of distance x from the surface of the semiconductor in the case of stationary generation at the surface.

1. *Excess minority carrier density* Δp under conditions of neutrality: This is the usual case in carrier generation by light or injection by a pn-junction. The excess hole density is neutralized by an equal excess electron concentration $\Delta n = n - n_0 = \Delta p(0)$ and possibly a charge of recombination centers. Since no external field is assumed and the space charge is zero, the electric field is also zero.

- Case 1a): Time decay of a homogeneous excess minority carrier density after excitation: This decay is obtained from Eq. (2.84) with div $\vec{J}_\mathrm{p} = 0$ and has been given already by Eq. (2.49). The minority concentration decays with the time constant τ_p.
- Case 1b): Stationary generation in a thin surface layer at $x = 0$, decrease of Δp with distance x from the surface: Since the field is zero, the hole particle current density is $J_\mathrm{p} = -D_\mathrm{p}\,\mathrm{d}p/\mathrm{d}x$. Hence Eq. (2.84) with $\partial p/\partial t = 0$ and $R_\mathrm{p} = \Delta p/\tau_\mathrm{p}$ turns into

$$-D_\mathrm{p}\frac{d^2p}{dx^2} + \frac{\Delta p}{\tau_\mathrm{p}} = 0$$

The solution is

$$\Delta p(x) = \Delta p(0)\cdot \mathrm{e}^{-x/l_\mathrm{p}} \text{ with } L_\mathrm{p} = \sqrt{D_\mathrm{p}\tau_\mathrm{p}} \tag{2.99}$$

The minority carrier concentration decreases with the decay length L_p which is called the minority carrier (here hole) diffusion length, indicating that the spreading of the minority carriers takes place by diffusion. L_p can be adjusted by the minority carrier lifetime τ_p. With $D_\mathrm{p} = kT/q\,\mu_\mathrm{p} = 12\ \mathrm{cm}^2/\mathrm{s}$ (for small doping density) a lifetime of 1 μs results in a diffusion length of 34.6 μm. The diffusion length of electrons is somewhat larger for a given lifetime because of the higher diffusion constant. With the appropriate lifetime, diffusion lengths of several hundred microns are possible. The diffusion lengths in devices are chosen often with reference to the base width w_B. Compared with the decay of a space charge considered in example (2b), the spreading of minority carriers from the point of generation or injection stretches much farther.

2. *Deviation from neutrality*: A small charge density $\rho = -q\cdot\delta n$ with $\delta n << n_0$ is assumed, caused by an increased electron concentration above the equilibrium density n_0. First the time decay of a homogeneous space charge is studied, then the spatial decrease of a stationary space charge caused at the surface $x = 0$.

- Case 2a): Time decay of a homogeneous space charge generated up to the time $t = 0$: From the equation of conservation of charge (2.87) one obtains with $j = q\cdot\mu_\mathrm{n}\cdot n_0\cdot \boldsymbol{E}$ and div $\boldsymbol{E} = \rho/\varepsilon$

$$\frac{q\cdot\mu_\mathrm{n}\cdot n_0}{\varepsilon}\rho + \frac{\mathrm{d}\rho}{\mathrm{d}t} = 0$$

This results in the time decay

$$\rho(t) = \rho(0) \cdot \mathrm{e}^{-t/\tau_{\mathrm{rel}}} \tag{2.100}$$

where the time constant is the relaxation time

$$\tau_{\mathrm{rel}} = \frac{\varepsilon}{q \cdot \mu_{\mathrm{n}} \cdot n_0} = \frac{\varepsilon_{\mathrm{r}} \varepsilon_0}{\sigma} \tag{2.101}$$

τ_{rel} is inversely proportional to the electrical conductivity σ and very small; in silicon at $n_0 = 1 \times 10^{15}\ \mathrm{cm}^{-3}$ ($\mu_{\mathrm{n}} = 1350\ \mathrm{cm}^2/\mathrm{Vs}$, $\sigma = 0.22$ A/(Vcm)) one obtains $\tau_{\mathrm{rel}} = 4.8$ ps. A homogeneous space charge left to itself can exist only very shortly.

- Case 2b): *Stationary* charge density $\rho(0) = -q \cdot \delta n(0)$ at the surface, caused, for example, by a positive voltage at an isolated gate electrode. The hole concentration $p < p_0 = n_{\mathrm{i}}^2/n_0$ and the hole components in the total current density j are then negligible. Since $d\rho/dt = 0$, one obtains from the equation of charge conservation (2.87) using Eq. (2.42) together with the Einstein relation (2.44)

$$0 = \frac{\mathrm{d}j}{\mathrm{d}x} = q\mu_{\mathrm{n}} \left(n \frac{\mathrm{d}E}{\mathrm{d}x} + \frac{\mathrm{d}n}{\mathrm{d}x} E + \frac{kT}{q} \frac{d^2 n}{dx^2} \right)$$

The term $\mathrm{d}n/\mathrm{d}x \cdot E$ as a product of two small quantities is negligible. Hence, using Eq. (2.89) and $\delta n = -\rho/q$, the equation turns into

$$\frac{n_0}{\varepsilon} \rho - \frac{kT}{q^2} \frac{d^2 \rho}{\mathrm{d}x^2} = 0$$

From this equation one obtains again an exponential decay with distance x:

$$\rho(x) = \rho(0) \cdot \mathrm{e}^{-x/L_{\mathrm{D}}} \tag{2.102}$$

But different from the minority carrier diffusion length, the decay length is now

$$L_{\mathrm{D}} = \sqrt{\frac{\varepsilon_{\mathrm{r}} \varepsilon_0 \cdot kT}{n_0 \cdot q^2}} = \sqrt{D_{\mathrm{n}} \tau_{\mathrm{rel}}} = 0.41 \cdot \sqrt{\frac{10^{14}/\mathrm{cm}^3}{n_0}}\ \mu\mathrm{m} \tag{2.103}$$

This decay length of a charge density is called "Debye length". It is inversely proportional to the equilibrium carrier density. The numerical expression on the right-hand side is obtained for silicon ($\varepsilon_{\mathrm{r}} = 11.7$) at $T = 300$ K. Even at a small concentration n_0 of $1 \times 10^{14}\ \mathrm{cm}^{-3}$, L_{D} amounts only to 0.41 μm. Hence the space charge decreases very rapidly toward zero. The Debye length is typically two or three orders of magnitude smaller than the diffusion lengths. The relation between L_{D} and the relaxation time is similar to the relation between the diffusion length and the minority carrier lifetime. The *majority* carrier diffusion constant appears in the relationship (2.103), because the spreading of the space charge is a majority carrier effect.

References

[Abb84] Abbas CC: "A theoretical explanation of the carrier lifetime as a function of the injection level in gold-doped silicon", IEEE Trans. Electron Devices, ED-31 pp. 1428–1432 (1984)

[Atk85] Atkinson CJ: "Power devices in gallium arsenide", IEE Proceedings **132**, Pt.I, pp. 264–271 (1985)

[Baf62] Baraff GA: "Distribution functions and ionization rates for hot electrons in semiconductors", Phys. Rev. 128, pp. 2507–2517 (1962)

[Bar09] Bartsch W, Schoerner R, Dohnke KO: "Optimization of Bipolar SiC-Diodes by Analysis of Avalanche Breakdown Performance", Proceedings of the ICSCRM 2009, paper Mo-P-56 (2009)

[Ben83] Bennett HS: "Hole and electron mobilities in heavily doped silicon: Comparison of Theory and Experiment", Solid-St. Electron., vol. 26, pp. 1157–1166 (1983)

[Bla62] Blakemore JS: Semiconductor Statistics, Pergamon Press. 1st ed 1962.

[Bri09] Briere MA: "GaN Based Power Conversion: A New Era in Power Electronics", Proc. PCIM Europe, Nuremberg (2009)

[Bul66] Bullis WM: "Properties of gold in silicon", Solid-St. Electron., vol. 9, pp. 143–168 (1966)

[Cau67] Caughey DM, Thomas RE: "Carrier Mobilities in Silicon Empirically Related to Doping and Field", Proceedings IEEE 23, pp. 2192–2193 (1967)

[Chy58] Chynoweth AG: "Ionization rates for electrons and holes in silicon", Phys. Rev., vol. 109, pp. 1537–1540 (1958)

[Con71] Conti M, Panchieri A: "Electrical properties of platinum in silicon", Alta Frequenza, vol. 40, pp. 544–546 (1971)

[Dan72] Dannhäuser F: "Die Abhängigkeit der Trägerbeweglichkeit in Silizium von der Konzentration der freien Ladungsträger - I", Solid-St. Electron., vol. 15, pp. 1371–1375 (1972)

[Dzi77] Dziewior J, Schmid W: "Auger Coefficients for Highly Doped and Highly Excited Silicon", Appl. Phys. Lett. 31, pp. 346–348 (1977)

[Dzi79] Dziewior J, Siber D: "Minority-carrier diffusion coefficients in highly doped silicon", Appl. Physics Let., vol. 35, pp. 170–172 (1979)

[Eng75] Engström O, Grimmeis HG: "Thermal activation energy of the gold acceptor level in silicon", J. Appl. Phys. 46, pp. 831–837 (1975)

[Eng78] Engström O, Alm A: "Thermodynamical analysis of optimal recombination centers in thyristors", Solid-St. Electron., vol. 21, pp. 1571–1576 (1978)

[Fai65] Fairfield JM, Gokhale BV: "Gold as a recombination centre in silicon", Solid-St. Electron. 8, pp. 685–691 (1965)

[Fri06] Friedrichs P: "SiC Power Devices - Recent and Upcoming Developments" IEEE ISIE 2006, July 9–12, Montreal, Quebec, Canada (2006)

[Ful67] Fulop W: "Calculation of Avalanche Breakdown Voltages of Silicon pn-Junctions", Solid-St. Electron., vol. 10, pp. 39–43 (1967)

[Gol01] Goldberg Y, Levinshtein ME, Rumyantsev SL: in: Properties of Advanced Semiconductor Materials GaN, AlN, SiC, BN, SiC, SiGe . Eds. Levinshtein et al., John Wiley & Sons, Inc., New York, pp. 93–148 (2001)

[Gra73] Grant WN: "Electron and hole ionization rates in epitaxial silicon at high electric fields", Solid-St. Electron. 16, pp.1189–1203 (1973)

[Gre90] Green MA: "Intrinsic concentration, effective densities of states, and effective mass in silicon", J. Appl. Phys. 67, pp. 2944–2954 (1990)

[Hag93] Hagmann G: Leistungselektronik, Aula-Verlag, Wiesbaden, 1993

[Hal52] Hall RN: "Electron-hole recombination in germanium", Phys. Rev. 87, p. 387, (1952)

[Ike08] Ikeda N, Kaya S, Jiang L, Sato Y, Kato S, Yoshida S: "High power AlGaN/GaN HFET with a high breakdown voltage of over 1.8 kV on 4 inch Si substrates and the suppression of current collapse", Proceedings of the ISPSD ′08 pp. 287–290 (2008)

[Jac77] Jacobini C, Canali C, Ottaviani G, Quaranta A: "Review of some charge transport properties of silicon", Solid-St. Electron. 20, pp. 77–89 (1977)

[Kla92] Klaassen DBM: "A unified mobility model for device simulation – I. Model equations and concentration dependence", Solid-St. Electron., vol. 35, pp. 953–959 (1992)

[Kla92b] Klaassen DBM: "A unified mobility model for device simulation – II. Temperature dependence of carrier mobility and lifetime", Solid-St. Electron., vol. 35, pp. 961–967 (1992)

[Kok74] Kokkas AG: "Empirical relationships between thermal conductivity and temperature for silicon and germanium", RCA Review, vol. 35, pp. 579–581 (1974)

[Kon98] Konstantinov AO, Wahab Q, Nordell N, Lindefelt U: "Study of avalanche breakdown and impact ionization in 4H silicon carbide", Journal of Electronic Materials 27, pp. 335–341 (1998)

[Kra72] Krause J: "Die Abhängigkeit der Träderbeweglichkeit in Silizium von der Konzentration der freien Ladungsträger - II", Solid-St. Electron., vol. 15, pp. 1377–1381 (1972)

[Kuz86] Kuzmicz W: "Ionization of impurities in silicon", Solid-St. Electron., vol. 29, pp. 1223–1227 (1986)

[Lag80] Lang DV, Grimmeis HG, Meijer E, Jaros M: "Complex nature of gold-related deep levels in silicon", Phys. Rev. B, vol. 22, pp. 3917–3934 (1980)

[Lan79] Lanyon HPD, Tuft RA: "Bandgap narrowing in moderately to heavily doped silicon", IEEE Trans. Electron Devices, ED-26, pp. 1014–1018 (1979)

[Lar54] Lark-Horovitz K: "The new Electronics", in: The present State of Physics (American Assn. for the Advancement of Science), Washington, 1954

[Lee64] Lee CA, Logan RA, Batdorf RL, Kleimack JJ, Wiegmann, W: "Ionization rates of holes and electrons in silicon", Phys. Rev. 134, pp. A761–A773 (1964)

[Lev01] Levinshtein ME, Rumyantsev SL, Shur MS: Properties of advanced semiconductor materials, John Wiley & Sons, New York, 2001

[Li78] Li SS: "The Dopant Density and Temperature Dependence of Hole Mobility and Resistivity in Boron Doped Silicon", Solid-St. Electron., Vol.21, pp. 1109–1117 (1978)

[Loh08] Loh WS, Ng B, Soloviev K et al.: "Impact ionization coefficients in 4H-SiC", IEEE Trans. Electron Devices ED-55, pp. 1984–1990 (2008)

[Lut94] Lutz J, Scheuermann U: "Advantages of the New Controlled Axial Lifetime Diode", Proceedings oft the 28th PCIM, pp. 163–169 (1994)

[Mad64] Madelung O: Physics of III-V Compounds, John Wiley & Sons, New York, 1964

[Mae90] Maes W, De Meyer K, Van Overstraaten R: "Impact ionization in silicon: A review and update", Solid-St. Electron. 33, pp. 705–718 (1990)

[Mas83] Masetti G, Severi M, Solmi S: "Modeling of carrier mobility against concentration in Arsenic-, Phosphorus-, and Boron-doped Silicon", IEEE Transaction on electron devices, ED-30, pp. 764–769 (1983)

[Mil76] Miller MD: "Differences Between Platinum- and Gold-Doped Silicon Power Devices", IEEE Trans. Electron Devices. Dev., vol. ED-23, pp. 1279–1283 (1976)

[Moe69] Mönch W: "On the physics of avalanche breakdown in semiconductors", Phys. Stat. Sol. 36, pp. 9–48 (1969)

[Mol64] Moll JL: Physics of semiconductors, McGraw Hill, New York, 1964

[Mon74] Monemar B: "Fundamental energy gap of GaN from photoluminescence excitation spectra", Phys. Rev. B 10, p. 676 (1974)

[Mue93] von Münch W: Einführung in die Halbleitertechnologie, B.G. Teubner, Stuttgart, 1993

[Ng03] Ng BK, David JPR, Tozer RC et al.: "Non-local effects in thin 4H-SiC UV avalanche photodiodes", IEEE Trans. Electron Devices, ED-50, pp. 1724–1732 (2003)

[Niw08] Niwa F, Misumi T, Yamazaki S, Sugiyama T, Kanata T, Nishiwaki K: "A Study of Correlation between CiOi Defects and Dynamic Avalanche Phenomenon of PiN Diode Using He Ion Irradiation", Proceedings of the PESC, Rhodos (2008)

[Oga65] Ogawa T: "Avalanche Breakdown and Multiplication in Silicon pin Junctions". Japanese J. of Applied Physics, vol. 4, pp. 473–484 (1965)

[Ove70] Van Overstraeten R, De Man H: "Measurement of the Ionization Rates in Diffused Silicon p-n junctions," Solid-St. Electron. 13, pp. 583–608 (1970)

[Pog64] Poganski S: "Fortschritte auf dem Gebiet der Selen-Gleichrichter", AEG-Mitteilungen 54, pp. 157–161 (1964)

[Ral78] Ralph HI: "The degeneracy factor of the gold acceptor level in Silicon", J. Appl. Phys. 49, pp. 672–675 (1978)

[Ras08] Rashid SJ, Udrea F, Twitchen DJ, Balmer RS, Amaratunga GAJ: "Single Crystal Diamond Schottky Diodes – Practical Design Considerations for Enhanced Device Performance", Proceedings of the ISPS, pp. 73–76, Prague (2008)

[Sah69] Sah CT, Forbes L, Rosier LI, Tasch AF Jr, Tole AB: "Thermal emission rates of carriers at gold centers in silicon", Appl. Phys. Letters 15, pp. 145–148 (1969)

[Scf69] Scharfetter DL, Gummel HK: "Large-signal analysis of a silicon Read Diode oscillator", IEEE Trans. Electron Devices, ED-16, pp. 64–77 (1969)

[Sco74] Schlangenotto H, Maeder H, Gerlach W: "Temperature dependence of the radiative recombination coefficient in silicon", Phys. Stat. Sol. (a) 21, pp. 357–367 (1974)

[Sco82] Schlangenotto H, Silber D, Zeyfang R: "Halbleiter-Leistungsbauelemente - Untersuchungen zur Physik und Technologie", Wiss. Ber. AEG-Telefunken 55, Nr. 1–2 pp. 7–24 (1982)

[Sco91] Schlangenotto H: Script of lectures on "Semiconductor power devices", Technical University Darmstadt, 1991 (in German)

[Scr94] Schäfer WJ, Negley GH, Irvin KG, Palmour JW: "Conductivity anisotropy in epitaxial 6H and 4H SiC", Proceedings of Material Res. Society Symposium, vol. 339, pp. 595–600 (1994)

[Scz66] Schultz W: "Rekombinations- und Generationsprozesse in Halbleitern", in Festkörperprobleme Band V, Vieweg & Sons, Braunschweig, pp.165–219 (1966)

[Sel84] Selberherr S: Analysis and simulation of semiconductor devices, Springer, Vienna, 1984

[Shi59] Shields J: "Breakdown in Silicon pn-Junctions", Journ. Electron. Control, vol. 6, pp. 132–148 (1959)

[Sho52] Shockley W, Read WT Jr: "Statistics of the recombinations of holes and electrons", Phys. Rev. 87, pp. 835–842 (1952)

[Sho59] Shockley W: Electrons and Holes in Semiconductors, Seventh printing, D. van Nostrand Company Inc, Princeton, 1959

[Sho61] Shockley W: "Problems related to p-n junctions in silicon", Solid-St. Electron. 2, pp. 35–67 (1961)

[Sie01] Siemieniec R, Netzel M, Südkamp W, Lutz J, "Temperature dependent properties of different lifetime killing technologies on example of fast diodes", IETA2001, Cairo, (2001)

[Sin59] Singh R, Baliga BJ: "Analysis and optimization of power MOSFETs for cryogenic operation", Sol. State Electronics, vol. 36, pp. 1203–1211 (1993)

[Slo76] Slotboom JW, De Graaff HC: "Measurements of bandgap narrowing in Si bipolar transistors", Solid-St. Electron., vol. 19, pp. 857–862 (1976)

[Slo77] Slotboom JW: "The pn-product in Silicon", Solid-St. Electron., vol. 20, pp. 279–283 (1977)

[Smi59] Smith RA: Semiconductors, University Press, Cambridge, 1959

[Spe58] Spenke E: Electronic semiconductors, McGraw Hill, New York, first edition, 1958

[Sue94] Südkamp W: DLTS-Untersuchung an tiefen Störstellen zur Einstellung der Trägerlebensdauer in Si-Leistungsbauelementen, Dissertation, Technical University of Berlin, 1994

[Swi87] Swirhun SE: Characterization of majority and minority carrier transport in heavily doped silicon, Ph.D. Dissertation, Stanford University, 1987

[Syn07] Advanced tcad manual, Synopsys Inc. Mountain View, CA. Available: http://www.synopsys.com (2007)

[Sze81] Sze SM: Physics of Semiconductor Devices, John Wiley & Sons, New York, 1981

[Sze02] Sze SM: Semiconductor Devices, Physics and Technology, 2nd Ed., John Wiley & Sons, New York, 2002

[Tac70] Tach AF Jr, Sah CT: "Recombination-Generation and optical properties of gold acceptor in silicon", Phys. Rev. B, vol. 1, pp. 800–809 (1970)

[Tho80] Thornber KK: "Relation of drift velocity to low-field mobility and high-field saturation velocity", J. Appl. Phys. 51, pp. 2127–2136 (1980)

[Thr75] Thurmond CD: "The Standard Thermodynamic Functions for the Formation of Electrons and Holes in Ge, Si, GaAs, and GaP", J. Electrochem. Soc., 122, pp. 1133–1141 (1975)

[Thu80] Thurber WR, Mattis RL, Liu YM, Filliben JJ: "Resistivity-Dopant Density Relationship for Phosporous-Doped Silicon", J. Electrochem. Soc., vol. 127, pp. 1807–1812 (1980)

[Thu80b] Thurber WR, Mattis RL, Liu YM, Filliben JJ: "Resistivity-Dopant Density Relationship for Boron-Doped Silicon", J. Electrochem. Soc., vol. 127, pp. 2291–2294 (1980)

[Ued05] Ueda D, Murata T, Hikita M, Nakazawa S, Kuroda M, Ishida H, Yanagihara M, Inoue K, Ueda T, Uemoto Y, Tanaka T, Egawa T: "AlGaN/GaN devices for future power switching systems", IEEE International Electron Devices Meeting, pp. 377–380 (2005)

[Val99] Valdinoci M, Ventura D, Vecchi M, Rudan M, Baccarani G, Illien F, Stricker A, Zullino L: "Impact-Ionization in silicon at large operating temperature", Proc. SISPAD'99, pp. 27–30, Kyoto, (1999)

[Var67] Varshni YP: "Temperature dependence of the energy gap in semiconductors", Physica 34, pp. 149–154 (1967)

[Vec76] Van Vechten JA, Thurmond CD: "Entropy of ionization and temperature variation of ionization levels of defects in semiconductors", Phys. Review B, vol. 14, pp. 3539–3550 (1976)

[Wfs60] Wolfstirn KB: "Hole and electron mobilities in doped silicon from radiochemical and conductivity measurements", J. Phys. Chem. Solids 16, pp. 279–284 (1960)

[Wof54] Wolff PA: "Theory of electron multiplication in silicon and germanium", Phys. Rev. 95, pp. 1415–1420 (1954)

[Wuf82] Wu RH, Peaker AR: "Capture cross sections of the gold and acceptor states in n-type Czochralski silicon", Solid-St. Electron., vol. 25, pp. 643–649 (1982)

[Wul60] Wul BM, Shotov AP: "Multiplication of electrons and holes in p-n junctions", Solid State Phys. in Electron. Telecommun., vol. 1, pp. 491–497 (1960)

Chapter 3
pn-Junctions

pn-Junctions are the basic element of nearly all power devices. They are formed when the type of conductivity changes from p-type to n-type within the same crystal. pn-junctions are rectifying, they conduct current only in one direction of the applied voltage, called forward direction, whereas in the opposite direction, the blocking direction, the current is extremely small. Although the function of pn-junctions has been described theoretically already in 1938 [Dav38], their full technological significance became obvious only after the invention of the transistor and major further advances in theory and technology [Sho49, Sho50]. This was a starting point of the enormous development of the semiconductor sector till now. Today, even the former poly-crystal rectifiers are thought for the most part to function by a pn-junction.

The rectifying effect can be simply understood qualitatively (see Fig. 3.1): If a positive voltage is applied to the p-region with respect to the n-region, then the free holes in the p-region and the free electrons in the n-region are driven toward the junction and are injected partly into the opposite region as excess minority carriers. Since there is no lack of carriers for the current flow, the pn-junction is conducting in this bias condition. If the voltage at the p-region is negative with respect to the n-region, then both types of majority carriers are withdrawn from the junction and cannot be supplied from the adjacent region of opposite conductivity, except for the few equilibrium minority carriers there. Hence only a very small current can flow, the pn-junction is biased in reverse or the blocking direction.

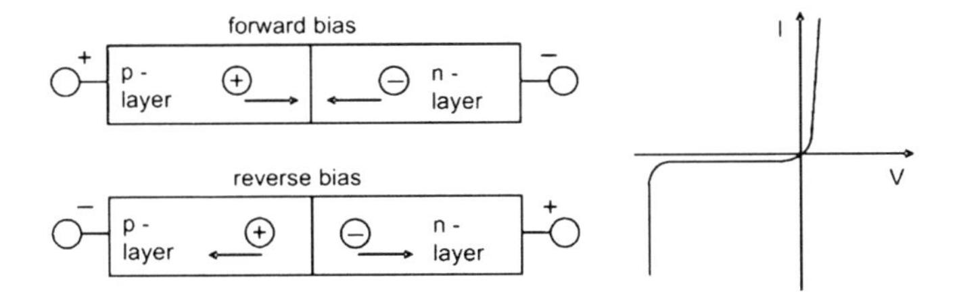

Fig. 3.1 pn-Junction in forward and in blocking direction

J. Lutz et al., *Semiconductor Power Devices*, DOI 10.1007/978-3-642-11125-9_3,

3.1 The pn-Junction in Thermal Equilibrium

First we consider a pn-junction in thermodynamic equilibrium, the case of zero external voltage and current. The formulae for this case can be transferred to a large extent to the case of applied voltage and hence are useful also for the *I–V* characteristics in forward and reverse direction. As discussed previously, the concentration of free holes in the bulk of the p-region is equal to the concentration of ionized acceptors, likewise deep in the n-region the electron density is given by the donor concentration, thus the charge of the impurities is neutralized by the carriers. This is not the case, however, near the transition between the p- and n-region, as is indicated in Fig. 3.2. Here the hole density in the p-region has a steep slope $-dp/dx$ resulting in a diffusion of holes toward the n-region. Since the fixed acceptors stay behind without compensation, a negative space charge arises in the p-region near the junction. From the n-region, in the same way electrons diffuse toward the p-region, so that a positive space charge of uncompensated donors remains in the n-region near the junction. Between both space charges an electric field is built up, which drives the holes toward the p-region and the electrons toward the n-region, i.e. in both cases in opposite direction to the particle diffusion currents. The thermal equilibrium is reached when the field current compensates the diffusion current both for electrons and holes. The field over the space charge region results in a built-in voltage $V_{\mathrm{bi}} = -\int_{x_\mathrm{p}}^{x_\mathrm{n}} E\,\mathrm{d}x$, where x_p, x_n are the boundaries of the space charge layer in the p- and n-region, respectively. The built-in potential holds the electrons in the n-region and the holes in the p-region. V_{bi} is often called "diffusion voltage, " because its primary cause is diffusion.

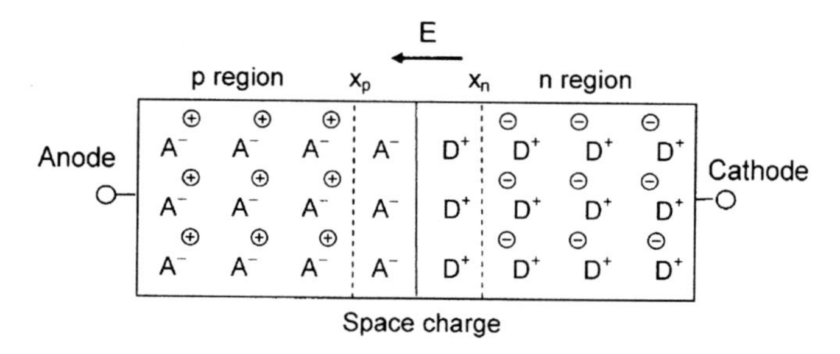

Fig. 3.2 pn-Junction in thermal equilibrium

For the built-in voltage a simple general relationship can be derived. In Sect. 2.7, the Einstein relation (2.44) has been derived from the Boltzmann distribution (2.45). Inversely, the experimentally confirmed Einstein relation results in the Boltzmann distribution. If one rewrites the current equations (2.42) and (2.43) using the Einstein relation

$$j_\mathrm{n} = q \cdot \mu_\mathrm{n} \left(n \cdot E + \frac{kT}{q} \frac{\mathrm{d}n}{\mathrm{d}x} \right) \tag{3.1}$$

$$j_\mathrm{p} = q \cdot \mu_\mathrm{p} \left(p \cdot E - \frac{kT}{q} \frac{\mathrm{d}p}{\mathrm{d}x} \right) \tag{3.2}$$

the Boltzmann distribution is obtained setting $j_\mathrm{p} = 0$ in Eq. (3.2) and integrating the bracket term:

$$p(x) = p(x_p) \cdot e^{-qV(x)/kT} \tag{3.3}$$

where the potential is given as $V(x) = -\int_{x_\mathrm{p}}^{x} E(x')\,\mathrm{d}x'$. Similarly, one obtains from Eq. (3.1) for the electrons

$$n(x) = n(x_\mathrm{p}) \cdot \mathrm{e}^{q \cdot V(x)/kT} \tag{3.4}$$

By these equations the space-dependent carrier densities in the space charge region are connected with that of the potential, which however is also not yet known as function of x. In Sect. 2.3 the relationship $np = n_\mathrm{i}^2$ has been derived for thermal equilibrium, without considering the present case of an existing electrical field. We note now that this relationship follows from Eqs. (3.3) and (3.4) at every point also in the space charge region where n and p each are strongly varying functions of x:

$$n(x) \cdot p(x) = n(x_\mathrm{p}) \cdot p(x_\mathrm{p}) = n_\mathrm{i}^2 \tag{3.5}$$

The built-in voltage V_bi is obtained from Eq. (3.3) or Eq. (3.4) as $V_\mathrm{bi} = V(x_\mathrm{n})$:

$$V_\mathrm{bi} = \frac{kT}{q} \ln \frac{p(x_\mathrm{p})}{p(x_\mathrm{n})} = \frac{kT}{q} \ln \frac{p(x_\mathrm{p}) \cdot n(x_\mathrm{n})}{n_\mathrm{i}^2} \tag{3.6}$$

$$\simeq \frac{kT}{q} \ln \frac{N_\mathrm{A}(x_\mathrm{p}) \cdot N_\mathrm{D}(x_\mathrm{n})}{n_\mathrm{i}^2} \tag{3.6a}$$

In the approximation (3.6a) complete ionization of the dopants is assumed together with neutrality at the boundaries of the space charge layer (see Sect. 2.5). According to this equation the built-in voltage is given by the doping densities at the boundaries of the space charge region.

How is now the space dependence of the potential and carrier concentrations, and how wide does the space charge layer reach into the p- and n-region? This can be calculated from the Poisson equation (2.92). Assuming complete ionization of the impurities and using Eqs. (3.3) and (3.4) one has without restriction to specific doping profiles

$$\begin{aligned}\frac{\mathrm{d}^2 V}{\mathrm{d}x^2} &= -\frac{\rho}{\varepsilon} = \frac{q}{\varepsilon}(n - p + N_\mathrm{A,tot}(x) - N_\mathrm{D,tot}(x)) \\ &= \frac{q}{\varepsilon}\left(n(x_\mathrm{p}) \cdot \mathrm{e}^{qV/kT} - p(x_\mathrm{p}) \cdot \mathrm{e}^{-qV/kT} - N(x)\right)\end{aligned} \tag{3.7}$$

where $N(x) \equiv N_\mathrm{D,tot} - N_\mathrm{A,tot}$. The donor and acceptor concentrations are indicated now additionally by an index "tot" to distinguish them from the net doping densities $N_\mathrm{A}(x) = N_\mathrm{A,tot}(x) - N_\mathrm{D,tot}(x)$ in the p-region and $N_\mathrm{D}(x) = N_\mathrm{D,tot} - N_\mathrm{A,tot}$ in the n-region, which are denoted like above in this chapter. The carrier concentrations at the p-sided boundary of the space charge region are then

$p_p(x_p) = N_A(x_p)$, $n(x_p) = n_i^2/N_A(x_p)$. Generally, the ordinary differential equation (3.7) has to be solved numerically, but for abrupt step junctions $V(x)$ can be expressed analytically by an integral (see Sect. 3.1.1). As will be seen, however, an approximate simpler calculation is often sufficient and even more useful. The exact solution of Eq. (3.7) will be used to test and partly correct the approximate formulae.

As obtained above, the voltage V_{bi} over the space charge region is much larger than the thermal voltage kT/q. Hence a small variation of the potential from the boundaries of the space charge region toward the interior is associated according to Eqs. (3.3) and (3.4) with a rapid decrease of the hole concentration near x_p and electron concentration near x_n. Hence apart from small transition layers the carrier concentrations in the space charge region may be neglected, and Eq. (3.7) reduces to

$$\frac{d^2 V}{dx^2} = -\frac{q}{\varepsilon} N(x) \tag{3.8}$$

Since the space charge region is assumed here to be completely depleted from carriers, the approach is called (abrupt) depletion approximation. It will be used extensively in following parts of the book.

3.1.1 The Abrupt Step Junction

The abrupt pn-junction or abrupt step junction is defined by a sharp, step-like doping transition between the p- and n-region and homogeneous doping densities within each of the two regions. Since N_A and N_D are independent of the positions x_p and x_n of the boundaries of the space charge layer, the built-in voltage is immediately determined by Eq. (3.6a) in this case. In Fig. 3.3, the built-in voltage of abrupt pn-junctions in silicon is plotted for two temperatures versus the doping concentration N_D of the n-region assuming a fixed doping $N_A = 1 \times 10^{19}\text{cm}^{-3}$ of the p-region. At 300 K, V_{bi} increases from 0.705 V to 1.00 V in the range 1×10^{13} to 1×10^{18} cm^{-3},

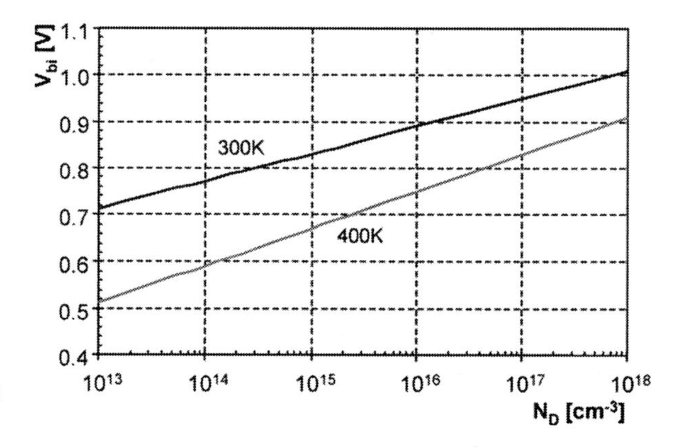

Fig. 3.3 Built-in voltage of abrupt pn-junctions in silicon as function of the doping concentration of the weakly doped side (N_D) for a fixed doping of 1×10^{19} cm^{-3}(N_A) of the highly doped region

at 400 K from 0.508 V to 0.905 V. The decrease of V_{bi} with increasing T is due to the strong increase of n_i. The dependency on T is approximately linear, because the pre-exponential factor of n_i^2 in Eq. (2.6), $N_c N_v$, is larger than $N_D N_A$ in the application range of the formula and its logarithm nearly constant like E_g. At high doping concentrations both of the p- and the n-regions the temperature dependence of V_{bi} is only weak.

We will now use the depletion approximation to calculate the potential and carrier concentrations in an abrupt junction as a function of x. Figure 3.4 illustrates the space dependence of (a) the doping and carrier concentrations, (b) the resulting charge density, (c) the electric field in the depletion approximation, (d) the potential, and (e) the corresponding band diagram. These qualitative plots are substantiated now by the following calculations.

Placing the origin of x at the transition between the acceptor and donor doping, the metallurgical junction, the doping profile $N(x)$ is given by

$$\begin{aligned} N(x) &= -N_A = \text{const for } x < 0, \\ N(x) &= +N_D = \text{const for } x \geq 0 \end{aligned} \tag{3.9}$$

In the region with acceptor doping the Poisson equation

$$\frac{d^2 V}{dx^2} = \frac{q}{\varepsilon} N_A \tag{3.10}$$

results in the field (see Fig. 3.4)

$$\frac{dV}{dx} = E(x) = -\frac{q}{\varepsilon} N_A \cdot (x - x_p) \quad \text{for } x_p \leq x \leq 0 \tag{3.11}$$

For the metallurgical n-region, integration of the Poisson equation $d^2 V/dx^2 = -q/\varepsilon N_D$ yields

$$-\frac{dV}{dx} = E(x) = -\frac{q}{\varepsilon} N_D \cdot (x_n - x) \quad \text{for } 0 \leq x \leq x_n \tag{3.12}$$

The continuity of E at $x = 0$ requires

$$N_A \cdot x_p = -N_D \cdot x_n \tag{3.13}$$

meaning that the charges on both sides of the junction are oppositely equal. Integration of Eqs. (3.11) and (3.12) leads to

$$V(x) = \frac{q}{2\varepsilon} \cdot N_A \cdot (x - x_p)^2 \quad \text{for } x_p < x \leq 0 \tag{3.14}$$

and

$$V(x) = \frac{q}{2\varepsilon} \left[-N_D \cdot (x_n - x)^2 + N_D x_n^2 + N_A x_p^2 \right] \quad \text{for } 0 \leq x < x_n \tag{3.15}$$

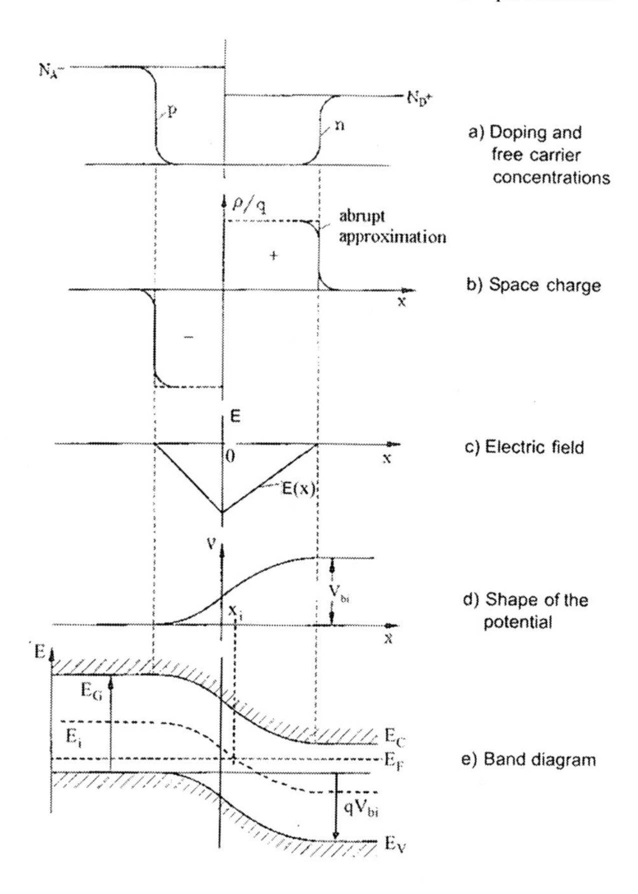

Fig. 3.4 Abrupt pn-junction in the depletion approximation

where the constant term in Eq. (3.15) is chosen to get continuity at $x = 0$. As is shown by Eqs. (3.11) and (3.12), and Fig. 3.4c, the field strength in each region is linear in x, the potential has a parabolic course. The ratio of the potential increase in the acceptor region to that in the donor region is obtained from Eqs. (3.13), (3.14), and (3.15) as

$$\frac{V_P}{V_N} = \frac{N_A \cdot x_p^2}{N_D \cdot x_n^2} = \frac{|x_p|}{x_n} = \frac{N_D}{N_A} \tag{3.16}$$

The penetration depths $-x_p$ and x_n can now be determined equating $V(x_n)$ to the built-in voltage as given by Eq. (3.6a). From Eqs. (3.15) and (3.13) one obtains

$$V_{bi} = \frac{q}{2\varepsilon} \cdot \left(N_D \cdot x_n^2 + N_A \cdot x_p^2\right) = \frac{q}{2\varepsilon} \cdot \left(N_D \cdot x_n^2 + \frac{N_D^2}{N_A} \cdot x_n^2\right) \tag{3.17}$$

$$x_n = \sqrt{\frac{2\varepsilon}{q} \cdot \frac{N_A/N_D}{(N_A + N_D)} \cdot V_{bi}} \quad |x_p| = \frac{N_D}{N_A} \cdot x_n \tag{3.18}$$

The total thickness of the space charge layer is

$$w_{sc} = x_n + |x_p| = \sqrt{\frac{2\varepsilon}{q} \cdot \frac{N_A + N_D}{N_A N_D} \cdot V_{bi}} \tag{3.19}$$

The maximum absolute field strength $E_m = |E(0)|$ is obtained from Eqs. (3.12) and (3.18) as

$$E_m = \sqrt{\frac{2q}{\varepsilon} \cdot \frac{N_A N_D}{N_A + N_D} \cdot V_{bi}} = \frac{2 \cdot V_{bi}}{w_{sc}} \tag{3.20}$$

For asymmetric junctions with very different concentrations N_A, N_D, as found often in devices, the formulae simplify. Referring in the notation to a p^+n-junction with $N_A >> N_D$, Eq. (3.19) turns into

$$w_{sc} = \sqrt{\frac{2 \cdot \varepsilon \cdot V_{bi}}{q \cdot N_D}} \tag{3.19a}$$

whereas Eq. (3.20) yields

$$E_m = \sqrt{\frac{2q}{\varepsilon} \cdot N_D \cdot V_{bi}} \tag{3.20a}$$

In the energy band diagram of Fig. 3.4e, the band edges vary inversely with the potential, and the Fermi level E_F is constant across the pn-junction. This follows because the conduction band edge E_c represents the potential energy $-qV(x)$ of the electrons, and using Eqs. (2.4) and (3.4) results in space-independent E_F. The built-in voltage multiplied with the elementary charge q is represented in the band diagram by the entire change of the band edges or of the intrinsic level E_i against the Fermi level. The point x_i where the intrinsic level crosses the Fermi level divides the region where $p > n$ from the region with $n > p$. This point differs generally from the metallurgical junction as is indicated in Fig. 3.4e. Instead of the above calculated voltage parts V_P, V_N in the metallurgical p- and n-region, the built-in voltage can be divided also into the potential difference ΔV_{p-i} between the neutral p-region and the intrinsic point x_i on one side and the potential difference ΔV_{i-n} between the intrinsic point and the neutral n-region on the other. Using Eqs. (2.11) and Eq. (2.10) these parts are obtained as

$$\begin{aligned} \Delta V_{p-i} &= \frac{1}{q}(E_i(x_p) - E_F) = \frac{kT}{q} \ln \frac{p(x_p)}{n_i} \simeq \frac{kT}{q} \ln \frac{N_A}{n_i} \\ \Delta V_{i-n} &= \frac{1}{q}(E_F - E_i(x_n)) = \frac{kT}{q} \ln \frac{n(x_n)}{n_i} \simeq \frac{kT}{q} \ln \frac{N_D}{n_i} \end{aligned} \tag{3.21}$$

The addition of both potentials yields the built-in voltage as given by Eqs. (3.6) and (3.6a).

As numerical example we consider a pn-junction in silicon with $N_A = 2 \times 10^{15}\text{cm}^{-3}$, $N_D = 1 \times 10^{15}\text{cm}^{-3}$. The built-in voltage is in this case 0.604 V. Using this, Eqs. (3.18) with the permittivity of silicon $\varepsilon_r = 11.7$ yield the penetration depths $x_n = 0.725\,\mu\text{m}$, $|x_p| = 0.363\,\mu\text{m}$. The maximum value of the field according to Eq. (3.20) is $E_m = 1.12 \times 10^4\text{V/cm}$. The parts of the built-in voltage in the metallurgical acceptor and donor region are $V_P = N_D/(N_A + N_D)V_{bi} = V_{bi}/3 = 0.203\text{V}$, $V_N = 2V_P = 0.407\,\text{V}$. In contrast to this, Eq. (3.20) yields for the voltage increase in the region with $p > n_i$ a value $\Delta V_{p\text{-}i} = 0.311\,\text{V}$, while the voltage in the region with $n > n_i$ is $\Delta V_{i\text{-}n} = 0.296\,\text{V}$. The considerably higher $\Delta V_{p\text{-}i}$ compared with V_P says that the intrinsic point is located not at the metallurgical junction but in the region with lower doping. Numerically Eq. (3.15) together with (3.21) yields $x_i = 0.106\,\mu\text{m}$.

The accuracy of Eqs. (3.10) to (3.20) can be tested by comparing with formulae following from the exact Poisson equation (3.7). By integrating Eq. (3.7) once,[1] the following exact relationship between the potential parts V_P, V_N in the acceptor and donor doped region, respectively, can be derived replacing the approximation (3.16):

$$\frac{V_P - kT/q}{V_N - kT/q} = \frac{N_D}{N_A} \tag{3.22}$$

Expressed explicity one has

$$\begin{aligned} V_P &= \frac{N_A - N_D}{N_A + N_D} \cdot \frac{kT}{q} + \frac{N_D}{N_A + N_D} \cdot V_{bi} \\ V_N &= V_{bi} - V_P \end{aligned} \tag{3.23}$$

For the maximum field the exact calculation yields

$$E_m = \sqrt{\frac{2kT}{\varepsilon} \cdot \left[N_A \cdot e^{-qV_P/kT} + N_D \cdot \left(\frac{qV_N}{kT} + e^{-qV_N/kT} - 1 \right) \right]} \tag{3.24}$$

As long as V_P, $V_N >> kT/q$, Eq. (3.22) is well approximated by Eq. (3.16), and Eq. (3.24) simplifies to the linear term in V_N leading to Eq. (3.20). If however the pn-junction is strongly asymmetric and satisfies the condition $N_A/N_D >> qV_{bi}/kT$ in the case of a p^+n-junction, V_P tends to kT/q according to Eq. (3.23), not to zero as would follow from Eq. (3.16). The N_A-term in Eq. (3.24) then is large against the N_D expression including the term leading to Eq. (3.20). *Hence E_m as given by* Eq. (3.24) *is much larger than according to the depletion approximation* (3.20a). The limiting expression of Eq. (3.24) for $N_A/N_D >> qV_{bi}/kT$ with V_P, V_N from Eq. (3.23) is

[1] After multiplication with $2 \cdot dV/dx$, Eq. (3.7) can be integrated analytically to obtain the field $E(x)$ as function of $V(x)$ in both regions. From the continuity of potential and field at the metallurgical junction the potential parts V_P, V_N are then determined.

$$E_\mathrm{m} = \sqrt{\frac{2kT}{\varepsilon} N_\mathrm{A}/\mathrm{e}} \quad \text{for}\, N_\mathrm{A}/N_\mathrm{D} >> qV_\mathrm{bi}/kT \tag{3.24a}$$

While the maximum field according to this equation depends only on the doping density of the highly doped region, the approximation (3.20a) contains only the doping concentration of the weakly doped region and additionally the built-in voltage. For the example $N_\mathrm{A} = 1 \times 10^{18}\mathrm{cm}^{-3}$, $N_\mathrm{D} = 1 \times 10^{14}\mathrm{cm}^{-3}$ Eq. (3.24) yields for Silicon $E_\mathrm{m} = 5.44 \times 10^4\mathrm{V/cm}$, whereas from Eq. (3.20a) with $V_\mathrm{bi} = 0.711$ V one obtains a field $E_\mathrm{m} = 4.69 \times 10^3\mathrm{V/cm}$ which is an order of magnitude too small.

The cause of this discrepancy becomes evident from Fig. 3.5 which shows the carrier distributions in the $\mathrm{p^+n}$-junction of the last example calculated by solving Eq. (3.7). It is seen that the hole cloud is not restricted to the acceptor region but reaches beyond the metallurgical junction into the n-region. In the first range the hole concentration is much higher than the doping density N_D and even the integrated hole charge in the n-region is large compared with the total charge of ionized donors in the space charge region. This explains why the field at the metallurgical junction is independent of N_D according to Eq. (3.24a). As will be shown in

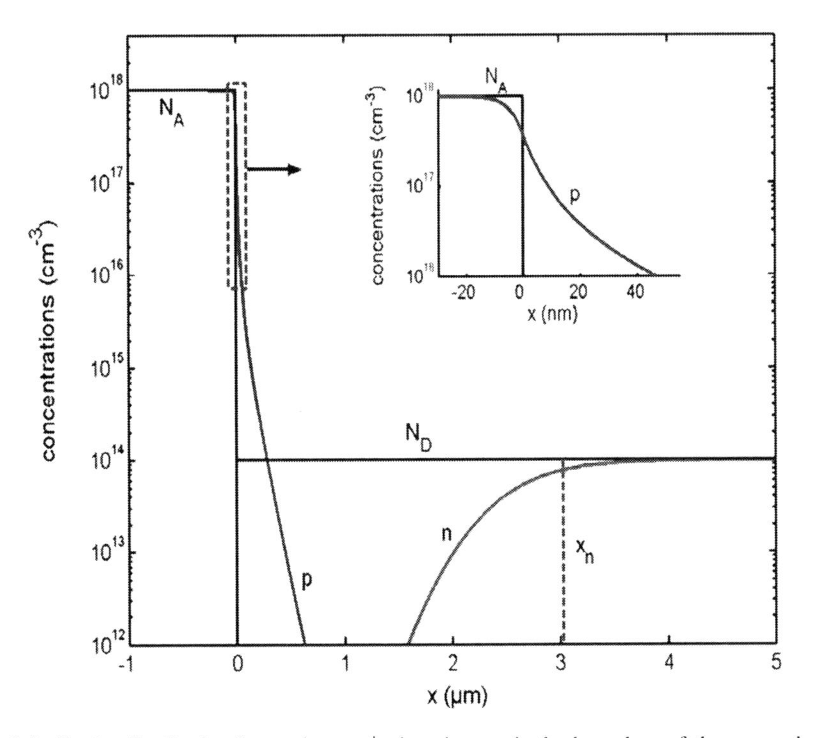

Fig. 3.5 Carrier distribution in an abrupt $\mathrm{p^+n}$-junction. x_n is the boundary of the space charge region in the depletion approximation

Sect. 3.5, the charge of mobile carriers in the space charge region can have a strong influence on the differential capacitance $c = \delta Q/\delta V$.

Practically, the perfect abruptness presupposed here will be realized to a sufficient extent mainly by junctions in wide-gap semiconductors like SiC, whose doping profiles are not smoothed out by diffusion. In silicon, only junctions prepared by low-temperature epitaxy (see Chap. 4) are so abrupt that Eq. (3.24) is applicable. In most "abrupt" asymmetric junctions in silicon, the doping concentration of the highly doped region ceases slowly enough that the carrier cloud does not strongly surmount the doping concentration and not reach beyond the metallurgical junction, but it is steep enough that the charge density of the weakly doped region is given with good approximation by the constant doping density. For such abrupt asymmetrical junctions the depletion approximation is a useful approach, especially if a reverse voltage is applied (see Sect. 3.3).

3.1.2 Graded Junctions

Very often the transition between the acceptor and donor doping is substantially graded. Particularly this holds for pn-junctions made by diffusion of the impurities, a technique described in detail in Chap. 4. If impurities are diffused into a semiconductor which contains already a doping of the opposite type and with a smaller density than the surface concentration of the diffused dopant, a pn-junction is formed at the point where the diffused impurity just compensates the back ground doping. The net doping density, $N(x) = N_D - N_A(x)$ for an acceptor diffusion, changes its sign at the junction. If the gradient of $N(x)$ is not too small, a significant space charge builds up as for abrupt junctions. In Fig. 3.6 the profile of a diffused acceptor with surface concentration $1 \times 10^{18}\text{cm}^{-3}$ added to a homogeneous donor doping with density $1 \times 10^{14}\text{cm}^{-3}$ is shown together with the resulting absolute net doping density and the electron and hole concentration calculated numerically for silicon. Near the pn-junction, the carrier concentrations are much smaller than the net doping density, indicating that the space charge there is a large. The net doping density shown on an expanded linear scale in the inset varies nearly over the whole space charge layer. The depletion approach neglecting the carrier charge in the space charge region can be used also in this case to a first approximation. Sufficiently near to the junction, the net doping concentration can be approximated by a linear dependence

$$N(x) = a \cdot x \tag{3.25}$$

the first term in the Taylor series, using the junction as origin of x.

In a *linearly graded junction* Eq. (3.25) holds by definition throughout the space charge region. In this case one obtains from Eqs. (3.8) and (3.25):

$$\frac{d^2 V}{dx^2} = -\frac{dE}{dx} = -\frac{q \cdot a}{\varepsilon} \cdot x \tag{3.26}$$

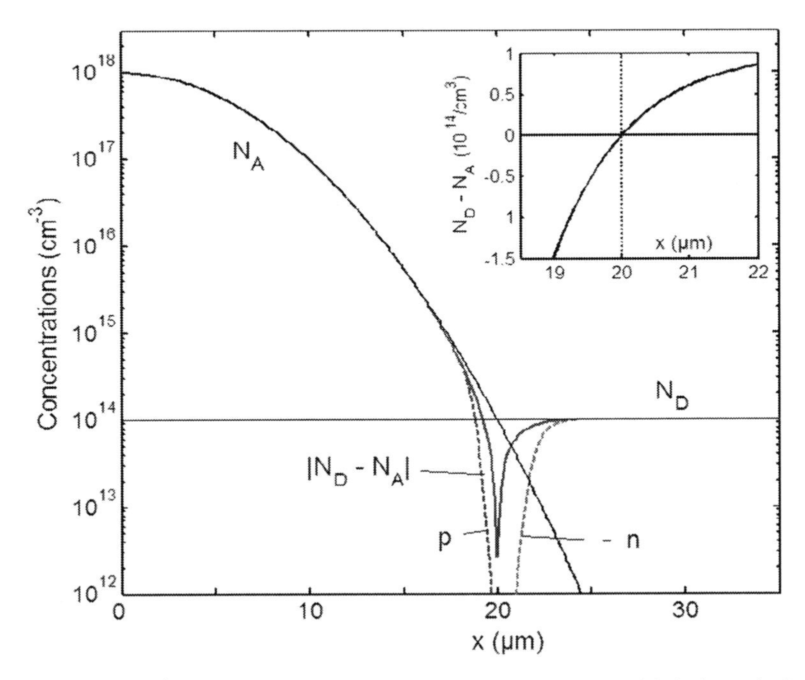

Fig. 3.6 Diffused p^+n-junction in silicon: Doping profile together with hole and electron distribution. A Gaussian function, $N_A \alpha \exp(-(x/L_A)^2)$, is assumed as diffusion profile

$$E(x) = \frac{q \cdot a}{2\,\varepsilon}\left(x^2 - w^2\right) \tag{3.27}$$

$$V(x) = \frac{q \cdot a}{2\,\varepsilon} \cdot \left(w^2 \cdot x - \frac{1}{3} \cdot x^3\right) \tag{3.28}$$

where w is the extension of the space charge layer in each of the two regions (half width of the total space charge layer) and the potential at the junction is set to zero. Since $V(w) - V(-w)$ is the built-in voltage, the half width w follows from Eq. (3.28) as

$$w = \left(\frac{3\,\varepsilon\,V_{bi}}{2\,q\,a}\right)^{1/3} \tag{3.29}$$

To calculate the built-in voltage this has to be inserted into Eq. (3.25) to obtain the net acceptor and donor density to be used in Eq. (3.6a):

$$\begin{aligned} V_{bi} &= \frac{kT}{q} \cdot \ln\left(\frac{-N(-w) \cdot N(w)}{n_i^2}\right) = \frac{kT}{q} \ln\left(\frac{(a \cdot w)^2}{n_i^2}\right) \\ &= 2\frac{kT}{q} \ln\left(\frac{a \cdot w}{n_i}\right) \end{aligned} \tag{3.30}$$

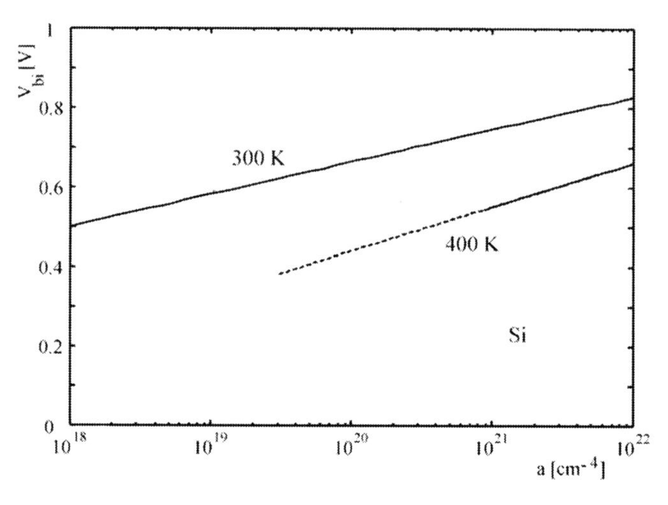

Fig 3.7 Built-in voltage of a linearly graded junction in Si as function of the doping gradient

Together with Eq. (3.29) this is an implicit equation for V_{bi} which can be solved by iteration to obtain the built-in voltage as function of doping gradient a. The result for silicon at 300 and 400 K is shown in Fig. 3.7. By comparing this result with exact numerical calculations it has been shown [Mol64, Mor60] that the depletion approximation is only good for impurity gradients

$$a > \sqrt{\frac{8\, n_{\mathrm{i}}^3\, q^2}{\varepsilon\, kT}} \times 10^4 \tag{3.31}$$

For silicon, the right-hand side amounts to $7.6 \times 10^{16}\mathrm{cm}^{-3}$ at 300 K, but $8.8 \times 10^{20}\mathrm{cm}^{-3}$ at 400 K. Below this value, the line for 400 K in Fig. 3.7 may be rather inexact.

For variable doping densities, also the (quasi-)neutral regions possess an inherent built-in field and built-in potential. Since the electron and hole currents are zero in thermal equilibrium, a field is necessary to compensate the diffusion current which arises from the concentration gradient. From Eq. (3.2) one obtains replacing the hole density by the net acceptor concentration

$$E(x) = \frac{kT}{q}\frac{\mathrm{d}\ln p}{\mathrm{d}x} = \frac{kT}{q}\frac{\mathrm{d}\ln N_{\mathrm{A}}}{\mathrm{d}x} \tag{3.32}$$

$$V(x) - V(x_0) = -\frac{kT}{q}\ln\frac{N_{\mathrm{A}}(x)}{N_{\mathrm{A}}(x_0)} \tag{3.33}$$

For n-regions the sign is opposite. The built-in potential of diffused regions in their quasi-neutral part is usually considerable. In the case of Fig. 3.6, the potential difference between the surface of the p-region and the boundary of the space charge region, where $N_{\mathrm{A}} = 3 \times 10^{14}\mathrm{cm}^{-3}$, amounts to 0.28 V at 400 K according to

Eq. (3.33). If also a diffused n-region follows on the homogeneous n-base, the whole built-in potential of the quasi-neutral regions comes near to that of the space charge region. For device characteristics, however, the built-in potential of the quasi-neutral regions is of minor significance. This follows because they stay (nearly) unchanged if an external voltage is applied, whereas the space charge region together with the voltage across it is essentially altered, as will be seen. If not further specified, the term "built-in voltage" refers therefore always to that of the space charge region. We have used here the term "quasi-neutral" because diffused regions are not completely neutral but have a charge density $\rho = \varepsilon \mathrm{d}E/\mathrm{d}x$ which can be calculated from Eq. (3.32). The border between the actual space charge region and the quasi-neutral regions is not precisely fixed by nature. It may be defined with some arbitrariness as the point where the majority carrier density falls below the net doping density by a certain percentage, for example 3%, or where the space charge is by a certain factor, for example five times, higher than according to Eq. (3.32). The precise limit has not a strong influence on the two built-in voltages.

If the p- and n-region are connected externally by a wire without applying a voltage, a current flow will not take place despite the presence of the built-in voltage, since this would contradict general laws of thermodynamics. Based on the constancy of the Fermi level throughout the contacted structure in thermal equilibrium, it can be concluded that contact potentials exist between the contacted two semiconductor regions and the metal which in sum are oppositely equal to the (whole) built-in voltage of the semiconductor.

As one of the results of this section we mention again that the extension of the space charge layer in thermal equilibrium is very small compared with the usual thickness of the (quasi-)neutral p- and n-regions enclosing it. If an external voltage is applied, the situation can be changed.

3.2 Current–Voltage Characteristics of the pn-Junction

Now a voltage V is applied to the p-region with respect to the n-region. If this voltage is positive, it is directed against the built-in voltage V_{bi}. Assuming that the caused current is small and the ohmic voltage drop over the neutral regions can be neglected, the voltage across the space charge region is now

$$\Delta V = V_{\mathrm{bi}} - V \tag{3.34}$$

For building up this voltage step, the required charges in the dipole layer at the junction are smaller or larger than without the external voltage depending on the sign of the voltage. Since the charge density on each side is given approximately by the doping density, the thicknesses of the space charge layer in the p- and n-region decrease or increase. Most important, however, is that the hole concentration in the n-region and electron concentration in the p-region are raised above the equilibrium minority densities for $V > 0$ and lowered for $V < 0$. This can be concluded from Eqs. (3.3) and (3.4) *assuming that the Boltzmann distribution is applicable also to cases away from thermal equilibrium*. This is a basic assumption of the whole device theory, justified by the fact that the deviation from equilibrium is usually weak, i.e.

the field and diffusion currents compensate each other largely in the space charge region.

In this chapter we assume furthermore that the minority carrier densities in the neutral regions are small compared with the doping densities. The junction is assumed to be abrupt. Using Eq. (3.34) in Eq. (3.4) replacing $V(x_n)$ in Eq. (3.3) by Eq. (3.34), one obtains for the minority carrier density p_n* higher as (3.35) in the n-region at the boundary $x = x_n$

$$p_n^* = N_A \cdot e^{-q(V_{bi}-V)/kT} \tag{3.35}$$

$$= p_{n0} \cdot e^{qV/kT} \tag{3.36}$$

Here the equilibrium hole density in the n-region is denoted by p_{n0}, and the relationship

$$p_{n0} = N_A \cdot e^{-qV_{bi}/kT} \tag{3.37}$$

was used which follows from Eq. (3.6). In the same way, for the electron density n_p^*; in the p-region at the boundary $x = x_p$ one obtains

$$n_p^* = n_{p0} \cdot e^{qV/kT} \tag{3.38}$$

where n_{p0} is the equilibrium electron concentration in the neutral p-region. The minority concentrations at the boundaries to the space charge region are raised respectively lowered by an exponential factor containing the applied voltage, the Boltzmann factor. In Fig. 3.8 this is illustrated for forward bias. The applied voltage raises the concentrations p_n^* and n_p^* here by eight orders of magnitude. On the chosen logarithmic scale the minority carrier concentrations decrease linearly with distance from the space charge layer. The figure anticipates here results of the following calculation.

For further visualization especially of the situation at reverse bias, the minority carrier densities are plotted in Fig. 3.9 on a linear scale. In the picture for reverse bias (lower part of the figure) the minority carrier concentrations at the boundaries to the space charge region are lowered already to zero. This is approached already for reverse voltages higher than a few times the thermal voltage kT/q, as follows from Eqs. (3.36) and (3.38). The diffusion of holes out of the n-region and electrons out of the p-region which determines the reverse current can then no longer be enhanced by a further increase of the reverse voltage. Because the equilibrium densities $p_{n0} = n_i^2/N_D, n_{p0} = n_i^2/N_A$ are very small and this transfers to the concentration gradients of the minority carriers, the blocking current is also very small.

As is noted, Eqs. (3.36) and (3.38) require only that the injection level of the *injecting* region is low, while (within these limits) the minority concentration n_p^* in the considered region may be arbitrary. For a p^+n-junction, Eq. (3.36) is applicable to arbitrary injection levels in the weakly doped n-region as long as the injection level in the p^+-region is low. The equations will appear later for pin-diodes with high level injection in the base region.

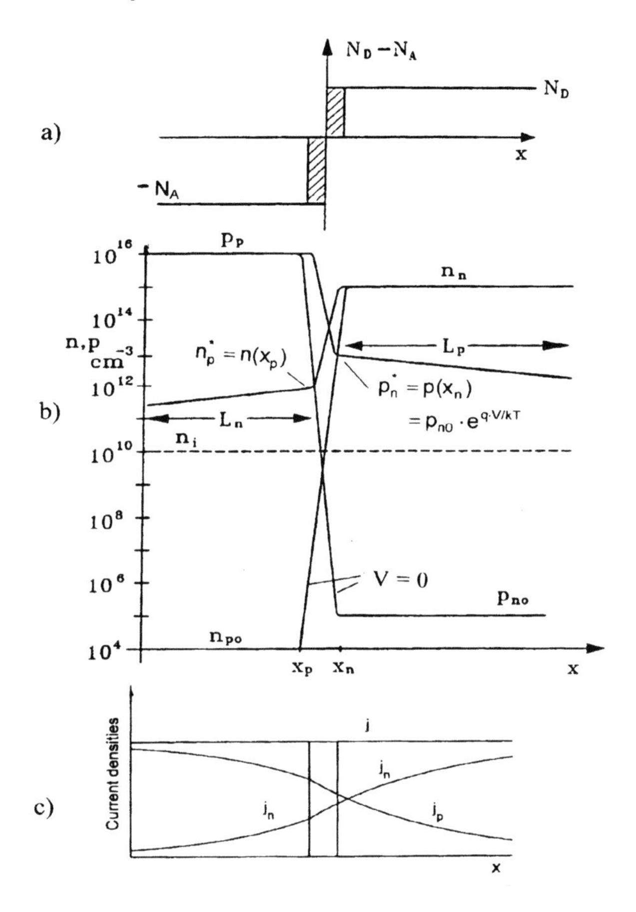

Fig. 3.8 Forward-biased pn-junction. (**a**) Net doping density, (**b**) carrier distribution for $V > 0$ and $V = 0$, (**c**) hole and electron current densities

From now on, we assume in this section that the injection level on both sides of the junction is low. From Eqs. (3.3) and (3.4) together with Eq. (3.38) one obtains then for the np-product in the space charge region

$$n(x) \cdot p(x) = n(x_\mathrm{p}) \cdot p(x_\mathrm{p}) = n_{\mathrm{p}0} \cdot \mathrm{e}^{qV/kT} \cdot p_0 = n_\mathrm{i}^2 \cdot \mathrm{e}^{qV/kT} \tag{3.39}$$

As without bias, the np-product in the space charge layer is independent of x, but it is increased or decreased depending on the sign of the voltage V by the exponential voltage factor. The deviation from equilibrium leads to net recombination or generation, respectively.

The I–V characteristic is governed by the minority carrier currents in the neutral regions. To calculate them one uses the continuity equation which for the holes in the n-region according to Eq. (2.84) reads

$$\frac{\mathrm{d}j_\mathrm{p}}{\mathrm{d}x} = -q \cdot R_\mathrm{p} = -q\,\frac{p - p_{\mathrm{n}0}}{\tau_\mathrm{p}} \tag{3.40}$$

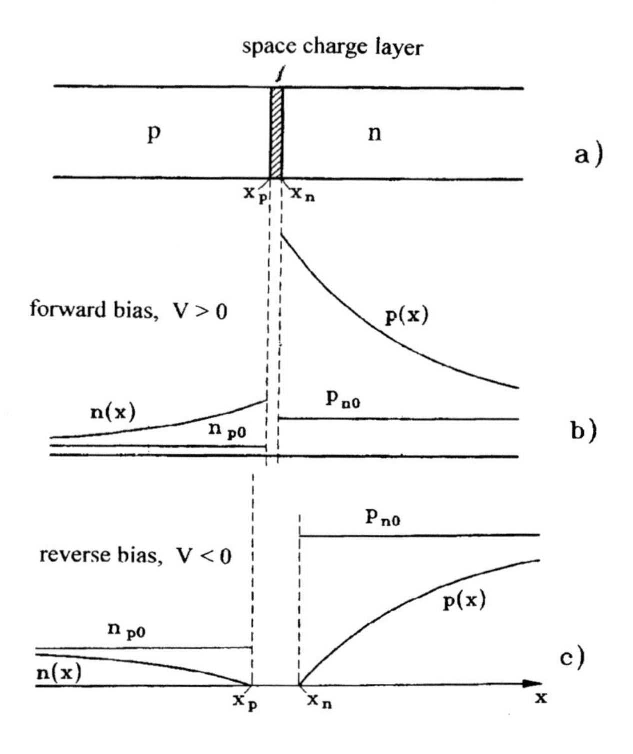

Fig. 3.9 Minority carrier distributions in a pn junction (**a**) under forward bias (**b**) and reverse bias (**c**) on a linear scale

Here the stationary case is assumed and the excess recombination rate R_p is expressed, according to Eq. (2.48), by the minority carrier lifetime and the excess hole concentration. Since the hole density p is assumed small compared with n, and additionally the field in the neutral region is small, the field term in the current equation (2.43) can be neglected. Hence we are dealing with the case 1b) in Sect. 2.10. Inserting

$$j_p = -q \cdot D_p \frac{dp}{dx} \tag{3.41}$$

into Eq. (3.40) one obtains

$$D_p \cdot \frac{d^2p}{dx^2} = \frac{p - p_{n0}}{\tau_p} \tag{3.42}$$

The solution of this differential equation with the boundary condition $p(x_n) = p_n^*$ is

$$p(x) - p_{n0} = (p_n^* - p_{n0}) \cdot e^{-(x-x_n)/L_p} \tag{3.43}$$

where L_p is the hole diffusion length:

$$L_p = \sqrt{D_p \cdot \tau_p} \tag{3.44}$$

Inserting p_{n}^{*} from Eq. (3.36) one has

$$p(x) - p_{\mathrm{n0}} = p_{\mathrm{n0}} \left(\mathrm{e}^{qV/kT} - 1 \right) \cdot \mathrm{e}^{-(x - x_{\mathrm{n}})/L_{\mathrm{p}}} \tag{3.45}$$

Using this hole distribution in Eq. (3.41) the hole current density at $x = x_{\mathrm{n}}$ is obtained as

$$j_{\mathrm{p}}(x_{\mathrm{n}}) = j_{\mathrm{ps}} \cdot \left(\mathrm{e}^{qV/kT} - 1 \right) \tag{3.46}$$

with

$$j_{\mathrm{ps}} = q \cdot p_{\mathrm{n0}} \cdot \frac{D_{\mathrm{p}}}{L_{\mathrm{p}}} = q \cdot \frac{n_{\mathrm{i}}^{2}}{N_{\mathrm{D}}} \cdot \frac{D_{\mathrm{p}}}{L_{\mathrm{p}}} \tag{3.46a}$$

Analogously the electron current density in the p-region at $x = x_{\mathrm{p}}$ is given by

$$j_{\mathrm{n}}(x_{\mathrm{p}}) = j_{\mathrm{ns}} \cdot \left(\mathrm{e}^{qV/kT} - 1 \right) \tag{3.47}$$

with

$$j_{\mathrm{ns}} = q n_{\mathrm{p0}} \cdot \frac{D_{\mathrm{n}}}{L_{\mathrm{n}}} = q \cdot \frac{n_{\mathrm{i}}^{2}}{N_{\mathrm{A}}} \cdot \frac{D_{\mathrm{n}}}{L_{\mathrm{n}}} \tag{3.47a}$$

where L_{n} is the electron diffusion length:

$$L_{\mathrm{n}} = \sqrt{D_{\mathrm{n}} \cdot \tau_{\mathrm{n}}} \tag{3.48}$$

As is seen, the minority carrier diffusion currents adopt the exponential voltage dependence of the minority carrier densities p_{n}^{*}, n_{p}^{*}. Apart from these quantities, the currents depend on the diffusion constants and the diffusion lengths L_{p}, L_{n} which are determined by the respective minority carrier lifetime in the neutral regions.

Whereas the recombination in the neutral regions is considered to be essential, *the recombination/generation in the thin space charge layer is neglected* in the ideal *I–V* characteristic. Since $dj_{\mathrm{n,p}}/\mathrm{d}x = \pm qR$, this means that the electron and hole currents are assumed constant across the space charge layer. With $j_{\mathrm{n}}(x_{\mathrm{n}}) = j_{\mathrm{n}}(x_{\mathrm{p}})$, $j_{\mathrm{p}}(x_{\mathrm{p}}) = j_{\mathrm{p}}(x_{\mathrm{n}})$ one obtains

$$j = j_{\mathrm{n}}(x_{\mathrm{n}}) + j_{\mathrm{p}}(x_{\mathrm{n}}) = j_{\mathrm{n}}(x_{\mathrm{p}}) + j_{\mathrm{p}}(x_{\mathrm{n}}) \tag{3.49}$$

Thus, the current density j is given by the sum of the minority carrier diffusion currents in the neutral regions at the borders to the space charge layer. Hence by adding Eqs. (3.46) and (3.47), the current–voltage characteristics of the pn-junction is obtained as

$$j = j_{\mathrm{s}} \cdot \left(\mathrm{e}^{qV/kT} - 1 \right) \tag{3.50}$$

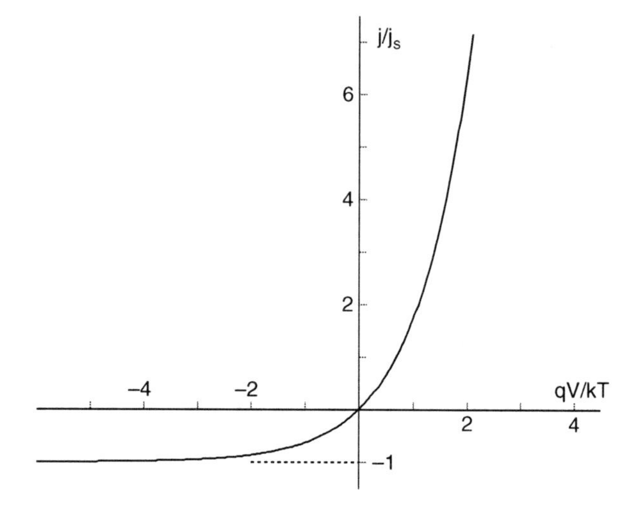

Fig. 3.10 Normalized ideal *I–V* characteristic of a pn-junction

with

$$j_s = q \cdot n_i^2 \cdot \left(\frac{D_p}{L_p \cdot N_D} + \frac{D_n}{L_n \cdot N_A} \right) \tag{3.51}$$

Equation (3.50) with (3.51) is the ideal current–voltage characteristic of the pn-junction as derived by Shockley [Sho49]. In Fig. 3.10 the characteristic is shown in the normalized form j/j_s versus qV/kT. The current increases exponentially with positive voltage, in the blocking direction it approaches quickly the saturation current which is very small. If for example

$$N_A = 1 \times 10^{16}\text{cm}^{-3},\ N_D = 1 \times 10^{15}\text{cm}^{-3},$$
$$L_n = L_p = 50\,\mu m,\ D_n = 30\,\text{cm}^2/\text{s},\ D_p = 12\,\text{cm}^2/\text{s}$$

one obtains for silicon with $n_i(300\,\text{K}) = 1.07 \times 10^{10}\text{cm}^{-3}$: $j_s = 5.5 \times 10^{-11}\,\text{A/cm}^2$. As is noted, the characteristic (3.50) has been published for a different model first by Wagner [Wag31] who showed that it describes the maximum possible rectification effect attainable by electronic conduction.

The factor $n_i{}^2$ in j_s includes the influence of the bandgap of the semiconductor material and contains the main part of the temperature dependence. In Fig. 3.11 the characteristics for several semiconductors at 300 K as given by Eqs. (3.50) and (3.51) are plotted directly and on a larger scale than in Fig. 3.10. The reverse current is scaled logarithmically. For all semiconductors the same values of N_A, N_D, and $D_{n,p}/L_{n,p}$ were used for j_s as in the above example for silicon, so that only the influence of the intrinsic concentration and the bandgap is shown (see Eq. (2.6) and Fig. 2.3). The values of the saturation current density obtained are in A/cm^2: 1.9×10^{-4} for Ge, 5.5×10^{-11} for Si, 2.1×10^{-18} for GaAs and 1.3×10^{-46} for 4H- SiC. The figure shows that up to a certain forward voltage, about 0.7 V for Si,

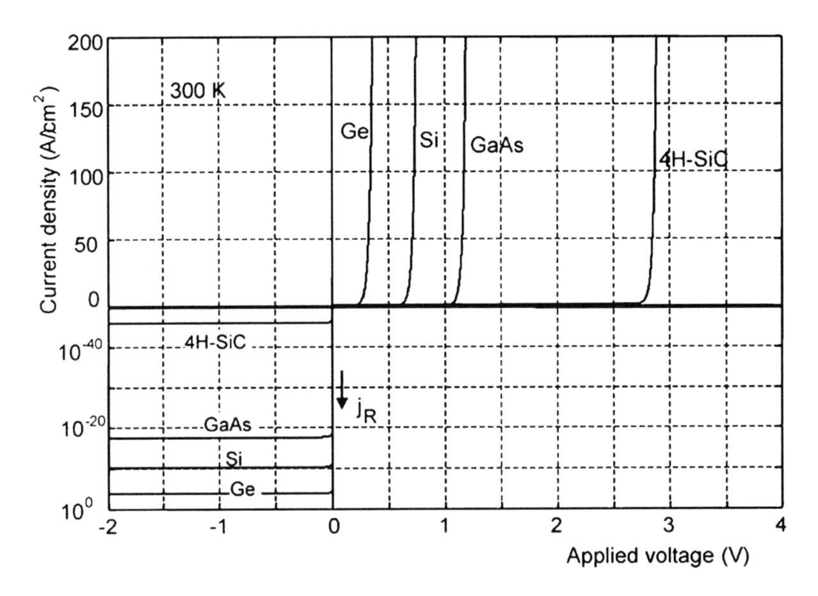

Fig. 3.11 Current density versus applied voltage for ideal pn-junctions in different semiconductors at 300 K

the current remains very small on the scale of normal conduction current densities, whereas above that voltage the current is soon strongly increasing. This threshold voltage can be defined with some arbitrariness as the voltage belonging to a current density $j_{\mathrm{thr}} = 5\mathrm{A/cm}^2$. Solving Eq. (3.50) for V

$$V = \frac{k \cdot T}{q} \ln\left(\frac{j}{j_{\mathrm{s}}} + 1\right) \tag{3.52}$$

and inserting $j_{\mathrm{thr}} = 5\ \mathrm{A/cm}$ the threshold voltage in the case of Fig. 3.11 is obtained to be 0.26 V for Ge, 0.65 V for Si[2], 1.09 V for GaAs, and 2.77 V for 4H-SiC. These values are close to the built-in voltage V_{bi} calculated by Eq. (3.6a). The threshold current, although near the lower end of normally used current densities, is already so high that the voltage balances largely the built-in voltage. The threshold voltage increases nearly linearly with increasing bandgap as follows from Eqs. (3.52) and (3.50) if Eq. (2.6) is inserted. Experimental results are sufficiently well reproduced by this theory.

Of course a small threshold voltage is advantageous since it implies small forward losses. On the other hand, however, the high j_{s} required for small V_{thr} means

[2] The threshold voltage given in data sheets of device manufacturers is about 0.2–0.5 V higher. Because it is defined there as the onset voltage V_{s} in the equivalent straight line $V_{\mathrm{F}} = V_{\mathrm{s}} + R_{\mathrm{diff}} I_{\mathrm{F}}$, the fitting to the mostly parabolic characteristics results in somewhat higher values than the threshold voltage defined above.

a high reverse leakage current. In the case of Ge this disadvantage outweighs strongly the advantage of small threshold voltage. Already at 100°C the leakage current grows so high that it results in a hardly controllable heating. For semiconductors with a wide bandgap on the other hand, the large threshold voltage is the more weighing disadvantage. Here, the large threshold voltage of pn-junctions is often avoided by using a metal–semiconductor junction and for switching devices a unipolar field effect transistor without junction in the current path.

In practical devices, the pn-junctions are mostly very asymmetric. With $N_A >> N_D$ Eq. (3.51) simplifies to

$$j_s = q \cdot n_i^2 \cdot \left(\frac{D_p}{L_p \cdot N_D} \right) \tag{3.53}$$

The saturation current is determined in these cases only by the minority carrier parameters in the weakly doped zone. Using this equation, we consider now the *temperature dependence* of the voltage V at a given forward current which is often used to indicate and control the temperature in integrated power devices. Inserting Eq. (2.6) for n_i^2 one obtains from Eqs. (3.53) and (3.52) for $V > 3\,kT/q$

$$V(j,T) = \frac{E_g}{q} - \frac{kT}{q} \cdot \ln \left(\frac{q \cdot D_p \cdot N_c \cdot N_v}{L_p \cdot N_D \cdot j} \right) \tag{3.54}$$

Because the term $qD_pN_cN_p/\left(L_pN_D\right)$ represents a very high current density whose temperature dependence due to the logarithm has little influence, the forward voltage V decreases nearly linearly with increasing T. In Fig. 3.12 the dependency (3.54)

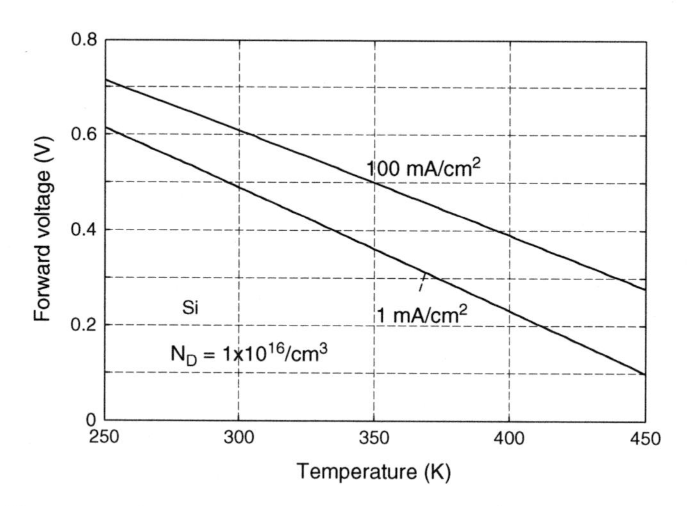

Fig. 3.12 Temperature dependence of forward voltage of a p⁺n-junction at constant current densities

with use of Eqs. (2.8) and (2.9) is plotted for two current densities for the example $N_D = 1 \times 10^{16}\text{cm}^{-3}$ and $\tau_p(T) = 1\,\mu s(T/300)^2$. This temperature dependence is found often and leads here to $D_p/L_p = \sqrt{(D_p/\tau_p)} \simeq 3.3 \times 10^3(300/T)^{3/2}$.

Regarding the conditions underlying the Shockley characteristic (3.50) and (3.51), the assumption of negligible recombination in the space charge region is sometimes not satisfied in silicon even at forward bias where the thickness of the space charge layer is very small. This can be seen, if at small forward current densities, for example in the range of 100 js to 1 A/cm^2, the measured current is plotted on a logarithmic scale versus voltage. The observed slope is often not q/kT as required by Eq. (3.50) but essentially smaller. The measured characteristic can be described then by

$$j = j_s \cdot \left(e^{qV/(nkT)} - 1\right) \tag{3.55}$$

where the coefficient n ranges often between 1.7 and 2. Because of the small current densities this cannot be attributed to a resistive voltage drop in the p- and n-region. The recombination in the space charge layer is given by the SRH equation (2.62) with $np = \text{const} >> n_i^2$ at forward bias (see Eq. (3.39)). The numerator of this equation has approximately the same magnitude as in the neighboring parts of the neutral regions. However, if the recombination level lies near the middle of a not too small bandgap and hence the concentrations n_r and p_r both are small, the denominator in Eq. (2.62) is on average small against the value in the neutral regions where n or p is large. Hence the recombination rate R is very high, and in spite of the small thickness of the space charge layer the integral recombination $\int R\mathrm{d}x$ can be significant. If the recombination levels, however, lie appreciably away from the middle of the bandgap, the assumption of negligible recombination in the space charge layer is justified, and the ideal characteristic is measured at sufficiently small current densities.

At higher current densities, where the resistive voltage drop V_{drift} caused by the majority carrier currents in the p- and n-region is considerable, the applied voltage V in the above equations and figures has to be interpreted as the junction voltage $V_j = V_F - V_{\text{drift}}$ where V_F signifies the total forward voltage. As long as the resistance is given by the doping concentration and thicknesses of the two semiconductor regions and hence is known and constant, V_F as function of j is obtained by adding the drift voltage $V_{\text{drift}} = rj\,(r = \text{resistance} \times \text{area})$ to the right-hand side of Eq. (3.52):

$$V_F = \frac{kT}{q} \cdot \left(\ln \frac{j}{j_s} + 1\right) + r \cdot j$$

However, the resistance can be simply calculated only if also the further assumption made for the derivation of Eq. (3.50), i.e. the condition of low injection level, is satisfied. This is the case over a considerable range of current densities only for relative high doping concentrations of the p- and n-region, as for example in the n-emitter and p-base region of thyristors and IGBTs. In many other cases this condition is not fulfilled. For power diodes with a p$^+$nn$^+$ structure, the injection level in the weakly doped n-base is at usual forward current densities not low but in fact high. Chapter 5 deals with the characteristics for this case.

3.3 Blocking Characteristics and Breakdown of the pn-Junction

3.3.1 Blocking Current

Especially for reverse voltages, Eqs. (3.50) and (3.51) agree with measurements only in limited ranges of voltage and temperature if at all. Often the measured blocking current is much higher than the saturation current (3.51) because of large generation in the space charge region. A second effect not included in Eqs. (3.50) and (3.51) is avalanche multiplication which at high reverse voltages leads to enhanced blocking current and at a certain voltage to complete breakdown of the blocking ability. This effect will be treated in the next section, in the present section we consider the blocking current as made up of the saturation current (3.51) and the current caused by generation in the space charge region.

Since a reverse voltage results in a negative excess carrier concentration (deviation from equilibrium), the thermal generation rate $G(x)$ is everywhere ≥ 0. The total reverse current density j_r owing to thermal generation is obtained by integrating the continuity equation $dj_p/\mathrm{d}x = -qR = qG$ from the left border of the p-region ($x = -\infty$) to the right border of the n-region ($+\infty$). This yields

$$q\int_{-\infty}^{\infty} G\,\mathrm{d}x = j_p(\infty) - j_p(-\infty) = -j_p(-\infty) = -j = j_r \tag{3.56}$$

since the hole current in the p-region at large distance from the junction equals the total current. For the calculation, one uses as before the abrupt depletion approximation with $n \simeq p \simeq 0$ in the space charge region and with the minority carrier distribution in the neutral regions as given for the holes in the n-region by Eq. (3.45), where now $V < 0$ (see Fig. 3.9c). The integral over the neutral regions delivers the reverse current of Eqs. (3.50) and (3.51), as can be easily verified. The minority carriers diffusing out of the neutral regions are *generated* there due to the negative excess concentration. For the current caused by generation in the *space charge region*, the formula (2.62) applies. In order that the terms with n and p can be neglected on the base of Eq. (3.39) and Eqs. (3.15) and (3.18), the reverse voltage V_r is assumed to be $> 3kT/q$. The generation rate is then constant and given by Eq. (2.69), and the reverse current generated in the space region (boundaries x_p and x_n, thickness $w = x_n - x_p$) is

$$j_{sc} = q\int_{x_p}^{x_n} G\,\mathrm{d}x = \frac{q \cdot n_i \cdot w}{\tau_g} = \frac{q \cdot N_r \cdot w}{1/e_n + 1/e_p} \tag{3.57}$$

Since the charge density on both sides of the junction is given by the doping concentration in the used approximation and is therefore independent of the voltage, the extension of the space charge region into the p- and n-region (and other relationships) is given by the same expressions as for zero bias except that V_{bi} has to be replaced by $V_{sc} = V_{bi} - V = V_{bi} + V_r$. Hence Eq. (3.19a) for an abrupt p$^+$n-junction generalizes to

$$w = \sqrt{\frac{2 \cdot \varepsilon \cdot (V_{bi} + V_r)}{q \cdot N_D}} \tag{3.58}$$

Inserting this and adding Eq. (3.57) to Eq. (3.53) the total blocking current density of a p$^+$n-junction is obtained as

$$j_r = j_s + j_{sc} = q \cdot \left(\frac{n_i^2}{N_D} \cdot \frac{L_p}{\tau_p} + \frac{n_i}{\tau_g} \cdot \sqrt{\frac{2 \cdot \varepsilon \cdot (V_{bi} + V_r)}{q \cdot N_D}} \right) \tag{3.59}$$

where D_p/L_p is written as L_p/τ_p to express the two current contributions in a similar manner by a concentration, a length, and a lifetime. Also for the diffusion current the condition $V_r > 3kT/q$ is used to be in saturation.

Via the width of the depletion region the blocking current increases now with voltage. Whereas the diffusion current is proportional to n_i^2, the space charge term increases only linear with n_i. Hence the diffusion term in proportion to the space charge current increases with intrinsic concentration, that is with decreasing bandgap and increasing temperature. Which part predominates depends, however, also on the lifetime τ_g as compared to τ_p as well as on the voltage V_r. If the generation lifetime τ_g is comparable with τ_p and w comparable with L_p, the space charge term in Eq. (3.59) predominates to the extent as $n_i >> p_{n0} = n_i^2/N_D$ or $n_i << N_D$. However, as has been discussed in Sect. 2.7.2, the generation lifetime depends exponentially on the recombination level and becomes very large if E_r is not located near the middle of the bandgap. According to Eq. (2.64), (2.58), (2.60), and (2.63) of the Shockley–Read–Hall model, the generation lifetime in the depletion region can be written as

$$\tau_g = \frac{n_i}{N_r} \left(\frac{1}{c_n \cdot n_r} + \frac{1}{c_p \cdot p_r} \right) = n_i \cdot \left(\frac{\tau_{n0}}{n_r} + \frac{\tau_{p0}}{p_r} \right) \tag{3.60}$$

whereas the low-level minority lifetime τ_p according to Eq. (2.66) is given by

$$\tau_p = \tau_{p0} + (\tau_{p0} \cdot n_r + \tau_{n0} \cdot p_r)/N_D \tag{3.61}$$

n_r and p_r are (except for the degeneracy factor) equal to the carrier concentrations obtained assuming the Fermi level equal to the recombination level (see Eqs. (2.58) and Eq. (2.60)). If E_r coincides with the intrinsic level E_i, one has $n_r = p_r = n_i$ and Eqs. (3.60) and (3.61) yield $\tau_g = \tau_{n0} + \tau_{p0}$ which is close to the minimum of τ_g. On the other hand, if the recombination level is distinctly distant from the middle of the bandgap, either n_r or p_r is small against n_i and hence $\tau_g >> \tau_p$ according to Eq. (3.60). In this case the space charge current can be small compared with the saturation current.

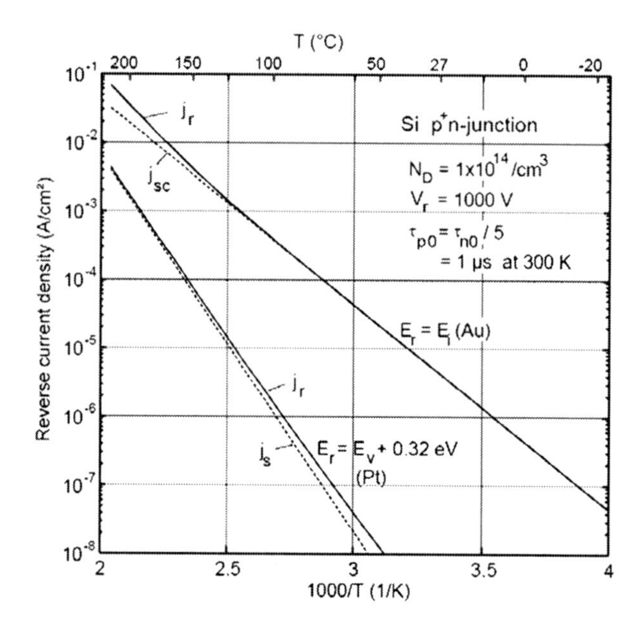

Fig. 3.13 Reverse current density of a p$^+$n-junction in Si as function of temperature for two recombination levels. In both cases the lifetime parameters τ_{n0}, τ_{p0} are assumed as $\tau_{p0} = \tau_{n0}/5 = 1\,\mu s(T/300)^2$

In Fig. 3.13 the reverse current density of a p$^+$n-junction in Si is plotted versus the reciprocal absolute temperature for $E_r = E_i$ as well as for $E_r = E_V + 0.32\,\text{eV}$. The reverse voltage is 1000 V. The case of $E_r = E_i$ reflects the behavior of gold-doped junctions since gold has an acceptor level very near the middle of the bandgap (see Fig. 2.18). The second case refers to the Pt level which is nearest to E_i and hence the active level in the depletion layer. As is seen, for $E_r = E_i$ the total reverse current is practically equal to the current generated in the depletion region, except in the upper temperature range. This holds even for a small reverse voltage. A predominant space charge current is also observed often in junctions which are not intentionally doped with deep traps, as is indicated by the temperature dependence. In the platinum case, on the other hand, the reverse current consists nearly totally of diffusion current j_s (dashed line in Fig. 3.13). With equal lifetime parameters τ_{n0}, τ_{p0} assumed in both cases, the generation lifetime for $E_r = E_i$ (and $g = 2$ in Eq. (2.58)) is 4.6 μs at 300 K whereas for $E_r = E_V + 0.32\,\text{eV}$ one obtains $\tau_g = 61$ ms.

The heat dissipation during reverse bias, $j_r V_r$, is usually much smaller than at forward conduction. However, because j_r increases strongly with temperature, the heat generation can increase stronger than the heat conduction to the sink. Since this can lead to an uncontrolled temperature increase (thermal run-away) and destruction, the reverse current density must be small enough. As an upper limit for diodes of the 1500 V class a value of about 10 mA/cm^2 at 150°C is typically used. Such a limit implies that the recombination center density has to be limited, which in the case of gold has the consequence that otherwise possible switching times cannot be attained.

3.3.2 Avalanche Multiplication and Breakdown Voltage

At a certain breakdown voltage, the reverse current increases abruptly and the blocking ability is lost. This is due to impact ionization or avalanche multiplication as described in Sect. 2.8. By the kinetic energy gained in the electric field a carrier raises an electron from the valence to the conduction band and creates an electron–hole pair, and these secondary particles again create electron–hole pairs, thus an avalanche process is initiated. The effect is significant only at high field strengths occurring under high reverse bias. With the avalanche generation rate (2.71) the continuity equations (2.83) and (2.84) in the stationary one-dimensional case take the form

$$\begin{aligned} \frac{\mathrm{d}j_\mathrm{p}}{\mathrm{d}x} &= \alpha_\mathrm{p} \cdot j_\mathrm{p} + \alpha_\mathrm{n} \cdot j_\mathrm{n} + q \cdot G \\ \frac{\mathrm{d}j_\mathrm{n}}{\mathrm{d}x} &= -\alpha_\mathrm{p} \cdot j_\mathrm{p} - \alpha_\mathrm{n} \cdot j_\mathrm{n} - q \cdot G \end{aligned} \tag{3.62}$$

where $\alpha_\mathrm{n}, \alpha_\mathrm{p}$ are the field-dependent impact ionization rates and G as before denotes the thermal generation rate. The total current density

$$j = j_\mathrm{n} + j_\mathrm{p} \tag{3.63}$$

is independent of x under stationary conditions (see Eq. (2.87)). The avalanche effect is expressed in Eqs. (3.62) by the proportionality of the generation rate to the current densities which at the relevant field strengths are proportional to the carrier densities. The course of the field strength, of the ionization rates and current densities is shown for a p$^+$n-junction in Fig. 3.14 (see the legend). To have positive current densities and field strength at reverse bias, the n-region is at the left and the p-region at the right-hand side. The strong variation of j_n, j_p in the high field region due to avalanche multiplication is illustrated qualitatively.

For a qualitative understanding we assume at first that the ionization rates are equal, $\alpha_\mathrm{n} = \alpha_\mathrm{p} = \alpha$, as is found e.g. in GaAs. Then Eqs. (3.62) with (3.63) simplify to

$$\frac{\mathrm{d}j_\mathrm{p}}{\mathrm{d}x} = \alpha \cdot j + q \cdot G = -\frac{\mathrm{d}j_\mathrm{n}}{\mathrm{d}x} \tag{3.64}$$

Integrating over the space charge region with the boundaries $x_\mathrm{n} = 0$ and $x_\mathrm{p} = w$ one obtains

$$j - j_\mathrm{ns} - j_\mathrm{ps} = j \cdot \int_0^w \alpha \mathrm{d}x \; + \; q \cdot w \cdot G$$

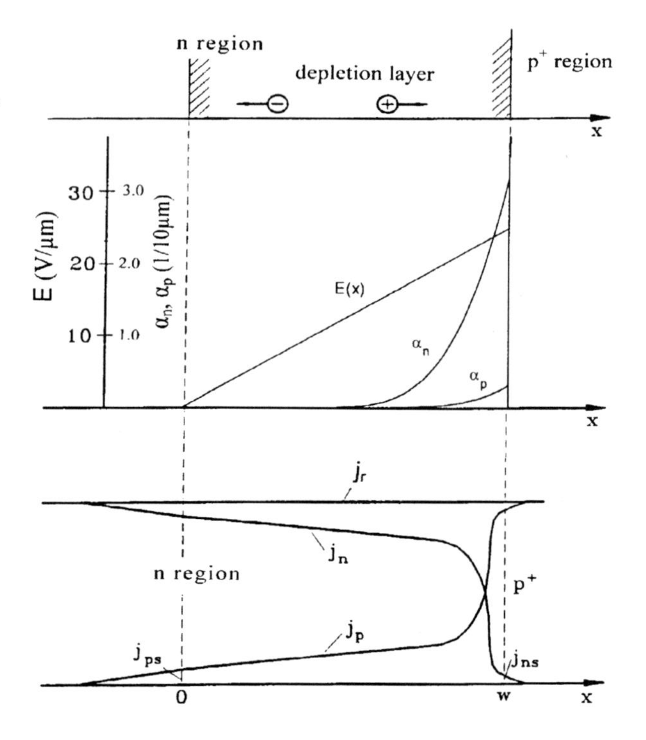

Fig. 3.14 Reverse biased p^+n-junction with avalanche multiplication: Field strength, ionization rates, and reverse current densities as function of x. The ionization rates in dependency of the field strength refer to Si

since $j_p(x_n) = j - j_{ns}$ and $j_p(x_n) = j_{ps}$, where j_{ns}, j_{ps} are the minority carrier saturation current densities entering the depletion layer from the neutral regions. The current density j follows as

$$j = \frac{j_{ns} + j_{ps} + j_{sc}}{1 - \int_0^w \alpha \, dx} \tag{3.65}$$

where the previous notation $j_{sc} = qwG$ is used. As shown by this equation, the effect of avalanche can be expressed by a multiplication factor

$$M = \frac{1}{1 - \int_0^w \alpha \, dx} \tag{3.66}$$

which enhances the three components of the thermal current density. If with increasing voltage the field via the ionization rate satisfies the equation

$$\int_0^w \alpha(E(x)) \, dx = 1 \tag{3.67}$$

the current increases to infinity. Hence this is the condition from which the breakdown voltage $V_r = V_B = \int_0^w E \, dx$ can be calculated. Below the breakdown voltage, Eq. (3.65) describes the enhancement of the reverse current by avalanche multiplication.

The considered case $\alpha_n = \alpha_p$ allows a simple calculation because the position where an impact ionization initiates the avalanche process has no influence on the total generated avalanche charge in this case: After generation of an electron–hole pair at a point x the electron moves to the neutral n-region and the hole to the p-region, so both together cover the whole width of the space charge region (high field region). This holds also for the pairs generated by the secondary carriers at a point $x\prime$. Hence if electrons and holes are equally effective, the total generated avalanche charge is independent of x. This is also the cause why all three current components j_{ns}, j_{ps}, and j_{sp} are multiplied by the same factor.

If $\alpha_n \neq \alpha_p$ as in the case of Si and most other semiconductors (see Figs. 2.20 and 2.21), the calculation is much more complicated. In Appendix B it is shown that the avalanche multiplication can be expressed generally by double integrals over the ionization rates. As expected, the three components of the thermal blocking current (numerator of Eq. (3.65)) are enhanced for $\alpha_n \neq \alpha_p$ each by a different multiplication factor and hence the leakage current density is generally given by

$$j = M_n \cdot j_{ns} + M_{sc} \cdot j_{sc} + M_p j_{ps} \tag{3.68}$$

From $\alpha_n > \alpha_p$ in the case of Si, it follows that $M_n > M_{sc} > M_p$. All three M′s tend to infinity at the same voltage, the breakdown voltage. For an abrupt p^+n-junction in Si the multiplication factors, as calculated numerically from the formulae in the appendix using the ionization rates of Eq. (2.78), are plotted in Fig. 3.15 versus the reverse voltage. The breakdown voltage for the assumed doping density $1 \times 10^{14} cm^{-3}$ of the n-region is 1640 V. Already at 1400 V, the current density j_{ns} is enhanced by $M_n = 5$, whereas the multiplication factor M_p amounts to 1.17 only. In the case of a p^+n-junction the higher M_n is not significant for the blocking current, since the electron saturation current j_{ns} is very small according to Eq. (3.51). For a transistor structure with a p^+n-junction between base and collector, however, the large M_n has a strong effect on the breakdown voltage, as will be seen.

Analytically, the multiplication factors as function of voltage are approximated often by the equation [Mil57]

$$M = \frac{1}{1 - (V/V_B)^m} \tag{3.69}$$

where the exponent m is used for fitting. To approximate the dependencies of Fig. 3.15, very different m-values are necessary for M_n and M_p. Fitting in both cases at $M_{n,p} = 2$, a value $m = 2.2$ is obtained for M_n and $m = 13.2$ for M_p. These values differ significantly from values in the literature where often $m = 4$ and 6 are used for M_n, M_p, respectively. Although Eq. (3.69) is only a rough approximation, considering the whole voltage range, it is very useful for dimensioning of transistor and thyristor structures. The large difference between M_n and M_p has a significant influence on the blocking behavior.

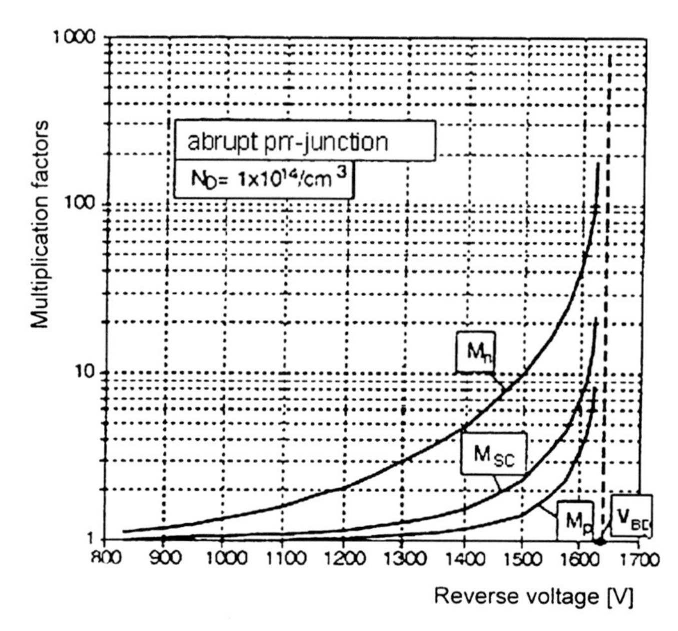

Fig. 3.15 Multiplication factors in dependence of the voltage for an abrupt p$^+$n-junction with $N_D = 1 \times 10^{14}\text{cm}^{-3} \cdot T = 300\,\text{K}$

If only the breakdown voltage of a (single) pn-junction is of interest, the complicated ionization integrals in Appendix B can be avoided. As noted already in Sect. 2.8 and shown in Appendix B, the breakdown voltage can be calculated using the effective ionization rate

$$\alpha_{\text{eff}} = \frac{\alpha_n - \alpha_p}{\ln(\alpha_n/\alpha_p)} \tag{3.70}$$

which at breakdown satisfies the condition [Oga65, WW60]

$$\int_0^w \alpha_{\text{eff}}(E(x)) = 1 \tag{3.71}$$

similarly as α in Eq. (3.67). This holds exactly only if the ratio α_n/α_p is independent of E. However, also if this is not well fulfilled over a large field range, Eq. (3.71) is often a good approximation, because only a relative small field range contributes significantly to the integral. This is the case also for silicon.

The maximum field strength in the depletion layer, at which breakdown occurs, is called the critical field strength E_c. If the width of the high-field region is varied via the doping density according to Eq. (3.58) for a p$^+$n-junction, this must be compensated according to Eq. (3.71) by an inverse variation of α_{eff}. However, due to the very strong increase of the α's with E, this requires only a small change of the field strength. Hence E_c depends only weakly on the thickness w of the depletion region, and for rough estimates E_c is assumed often as constant.

Without restriction we assume at first that the critical field (in dependence of w and N_D) is known and ask how the width and doping concentration of the weakly doped region(s) of a pn-junction are to be chosen to reach a desired breakdown voltage V_B. Considering an abrupt p^+n-junction as shown in Fig. 3.14 the breakdown voltage is

$$V_B = \int_0^w E \, dx = \frac{1}{2} w E_c \tag{3.72}$$

Within the approximation of constant E_c, the breakdown voltage is proportional to the width w of the space charge region. According to the Poisson equation

$$\frac{dE}{dx} = \frac{E_c}{w} = \frac{q \cdot N_D}{\varepsilon} \tag{3.73}$$

w depends inversely on the doping density N_D. Expressing the width w in Eq. (3.72) by N_D one obtains

$$V_B = \frac{\varepsilon \cdot E_c^2}{2 \cdot q \cdot N_D} \tag{3.74}$$

Hence, within the approximation of constant E_c, the breakdown voltage is inversely proportional to the doping density of the n-region. However, as mentioned, this is only a very rough approximation.

To calculate the critical field in dependence of w or N_D in analytical form, we use the power approach (2.76) in the form [Ful67, Shi59]

$$\alpha_{eff} = B \cdot E^n \tag{3.75}$$

With the coordinates of Fig. 3.14, the field strength is

$$E(x) = \frac{q \, N_D}{\varepsilon} x \tag{3.76}$$

Inserting Eq. (3.75) together with $dx = \varepsilon/(qN_D) \, dE = w/E_c dE$ into the condition (3.71) one gets

$$\frac{B \, w}{E_c} \int_0^{E_c} E^n \, dE = B \, w \frac{E_c^n}{n+1} = 1 \tag{3.77}$$

$$E_c = \left(\frac{n+1}{Bw} \right)^{\frac{1}{n}} = \left(\frac{q \cdot (n+1) \cdot N_D}{B \cdot \varepsilon} \right)^{\frac{1}{n+1}} \tag{3.78}$$

The last expression follows from Eq. (3.77) with $w = \varepsilon E_c/(qN_D)$ (Eq. (3.73)). As expected, E_c decreases slightly with increasing w and decreasing

N_D. Inserting Eq. (3.78) into (3.74) the breakdown voltage as function of doping density is obtained as

$$\begin{aligned} V_B &= \frac{\varepsilon}{2 \cdot q \cdot N_D \cdot} \left(\frac{q \cdot (n+1) \cdot N_D}{B \cdot \varepsilon} \right)^{\frac{2}{n+1}} \\ &= \frac{1}{2} \cdot \left(\frac{n+1}{B} \right)^{\frac{2}{n+1}} \cdot \left(\frac{\varepsilon}{q \cdot N_D} \right)^{\frac{n-1}{n+1}} \end{aligned} \tag{3.79}$$

Equations (3.72) and (3.78) yield

$$V_B = \frac{1}{2} \cdot \left(\frac{n+1}{B} \cdot w^{n-1} \right)^{\frac{1}{n}} \tag{3.80}$$

The width w of the depletion region as function of N_D at breakdown follows from Eqs. (3.73) and (3.78) as

$$w = \left(\frac{n+1}{B} \right)^{\frac{1}{n+1}} \cdot \left(\frac{\varepsilon}{q \cdot N_D} \right)^{\frac{n}{n+1}} \tag{3.81}$$

Due to the variation of the critical field, the increase of V_B with w and with $1/N_D$ is sub-linear.

To adjust Eq. (3.75) to the Chynoweth law, the constants n and B are to be chosen according to Eqs. (2.77), (2.76) and (3.75). With the numerical equation (2.74) for Si at 300 K, one obtains in dependence of the field E_0 where the adjustment is carried out

$$n = \frac{1.68 \times 10^6 \text{ V/cm}}{E_0} \tag{3.82}$$

$$B = \frac{1.06 \times 10^6 \text{ /cm}}{E_0^n \cdot \exp\ (n)} \tag{3.83}$$

Following Shields [Shi59], most authors set the exponent to $n = 7$ in silicon. According to Eq. (3.82) this is obtained at $E_0 = 2.4 \times 10^5 \text{V/cm}$. The corresponding B-value according to Eq. (3.83) is $B = 2.107 \times 10^{-35} \text{cm}^6/\text{V}^7$. Using these constants, Eqs. (3.80) and (3.81) can be written as

$$V_B = \frac{1}{2} \cdot \left(\frac{8}{B} \right)^{\frac{1}{4}} \cdot \left(\frac{\varepsilon}{q N_D} \right)^{\frac{3}{4}} = 563 \text{ V} \cdot \left(\frac{4 \times 10^{14}/\text{cm}^3}{N_D} \right)^{\frac{3}{4}} \tag{3.84}$$

$$w = \left(\frac{8}{B} \right)^{\frac{1}{8}} \cdot \left(\frac{\varepsilon}{q \cdot N_D} \right)^{\frac{7}{8}} = 44.6 \text{ μm} \left(\frac{4 \times 10^{14}/\text{cm}^3}{N_D} \right)^{\frac{7}{8}} \tag{3.85}$$

In the numerical expressions on the right-hand side, the doping density is related to the value $4 \times 10^{14} \text{cm}^{-3}$ because this is roughly the point at which the approximation is adapted and around which it will be very accurate. Here it is taken into account that $\alpha_{eff}(E)$ in the triangular field distribution of the junction (see Fig. 3.14)

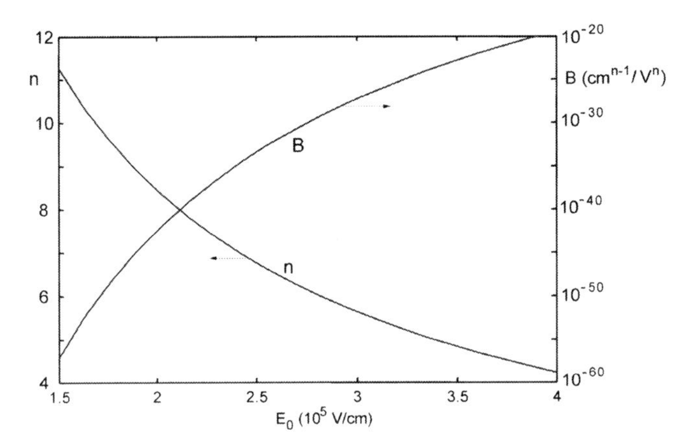

Fig. 3.16 Constants in the power law $\alpha_{\text{eff}} = BE^n$ as functions of the field strength at which it is fitted to $\alpha_{\text{eff}} = a/\exp(b/E).a = 1.06 \times 10^6/\text{cm}, b = 1.68\,\text{V/cm}$ (Si at 300 K)

is best approximated choosing E_0 near $0.9E_c$. This can be seen by maximizing the calculated breakdown voltage. Hence, with the choice $n = 7$ the function (3.75) is matched to Eq. (2.74) at the point where the maximum field in the junction is $E_c = 2.4 \times 20^5/0.9\,\text{V/cm} = 2.667 \times 10^5\text{V/cm}$. This field corresponds to the doping density $N_D = 4.36 \times 10^{14}\text{cm}^{-3}$ according to Eq. (3.78).

At much higher and lower voltages the accuracy suffers from the deviation of the power approximation from the Chynoweth law as illustrated in Fig. 2.20. In Fig. 3.16, the exponent n and the constant B are plotted versus the field E_0 where the fitting is carried out. n varies nearly by a factor 3 in the relevant field range. The extreme increase of B with E_0 is caused according to Eq. (3.83) by the decrease of n and the resulting rapid decrease of E_0^n. To take into account these variations, it is suitable to choose the matching point E_0 for each doping density separately as about $0.9E_c(N_D)$. Inserting Eqs. (3.82) and (3.83) with this E_0 into the right-hand side of Eq. (3.78), one can solve the equation for E_c by iteration. The breakdown voltage and width of the space charge layer are given by Eqs. (3.74) and (3.73) or can be calculated from Eqs. (3.79) and (3.81). The breakdown voltage and width w at breakdown obtained in this way are plotted in Fig. 3.17 as functions of N_D. These results are very close to those calculated from the exact ionization integral given in Appendix B. In the figure also $N_D^{-3/4}$ according to Eq. (3.84) is shown (dotted line). As is seen, this simple formula is a good approximation over a wide range. For doping densities $N_D < 10^{14}$ and $> 10^{15}\text{cm}^{-3}$ it underestimates the breakdown voltage. The used approach allows an analytical integration of Eq. (3.71) also in other cases where the field strength depends linearly on x. For p^+nn^+-diodes this is utilized later in Chap. 5.

For diffused junctions, the approach (3.75) is useless because the integration of Eq. (3.71) or the ionization integral can be carried out even with this approximation only numerically. If the diffusion profile is very steep, the results for abrupt

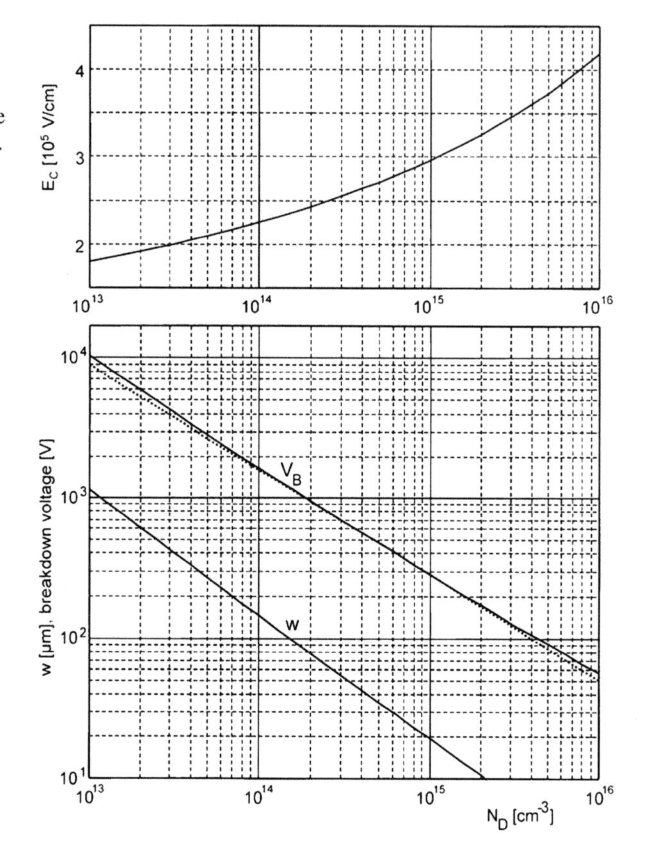

Fig. 3.17 Critical field strength, breakdown voltage, and depletion width at breakdown as functions of the doping density N_D for abrupt. p^+n-junctions in Si at 300 K. The dotted straight line represents the approximation (3.84)

junctions can be used as an approximation. Generally diffused junctions have a higher breakdown voltage than abrupt ones with equal doping density of the lightly doped region. This follows because (i) the maximum field at the junction is reduced and (ii) the field extends considerably into the highly doped diffused region near the junction. In Fig. 3.18 the numerically calculated breakdown voltage of diffused p^+n-junctions is plotted versus the background doping density N_D for two examples of steepness of the diffusion profile. The acceptor density around the junction at $x = 0$ is approximated here by an exponential decrease $N_A(x) \sim \exp(-x/\lambda)$ (orientation as in Fig. 3.6). Hence, λ is the decay length of the diffusion profile near the junction. Since $N_A(x) = N_D$ at $x = 0$, the net doping density in the space charge region is

$$N(x) = N_D - N_A(x) = N_D \cdot (1 - e^{-x/\lambda}) \tag{3.86}$$

and the doping gradient at the junction is $dN/dx(0) = -N_D/\lambda$. The case $\lambda = 3\,\mu m$ is typical for pn^-p structures in thyristors diffused with gallium or boron, the case

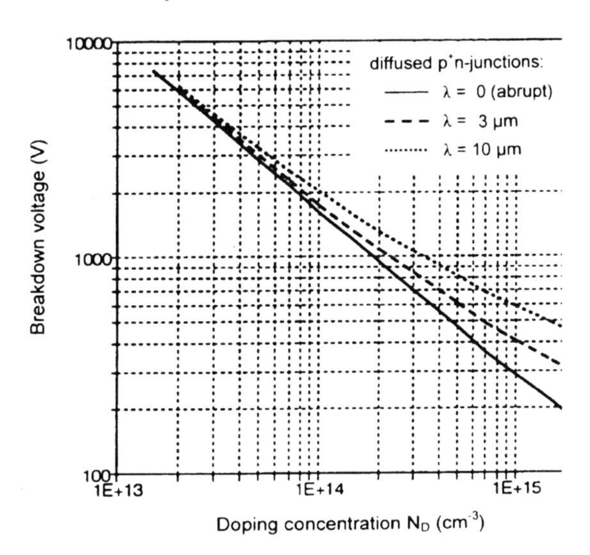

Fig. 3.18 Breakdown voltage of diffused p^+n-junctions in Si. λ is the decay lengths of the diffusion profile in the depletion layer

$\lambda = 10\,\mu$m is representative for high voltage pnp-structures realized with very deep aluminum diffusion.

Temperature Dependence

As described in Sect. 2.8, the ionization rates decrease with increasing temperature; for the effective ionization rate this is expressed numerically by Eq. (2.79). Hence the critical field strength and avalanche breakdown voltage increase slightly with T. Using again the power approach (3.75) to give an analytical description, the exponent n and constant B are needed as function of temperature. This is obtained from Eq. (2.80) together with Eq. (3.75):

$$\begin{aligned} n(T) &= \frac{b_{\mathrm{eff}}(T)}{E_0} = \frac{1.68 \times 10^6 + 1100 \cdot (T - 300)}{E_0} \\ B(T) &= \frac{C(T)}{E_0^{n(T)}} = \frac{a_{\mathrm{eff}}}{E_0^{n(T)} \cdot \exp(n(T))} = \frac{1.06 \times 10^6}{E_0^{n(T)} \cdot \exp(n(T))} \end{aligned} \tag{3.87}$$

where E_0 has to be inserted in volts per centimeter. Substituting these parameters in Eq. (3.79), the breakdown voltage for a triangular field distribution is given analytically as function of temperature. If for example the doping concentration is $N_{\mathrm{D}} = 1 \times 10^{14}\mathrm{cm}^{-3}$, a good choice for E_0 is $2.2 \times 20^5\mathrm{V/cm}$ (see Fig. 3.17). Then Eq. (3.87) at 300 K yields $n = 7.64$, $B = 7.7810^{-39}\mathrm{cm}^{\mathrm{n}-1}/\mathrm{V}^{\mathrm{n}}$. With these values Eq. (3.79) gives $V_{\mathrm{B}} = 1630\,\mathrm{V}$. At $T = 400\,\mathrm{K}$ one obtains $n = 8.14$, $B = 1.01 \times 10^{-41}\,\mathrm{cm}^{\mathrm{n}-1}/\mathrm{V}^{\mathrm{n}}$, and Eq. (3.79) results in $V_{\mathrm{B}} = 1835\,\mathrm{V}$, a 12.6% higher value than at 300 K. Due to a considerable decrease of V_{B} from room temperature downward, one has to dimension so that the targeted blocking ability is maintained still at the lower limit of operation temperature ($\approx 250\,\mathrm{K}$)[3]. This holds for diodes. For thyristors, the temperature dependency is weaker and can be inverted, because the blocking voltage is supplied here by a transistor structure and depends not only on avalanche multiplication but also on the current amplification factor of a transistor, which increases with temperature

3.3.3 Blocking Capability with Wide-Gap Semiconductors

The presented theory is of course applicable also to other semiconductors, particularly to those which have a wider bandgap than Si. Owing to the higher energy required to carry an electron from the valence into the conduction band, the critical field strength increases with bandgap. A summary of raw values of the critical field is given in Table 3.1. Using these values in Eqs. (3.72), (3.73), and (3.74) a rough estimate can be obtained about the needed width w of the depletion region and the allowed maximum doping density for a given blocking ability.

Table 3.1 Raw values of the critical field strength for Si and semiconductors with a wider bandgap

	E_C(V/cm)
Si	2×10^5
GaAs	4×10^5
4H-SiC	2×10^6
GaN	$> 3 \times 10^6$
C (diamond)	$1 - 2 \times 10^7$

To attain for example a breakdown voltage of 10 kV with a p^+n or pn^+-junction in 4H-SiC, a width $w = 2V_B/E_c = 100\,\mu\mathrm{m}$ and a maximum doping density $N_D = 1.07 \times 10^{15}\mathrm{cm}^{-3}$ are obtained using the E_c value in the table. For a more precise calculation the dependence of the critical field strength on the thickness of the depletion region and the doping density determining it is necessary. For 4H-SiC, this can be obtained from Eq. (2.82b) inserting the constants $n = 8.03$ and $B = 2.18 \times 10^{-48}\mathrm{cm}^{7.03}V^{-8.03}$ [Bar09] into Eq. (3.78):

$$E_c = 2.58 \cdot \left(\frac{N_D}{10^{16}/\mathrm{cm}^3}\right)^{0.111} \mathrm{MV/cm} \tag{3.88}$$

[3] Some manufacturers' data sheets specify V_B at +25°C, and for application at lower temperatures one must be aware of said decrease of V_B with temperature.

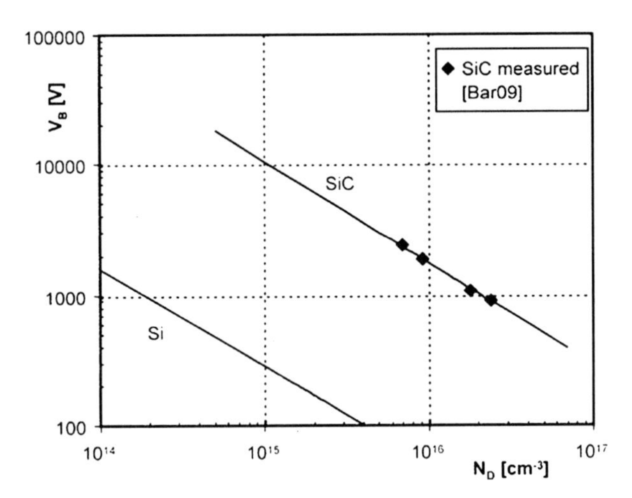

Fig. 3.19 Breakdown voltage of asymmetrical junctions in 4H-SiC and Si as function of the doping concentration of the weakly doped region

For the blocking capability Eq. (3.79) gives

$$V_{\mathrm{B}} = 1770 \left(\frac{10^{16}/\mathrm{cm}^3}{N_{\mathrm{D}}} \right)^{0.78} \mathrm{V} \tag{3.89}$$

The breakdown voltage calculated in this way for an asymmetrical junction in 4H-SiC is shown in Fig. 3.19 together with that of silicon. As can be seen, the maximum doping concentration for a given blocking voltage, for example 1000 V, is for SiC two decades higher than for Si. This results because the critical field strength is an order of magnitude higher than in Si, and according to Eq. (3.74) $N_{\mathrm{D}} \sim E_{\mathrm{c}}^2$ for a given V_{B}. The thickness w is about an order of magnitude smaller since w depends only linearly on $1/E_{\mathrm{c}}$ according to Eq. (3.72). Owing to the smaller possible base width and the much higher doping concentration SiC devices show superior characteristics regarding conduction and switching losses, as will be shown in detail later.

Several years it was practice in the design of SiC-devices to limit the electric field to a maximum value of about 1.5 MV/cm. The devices were specified for a lower blocking voltage than the breakdown voltage resulting from doping and thickness dimensions, because the leakage current otherwise surmounts the allowed range. The reasons for the enhanced reverse current are crystal defects. In recent time, the SiC crystal quality has been improved strongly and former limits are overcome. Since the SiC technology is still developing, Eqs. (3.88) and (3.89) must be considered to some extent as preliminary.

3.4 Injection Efficiency of Emitter Regions

In preceding sections mainly asymmetric pn-junctions have been considered. It turned out that to the extent as the doping ratio $N_{\mathrm{A}}/N_{\mathrm{D}}$ is very large or small the influence of the properties of the highly doped region on the characteristics vanishes, in case of blocking as well as in forward direction. Since the density of injected minority carriers at forward bias is inversely proportional to the doping density according to Eqs. (3.36) and (3.38), mainly the highly doped region injects carriers into the weakly doped region and not vice versa. Because this is similar to the function of the emitter and base regions in transistors, the highly and weakly doped regions of asymmetrical junctions are also called emitter and base region, respectively. The injection efficiency, defined as ratio of minority carrier current in the base region to the total current, is called emitter efficiency. A quantitative comparison of theory and experiment shows now that the measured emitter efficiency is considerebly smaller than calculated on the base of the equations in Sect. 3.2. This manifests itself immediately in the current amplification factor of transistors. For power diodes with a p^+nn^+-structure, the current injected from the base into the emitter regions becomes much earlier significant with increasing injection level than according to the given "classical" equations. A consequence is that the forward voltage drop of diodes and other power devices is significantly enhanced at high current densities compared with characteristics obtained theoretically taking into account high injection levels in the base region. The cause of the unrealistic emitter efficiency obtained from theory is that bandgap reduction in the highly doped region(s) has been neglected till now.

In the present section we examine the injection properties of emitter regions with consideration of bandgap narrowing. Furthermore, Auger recombination in the emitter is considered and the injection level in the base region is allowed to be arbitrary. According to the continuity equation (3.40), the minority carrier current entering the emitter is absorbed by recombination in the bulk and, if the thickness of the region is smaller than a few minority carrier diffusion length, at the contact. Although the contact recombination is utilized in novel devices to improve switching properties, we assume at first that only recombination in the volume is significant.

We consider an abrupt n^+p-junction which is forward biased with low injection level in the emitter region, while – within this limit – the injection level in the base region may be arbitrary (see Fig. 3.20). At first we expand the pn-theory of Sect. 3.2 to the case of arbitrary injection level in the base. According to Eqs. (3.3) and (3.4), the ratio of carrier densities on one side to that on the other side of the space charge region are given by the Boltzmann factor with the potential difference (3.34):

$$\frac{p_n^*}{p_B^*} = \frac{n_B^*}{n^+} = e^{-q \cdot \Delta V / kT} \tag{3.90}$$

As illustrated in Fig. 3.20, the star indicates the carrier concentrations in the neutral regions at the boundaries to the space charge layer. The concentration n^+ in the emitter is equal to the donor density N_D due to the assumed low injection level there. For the minority carrier density in the emitter region Eq. (3.90) gives

$$p_n^* = \frac{n_B^* \cdot p_B^*}{n^+} = \frac{n_B^* \cdot (N_B + n_B^*)}{n^+} \tag{3.91}$$

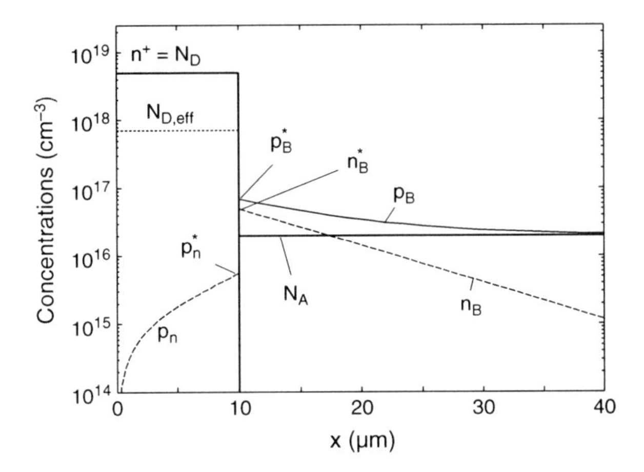

Fig. 3.20 Forward-biased n$^+$p-junction with intermediate injection level in the p-base region

where the equation $p_B^* = N_B + n_B^*$ follows from neutrality. The base doping N_A is replaced here by N_B to include the case of an n$^+$n-junction, occurring in power diodes, by defining $N_B = -N_D$ in case of an n-base. For power diodes which have a low base doping, Eq. (3.91) is mainly significant for high injection levels in the base region where $p_B^* \simeq n_B^*$, in bipolar transistors and thyristors the injection level in the p-base region is typically intermediate in the interesting current range. As shown earlier (see Eqs. (3.46) and (3.43)) the injected holes result in the minority current density in the n$^+$-region:

$$j_p(x_n) = q \cdot \frac{D_p}{L_p} \cdot (p_n^* - p_{n0}) \tag{3.92}$$

Neglecting the very small equilibrium concentration p_{n0} this can be written using Eq. (3.91) as

$$j_p(x_n) = q \cdot h_n \cdot n_B^* \cdot p_B^* \tag{3.93}$$

where the properties of the n emitter are comprised in the constant:

$$h_n = \frac{D_p}{N_D \cdot L_p} = \frac{1}{N_D} \cdot \sqrt{\frac{D_p}{\tau_p}} \tag{3.94}$$

This parameter characterizes the emitter recombination which reduces the injection efficiency. It equals the saturation current density j_{ps} of the emitter defined in Eq. (3.46a) except that the factor qn_i^2 is omitted. It is useful to introduce the h-quantity, because the intrinsic concentration with its strong temperature dependence is not relevant for the injection efficiency for a given current. The h-parameter is nearly temperature independent. Equation (3.93) defining the h-parameter by the proportionality between $j_p(x_n)$ and the product $n_B^* \cdot p_B^*$ is usable also for diffused emitter regions. The experimental h-values determined on the basis of Eq. (3.93)

[Bur75, Sco69, Sco79] are not in close accordance with the saturation current density determining the *I–V* characteristics at low current densities.

For the emitter efficiency of the n^+p-junction, defined as the electron current in the p-base near the space charge layer in proportion to the total current

$$\gamma = j_n(x_p)/j \tag{3.95}$$

one obtains with $j_n(x_p) = j - j_p(x_p) \simeq j - j_p(x_n)$ by insertion of Eq. (3.93)

$$\gamma = 1 - j_p(x_n)/j = 1 - q \cdot h_n \cdot n_B^* \cdot p_B^*/j \tag{3.96}$$

As is seen, a small h_n-value is required to get a high emitter efficiency and current amplification.

According to Eq. (3.94) h_n can be made small by choosing a high doping density N_D. The explicit N_D dependence is however counteracted by the decrease of the minority carrier lifetime due to Auger recombination described in Sect. 2.7. For $N_D > 5 \times 10^{18}\text{cm}^{-3}$ the lifetime is often solely determined by this recombination process, hence according to Eq. (2.53) one has $1/\tau_p = c_{A,n} \cdot N_D^2$. Inserting this, Eq. (3.94) yields the following "limiting" h-value for high N_D

$$h_{n,\text{lim}} = \sqrt{D_p \cdot c_{A,n}} \tag{3.97}$$

which decreases however still slightly with increasing n^+ owing to the decrease of the diffusion constant D_p. Using the Auger coefficient of Eq. (2.54) and $D_p = 1.9\,\text{cm}^2/\text{s}$ for a doping concentration of $10^{19}\,\text{cm}^{-3}$, Eq. (3.97) gives $h_{n,\text{lim}} = 7.3 \times 10^{-16}\text{cm}^4/\text{s}$.

Experimentally, the lowest h-values are an order of magnitude higher [Sco69, Bur75, Sco79, Coo83]. This is attributed to the bandgap narrowing in the emitter region due to high doping, which has been neglected above. Actually the bandgap narrowing has been determined essentially from this discrepancy to explain the experimental current amplification [Slo76]. As discussed in Sect. 2.5 (see Fig. 2.11), a cause of bandgap narrowing consists in the decrease of the energy of the electrostatic field which is built-up by a minority carrier in the environment of majority carriers [Lan79]. It is obvious that the energy ΔE_g saved by this phenomenon during injection must be subtracted from the potential barrier $q\Delta V$ in the Boltzmann factor for the minority carrier density p_n^*. Hence it follows that

$$\frac{p_n^*}{p_B^*} = e^{-(q \cdot \Delta V - \Delta E_g)/kT} \tag{3.98}$$

whereas because of the lower doping in the base region the electron concentration n_B^* is given by the classical relationship

$$\frac{n_B^*}{n^+} = e^{-q \cdot \Delta V/kT} \tag{3.99}$$

Dividing the two equations one obtains in place of Eq. (3.91)

$$p_n^* = \frac{n_B^* \cdot p_B^*}{n^+} \cdot e^{\Delta E_g/kT} \tag{3.100}$$

This holds also in case of the ΔE_g resulting from band tails [Slo76]: In thermal equilibrium one has according to Eq. (2.27):

$$n^+ \cdot p_n^* = n_{i0}^2 \cdot \exp(\Delta E_g/kT) = n_B^* \cdot p_B^* \exp(\Delta E_g/kT)$$

If a forward voltage is applied, the lowered potential difference across the junction results in an enhancement of p_n^*/p_B^* and n_B^*/n^+ by the same factor, hence Eq. (3.100) follows generally. Defining an effective doping concentration $N_{D,eff} = N_D \cdot \exp(-\Delta E_g/kT)$, one has for the concentration ratios at opposite sides of the space charge layer for all doping densities $p_n^*/p_B^* = n_B^*/N_{D,eff}$ (see Fig. 3.20). This effective doping concentration increases only very slightly with increasing N_D. It serves only to illustrate the effect of bandgap narrowing on the injection efficiency and has no whatever significance for the majority carrier concentration $n^+ = N_D$. Inserting Eq. (3.100) into Eq. (3.92) the h_n quantity as defined by Eq. (3.93) takes the form

$$h_n = \frac{1}{N_D} \cdot \frac{D_p(N_D)}{L_p(N_D)} \cdot e^{\Delta E_g(N_D)/kT} \tag{3.101}$$

The explicit dependence on N_D in the denominator can be overcompensated now by the increase of ΔE_g with increasing N_D, which adds to decrease of $L_p = \sqrt{(D_p \tau_p)}$ and results in an increase of h_n at high N_D.

As mentioned, the highly doped region was assumed till now thick enough that recombination at the contact can be neglected. Often however it is considerable or made large even intentionally as noted also. Fortunately, the contact recombination can be taken into account simply by replacing the diffusion length L_p in Eqs. (3.92) and (3.101) by a length $L_{p,eff}$ defined for this purpose. Assuming a contact with infinite recombination velocity the minority carrier concentration at the surface is pressed down to the equilibrium value p_{n0} so that $p \simeq p - p_{n0} = 0$. With this boundary condition at $x = 0$ the solution of the differential equation (3.42) is $p_n(x) = p_n^* \cdot \sinh(x/L_p)/\sinh(w_n/L_p)$ (w_n width of the emitter). The gradient of this carrier distribution at $x = w_n$ is

$$dp/dx(x_n) = p_n^*/(L_p \cdot \tanh(w_n/L_p))$$

which for $w_n \to \infty$ tends to the previous form p_n^*/L_p. Hence for small w_n the length to be used instead of L_p is

$$L_{p,eff} = L_p \cdot \tanh(w_n/L_p) \tag{3.102}$$

Thus the emitter parameter is finally given in this model by

$$h_{\mathrm{n}} = \frac{D_{\mathrm{p}}}{N_{\mathrm{D}} \cdot L_{\mathrm{p}} \cdot \tanh(w_{\mathrm{n}}/L_{\mathrm{p}})} \cdot \mathrm{e}^{\Delta E_{\mathrm{g}}/kT} \tag{3.103}$$

where the dependency on the doping and carrier concentration $N_{\mathrm{D}} = \mathrm{n}^{+}$ is contained to an essential part in $L_{\mathrm{p}} = \surd(D_{\mathrm{p}}\tau_{\mathrm{p}})$, ΔE_{g}, and D_{p}. For very high N_{D} where τ_{p} is determined by Auger recombination and $L_{\mathrm{p}} = \surd(D_{\mathrm{p}}/c_{\mathrm{A,n}})/N_{\mathrm{D}}$ is smaller than about $w_{\mathrm{n}}/3$, Eq. (3.103) yields instead of Eq. (3.97)

$$h_{\mathrm{n}} = \sqrt{D_{\mathrm{p}} \cdot c_{\mathrm{A,n}}} \cdot \mathrm{e}^{\Delta E_{\mathrm{g}}/kT} \tag{3.104}$$

Formulae for a p-emitter are obtained by exchanging n and p and N_{D} and N_{A}. Eq. (3.103) for a p-emitter reads

$$h_{\mathrm{p}} = \frac{D_{\mathrm{n}}}{N_{\mathrm{A}} \cdot L_{\mathrm{n}} \cdot \tanh(w_{\mathrm{p}}/L_{\mathrm{n}})} \cdot \mathrm{e}^{\Delta E_{\mathrm{g}}/kT} \tag{3.105}$$

For later use, we note also the formula for a small width of the p^{+}-region: If $w_{\mathrm{p}} < L_{\mathrm{n}}/3$, Eq. (3.105) turns into

$$h_{\mathrm{p}} = \frac{D_{\mathrm{n}}}{N_{\mathrm{A}} \cdot w_{\mathrm{p}}} \cdot \mathrm{e}^{\Delta E_{\mathrm{g}}/kT} \tag{3.106}$$

This equation holds also for variable emitter doping if $N_{\mathrm{A}}w_{\mathrm{p}}$ is replaced by the integral $\int N_{\mathrm{A}}(x)\mathrm{d}x$. For ΔE_{g} and D_{n} one has to use then suitable mean values. In Fig. 3.21 the h-parameter of a p^{+}-region as calculated from Eq. (3.105) is plotted versus the doping density N_{A} for a very small and a larger width w_{p}. The minority carrier lifetime τ_{n} was calculated from the Shockley–Read–Hall lifetime τ_{SRH} assumed to be 2 μs and Auger recombination according to $1/\tau_{\mathrm{n}} = 1/\tau_{\mathrm{SRH}} + c_{\mathrm{A,n}} \cdot N_{\mathrm{A}}^{2}$, where the Auger constant $c_{\mathrm{A,p}}$ of Eq. (2.54) was inserted. As

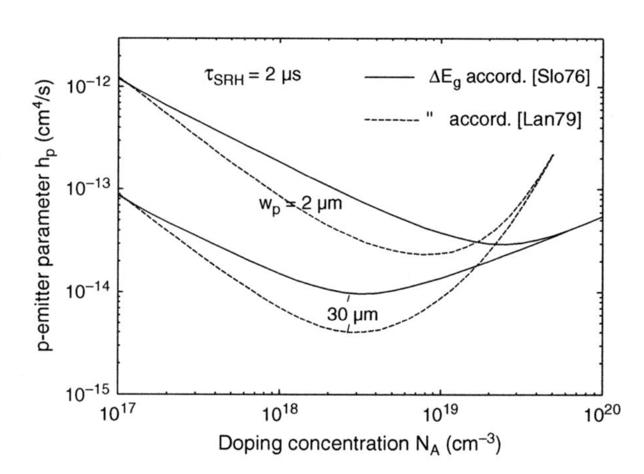

Fig. 3.21 h-Parameter of a p^{+}-region as a function of doping concentration according to Eq. (3.105)

bandgap narrowing $\Delta E_g(N_A)$ Eq. (2.25) of [Slo76] as well as Eq. (2.26) of Lanyon and Tuft [Lan79] was used. The minority diffusion constant $D_n(N_A)$ was calculated from the mobility μ_n using for simplicity the same dependence on doping concentration as in an n$^+$-region: $D_n(N_A) = kT/q\,\mu_n(N_D = N_A)$ (see Sect. 2.6 and Appendix A). As expected from Fig. 2.11, the ΔE_g of [Lan79] results in considerably smaller h_p below 10^{19} cm^{-3} than that of Slotboom and De Graaf. Experimental h-values obtained with power diodes are mostly in rough agreement with the $\Delta E_g(N)$ dependency of [Slo76]. At small N_A corresponding curves tend to the same value, since ΔE_g then approaches zero in both cases. As is shown already by the formulae, h can be strongly enhanced by choosing a small thickness of the emitter. The curves for $w_p = 2\,\mu$m are described up to about 1×10^{19}cm^{-3} by Eq. (3.106) showing that recombination at the contact predominates. For $w_n = 30\,\mu$m on the other hand, recombination in the emitter volume is significant in the whole range. Due to the decrease of the lifetime because of Auger recombination, adding to the increase of ΔE_g with N_A, h_p has a minimum at 3×10^{18}cm^{-3}.

Concerning n-emitter regions, it can be seen, that due to the smaller minority diffusion constant D_p and inspite of the higher Auger recombination constant $c_{A,n}$ the numerical h_n-values will be somewhat smaller than h_p under most conditions. For a small emitter thickness the formula corresponding to Eq. (3.106) yields a h_n which is by the factor D_p/D_n smaller than h_p provided the integrated doping density is equal in both cases. At very high doping densities the parameters h_n and h_p are very similar according to Eq. (3.104) since the products $D_n c_{A,p}$ and $D_p c_{A,n}$ do not differ much. The bandgap narrowing is generally assumed to be the same in n$^+$ and p$^+$-regions

Although not all consequences of the theoretical model have been verified precisely and the bandgap narrowing itself is not yet completely understood quantitatively, the model has been proved to be quite useful for describing the main influence of design measures of emitter regions.

For the experimental determination of emitter properties, diodes with a p$^+$nn$^+$- or p$^+$pn$^+$-structure have been used in a forward current range where the injection levels in the base region is high. Hence Eq. (3.93) was used in the form

$$j_p(x_n) = q \cdot h_n \cdot n_B^{*2} \tag{3.107}$$

Typical highly doped emitter regions have h-values in the range 1×10^{-14}cm^4/s to 3×10^{-14}cm^4/s, and the lowest values observed are about 7×10^{-15}cm^4/s [Sco69, Bur75, Sco79, Coo83]. As will be treated in detail in Chap. 5, the h-values of the two emitter regions of pin-diodes and thyristors influence largely their forward voltage drop from moderate current densities upward. For low-frequency power diodes which are designed to show a forward voltage drop as low as possible, small h-values are required. The injection efficiency of the junctions then is high. If $h = 1 \times 10^{-14}$cm^{-14}/s and $n_B^* = p_B^* = 1 \times 10^{17}$/cm^3 at a current density of 200 A/cm^2, one obtains from Eq. (3.96) $\gamma = 0.92$. On the other hand, the h-value of the p-emitter of fast power diodes and IGBTs is made often very high by choosing a small thickness and a relatively low doping density. In a typical example one may have $h_p = 1 \times 10^{-12}$cm^4/s, and the carrier concentration in the base region at the p-emitter side may be $1.8 \times 10^{16}\text{cm}^{-3}$ at a forward current density of 150 A/cm^2. In this case Eq. (3.96) yields $\gamma = 0.65$. The electron particle current which enters the p$^+$-region and ends at the contact by recombination, amounts then to more than a third of the total current.

An observation which is not expected from the above theory is that the minority carrier current flowing into the emitter regions contains besides the quadratic term (3.107) also a significant part which increases proportional with the concentration n_B^* [Sco79,Coo83]. Hence the minority carrier current density is actually given by

$$j_p(x_n) = q \cdot (h_n \cdot n_B^{*2} + s_n \cdot n_B^*) \tag{3.108}$$

This was found for all samples investigated regardless whether prepared by diffusion, epitaxy, or alloying. The linear term in Eq. (3.108) is often considerably higher than the recombination in the volume of the base. For gold doped pin-diodes with diffused junctions, Cooper [Coo83] found that the linear constant s_n in Eq. (3.108) is nearly proportional to the gold concentration. The linear recombination is caused possibly by an accumulation of recombination centers at the junctions. Its existence follows because the high-level lifetime in the base region determined from the carrier distribution is appreciably higher than the total linear recombination lifetime determined from stored charge measurements. (See also Sect. 5.5)

3.5 Capacitance of pn-Junctions

In this section, we consider the capacitance of a pn-junction for small oscillations around a given stationary state. The actual switching behavior of power devices consisting of large current and voltage changes between forward and reverse states will be discussed in later chapters. The small-signal capacitance in conjunction with an external stray inductance for example can be a source of disturbing oscillations in power electronic circuits (see Chap. 13). On the other hand, the capacitance is used as a tool for determining the doping concentration of the weakly doped base region of a junction. It is also used in different methods to investigate the properties of deep-level impurities which will be discussed in Sect. 4.9. In the present section, however, the concentration of deep impurities is neglected against the density of the normal doping. The doping atoms are assumed to be completely ionized.

In a reverse biased pn-junction, the capacitance per unit area is defined as $c_j = dQ/dV_r$, where dQ is the variation of charge per unit area caused by an incremental change dV_r of the reverse voltage. Assuming a p$^+$n-junction the charge in depletion approximation is given by $dQ = qN_D dw$, where N_D denotes the doping density of the base region and dw the change of the width w of the space charge layer caused by the voltage dV_r. The latter can be expressed according to Eq. (3.58) as $dV_r = q \cdot N_D \cdot w \cdot dw/\varepsilon$. This results in $c_j = dQ/dV_r = \varepsilon/w$, a well-known formula for a planar parallel plate capacitor. Substituting Eq. (3.58) for w, one obtains the capacitance as function of the reverse voltage

$$c_j = \sqrt{\frac{q \cdot \varepsilon \cdot N_D}{2 \cdot (V_{bi} + V_r)}} \tag{3.109}$$

or

$$\frac{1}{c_j^2} = \frac{2 \cdot (V_{bi} + V_r)}{q \cdot \varepsilon \cdot N_D} \tag{3.110}$$

According to this equation, a plot of $1/c_j^2$ versus V_r yields a straight line, whose slope can be used to determine the doping concentration N_D. This measuring method has the advantage compared with other methods that it allows the direct determination of donor and acceptor doping concentrations down to $1 \times 10^{13}\text{cm}^{-3}$, without using any other quantities and material characteristics. Because the capacitance according to Eq. (3.109) refers to a depleted space charge region, it is called depletion capacitance.

If we assume for example $N_D = 1 \times 10^{14}\text{cm}^{-3}$, $V_r = 600\,\text{V}$ and a junction area $A = 2\,\text{cm}^2$, the capacitance $C_j = Ac_j$ amounts to 235 pF. With a stray inductance $L_{par} = 50\,\text{nH}$ in series, this capacitance gives rise to oscillations with frequency $f = 1/(2\pi\sqrt{L_{par} \cdot C_j}) = 46.4\,\text{MHz}$. Note that the switching frequency of power electronic circuits is typically in the order of 10 kHz.

As illustrated in Fig. 3.5, the charge of mobile carriers in the space charge region can be significant or even dominating in abrupt asymmetrical junctions. The exact expression for the maximum field strength is given by Eq. (3.24). Since the charge of one sign in the space charge region is $Q = \int \rho dx = \varepsilon E_m$ according to the Poisson equation, Eq. (3.24) yields

$$Q = \sqrt{2 \cdot \varepsilon \cdot kT \cdot \left[N_A \cdot e^{qV_P/kT} + N_D \cdot \frac{qV_N}{kT} \right]} \tag{3.111}$$

The exponential term in V_N and the term "1" in Eqs. (3.24) have been neglected, because the condition $N_D << N_A$ implicates that V_N (voltage over the space charge layer in the n-region) is large compared with kT/q. V_N and V_P are given by Eqs. (3.23), where the built-in voltage has to be replaced by $V_{bi} + V_r$ at reverse bias. Hence, Q is given by Eq. (3.111) together with Eq. (3.23) as an explicit function of V_r which by differentiation yields the capacitance. Using the inequality $(V_{bi} + V_r)/kT >> 1$ in addition to $N_D/N_A << 1$, but taking into account the product

$$\gamma = \frac{N_D}{N_A} \cdot \frac{q}{kT} \cdot (V_{bi} + V_r) \tag{3.112}$$

the voltages can be expressed as $V_P \simeq kT/q(1+\gamma)$, $V_N \simeq V_{bi} + V_r$. After some conversions the capacitance per unit area is obtained then as

$$c_{\mathrm{j}} = \frac{\mathrm{d}Q}{\mathrm{d}V_{\mathrm{r}}} = \sqrt{\frac{q \cdot \varepsilon \cdot N_{\mathrm{D}}}{2 \cdot (V_{\mathrm{bi}} + V_{\mathrm{r}}) \cdot}} \frac{1 - e^{-1-\gamma}}{\sqrt{1 + e^{-1-\gamma}/\gamma}} \tag{3.113}$$

For $\gamma > 2$, i.e. for a sufficiently high voltage V_{r}, Eq. (3.113) approaches the depletion formula (3.109). For concentrations $N_{\mathrm{A}} = 1 \times 10^{18}\,\mathrm{cm}^{-3}$ and $N_{\mathrm{D}} = 1 \times 10^{14}\,\mathrm{cm}^{-3}$ for example, the condition $\gamma > 2$ leads to $V_{\mathrm{r}} > 516\,\mathrm{V}$ (with $\varepsilon_{\mathrm{r}} = 11.7$ for silicon). For smaller γ, Eq. (3.113) yields a significantly lower capacitance than Eq. (3.109). For $V_{\mathrm{r}} = 0$, the factor between both equations is

$$\frac{c_{\mathrm{j}}(V_{\mathrm{r}} = 0)}{c_{\mathrm{jdpl}}(V_{\mathrm{r}} = 0)} = 1.042 \cdot \sqrt{\frac{N_{\mathrm{D}}}{N_{\mathrm{A}}} \cdot \frac{qV_{\mathrm{bi}}}{kT}} \tag{3.114}$$

where the depletion capacitance is indicated by the subscript "dpl". With $N_{\mathrm{D}}/N_{\mathrm{A}} = 10^{-4}$ and $V_{\mathrm{bi}} = 0.7\,\mathrm{V}$ this yields $C_j/C_{jdpl} = 0.0542$ at 300 K: the capacitance is nearby a factor 20 smaller than calculated from Eq. (3.109) in thermal equilibrium. It is to be noted, however, that the formula (3.113) applies to very abrupt junctions, whose doping transition occurs over a smaller length than a Debye length of the highly doped region. As mentioned already at the end of Sect. 3.1.1, such very abrupt doping transitions are shown usually only by pn-junctions in SiC and other wide-gap semiconductors whose impurity diffusion constants are extremely small. Typical "abrupt" junctions in Si have a somewhat smoother doping transition, so Eqs. (3.109) and (3.110) will be more useful than Eq. (3.113) in these cases, although some influence of the mobile charges may remain.

At forward voltages above a few kT/q, the capacitance is dominated by the charge of minority carriers in the neutral regions. Concerning this "diffusion capacitance" we refer to the literature [Sze02]

References

[Bad97] Badila M, Brezeanu G, Tudor B, Bica G, Lungu P, Millan J, Godignon P, Locatelli ML, Chante JP: "Electrical behavior of 6H-SiC pn diodes" Int. Semiconductor Conf. 20th edition CAS'97 Proceedings NY, IEEE (1997)

[Bar09] Bartsch W, Schoerner R, Dohnke KO: "Optimization of Bipolar SiC-Diodes by Analysis of Avalanche Breakdown Performance", Proceedings of the ICSCRM 2009, paper Mo-P-56 (2009)

[Bur75] Burtscher J, Dannhäuser F, Krasse J: "Die Rekombination in Thyristoren und Gleichrichtern aus Silizium: Ihr Einfluß auf die Durchlaßkennlinie und das Freiwerdezeitverhalten", Solid-St. Electron 18, pp. 35–63 (1975)

[Coo83] Cooper RN: "An investigation of recombination in Gold-doped pin rectifiers", Solid-St. Electron. 26, pp. 217–226 (1983)

[Dav38] Davydov B: "The rectifying action of semiconductors", Techn. Phys. UdSSR 5, pp. 87–95 (1938)

[Ful67] Fulop W: "Calculation of Avalanche Breakdown Voltages of Silicon pn-Junctions", Solid-St. Electron. 10, pp. 39–43 (1967)

[Lan79] Lanyon HPD, Tuft RA: "Bandgap narrowing in moderately to heavily doped silicon" IEEE Trans. Electron Devices. ED-26, pp. 1014–1018 (1979)

[Mil57] Miller SL: "Ionization rates for holes and electrons in silicon", Phys. Rev., 105, pp.1246–1249 (1957)

[Mol64] Moll JL, Physics of semiconductors, McGraw Hill, New York, 1964

[Mor60] Morgan SP, Smits FM, "Potential distribution and capacitance of a graded p-n junction", Bell System Tech. J. vol. 39, pp. 1573–1602, 1960

[Oga65] Ogawa T; "Avalanche Breakdown and Multiplication in Silicon pin Junctions". Japanese J. of Applied Physics, 4, pp. 473–484 (1965)

[Sco69] Schlangenotto H, Gerlach W: "On the effective carrier lifetime in psn-rectifiers at high injection levels", Solid-St. Electron., Vol 12, pp. 267–275 (1969)

[Sco79] Schlangenotto H, Maeder H: "Spatial Composition and Injection Dependence of Recombination in Silicon Power Device Structures", IEEE Trans. Electron Devices, Ed-26, pp. 191–200 (1979)

[Shi59] Shields J: "Breakdown in Silicon pn-Junctions", Journ. Electron. Control 6, pp. 132–148 (1959)

[Sho49] Shockley W: "The theory of p-n junctions in semiconductors and p-n junction transistors", Bell System Techn. J. 28, pp. 435–489 (1949)

[Sho50] Shockley W, "The Theory of pn Junctions in Semiconductors", in Electrons and Holes in Semiconductors, D. van Nostrand Company Inc, Princeton 1950

[Slo76] Slotboom, J.W. and De Graaff, H.C.: "Measurements of bandgap narrowing in Si bipolar transistors", Solid-St. Electron. 19, pp. 857–862 (1976)

[Sze02] Sze SM: Semiconductor Devices, Physics and Technology, 2nd Ed., John Wiley & Sons, New York, 2002

[Wag31] C. Wagner, Phys. Zeits. 32, pp. 641–645, 1931

[Wul60] Wul, B. M. and Shotov, A. P., "Multiplication of electrons and holes in p-n junctions", Solid-St. Phys. in Electron. Telecommun., vol. 1, pp. 491–497, 1960

Chapter 4
Short Introduction to Power Device Technology

In the following some basic aspects of power device production technology will be described. The selection was done with the aim to describe the process steps which are important for the understanding of the power device operation and limitations.

4.1 Crystal Growth

The material for silicon devices must be of very high purity. Metallurgical silicon is converted to trichlorosilane, $SiHCl_3$, which is liquid and purified by fractional distillation. Especially chlorides of metals must be eliminated. By reduction of $SiHCl_3$ in a hydrogen atmosphere, polycrystalline rods of pure silicon are formed. For more details, see [Ben99].

The semiconductor material used for power device manufacturing must be monocrystalline. To produce such monocrystals there are two important processes:

In the *Czochalski process* (*CZ*) the crystal growth is performed in a crucible in which molten silicon is kept at a defined temperature. The required p- or n-type dopants are added to the melt. A small monocrystalline seed crystal is then dipped into the melt (Fig. 4.1). During slow rotation of the seed crystal, and of the crucible in the opposite direction, monocrystalline layers of silicon are deposited on the seed crystal maintaining the crystal structure of the seed. The growing bar is simultaneously pulled slowly upward.

With the CZ process very large single crystals can be grown. Si cylinders of a length of several meters and a diameter of more than 30 cm for fabrication of 300 mm wafers are produced on an industrial scale. However, the purity and quality of single crystals is limited with the CZ process, since the melt is in contact with the crucible during crystal growth. The oxygen content in CZ silicon is typically $>10^{17}$ cm^{-3} and also the content of impurity carbon is in the same range. CZ wafers are mainly used as substrates for epitaxial wafer growth (see Sect. 4.3), from which integrated circuits, etc., are produced. Some power devices like MOSFETs are also produced from epitaxial wafers using CZ substrates. For power devices in which the volume of the wafer is used, in most cases the purity of CZ crystals is not sufficient.

J. Lutz et al., *Semiconductor Power Devices*, DOI 10.1007/978-3-642-11125-9_4,

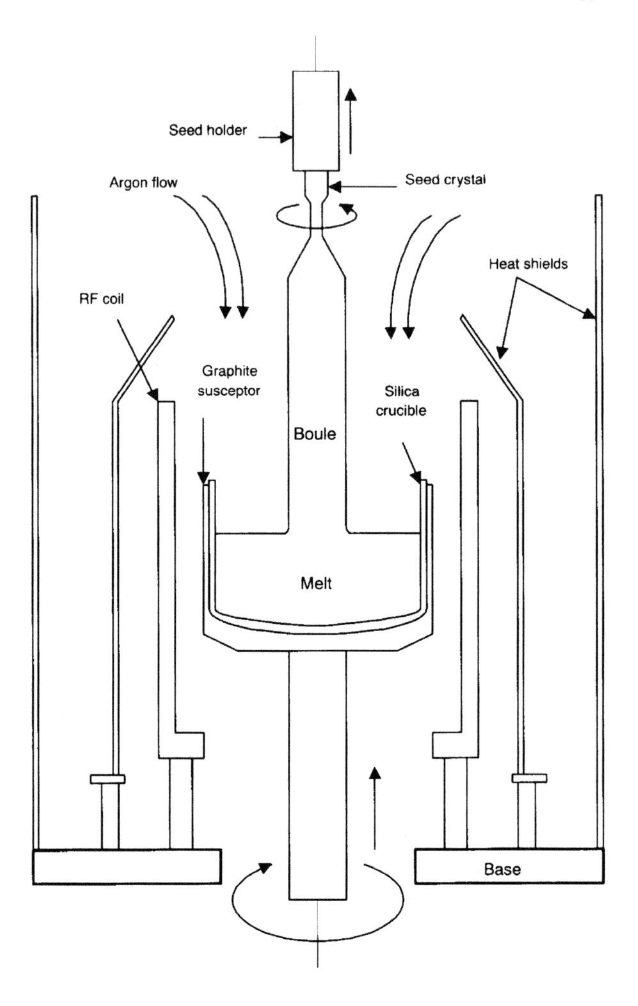

Fig. 4.1 Czochalski process for production of single crystalline silicon rods. Figure taken from [Ben99] reprint with permission of John Wiley & Sons, Inc.

The *Float Zone* (*FZ*) *process* is a crucible-free crystal growth method. The starting material is a rod of high-purity polysilicon. A seed crystal is clamped in contact with the end of the rod (Fig. 4.2). An induction heating coil is placed around the rod, melting a small zone next to the seed crystal. While slowly moving the coil along the bar, the molten zone, starting at the interface between seed crystal and polysilicon rod, passes slowly along the length of the bar. The polysilicon melts and monocrystalline silicon is regrown with the same orientation as the seed crystal. Because of the absence of a crucible, crystals with higher purity and higher quality can be grown with the FZ process. Also since the solubility of the impurities is higher in the melted zone they tend to be accumulated there. The carbon content is $< 5 \times 10^{15}$ cm^{-3} and the oxygen content is $< 1 \times 10^{16}$ cm^{-3}. With the FZ method, monocrystalline silicon rods with a diameter of more than 15 cm can be produced.

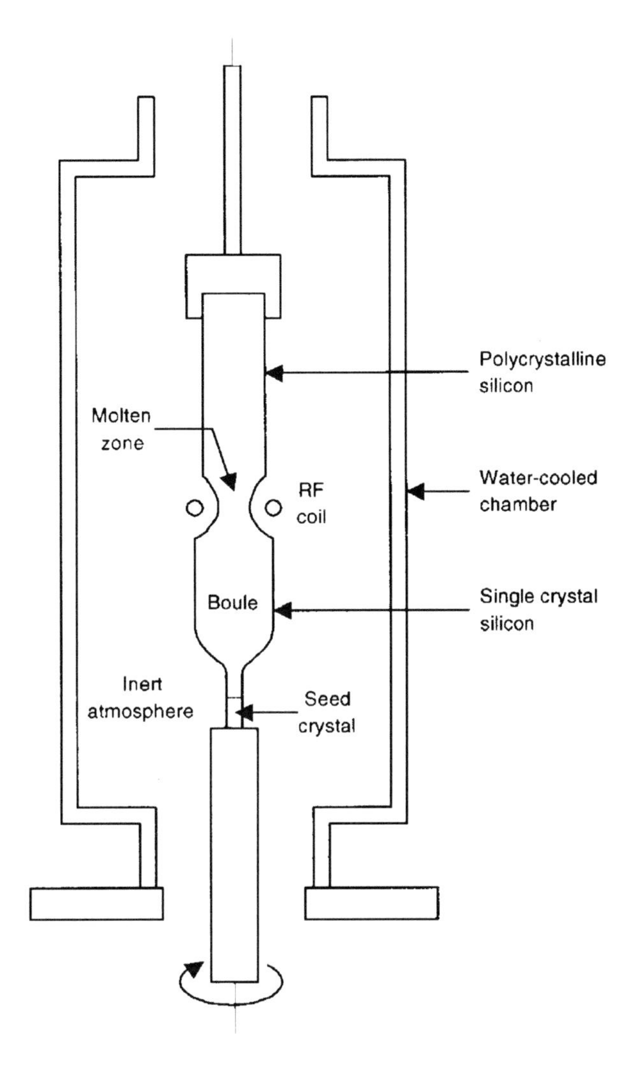

Fig. 4.2 Float-zone (FZ) process for crystal growth. Figure taken from [Ben99] reprint with permission of John Wiley & Sons, Inc.

Power devices, which use the whole volume of the wafer, are mostly fabricated from FZ silicon because of the higher crystal quality and excellent purity. Only power devices which are manufactured with epitaxial techniques are using CZ silicon as material for the substrates.

In the FZ as well as in the CZ process, the doping of the crystal is done by adding dopants to the melt. Power devices normally require a thick low-doped middle region to support a large voltage. Most often an n-type region is preferred. The value of this doping depends on the voltage for which the power device is designed; see further Eq. (3.84) and Fig. 3.17. The doping processes will be discussed in more detail below.

After crystal growth follows a sawing process to slice the rod in single wafers. To remove the surface damage caused by sawing and to obtain a clean and undisturbed surface, the wafer surface layer is lapped mechanically and in some cases additionally etched chemically. For some semiconductor processes also wafers with single-side polished surface are necessary.

4.2 Neutron Transmutation for Adjustment of the Wafer Doping

The homogeneity of the doping concentration is extremely important for power devices. If variations in doping concentration (or local defects) exist in the wafers, the current may be unevenly distributed, especially at avalanche breakdown. This can in turn result in local overheating and device failure. However, even with the FZ process, which alone is suited for the preparation of the extremely pure silicon crystals needed for power devices, fluctuations in doping cannot be avoided. The obtained wafers show periodic concentric rings with doping variations called "striations". An example for the resistivity profile measured across a wafer is shown in Fig. 4.3.

According to Eq. (2.34), the resistivity variations correspond to doping fluctuations in an inverse manner. Since the doping directly determines the blocking

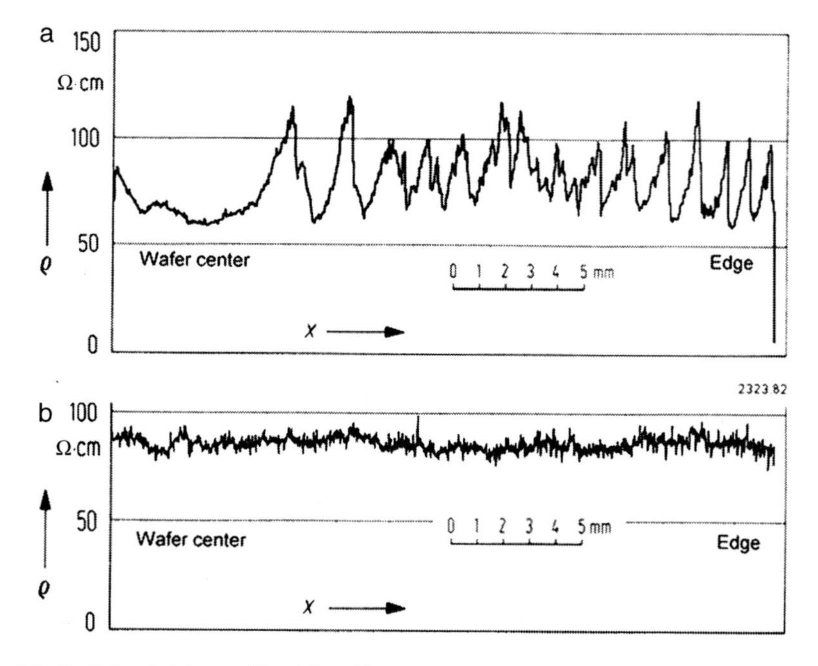

Fig. 4.3 Radial resistivity profile achieved by conventional doping (*top*) and by neutron transmutation doping (*bottom*). Figure from [Sco82]

voltage, it is not possible to produce high-voltage power devices with the required narrow design, if the doping of the starting material is fluctuating in the amount as shown in Fig. 4.3a. Only with the method of neutron doping (Fig. 4.3b), it became possible to realize for instance thyristors with a blocking voltage of >2000 V.

The neutron transmutation doping (NTD) of silicon was proposed and investigated first by Lark-Horovitz [Lar51] and Tannenbaum and Mills [Tan59, Tan61]. For the fabrication of 5 kV thyristors with an acceptable recovery time, the method was reinvented and introduced on industrial scale in 1973–1974 [Scl74, Haa76, Jan76]. It is based on the radioactive transmutation of the silicon isotope ${}^{30}_{14}\mathrm{Si}$ to phosphorous ${}^{31}_{15}\mathrm{P}$ by thermal neutrons. This isotope is contained in natural silicon to 3.09%, besides ${}^{29}_{14}\mathrm{Si}$ with a share of 4.67% and the main isotope ${}^{28}_{14}\mathrm{Si}$ with 92.23%. As starting material a crystal bar is used which is grown by the FZ method and doped very low, typically below 5×10^{12} cm^{-3}. If the silicon rod is placed near the core of a nuclear reactor, the ${}^{30}_{14}\mathrm{Si}$ atoms capture partly a thermal neutron and turn under emission of a γ quant into the unstable isotope ${}^{31}_{14}\mathrm{Si}$ which by emission of an electron (β particle) decays then into the stable phosphorous isotope ${}^{31}_{15}\mathrm{P}$:

$${}^{30}_{14}\mathrm{Si}\quad (n,\gamma)\rightarrow\ {}^{31}_{14}\mathrm{Si}\rightarrow\ {}^{31}_{15}\mathrm{P}+\beta$$

The decay of ${}^{31}\mathrm{Si}$ to ${}^{31}\mathrm{P}$ has a half-time of 2.6 h. Thermal neutrons exhibit a decay length in Si of 19 cm, while the diameter of the bars is usually 10 or 15 cm. By rotation of the bar during the irradiation, however, the mean neutron flux in the Si bar becomes nearly homogeneous, the difference between the value at the axis to that at the periphery being less than 4% even for a rod diameter of 15 cm [Jan76]. The final phosphorous concentration obtained by these processes is given by

$$N_{\mathrm{Phos}}=c\cdot\Phi\cdot t \tag{4.1}$$

where Φ denotes the neutron flux in cm^{-2}s^{-1} and t the time. With the cross-section data given in [Jan76], the constant c for natural silicon is obtained to be 2.0×10^{-4} cm^{-1}. The typical flux density Φ of thermal neutrons at the used positions is in the range 10^{13}–10^{14} cm^{-2}s^{-1} [Amm92]. At a flux of 2×10^{13} cm^{-2}s^{-1}, for example, an irradiation time of 3.5 h is necessary according to Eq. (4.1) to obtain a phosphorous doping of 5×10^{13} cm^{-3}. The phosphorous concentration can be adjusted by this method very exactly within limits of only $\pm$ 3%.

After the neutron irradiation a storage period is necessary for the ongoing transmutation of the ${}^{31}_{14}\mathrm{Si}$ into the ${}^{31}_{15}\mathrm{P}$ and particularly for the decay of the activity due to the β emission. The time for the decay of the activity to an undetectable level is 3–5 days. As secondary process, the reaction of the generated ${}^{31}_{15}\mathrm{P}$ atoms with the neutrons and the decay of the resulting ${}^{32}_{15}\mathrm{P}$ according to

$${}^{31}_{15}\mathrm{P}(n,\gamma)\rightarrow\ {}^{32}_{15}\mathrm{P}\rightarrow\ {}^{32}_{16}S+\beta$$

has to be considered. The decay of $^{32}_{15}P$ into $^{32}_{16}S$ has a half-time of 14.3 days and will prolong the time required for the activity decay if the phosphorous concentration generated is higher than about $1\times10^{15}\text{cm}^{-3}$ [Jan76]. To eliminate lattice defects induced by the radiation, the crystal bar is annealed at a temperature of 800°C, before it is further processed to wafers.

Because the neutron doping method is rather costly there have been major efforts of silicon wafer suppliers to improve the FZ crystal growth process, and essential advances have been reached in the last 20 years. The doping tolerances of wafers with a diameter of 200 mm are decreased now to ± 12% [SIL06]. Nevertheless, the NTD method is still indispensable.

4.3 Epitaxial Growth

Epitaxial growth is an alternative method to produce high-purity single crystalline layers. As a starting material or substrate for the epitaxial growth process, a single-side polished CZ wafer is used. On this wafer a layer with higher purity and higher crystal quality is then grown in the same crystal orientation as the substrate, see Fig. 4.4. The epitaxy process is done significantly below the melting temperature of the semiconductor material.

The growth of epitaxial layers from silicon is made in an enclosed chamber, a reactor, using a vapor phase process. There are different possible processes which use one of the following chemical reactions [Ben99]:

$$
\begin{array}{lll}
SiCl_4 + 2\,H_2 & \rightarrow Si + 4\,HCl & \text{at } 1150-1220°C \\
SiHCl_3 + H_2 & \rightarrow Si + 3\,HCl & \text{at } 1100-1175°C \\
SiH_2Cl_2 & \rightarrow Si + 2\,HCl & \text{at } 1025-1100°C \\
SiH_4 & \rightarrow Si + 2\,H_2 & \text{at } 1150-1220°C
\end{array}
$$

Prior to the growth an intensive mechanical and chemical cleaning of the wafer surfaces is done. The substrates are also cleaned and etched in the process reactor at a temperature of 1140–1240°C with HCl. The growth is executed in a H_2 atmosphere. For doping, PH_3 (phosphine) or B_2H_6 (diborane) for n-type or p-type, respectively, are added to the hydrogen gas in controlled proportions.

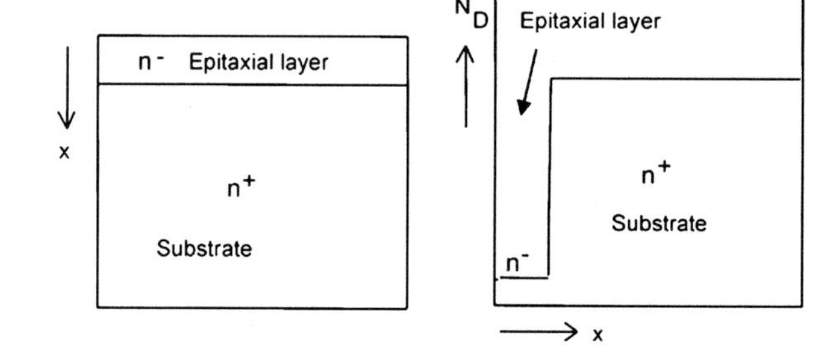

Fig. 4.4 Epitaxially grown wafer in cross section. Example n^+-substrate with n^--epitaxial layer

Epitaxial layers in silicon can be fabricated with high purity, especially with very low impurity content of carbon and oxygen.

For the production of SiC monocrystals a high effort is necessary. The SiC substrate crystal growth is done at a temperature of 2300°C. It is very difficult to achieve monocrystals of low defect density. For SiC power devices, a following epitaxy process is always necessary to reach the desired material quality. This epitaxial process is done at 1400–1600°C under H_2 atmosphere using SiH_4 and CH_4 or C_3H_8 as reaction gases. The growth rate is typically in the range of only a few micrometers per hour. However, there are still problems with crystal defects in the SiC substrate that propagate into the epitaxial layer, and particularly large area SiC wafers of high quality are therefore very expensive compared to Si.

4.4 Diffusion

One important way to introduce dopants into semiconductors is by diffusion. Particle diffusion was already introduced with respect to electrons and holes in Sect. 2.6, and the following description of dopant atoms is clearly similar.

Considering a gradient in the particle density $N(x)$ according to Fig. 4.5a, this gradient will cause a particle current density j. The magnitude of this particle current density is proportional to the slope of the particle density grad N, and its direction is opposite to grad N (first Fick's law)

$$\vec{j} = -D \times \operatorname{grad} N \tag{4.2}$$

or in one-dimensional formulation

$$j = -D \cdot \frac{\mathrm{d}N}{\mathrm{d}x}$$

The diffusion constant D is a proportionality factor between the density gradient and particle current density j.

If j flowing out of a volume element of the thickness dx is higher than j flowing in, i.e. $\partial j/\partial x > 0$, then the density of particles in this volume element is decreasing

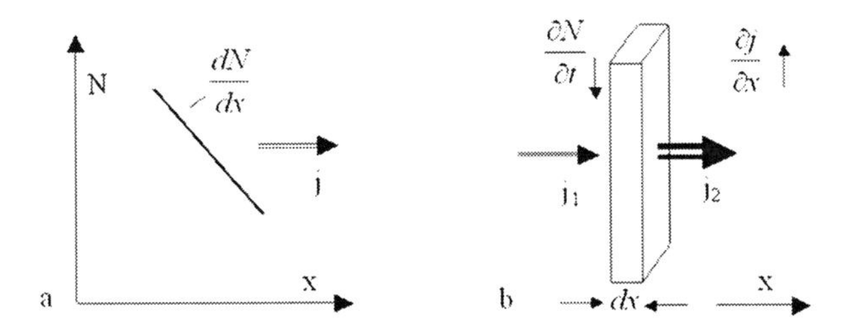

Fig. 4.5 Illustration of the relation between (**a**) gradient of a particle density and particle current density (first Fick's law) (**b**) divergence of the particle current density and change of the particle density (continuity equation, second Fick's law)

(Fig 4.5b). The rate of change in particle density will be $-\partial N/\partial t = \partial j/\partial x$, or in a general formulation

$$-\frac{\partial N}{\partial t} = \mathrm{div}\,\vec{j} \tag{4.3}$$

The source of the particle current density j, expressed as div j, arises from the decrease of particle density in the volume element. By the same way the continuity equation for electrons and holes was derived.

For a constant D, Eq. (4.2) inserted into Eq. (4.3) leads to Fick's second law

$$\frac{\partial N}{\partial t} = \mathrm{div}\left(D \cdot \mathrm{grad} N\right) \tag{4.4}$$

or in one-dimensional expression

$$\frac{\mathrm{d}N}{\mathrm{d}t} = D \cdot \frac{\mathrm{d}^2 N}{\mathrm{d}x^2}$$

A solution of Eq. (4.4) gives the density profiles $N(x, t)$ (in one dimension).

We will now examine the diffusion of dopant atoms in a semiconductor and consider two situations.

a) Constant total amount of impurities: *Diffusion profile of the Gauss type.* With the boundary condition that the integrated total number of diffusing dopant atoms is constant in time and equal to the initial doping amount S per area

$$\int_0^\infty N(x, t)\mathrm{d}x = S = \mathrm{const} \tag{4.5}$$

where $x = 0$ indicates the semiconductor surface, resulting in the solution

$$N(x, t) = \frac{S}{\sqrt{\pi \cdot D \cdot t}} \cdot \mathrm{e}^{-\frac{x^2}{4Dt}} \tag{4.6}$$

If the penetration depth is increasing by diffusion, then the density of particles at the surface N_S must decrease with diffusion time:

$$N_\mathrm{S} = \frac{S}{\sqrt{\pi \cdot D \cdot t}} \tag{4.7}$$

Using the diffusion length L_D – the distance, at which the density has decreased to 1/e of N_S

$$L_\mathrm{D} = 2 \cdot \sqrt{D \cdot t} \tag{4.8}$$

Equation (4.6) becomes

$$N(x, t) = N_\mathrm{S} \cdot \mathrm{e}^{-\frac{x^2}{L_\mathrm{D}^2}} \tag{4.9}$$

The Gaussian diffusion profile is always attained when no additional supply of particles from outside during the diffusion process occurs. This type of solution, a Gaussian diffusion profile, develops after an initial deposition of dopants, for instance by diffusion from a finite particle source, or an initial ion implantation. During the subsequent diffusion process no source of dopants from outside is supplied. A result of this often used diffusion process is shown in Fig. 3.6. This figure shows a p-type dopant diffusion profile of boron in a fast diode. The penetration depth – the depth at which the profile $N_A(x)$ is reaching the value of the constant n-type background doping N_D – amounts to 20 μm.

b) Constant surface concentration: The *diffusion profile of* erfc-*type* develops, if during diffusion there is always a source of the doping atoms present, i.e. diffusion from an infinite source. Therefore, the surface concentration N_S is constant and usually given by the solubility of the dopant in silicon at the diffusion temperature. The solution of Eq. (4.4) is given by

$$N(x,t) = N_S \cdot \mathrm{erfc}\left(\frac{x}{2 \cdot \sqrt{D \cdot t}}\right) = N_S \cdot \mathrm{erfc}\left(\frac{x}{L_D}\right) \tag{4.10}$$

with the complementary error function

$$\mathrm{erfc}(u) = \frac{2}{\sqrt{\pi}} \cdot \int_u^{\infty} \mathrm{e}^{-u^2} \mathrm{d}u \approx \mathrm{e}^{-1.14u - 0.7092u^{2.122}} \tag{4.11}$$

The right-hand side in Eq. (4.11) gives an approximation for the erfc function with a maximal deviation of $2^{\mathrm{o}}/_{\mathrm{oo}}$.

The erfc profile is created, for example, during the diffusion from the vapor phase, where the semiconductor wafer is exposed to an atmosphere which contains the dopant. Another possibility is to deposit a layer containing the dopant in high concentration on the semiconductor surface. Only a small amount of the dopants can penetrate into the semiconductor due to the solubility limit, so that the concentration inside the semiconductor can be continuously augmented by the deposited layer.

Both the specific diffusion constant and the specific solubility determine the diffusion, and both are functions of temperature. The temperature dependence of the diffusion constant D can be expressed in a first approximation by an Arrhenius function

$$D(T) = D_0 \cdot \mathrm{e}^{-E_A/kT} \tag{4.12}$$

where E_A represents the activation energy of the diffusion process. Figure 4.6 shows the temperature dependence of the diffusion constants for dopants used for n- and p-doping in silicon. In a more detailed investigation, the diffusion constant also depends on the dopant density, on the concentration of other dopants and, in some cases, on the atmosphere. However, Eq. (4.12) is a good approximation in most

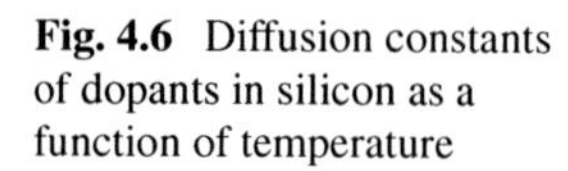

Fig. 4.6 Diffusion constants of dopants in silicon as a function of temperature

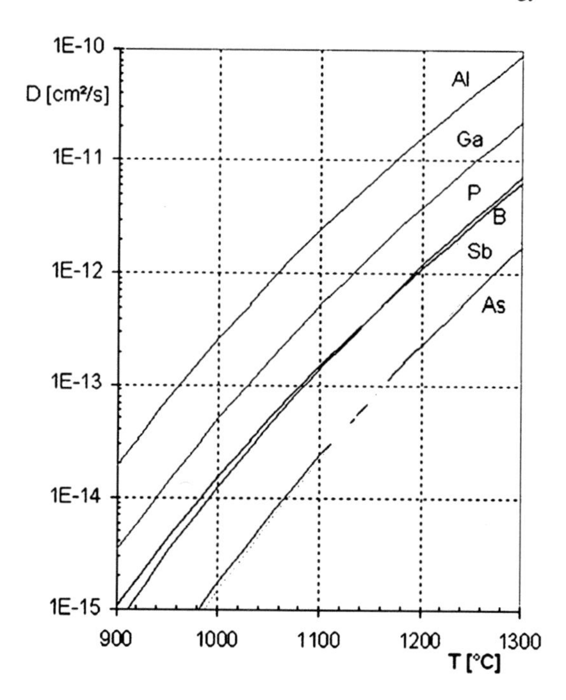

Table 4.1 Parameters for calculation of the diffusion constant in Si with Eq. (4.12)

Element	D_0 (cm^2/s)	E_A (eV)	Reference
B	0.76	3.46	[Sze88]
Al	4.73	3.35	[Kra02]
Ga	3.6	3.5	[Ful56]
P	3.85	3.66	[Sze88]
As	8.85	3.971	[Pic04]
Sb	40.9	4.158	[Pic04]

cases. The parameters for the calculation of Fig. 4.6 are given in Table 4.1. The values in the literature are sometimes different; however, the values applied in Fig. 4.6 and listed in Table 4.1 are confirmed by own experience.

The elements commonly used for doping of silicon are B, Al, Ga for p-type doping and P, Sb, As for n-type doping. These elements occupy silicon lattice positions, so-called substitutional sites, as well during the diffusion process as in their final position. The diffusion on substitutional positions is relatively slow. Al is the fastest diffusing element of them; it is used for creating deep pn-junctions especially for thyristors in which pn-junctions of a depth up to 70–100 μm are used. The temperature of the diffusion process is normally higher than 1200°C to achieve the desired penetration depth in a reasonable time. However, of all doping elements, Al has the lowest solubility in silicon. The dependence of the solubility on temperature

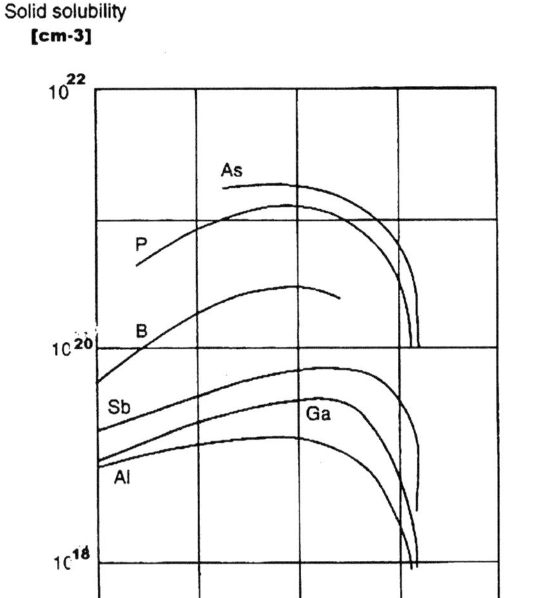

Fig. 4.7 Solubility of doping elements in silicon. Figure according to [Ben99] reprint with permission of John Wiley & Sons, Inc.

is shown in Fig. 4.7. The low Al solubility is often not sufficient to obtain a surface concentration high enough for good ohmic contacts. Therefore, several p-type diffusion steps are performed subsequently for some thyristors. After the Al diffusion, a Ga diffusion follows and some manufacturers even use a third diffusion step with boron. In Fig. 8.3, a diffusion profile of a thyristor manufactured in this way is shown.

Profiles of p-type with steep gradients and a penetration depth in the range of 20 μm are created preferably with boron, because of its higher solubility. Profiles of n-type are almost solely generated with phosphorus. P is the only n-doping element in silicon, which has a sufficient solubility and an acceptable diffusion constant. Nevertheless very deep n-profiles, for instance used for bipolar transistors, require a high temperature and a long diffusion time. For the 120 μm deep profile of the collector layer of a npn-bipolar transistor (Fig. 7.3), a process time of about 140 h at a temperature > 1250°C is necessary.

The high solubility of As in silicon is not used for diffusion processes. Arsenic is used for doping substrates of epitaxial wafers. With As, a very low specific resistivity of less than 5 mΩcm can be reached, while with Sb only values <15 mΩ cm are achievable. Since in many power devices the resistivity of the substrate is acting as a series resistor, a very low resistance is important for low conduction losses.

Other atoms, as the heavy metals Au, Pt, Ni, and Ag, can occupy lattice places as well as interstitial positions in the silicon crystal. These impurities are very important for controlling the charge carrier lifetime in an indirect semiconductor as silicon. Placed at interstitial positions the diffusion process is very much faster. The diffusion constant of Au in silicon at 900°C is seven orders of magnitude higher than for Al at the same temperature. For this type of heavy metal impurities, already a temperature of 850°C and a time of 10 min are sufficient to reach the backside of a 300 μm thick wafer. Furthermore, these elements are preferably implemented in regions in which the lattice is under stress because of a high concentration of boron or phosphorus. The diffusion mechanism is strongly influenced by lattice disorder like vacancies and interstitials, and these heavy metals prefer to accumulate at dislocations and other defects. As a result, the diffusion profile is no longer of Gauss type, as in Fig. 3.6, or of erfc type. More likely a "bath tub" profile will arise, where the concentration of impurities is lowest in the middle region and higher close to the wafer surfaces, see later in Fig. 4.23. Due to the rapid diffusion and to the interaction with vacancies, interstitials, and other lattice disorder, it is very difficult to achieve reproducible doping profiles, even with modern process techniques with fast heating and very fast cooling rates.

Other impurity atoms like H, Li, and Na diffuse solely on interstitial positions, and they diffuse very fast. Regarding alkaline and alkaline earth-metal elements, it is necessary to exclude a contamination of the silicon wafers, and therefore very clean process conditions are necessary.

While in silicon the diffusion process is used widely, this is not possible in SiC. All elements, which can be used as dopants feature very low diffusion constants even at very high temperatures. If pn-junctions and other device structures cannot be produced by an epitaxial process, the only possibility to introduce dopants is ion implantation.

4.5 Ion Implantation

In the ion implantation process, atoms of the dopants are ionized and accelerated in an electric field to form a beam of monoenergetic ions. The focused beam is scanned vertically and horizontally and the particles are injected into the wafer homogenously distributed across the wafer surface. The ions will slow down and stop in the semiconductor. The kinetic energy of the impacting ions is lost to the wafer by two types of collisions: with the cores of the lattice atom – elastic nuclear collisions – and by retardation in the electron shells of the lattice atoms – electronic deceleration. The nuclear collisions will also scatter the ions randomly, giving rise to a distribution of implanted ions where the mean penetration depth is mainly determined by the initial energy of the ions.

The dose of the implanted ions and thereby the quantity of the dopants can be controlled very exactly and it is also possible to mask different areas of the wafer to prevent ions from reaching the semiconductor at these locations. Power semiconductors with challenging technology are often doped using ion implantation.

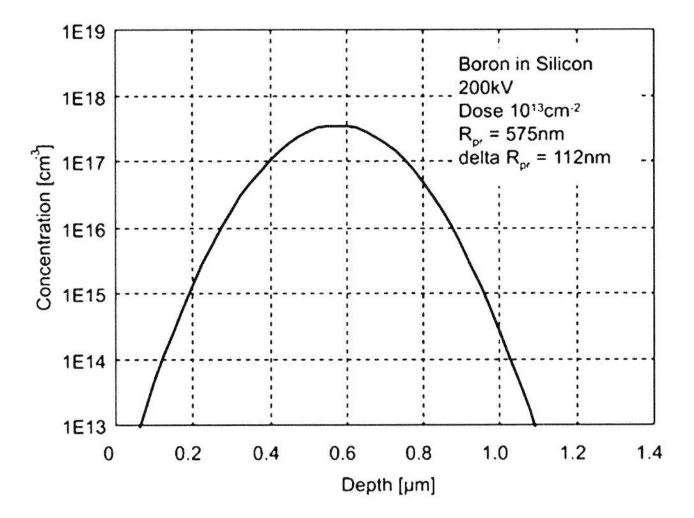

Fig. 4.8 Calculated doping profile of an ion implantation of boron in silicon

Typically, all semiconductor devices with MOS structures at the surface, as power MOSFET and IGBT, use ion implantation to form the p- and n-type regions.

The generated profile of the dopants can be described in a simplified way with a Gauss function

$$N(x) = \frac{S}{\sqrt{2\pi} \cdot \Delta R_{\mathrm{pr}}} \cdot \mathrm{e}^{-\frac{(x-R_{\mathrm{pr}})^2}{2\Delta R_{\mathrm{pr}}^2}} \tag{4.13}$$

where R_{pr}, the projected range, corresponds to the mean depth of the implanted ions, i.e. the peak of the profile in this simplified description. ΔR_{pr} the projected range straggling, is the statistical variation around this mean value, similar to the standard deviation. The integral amount of doping atoms S corresponds to the implanted dose. A doping profile calculated with Eq. 4.13 is drawn in Fig. 4.8.

The projected range depends primarily on the implantation energy, this relation is shown for boron in Fig. 4.9. Ions that are heavier than boron will have a smaller R_{pr} for the same energy.

With increasing energy, the peak height of the doping profile will decrease due to the increasing straggling. This is shown for boron in Fig. 4.10, where a linear scale is used for the concentration. Most often a log-scale is used, since the concentration varies over many orders of magnitude. Tables for the projected range and projected range straggling for the different ions are found in the literature, e.g., in [Rys86]. The ion distributions can also be simulated using for instance the SRIM software, which is a widely used program available on www.SRIM.org [Zie06].

The description discussed up to now has assumed a solid target with unorganized distribution of atoms, i.e. an amorphous target. But in a single crystalline target the atoms are regularly ordered. The elementary cell of the silicon lattice has been

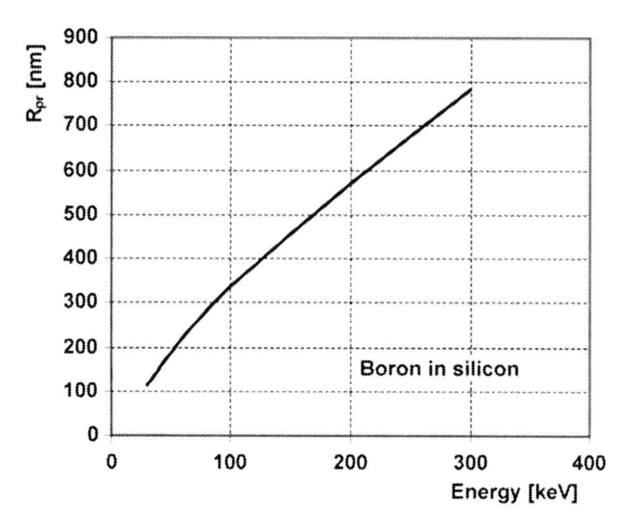

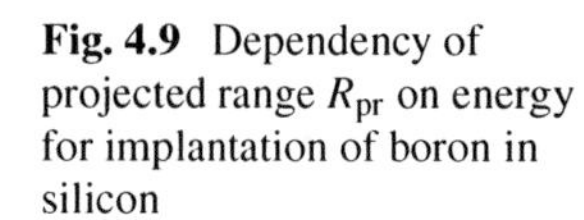

Fig. 4.9 Dependency of projected range R_{pr} on energy for implantation of boron in silicon

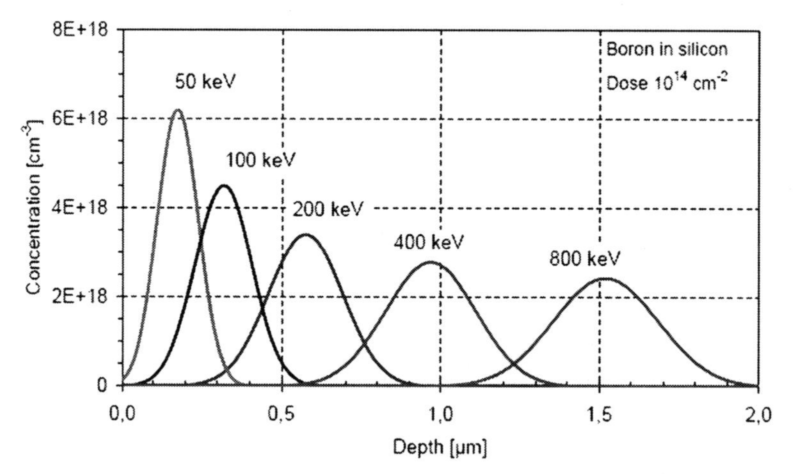

Fig. 4.10 Calculated doping profiles according to Eq. (4.13) after implantation of boron in silicon with different energies. R_{pr} and ΔR_{pr} are calculated with SRIM [Zie06]

already shown in Fig. 2.1. In Fig. 4.11 a drawing of the arrangement of atoms at a view along the [110] direction into a silicon crystal is shown.

If the direction of the penetrating ion beam is aligned with a major crystal orientation, it can penetrate deeper. In the so-called channels, nuclear collisions are very rare and many of the ions are not deflected from their original path. The retardation occurs only due to inelastic impacts with the electron shells of the lattice atoms, i.e. due to the electronic deceleration. The penetration depth in the channels can therefore be 10 times higher than the projected range in an amorphous solid object.

Normally one tries to avoid the channeling, and to counteract this effect the wafer is tilted, so that the implantation occurs in a direction deviating from the

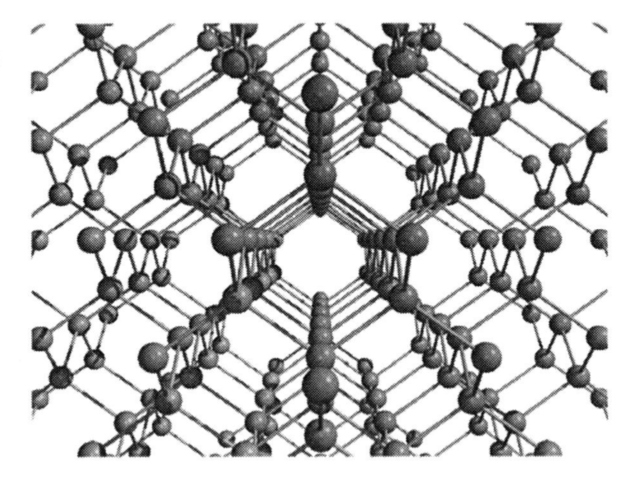

Fig. 4.11 Atomic structure of silicon: viewed in the [110] direction through a silicon crystal of 3×3×3 unit cells. Figure from [Pic04] © 2004 Springer

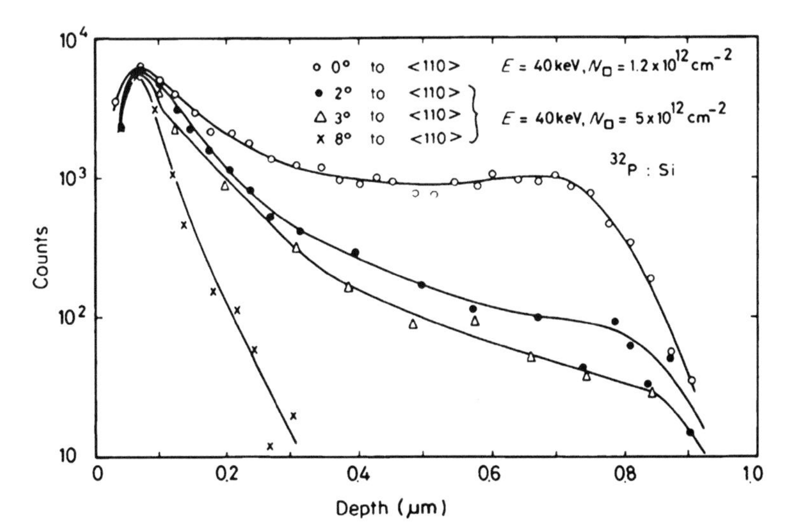

Fig. 4.12 Dependency of the channeling effect on the tilt angle between the ion beam direction and the wafer surface normal for a phosphorus implantation in silicon [Dea68]. Copyright 1968 National Research Council of Canada

major crystal axis. In Fig. 4.12 the dependence of the channeling effect on the tilt angle is shown. Even at a tilt angle of 8° a partial channeling is still visible. A further increase of the tilt angle would cause channeling into other channels. Usually semiconductor wafers are tilted by 7° during ion implantation.

Another way to reduce the channeling is by covering the crystal with an amorphous layer, for instance SiO_2. In this amorphous layer the ions are scattered and the ions will no longer travel in parallel paths. Oxides with a thickness of just 10–20 nm reduce effectively the channeling effect.

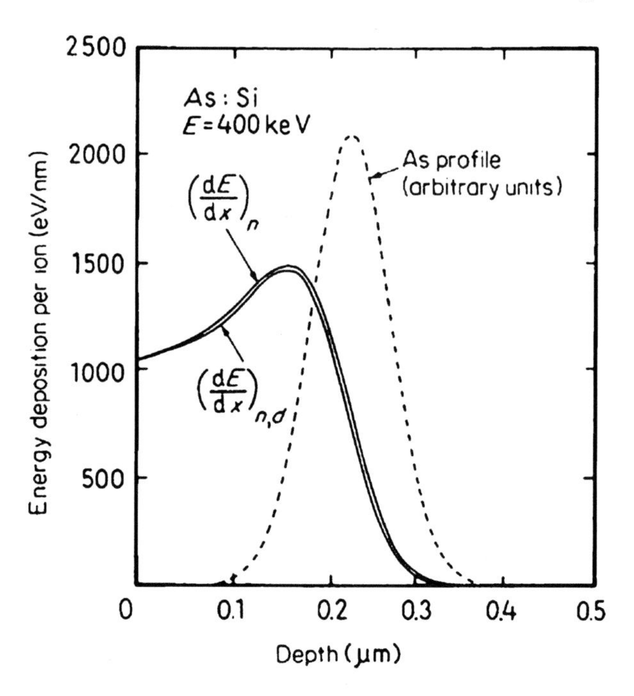

Fig. 4.13 Energy deposition for an implantation of arsenic in silicon in comparison to the distribution of implanted atoms. The profile of lattice defects follows the profile of energy deposition. Figure taken from [Rys86]. Original copyright 1971 Phs. Society Japan JPSJ Vol. 31, pp. 1695–1711

An increased target temperature results in increased amplitude of the lattice oscillations and this will also increase the number of ions that are scattered at the surface. Therefore, the channeling effect is reduced with increased target temperature.

However, none of these countermeasures is sufficient to avoid the channeling effect completely [Rys86]. The best way to reduce the channeling is to use a previously amorphized target. If the silicon wafer is first implanted with Si ions at a high dose, an amorphous Si layer of sufficient thickness is created at the surface, which will effectively prevent the channeling.

When dopants are introduced by ion implantation, one serious complication is the generation of lattice defects by the elastic collisions. These defects consist of atoms knocked out of their lattice positions into interstitial sites, with vacant lattice sites left behind. Combinations of these defects are also known, for instance the Si divacancy, formed by two neighboring vacancies. The maximum amount of ion beam induced lattice defects is located just before the ion comes to rest, e.g., at an implantation of boron in Si at $0.8R_{\mathrm{pr}}$. The profile of lattice defect is reaching up to the semiconductor surface. The distribution of lattice defects in comparison to the distribution of implanted atoms is shown in Fig. 4.13 on example of an implantation of heavy As ions into silicon.

If the implantation dose is sufficiently high, an amorphous layer can be generated due to the large number of lattice defects. For every ion there is a critical dose for silicon amorphization, which depends on temperature. The higher the mass of the ion, the lower is the critical dose. However, there is a balance between defect build-up by the incoming ions and the recombination of Si interstitials and vacancies, and this balance depends very much on the target temperature and the flux of incoming ions. For boron implantation at room temperature for instance, no amorphous layer can be generated.

The implantation process is always followed by a thermal annealing process and there are two reasons for it:

The lattice must be restored to its crystalline form and remaining point defects should recombine.

The implanted donor ions or acceptor ions must be positioned at substitutional sites where they become electrically active, i.e. contribute with an electron to the conduction band or a hole to the valence band.

The annealing of lattice defects starts already at room temperature, but defect complexes of higher order anneal only at a higher temperature. Mostly used in Si processing is a fast annealing process – rapid thermal annealing, RTA, to keep the effect of diffusion of the implanted ions small during annealing. In an RTA process the wafer is heated in a short time to a high temperature by using very intensive light radiators. The temperature is typically above 1000°C and the annealing time is only some seconds or half a minute. A fast cooling step follows. With this process an effective electrical activation of the doping atoms and an effective annealing of lattice defects is possible without a significant increase of the penetration depth of the doping profile. Even amorphous layers can be restored to monocrystalline layers. Careful optimization of the implantation and annealing processes is necessary to achieve the desired device structures in the field of micro- and nanoelectronics. For power devices, higher penetration depths than can be reached by conventional implanters are often necessary. In these cases, a diffusion step follows the ion implantation step. However, the process of ion implantation is frequently used because of the possibility of an exact control of the dose and thereby a very exact adjustment of the resulting profile. Modern implanters can produce doping profiles with a deviation of less than 1% across 8 in. wafers.

4.6 Oxidation and Masking

One of the great advantages of silicon compared with other semiconductors is the ability to form an oxide of high quality. Silicon dioxide, SiO_2, is widely used for thin insulating layers (e.g., gate insulation in MOSFETs), and for protecting or masking certain areas of a wafer during processing. SiO_2 features a disordered amorphous structure. For the oxidation of silicon two processes are used.

Dry oxidation:

$$Si + O_2 \rightarrow SiO_2$$

This process gives a low growth rate of the oxide, but the quality is very good. It is used for fabrication of thin oxide layers, which are needed as scattering layer for ion implantation to prevent channeling, and also for the creation of gate oxide in field controlled devices, so-called metal oxide semiconductor (MOS) structures.

Wet oxidation:

$$Si + 2H_2O \rightarrow SiO_2 + 2H_2$$

The growth rate of the oxide is higher with this process. It is used to fabricate oxides for masking purposes and oxides as passivation layers. After the wet oxidation step, a dry oxidation step to improve the surface quality of the oxide layer follows in many applications.

The thickness of the oxide layer, d_{ox}, can be calculated with [Ben99]

$$d_{ox} = d_0 + A \cdot t \quad \text{for thin oxide}$$
$$d_{ox} = B \cdot \sqrt{t} \quad \text{for thick oxide.}$$

The constants A and B are temperature dependent. For instance, an oxide layer of the thickness of 1.2 μm for masking purpose is created with a wet oxidation at 1120°C and a time duration $t = 3$ h. The process is finished with a dry oxidation in the same reactor for 1 h at the same temperature, which improves the oxide quality.

The diffusion constants of the dopants B, P, As, and Sb in SiO_2 are several decades lower than their diffusion constants in crystalline Si. This is used in the manufacturing process. The oxide layer is structured with a photolithographic process. The photoresist is exposed to ultraviolet light through a photomask with the intended pattern. After developing and hard baking of the photoresist, the SiO_2 layer is etched in buffered hydrofluoric acid NH_4F/HF, and finally the photoresist is removed ("stripped"). After removal of the photoresist the wafer has a pattern of SiO_2 layers as shown in Fig. 4.14a.

After cleaning, this pattern can now be used as a mask for a diffusion process. A phosphorus diffusion is typically executed from the vapor phase in a diffusion

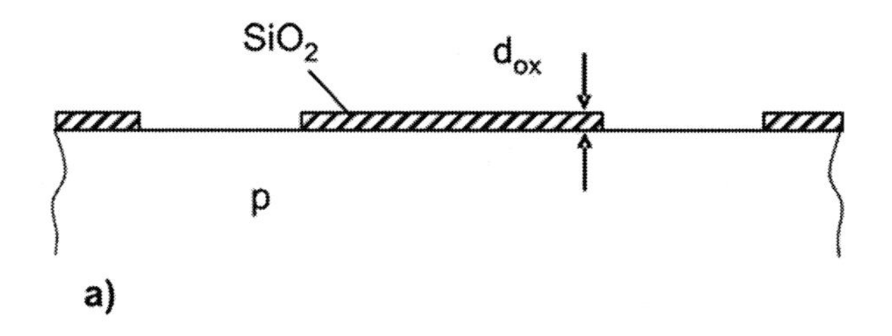

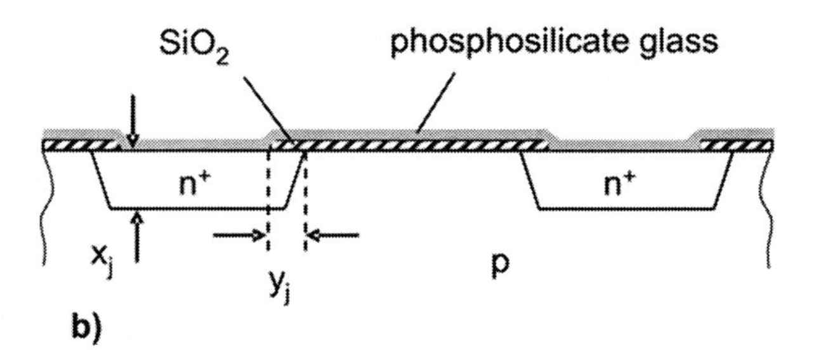

Fig. 4.14 Masking, example of fabrication of n^+-structures in a p-layer. (**a**) p-layer, SiO_2 structured with a photolithographic process. (**b**) Phosphorus diffusion with a penetration depth x_j

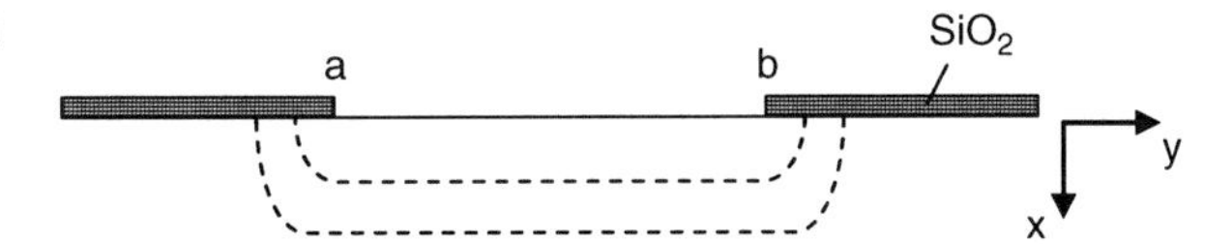

Fig. 4.15 Vertical and lateral diffusion

furnace. With the parameters temperature and time the penetration depth x_j is adjusted. However, the time can only be so long as the phosphorus does not penetrate the masking oxide layer with the thickness d_{ox}. With an oxide thickness of 1.2 μm a penetration depth x_j in the range of 10 μm can be achieved. During diffusion a layer of phosphor silicate glass is growing at the surface of the silicon and of the oxide, it acts as source of diffusion atoms. This layer is subsequently removed.

During the vertical diffusion process a simultaneous lateral diffusion under the oxide takes place. This lateral diffusion extends to a length y_j, which is always smaller than x_j. In the literature is found a relation in the range $y_j/x_j \cong 0.6 - 0.9$ [Sce83]. However, also an exact solution of Eq. (4.4) is possible for the given two-dimensional problem. If there is initial deposition of dopants with constant integral particle density per area S (see Eq. 4.5), and if a constant D is presumed, then the exact solution is

$$N(x, y, t) = \frac{S}{\sqrt{\pi} L_D} \cdot e^{-\frac{x^2}{L_D^2}} \left(\mathrm{erf}\frac{b-y}{L_D} + \mathrm{erf}\frac{y-a}{L_D} \right) \tag{4.14}$$

with $L_D = 2 \cdot \sqrt{D \cdot t}$. Here the edges of the SiO_2 mask are at $y = a$ and $y = b$, see Fig. 4.15. erf(u) is the normal error function, which is related to the more frequently used complementary error function erfc(u) by

$$\mathrm{erf}(u) = 1 - \mathrm{erfc}(u)$$

and for erfc(u) the approximation in Eq. (4.11) can be used for $u > 0$. For negative values in erf(u) holds

$$\mathrm{erf}(-|u|) = -\mathrm{erf}(|u|)$$

According to Eq. (4.14), the surface concentration between a and b is not constant as often assumed, but decreases at $y = a$ and $y = b$ to ≈ 50% of the density in the middle.

Equation (4.14) can be simply extended to describe profiles obtained by ion implantation. In Eqs. (4.10) and (4.14), a loss of doping by diffusion out of the surface during the drive-in diffusion has been neglected. Additionally, during drive-in diffusions often an oxide layer grows at the surface and consumes some of the semiconductor. More detailed calculations of this process can be done with process simulation tools. Nevertheless, Eq. (4.14) is an approximation which is found to be closely correlated to experimental results.

The lateral diffusion must be taken into account in the design of structures and the layout of the particular masks. The lateral diffusion is not in all cases a disadvantage. In fact, it can also be used for the formation of special structures in power devices. It is, for example, used to adjust the channel length in MOSFETs, see Chap. 9.

Masking with SiO_2 is not possible for the dopants Ga and Al, since for these impurities the diffusion constants in SiO_2 are too high.

Ion implantation is also often masked with SiO_2. The penetration depth of the ions into the oxide layer is in the same range as in Si because of the similar density. Therefore, the thickness of the oxide must be chosen accordingly. Other coating layers are also used for laterally masking during ion implantation, e.g., Si_3N_4 and even photoresist layers can be used, as long as the temperature during ion implantation is kept low.

4.7 Edge Terminations

The one-dimensional investigation of the blocking behavior in Chap. 3 is valid only as long as the semiconductor body is unlimited in lateral direction. In reality, however, the device structures have a finite dimension and an edge termination must be applied to lower the electric field at the edges. Edge termination structures may be divided in two main groups:

1. Edge structures with *beveled termination structures*. By beveling a defined angle can be adjusted between a lateral pn-junction and the surface, and thereby the edge is relieved from high electric fields. An overview is given in [Ger79].
2. Edge structures with a planar semiconductor surface are denoted as *planar termination structures*. An overview is given in [Fal94].

4.7.1 Bevelled Termination Structures

The beveled edge contour is produced by mechanical grinding. The angle α is defined in relation to the junction from the high doped side to the low doped side. A beveled edge with negative bevel angle is shown in Fig. 4.16. The effect can be explained in a simplified way as follows. The equipotential lines of the space charge

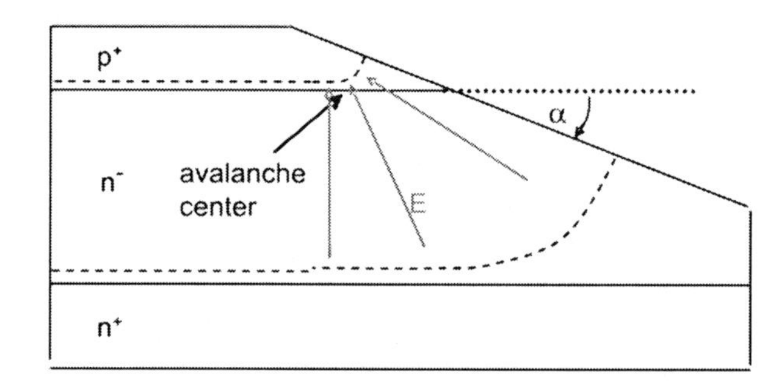

Fig. 4.16 Edge termination with negative bevel angle

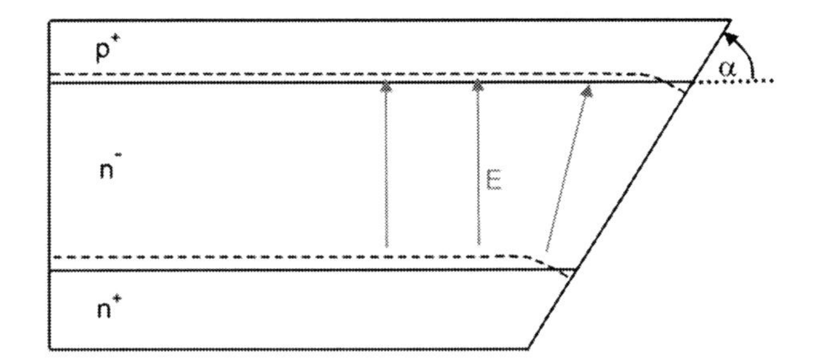

Fig. 4.17 Edge termination with positive bevel angle

must intersect with the surface orthogonally, if there are no surface charges. This forces the space charge region to widen at the edge, and thereby the electric field strength is lowered at the surface.

In a structure with negative bevel angle, a compression of the field lines occurs close to the junction on the p-side near the beveled surface. Therefore, the electric field is increased at this location. To counteract this, an edge structure with negative bevel angle is always fabricated with a very shallow angle, usually between 2° and 4°. For this case, approximately 90% of the volume blocking capability can be achieved. The avalanche breakdown always starts at the edge close below the semiconductor surface at the location which is marked in Fig. 4.16 as avalanche center.

With a beveled edge termination using a positive bevel angle, as shown in Fig. 4.17, the distance between the equipotential lines is also increased at the surface. Especially close to the pn-junction, where the electric field is high, the field lines are widened at the edge. Therefore no avalanche center occurs and 100% of the volume breakdown voltage can be achieved with this termination structure. The angle α can be chosen in a wide range between 30° and 80°.

The etched structure shown in Fig. 4.18 is also a junction termination structure with positive bevel angle. The semiconductor wafer is etched starting from

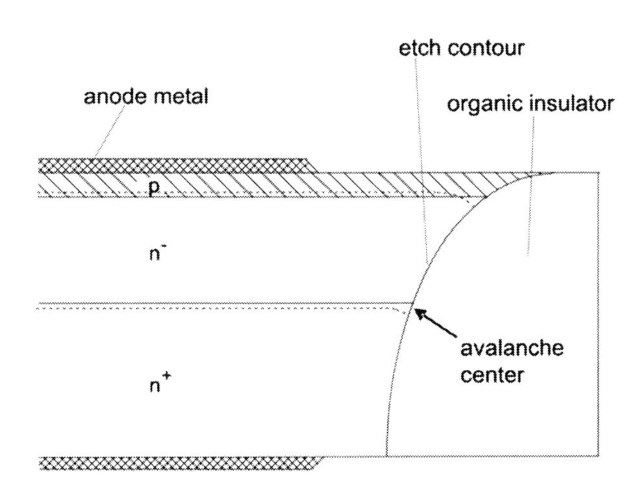

Fig. 4.18 Etched edge termination structure with positive bevel angle at the pn-junction

the n^+-side. Also for this structure, the breakdown will take place in the volume and the breakdown voltage is not lowered compared to the one-dimensional calculation. However, if the space charge penetrates the n^+-layer, an avalanche center may occur at the nn^+-junction at the location marked in Fig. 4.18.

If an avalanche center at the nn^+-junction is avoided, the structure according Fig. 4.18 proves to be very insensitive against surface charges. For the long time stability in this case a passivation layer of silicone gel is sufficient. The sharp corner at the anode side is mechanically very susceptible and such devices can easily be damaged. Therefore, this edge structure is not suited for modern devices with shallow penetration depth of the p-layer.

4.7.2 Planar Junction Termination Structures

Planar structures are mechanically much more insensitive. The structure with floating potential rings, as shown in Fig. 4.19, can be fabricated in a single mask step together with the p-type anode layer. The potential rings lead to a widening of the space charge at the semiconductor top surface. The potential ring structure was first suggested by Kao and Wolley [Kao67].

The maxima of the electric field are marked in Fig. 4.19, they can be lowered by selecting the optimal distance between the potential rings. With numerical simulation using the Poisson equation in a two-dimensional grid, the optimal arrangement of potential rings can be calculated, as shown by Brieger and Gerlach [Bri83]. Nevertheless field maxima cannot be completely avoided and the avalanche breakdown will occur in the region of the junction termination. Around 85–95% of the volume breakdown voltage can be achieved. A big advantage of this structure is that no additional photolithographic step is necessary in production; it accrues simultaneously with the fabrication of the p-layer, which is used as anode layer for a diode or p-base for a transistor. Therefore, this structure is the most frequently used edge termination structure. A disadvantage is the large space requirement.

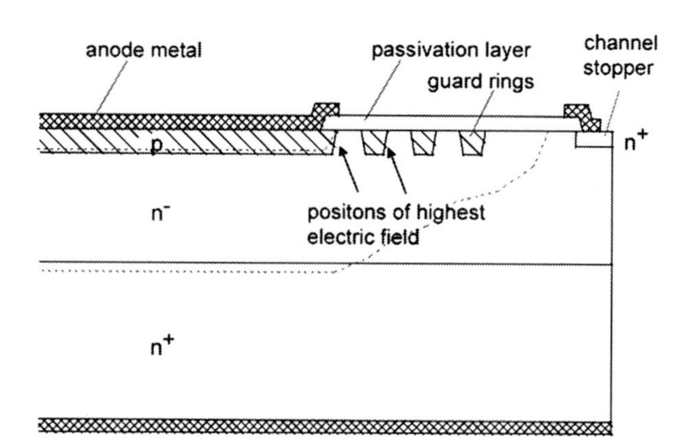

Fig. 4.19 Planar junction termination with floating potential rings

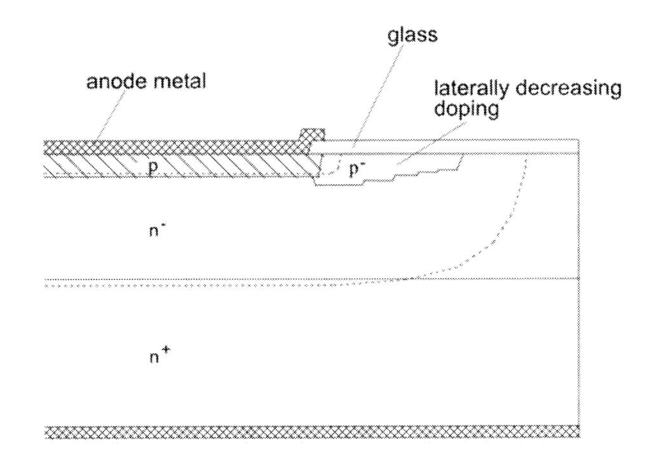

Fig. 4.20 Junction termination structure with laterally decreasing doping (VLD)

With a very lowly doped p^--zone, the so-called junction termination extension (JTE) structure, it is also possible to reach the volume breakdown voltage using planar structures. The variation of lateral doping (VLD) structure shown in Fig. 4.20 is one of the possible varieties of the JTE structure, it was first suggested by Stengl and Gösele [Ste85]. A p^--zone, in which the doping is decreasing outward to the device edge, is connected to the p-anode layer.

The structure is manufactured in such a way that the mask for the p^--layer has stripe-shaped openings, with a pitch in a form that the share of open area decreases from the anode region toward the edge. The deposition of the p-dopant is executed in these stripes, and during the following drive-in diffusion the p^--zones are joined by lateral diffusion. This process results in a profile with decreasing doping concentration and decreasing depth toward the edge as shown in Fig. 4.20. With an optimal design the breakdown takes place in the volume. Structures fabricated with implantation of Al reach 100% of the volume blocking voltage [Scu89]. Because of the lower solubility of Al, low-doped regions are easier to realize with Al.

Compared to the structure with floating potential rings, the VLD structure features a smaller space requirement and it is also much more insensitive against surface charge states [Scu89]. Because of the small tolerance window of the doping in the edge region, an ion implantation is necessary for controlling the deposited amount of B or Al. Regarding other parameters, e.g., the penetration depth, the VLD structure is less sensitive.

With field plate structures, the metallization of the p-layer is extended above the insulating passivation layer at the edge of the device. Figure 4.21a shows the effect of a one-step field plate. Also with this structure the space charge is elongated at the edge. A one-step field plate is rarely sufficient to reach blocking voltages close to the volume breakdown voltage. For the fabrication of field-controlled devices, several insulating layers are necessary in the cell structure, and using a combination of these layers several steps at the edge can be realized as shown in Fig. 4.21b. The determination of the position of the specific steps is done by numeric simulation

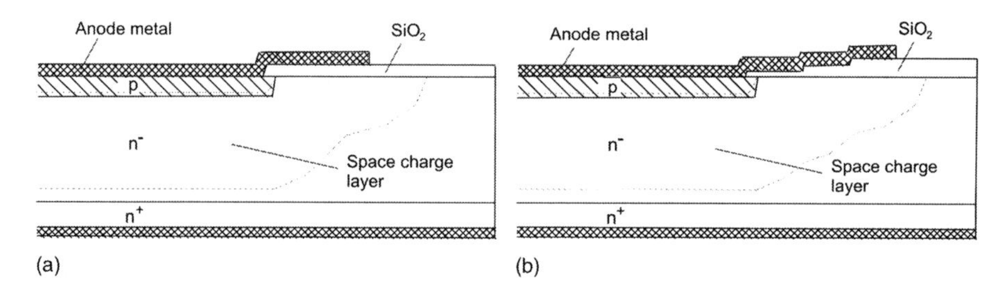

Fig. 4.21 Field plate junction termination. (**a**) One-step field plate (**b**) multi-step field plate

solving the Poisson equation, similar to the calculation of field rings, mentioned earlier. Field plate structures are often used in MOSFETs and IGBTs.

It is also possible to combine guard rings and field plates in order to spare some potential rings and thus to reduce the area requirement.

For economic reasons, it is necessary to keep the edge termination area as small as possible, since this area cannot be used for current conduction. On the other hand, additional photolithographic steps should also be avoided. In several ways the edge is the weak point of a power semiconductor device and the development of protecting junction terminations is one of the most important tasks in the development of stable and rugged power devices.

4.7.3 Junction Termination for Bidirectional Blocking Devices

Bidirectional blocking devices like the thyristor have two blocking pn-junctions, where the low-doped layer n^--layer is commonly used for both directions. Thyristors are fabricated, for example, with two structures of negative bevel angle like in Fig. 4.16, or with one structure of negative and one with a positive bevel angle as shown Fig. 4.17. Also other structures like etched mesa grooves from both sides are applied.

An attractive solution for a bidirectional blocking device is the structure with deep edge diffusion as shown in Fig. 4.22. The deep p-diffusions on the right side

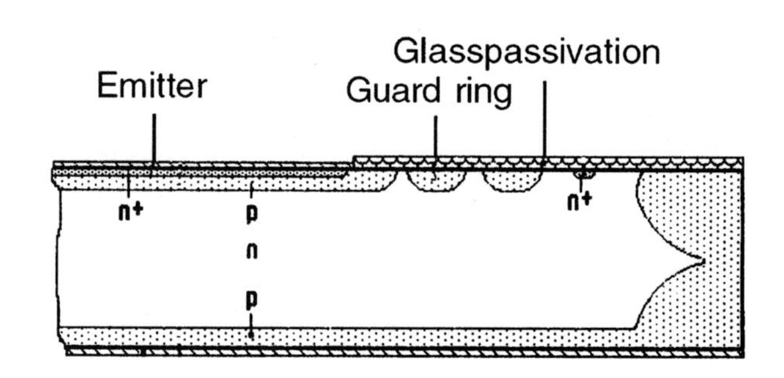

Fig. 4.22 Thyristor with edge diffusion. Figure from IXYS semiconductors AG

are carried out from both sides as one of the first process steps. In a final step the wafer is cut in the regions of the deep p-diffusions. The upper pn-junction has a planar junction with guard rings and a channel stopper as in Fig. 4.19. The lower pn-junction is connected to the top side of the wafer via the deep p-layers. The deep p-diffusion diffuses additionally to the side and thus a p^--structure as used in JTE structures is achieved. No further structures are necessary.

The advantage of the edge diffusion structure is that the wafer is compatible with processes of modern semiconductor technology, in which photolithographic processes can usually be applied only to one side of a wafer. Bidirectional blocking IGBTs are using a similar junction termination, see Sect. 10.7.

4.8 Passivation

The surface of a semiconductor is very sensitive to high electric fields. In addition to the edge terminations just described, it is also necessary to treat the surface in some way to obtain a well-defined surface and to terminate the free bonds of the silicon atoms at the surface. This treatment is called surface passivation and involves cleaning processes and subsequent deposition of an insulating material or a material with high resistivity.

For conventional devices with beveled edge an organic passivation layer is often used which consists of silicon rubber or polyimide. For junction terminations like the beveled structure with positive angle (see Figs. 4.17 and 4.18) the passivation layer is not critical, since no field peaks occur at the interface.

In junction terminations such as the planar structure with potential rings, field peaks occur at the surface and the passivation becomes crucial. The resulting blocking capability of the device is very sensitive to charges in the passivation layer and the charge state of the passivating material must be taken into account in the calculation of the breakdown voltage.

Often SiO_2 is used as passivation layer. After diffusion processes, an oxidized semiconductor surface is attained often and no additional process for passivation is necessary. This oxide layer must have a very high purity and this requirement becomes more critical for lower background doping levels N_D, since at low N_D already a very small density of surface charges is sufficient to generate an inversion layer at the surface. Such inversion layers lead to increased leakage currents and long-term instability of the device. Instead of SiO_2 also different glass layers are used, which consist of SiO_2 and additional elements.

Also semi-insulating layers are possible to use as a combined passivation and edge termination. By adjusting the electric conductivity of a semi-insulating layer, a continuous decrease of the potential at the surface can be achieved.

One commonly used evaluation method and selection criterion for the quality of a passivation layer is the hot reverse test (see Sect. 11.6), where devices are subjected to the maximal allowed temperature and the maximum allowed voltage, applied as

DC voltage, for 1000 h. If there are mobile ions or other mobile charges in the passivation material, they will move, driven by the high electric field, and accumulate at unfavorable locations where an inversion layer may be created. In this case a significant leakage current increase will be observed and the device will be rejected.

Most challenging is the passivation of very high-voltage devices between 5 and 10 kV, because of the very low doping which is mandatory for such devices. For these very demanding conditions a passivation layer of amorphous hydrated carbon (a-C:H) is used in some applications [Bar99]. The mechanical and chemical properties of a-C:H feature a diamond-like characteristic, although the bandgap is smaller than for diamond, it is only in the range of 1 to 1.6 eV. One positive property of a-C:H is that mirror charges can be induced in the bandgap, which have the capability to compensate disturbing charges. They can even reduce field peaks at the surface. Layers of a-C:H are very stable under the condition of hermitically sealed housings.

For SiC devices, 10 times higher fields are allowed in the bulk, as compared to silicon. This imposes difficult challenges for SiC passivation layers and new solutions are required both for surface processing and for the insulating layers.

4.9 Recombination Centers

Fundamental device characteristics depend on the charge carrier lifetime. The carrier lifetime in silicon, as well as all other indirect bandgap semiconductors, is adjusted by recombination centers. Recombination centers always decrease the carrier lifetime and therefore increase the voltage drop at forward conduction of a device, which leads to higher on-state losses. On the other hand, recombination centers decrease many parameters important for the switching behavior, for instance the switching time and the reverse recovery charge. This lowers the losses associated with the turn-off. The following part of this chapter deals with the particularities of different recombination centers, for instance their temperature dependence and how they can affect device characteristics. For power devices it is very important to know the differences between different recombination center technologies and how they can be used to optimize the trade-off between on-state and switching losses.

4.9.1 Gold and Platinum as Recombination Centers

Gold is the earliest used recombination center in silicon [Far65]. Platinum is used as a recombination center since middle of 1970s as an alternative to gold [Mil76]. Both recombination centers, their energy levels (Fig. 2.8) and physical characteristics, have been discussed in Sect. 2.7.

Gold as well as platinum are diffused into silicon, and both have similar characteristics in their diffusion mechanism. In silicon they can occupy interstitial and substitutional positions; however, the solubility of the heavy metal is higher at the substitutional sites. On the other hand, gold and platinum on interstitial sites are much more mobile and diffuse rapidly. The diffusion from substitutional heavy

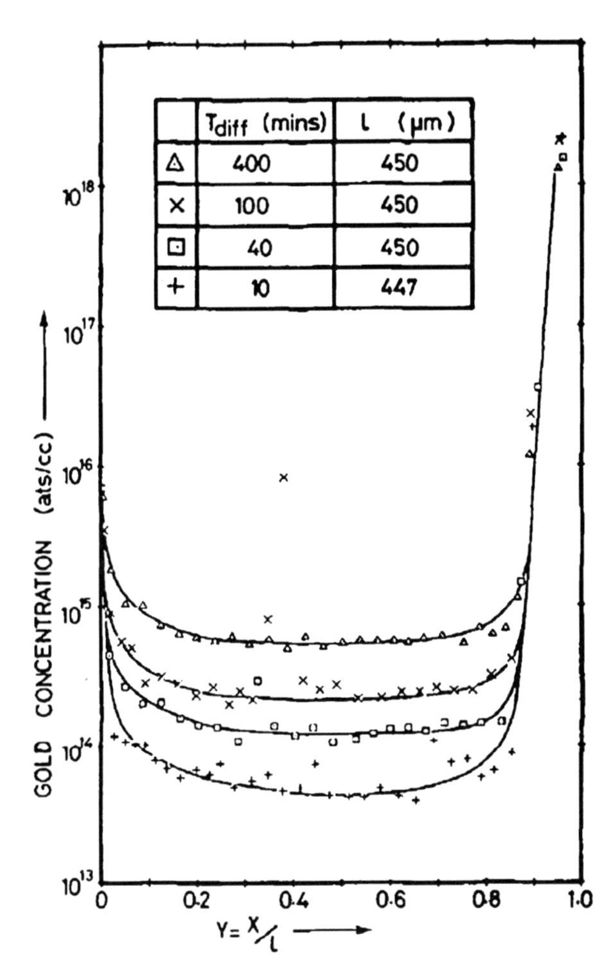

Fig. 4.23 Concentration profiles of gold diffused into FZ silicon slices at 900°C. Gold deposition was executed before diffusion on the right-hand side. The very high concentrations on this side are not valid. Figure from [Hun73] © 1973 The Electrochemical Society

metals can in fact be neglected, but there is always a relatively small barrier for interstitial to substitutional sites which results in a very fast diffusion. For instance, at a diffusion temperature of 850°C, a substantial part of the heavy metals can be found at the opposite side of the semiconductor wafer already after a diffusion time of only 10 min. This fast diffusion leads to an U-shaped concentration versus depth, or a bathtub-shaped profile with increased density of heavy metal atoms close toward the wafer surfaces, as shown in Fig. 4.23.This profile also means that the density is increased close to the pn- and nn^+-junctions in devices.

Figure 4.23 does not contain highly doped zones, which are produced in power devices before the gold diffusion. Highly boron-doped p-layers will have more stress in the lattice and will exhibit an increased gold concentration. Highly phosphorous-doped n^+-layers will collect gold atoms (gettering), a gold diffusion through such layers is not possible.

It is difficult to control the diffusion profile of gold and platinum. Because of the interaction of the diffusion mechanism with other defects in the crystal, it is also difficult to obtain a good reproducibility of the diffused heavy metal profiles. Over many years of application of gold and platinum diffusion to control the charge carrier lifetimes, a wide spread in the device characteristics even within the same batch, as well as a bad yield, had to be accepted. Much due to the problems to control the gold and platinum diffusion, a large difference is found between the real parameters and the maximal allowed values in data sheets of many fast diodes of older generations.

While the diffusion mechanism is very similar for gold and platinum, the characteristics of gold- and platinum-diffused devices are very different.

Platinum features a decrease of the lifetime with the injection level, see Fig. 2.17, this holds not for gold. Using gold, it is possible to achieve a relative good trade-off between forward voltage drop and reverse recovery charge extracted at turn-off. In this respect, gold would be a suitable recombination center. However, one of the Au levels is located almost exactly in the middle of the bandgap, see Fig. 2.18. This has the consequence that it acts also very effectively as generation center during the blocking state and this generation creates a high reverse leakage current at elevated temperature, for further details see Fig. 3.13. The leakage current of devices, where gold has been diffused to control the recombination rate, are at 150°C a factor of 50 higher than for platinum-diffused devices.

The leakage current for different recombination center technologies are compared in Fig. 4.24. For devices with a high density of gold recombination centers

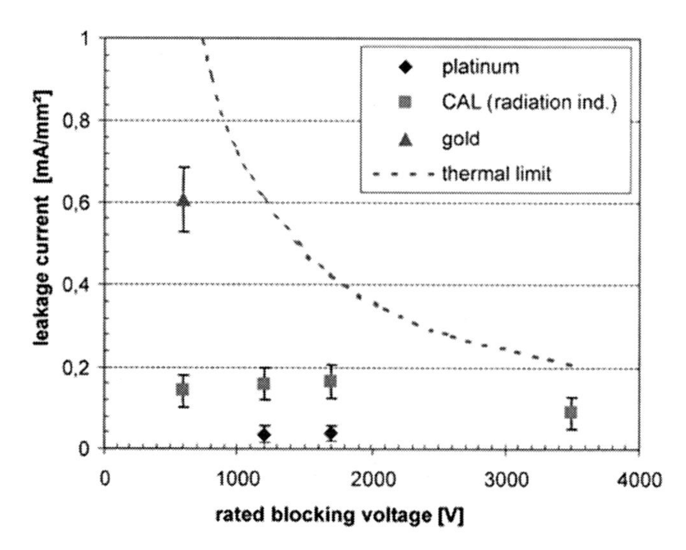

Fig. 4.24 Leakage current of diodes fabricated with different recombination centers at $T = 150°C$. *X*-axis: Rated voltage of the respective device. The density of recombination centers is chosen so that requirements for freewheeling diodes for IGBTs with the specified voltage are fulfilled. Figure from [Lut97] © 2007 EPE

as necessary for instance for fast freewheeling diodes and for a rated voltage over 1000 V, at a temperature of 150°C such high reverse blocking losses occur that they will, in fact, result in thermal instability. Gold must therefore be excluded as recombination center for fast freewheeling diodes in this voltage range. As a thermal limit in Fig. 4.24, it is assumed that the temperature increase by the leakage current is allowed to be at maximum $\Delta T = 15$ K at a DC voltage of two-thirds of the specified blocking voltage, and a thermal resistance $R_{th(j-c)} = 31$ KW^{-1}mm^2 is assumed which is, for example, a typical value for a power module.

For gold-diffused devices at a voltage above 1000 V, the maximum allowed junction temperature is limited, typically to 125°C. For a DC voltage load it must be further reduced and limited to only 100°C.

In a space charge region the acceptor level of the gold center is negatively charged. If the density of the gold atoms is in the range of the background doping – this is the case for very fast power diodes – then a compensation effect occurs and the device behaves according to the reduced background doping [Mil76, Nov89]. This affects also the turn-on behavior. The voltage peak V_{FRM}, occurring at turn-on of a device, is a function of the resistance of the low-doped layer – see Chap. 5, Eq. (5.66). For gold-diffused devices, V_{FRM} can amount to the multiple of the voltage peak of diodes completely without, or with other recombination centers.

For platinum, the trade-off between forward voltage drop and reverse recovery charge is significantly less convenient than that for gold. On the other hand, for platinum there is no energy level close to the middle of the bandgap, and the small leakage current of a platinum-diffused device can hardly be distinguished from a device without recombination centers. Therefore higher junction temperatures can be realized with platinum, for instance 150 or 175°C and in the future probably 200°C.

For platinum-diffused fast diodes the reverse recovery charge is strongly increasing with temperature. The effect of the recombination center is therefore decreasing with increasing temperature. As a consequence platinum-diffused diodes, in which the p-emitter is strongly doped and the emitter recombination is low, show a negative temperature dependency of the forward voltage. The temperature dependency of capture coefficients of different recombination centers is summarized in [Sie01].

4.9.2 Radiation-Induced Recombination Centers

The characteristics of the most important centers generated by radiation have been discussed in Sect. 2.7. This technology has a much better reproducibility, and it is easy to control the lifetime by adjusting the radiation dose. The process of irradiating silicon power devices with high-energy electrons to induce lattice defects, which act as recombination centers, has been in use since the 1970s. Electron and also γ-radiation create a homogeneous distribution of defects by knocking out lattice atoms in random collisions with low probability. This technique has many advantages over diffusion technology, but the resulting constant distribution of recombination centers is often a disadvantage for the switching characteristics of devices. More recently, implantation of protons or helium ions has been introduced for charge carrier lifetime control [Won87, Lut97]. These particles also collide with the host atoms and knock them out of their positions, but the collisions occur with large probability toward the end of the ion track. As a result, low mass ion implantation creates a localized region of recombination centers, whose position can be adjusted by the energy of the incoming particles. This process has become widely used for the adjustment of charge carrier lifetime (see also Sect. 5.7, CAL diode). Irradiation processes feature high accuracy and high reproducibility and, furthermore, the irradiation techniques offer the possibility to perform the carrier lifetime control at the end of the fabrication process after the devices have been metalized and passivated.

The vacancies and interstitials that are generated by the collisions diffuse and form stable complexes with each other and with impurity atoms which are present even in high-purity silicon, such as carbon, oxygen, and phosphorus. The irradiation process is followed by a mild annealing step to eliminate the thermally unstable centers and to ensure a long-term stability of the device characteristics. The annealing process is also needed to reduce the effects of further processing steps, such as soldering, that can alter the defect properties.

Devices for packaging in presspack technology are annealed in the range of 220°C and they will not be exposed to a higher process temperature after the irradiation. Devices which are exposed to a soldering process during packaging and other possible thermal treatments should be annealed at temperature in the range 340–350°C to ensure that no further defect annealing occurs in subsequent soldering processes, whose temperatures can be quite high.

A lot of work has been done to determine the characteristics of the radiation-induced centers [Sue94, Hal96, Ble96]; a summarizing work of recent results is given in [Sie06] and [Haz07]. The most important centers are shown in Fig. 2.19. Figure 4.25 shows additionally the centers which are still present after annealing at 220°C and which impact the carrier lifetime or other device characteristics [Sie02, Sie06]. The notations in the base line E(90 K) denote the signal of the center found in Deep Level Transient Spectroscopy (DLTS). This notation is often used, since the determination of the specific atomic structure is difficult and the literature on it is not always consistent.

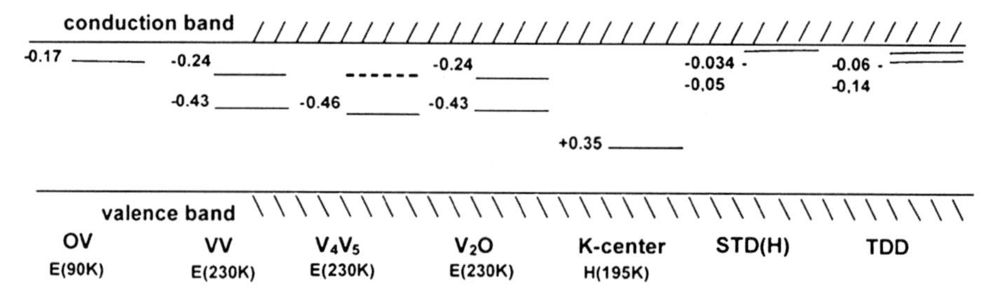

Fig. 4.25 Energy positions in the bandgap for the most prominent irradiation-induced defects in silicon

The OV center or A center $E(90\text{ K})$ is a vacancy–oxygen complex. This center starts to anneal at temperatures above 350°C and vanishes almost completely at an annealing temperature of 400°C [Won85]. As described in Sect. 2.7, the high-injection lifetime τ_{HL} and the resulting forward characteristics as well as switching properties are mainly controlled by the OV center.

The K-center $H(195\text{ K})$ was found as a hole trap at +0.35 eV above the valance band. Different assignments on its origin are found in the literature, often a COVV-complex involving a carbon impurity, an oxygen atom and two vacancies is assumed. Recent work using the cathode luminescence method identified the relevant center as C_iO_i, an association of an interstitial carbon atom and an interstitial oxygen atom [Niw08]. The K-center starts to anneal at a temperature in the range of 370– 400°C and vanishes after annealing at temperatures over 450°C [Niw85]. It has an energy level close to the middle of the bandgap, but the contribution to recombination processes is rather small due to low probability to capture charge carriers, i.e. small capture cross sections. Under high-injection condition, which is given in a forward-biased power diode, for example, this center is positively charged. When the diode is turned-off sufficiently fast, the defects remain positively charged for a certain time, typically some 100 ns or 1 μs, due to the relatively low electron-capture rate. The positively charged K-centers will act as donors, and they will increase the effective doping concentration. This reduces temporarily the breakdown voltage at the pn-junction according to Eq. (3.84), resulting in avalanche breakdown at voltages far below the static breakdown voltage [Lut98]. These detrimental effects are further described in Sect. 13.3. It is thus important to avoid high densities of K-centers, and the maximum irradiation dose is therefore limited for some applications.

The divacancy, VV $E(230\text{ K})$, has as many as three different charge states, that give rise to three separate levels in the bandgap. Most important for charge carrier dynamics is the level at 0.43 eV below the conduction band edge. It acts as a recombination center, but since this energy level is so close to the middle of the bandgap, it also acts as a generation center and will determine the generation lifetime and leakage current. However, its effect as generation center is fortunately weaker than gold.

Divacancies anneal out at temperatures of about 300°C [Bro82, Won85], but when using proton- or $^4He^{2+}$ irradiation, this level is still found even after annealing temperatures in excess of 350°C. This can be explained by a transformation of the defect to another defect with nearly the same bandgap position, and recently the remaining centers have been assigned to the singly and doubly charged states of the V_2O defect [Mon02]. Another possible origin of this $E(230\text{ K})$ peak is given in [Gul77], which suggests that it arises from a V_4 or V_5 complex. Because of their low concentration, the influence of the V_2O defects on τ_{HL} is rather small, but this center is responsible for the fact that devices implanted with helium ions exhibit a higher leakage current than platinum-diffused devices (see Fig. 4.24). The leakage current of helium-implanted devices is about 20% of the leakage current of comparable gold-diffused devices and poses no problem for the thermal stability at junction temperatures up to 150°C.

The $E(230\ \text{K})$ defect also shows a significant compensation effect on the doping in n-type silicon, where the Fermi level is above the trap level W_C −0.43 or −0.46 eV. After annealing at a temperature of 350°C, one can find a decrease of the effective doping due to the charged acceptor states of $E(230\ \text{K})$ in the region affected by the helium implantation. This effect can be used to increase or to adjust the blocking voltage of a device after the main manufacturing steps [Sie06].

Annealing temperatures significantly above 350°C may result in the formation of so-called thermal double-donors TDD. The maximum TDD concentration was found after annealing at $T \approx 450°\text{C}$. The TDDs increase the doping concentration in n-type silicon but compensate the doping in p-type silicon. In n-type silicon, thermal double donors can be used to create buried buffer layers with shallow doping.

After proton irradiation, additional effects arise since the protons (hydrogen ions) may also participate in the defect complexes. The hydrogen-related shallow thermal donors STD(H) are an example for this. These centers are found after annealing of a proton implanted sample above 200°C [Won85]. Their maximum density is found in the region of the projected range R_{pr} of the protons, a clear indication that the defects involve the implanted hydrogen. In contradiction to the K-center, the doping effect is not temporary. The STD(H) is a stable donor with an energy level close to the conduction band. If proton implantation is used to reduce the carrier lifetime, the dose must be limited to keep the density of these hydrogen related centers below the background doping. Otherwise the effect will be that deep buried n-layers are created. However, this effect may also be used in device design. If proton implantation is executed in conjunction with a subsequent annealing step at temperatures between 300 and 500°C, "buffers" which lead to a trapezoidal field in the device (PT dimensioning, Chap. 5) can be created. This technology is applied for some IGBTs of the latest generations. Furthermore, proton irradiation is used to adjust the threshold voltage of protection devices [Sie06].

While radiation-induced centers are used widely, their atomic structure and the charge carrier capture and emission processes are still not fully understood. An overview with a detailed treatment of the characteristics of the specific centers can be found in [Sie06]. Many important details, for instance the temperature dependency of the center characteristics, are still an object of active research.

4.9.3 Radiation-Enhanced Diffusion of Pt and Pd

As described, the diffusion mechanism as well as the built-in mechanism of Au and Pt into the Si lattice interacts with crystal defects. Therefore the final profile can be influenced if there is defined crystal damage at a defined penetration depth as it is created by radiation with particles such as H^+ and He^{2+} ions. After a He^{2+} irradiation, Pt can be diffused at a temperature much lower than usual for Pt diffusions into the position of maximal radiation damage caused by the He^{2+} implantation [Vob02]. If the temperature is high enough to anneal the radiation-induced defects, they are replaced by a local profile of Pt atoms. Since not all of the radiation-induced defects have optimal properties, the final device characteristics can be further improved by the radiation enhanced diffusion. A level close to the middle of the bandgap, which leads to increased leakage current, is avoided if Pt or Pd is used.

The energy levels of Pd are at positions in the bandgap close to that of Pt, so that the electronic structure of a substitutional Pd atom seems to be quite similar to a substitutional Pt atom. A wider usable temperature range was found for the radiation enhanced diffusion of Pd [Vob07]. Depending on the temperature, the formation of an acceptor is observed and a buried p^--layer can be formed at the penetration depth of the primary He-implantation [Vob09]. This p^--layer increases the static breakdown voltage and reduces peaks of the electric fields and dynamic avalanche at fast switching events of diodes. Diodes with increased ruggedness have been presented using this technique [Vob09]. Further work is of interest to understand the details of the created centers.

References

[Amm92] von Ammon W. "Neutron transmutation doped silicon - technological and economic aspects" Nuclear Instruments and Methods in Physics Research B63 pp. 95–100 (1992)

[Bar99] Barthelmeß R, Beuermann M, Winter N: "New Diodes With Pressure Contact For Hard-Switched High Power Converters", Proceedings of the EPE '99, Lausanne (1999)

[Ben99] Benda V, Govar J, Grant DA, Power Semiconductor Devices, John Wiley & Sons, New York 1999

[Ble96] Bleichner H, Jonsson P, Keskitalo N. Nordlander E.: "Temperature and injection dependence of the Shockley–Read–Hall lifetime in electron irradiated n-type silicon", J. Appl. Phys. 79, 9142 (1996)

[Bri83] Brieger KP, Gerlach W, Pelka J: "Blocking Capability of Planar Devices with Field Limiting Rings", Sol. State Electron. 26, 739 (1983)

[Bro82] Brotherton SD, Bradley P: "Defect production and lifetime control in electron and ?-irradiated silicon", Journal of Applied Physics 53 (8) 5720–5732, (1982)

[Coo83] Cooper RN: "An investigation of recombination in Gold-doped pin rectifiers", Solid-St. Electron., Vol. 26, 217–226 (1983)

[Dea68] Dearnaley G et al, Can. Journ. Phys. 4 S. 587 (1968)

[Fal94] Falck E, Untersuchung der Sperrfähigkeit von Halbleiter-Bauelementen mittels numerischer Simulation, Dissertation, Berlin 1994

[Far65] Farfield JM, Gokhale BV: "Gold as Recombination Center in Silicon", Solid St. Electr., Vol 8, pp6 85–691 (1965)

[Ful56] Fuller und Ditzenberger, J. Appl. Phys., Vol. 27, p. 544–553 (1956)

[Ger79] Gerlach W: Thyristoren, Springer Berlin 1979

[Gul77] Guldberg J: "Electron trap annealing in neutron transmutation doped silicon", Appl. Phys. Lett., 31 (9):578, (1977)

[Haa76] Haas EW.; Schnoller M.S "Phosphorus doping of silicon by means of neutron irradiation", IEEE Trans. Electron Devices, vol 23, Issue 8, Pages 803–805 (1976)

[Hal96] Hallén A, Keskitalo N, Masszi F, Nágl V: "Lifetime in proton irradiated silicon", J. Appl. Phys. 79, p. 3906 (1996)

[Haz07] Hazdra P, Komarnitskyy V: "Local lifetime control in silicon power diode by ion irradiation: introduction and stability of shallow donors"; IET Journal Circuits, Devices & Systems, Volume 1, Issue 5, Page(s):321–326 (2007)

[Hun73] Huntley FA, Willoughby AFW, "The Effect of Dislocation Density on the Diffusion of Gold in Thin Silicon Slices", J. Electrochem. Soc., Volume 120, Issue 3, pp. 414–422 (March 1973)

[Jan76] Janus HM Malmros O, "Application of thermal neutron irradiation for large scale production of homogeneous phosphorous doping of floatzone silicon", IEEE Trans. Electron Devices 21, pp. 797–805 (1976)

[Kao67] Kao YC, Wolley ED: "High Voltage Planar pn-Junctions", IEEE Trans. Electron Devices 55, 1409 (1967)

[Kra02] Krause O, Pichler P, Ryssel H: "Determination of aluminum diffusion parameters in silicon", Journ. Appl. Phys vol 91, No 9 (2002)

[Lar51] Lark-Horovitz K, "Nuclear-bombarded semi-conductors,'' in Semiconductor Materials, Proc. Conf. Univ. Reading. London: Butterworths, 1951, pp 47–69.

[Lut97] Lutz J: "Axial recombination centre technology for freewheeling diodes" Proceedings of the 7th EPE, Trondheim, 1.502 (1997)

[Lut98] Lutz J, Südkamp W, Gerlach W: "IMPATT Oscillations in Fast Recovery Diodes due to Temporarily Charged Radiation Induced Deep Levels" Solid-St. Electron., Vol 42 No. 6, 931–938 (1998)

[Mil76] Miller MD: "Differences Between Platinum- and Gold-Doped Silicon Power Devices", IEEE Trans. Electron Devices, Vol. ED-23, No. 12 (1976)

[Mol64] Moll JL, Physics of semiconductors, McGraw Hill, New York, 1964

[Mon02] Monakhov EV, Avset BS, Hallen A, Svensson BG: "Formation of a double acceptor center during divacancy annealing in low-doped high-purity oxygenated Si," Phys. Rev. B, 65:233207 (2002)

[Niw08] Niwa F, Misumi T, Yamazaki S, Sugiyama T, Kanata T, Nishiwaki K: "A Study of Correlation between CiOi Defects and Dynamic Avalanche Phenomenon of PiN Diode Using He Ion Irradiation", Proceedings of the PESC, Rhodos (2008)

[Nov89] Novak WD, Schlangenotto H, Füllmann M: "Improved Switching Behaviour of Fast Power Diodes", PCIM Europe (1989)

[Pic04] Pichler P, Intrinsic Point Defects, Impurities, and Their Diffusion in Silicon, Springer Wien New York 2004

[Rys86] Ryssel H, Ruge I, Ion Implantation, John Wiley & Sons, New York 1986

[Sce83] Schade K, Halbleitertechnologie Bd 2, VEB Verlag Berlin 1983

[Scl74] Schnöller MS, "Breakdown behaviour of rectifiers and thyristors made from striation-free silicon", IEEE Trans. Electron Devices ED 21 pp. 313–314 (1974)

[Sco82] Schlangenotto H, Silber D, Zeyfang R: "Halbleiter-Leistungsbauelemente - Untersuchungen zur Physik und Technologie", Wiss. Ber. AEG-Telefunken 55 Nr. 1–2 (1982)

[Scu89] Schulze HJ, Kuhnert R: "Realization of High Voltage Planar Junction Termination for Power Devices", Solid-St. Electron., Vol.32, S. 175 (1989)

[Sie01] Siemieniec R, Netzel M, Südkamp W, Lutz J, "Temperature dependent properties of different lifetime killing technologies on example of fast diodes", IETA2001, Cairo, (2001)

[Sie02] Siemieniec R, Südkamp W, Lutz J: "Determination of parameters of radiation induced traps in silicon", Solid-St. Electron., Vol 46, 891–901 (2002)

[Sie06] Siemieniec R, Niedernostheide FJ, Schulze HJ, Südkamp W, Kellner-Werdehausen U, Lutz J: "Irradiation-Induced Deep Levels in Silicon for Power Device Tailoring" Journal of the Electrochemical Society, 153 (2) G108–G118 (2006)

[SIL06] Siltronic AG, "FLOAT ZONE SILICON AT SILTRONIC" www.siltronic.com/int/media/publication/.../Leaflet_Floatzone_en.pdf (2006)

[Ste85] Stengl R, Gösele U: "Variation of Lateral Doping - a new Concept to Avoid High Voltage Breakdown of Planar Junctions", IEEE IEDM 85, S.154 (1985)

[Sue94] Südkamp W: DLTS-Untersuchung an tiefen Störstellen zur Einstellung der Trägerlebensdauer in Si-Leistungsbauelementen, Dissertation, Technical University of Berlin, 1994

[Sze88] Sze SM, VLSI Technology, McGrawHill, New York 1988

[Tan59] Tannenbaum M: Uniform n-type silicon, U.S. patent 3076732, filled 1959/12/15

[Tan61] Tannenbaum M, Mills AD, "Preparation of uniform resistivity n-tvpe silicon by nuclear transmutation", J. Electrochem. SOC., vol. 108, pp. 171–176, (1961)

[Vob02] Vobecký, J.; Hazdra, P: "High-power P-i-N diode with the local lifetimecontrol based on the proximity gettering of platinum" IEEE Electron DeviceLetters, vol 23, Issue 7, pp 392–394 (2002)

[Vob07] Vobecký J, Hazdra, P: "Radiation-Enhanced Diffusion of Palladium for a Local Lifetime Control in Power Devices", IEEE Trans. Electron Devices, vol 54, Issue 6, pp 1521–1526 (2007)

[Vob09] Vobecký J, Záhlava V, Hemmann K, Arnold M, Rahimo M, "The Radiation Enhanced Diffusion (RED) Diode - Realization of a Large Area $p^+p^-n^-n^+$ Structure with High SOA", Proceedings of the 21st ISPSD, Barcelona, pp 144–147 (2009)

[Won85] Wondrak W, Erzeugung von Strahlenschäden in Silizium durch hochenergetische Elektronen und Protonen, Dissertation, Frankfurt 1985

[Won87] Wondrak W, Boos A, "Helium Implantation for Lifetime Control in Silicon Power Devices," Proc. of ESSDERC 87, Bologna, pp. 649–652, (1987)

[Zie06] Ziegler JF, Biersack JP, "The Stopping and Range of Ions in Matter". [Online]. http://www.srim.org/SRIM/SRIMINTRO.htm (accessed 1/4/2006)

Chapter 5
pin-Diodes

Most power diodes are pin-diodes, i.e. they possess a middle region with a much lower doping concentration than the outer p- and n-layers enclosing it. Compared with unipolar devices (see Chap. 6), pin-diodes have the advantage that the on-resistance is strongly reduced by high-level injection in the base region, which is known as conductivity modulation. Hence pin-diodes can be used up to very high blocking voltages. The base region is not intrinsic, as suggested by the name. The intrinsic case – doping in the range of $< 10^{10}\,\text{cm}^{-3}$ – would not only be difficult to attain by technology, extremely low doping would cause essential disadvantages in the turn-off behavior and other properties. Power diodes usually have a $p^+n^-n^+$-structure, hence the so-called i-layer is actually an n^--layer. Since it is several orders of magnitude lower than the doping of the outer layers, the name pin-diode has become the usual denotation in almost every case,

From the viewpoint of application, power diodes can be distinguished into two main types:

Rectifier diodes for grid frequency of 50 or 60 Hz: the switching losses play a subordinate role, and there is a high carrier lifetime in the middle layer.

Fast recovery diodes that work as freewheeling diodes for a switching device or that are in the output rectifier after a high-frequency transformer. They have to be generally capable of switching frequencies of up to 20 kHz and in switch-mode power supplies of 50–100 kHz and more. In fast diodes manufactured from silicon, the charge carrier lifetime in the middle low-doped layer has to be reduced to a defined low value.

5.1 Structure of the pin-Diode

With respect to structure and technology, pin-diodes can be classified into two types. For pin-diodes using epitaxial technology (epitaxial diodes, Fig. 5.1a), first, an n^--layer is deposited by epitaxy on a highly doped n^+-substrate. Then, the p-layer is diffused. With this process a very small base width w_B down to some micrometers can be created, whereby the silicon wafer is thick enough by the substrate to allow production with low wafer breaking and at high yield. By implementing recombination centers – in most cases by gold diffusion – very fast diodes can be

J. Lutz et al., *Semiconductor Power Devices*, DOI 10.1007/978-3-642-11125-9_5,

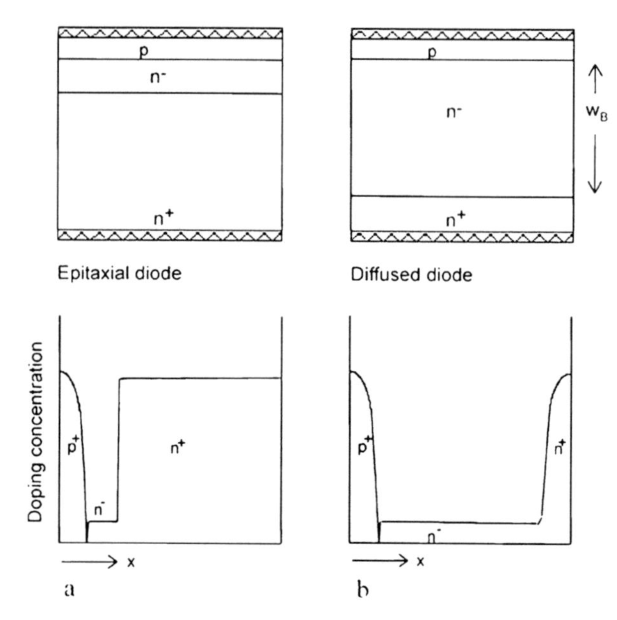

Fig. 5.1 Structure of pin power diodes. (**a**) Epitaxial diode. (**b**) Diffused diode

realized. Since w_B is kept very small, the voltage drop across the middle layer is low. Epitaxial (epi-) diodes are mainly applied for blocking voltages of between 100 and 600 V; however, some manufacturers also produce 1200 V with epi-diodes.

Because the costs of the epitaxy process are notable, diodes for higher blocking voltages – usually 1200 V and above – are fabricated by diffusion. For a diffused pin-diode (Fig. 5.1b), one starts with a low-doped wafer in which the p^+-layer and the n^+-layer are created by diffusion. The thickness of the wafer now is determined by the thickness w_B of the middle n^--layer and the depths of the diffusion profiles. The required w_B is small for lower voltages. With deep n^+-and p^+-layers the wafer thickness can be increased again, but deep p-layers have disadvantages regarding the reverse recovery behavior. The processing of such thin wafers is challenging. Infineon has introduced a technology for handling very thin wafers, down to a thickness of 80 μm in the manufacturing process. With this technology, also freewheeling diodes for 600 V with shallow p- and n^+-border layers can be fabricated as diffused diodes.

5.2 *I–V* Characteristic of the pin-Diode

The *I–V* characteristic of a fast 300 V pin-diode measured at 25°C as well as some definitions for parameters of the *I–V* characteristic are shown in Fig. 5.2. In the figure different scales in forward and reverse directions are applied. In forward bias, the characteristic associates a defined current I_F to the voltage drop V_F. This has to be distinguished from the maximally allowed voltage drop V_{Fmax} specified in manufacturers' data sheets. V_{Fmax} is the maximum forward voltage drop that can

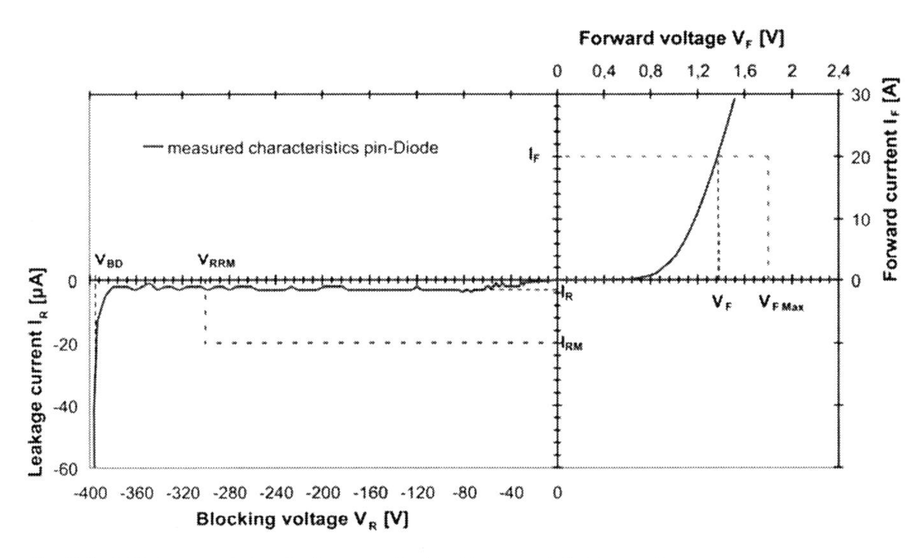

Fig. 5.2 *I–V* characteristic of a fast pin-diode with some definitions of parameters

occur at a diode of this type under specified conditions. In most cases this value is significantly higher than the value measured for an individual sample due to tolerances of parameters during production, e.g. of the base width w_B. In fast diodes, the carrier lifetime strongly effects V_F. For fast gold- or platinum-diffused diodes of older generations, relatively high variations in the carrier lifetime are typical due to the difficulties to control these technologies, see Chap. 4. Some manufacturers also specify typical values, but this is not guaranteed for an individual sample.

In reverse bias, V_{BD} is the physical breakdown voltage of a given sample. I_R denotes the leakage current measured at a defined reverse voltage. This has to be distinguished from V_{RRM}, the maximal reverse voltage that is specified in the data sheet, as well as from I_{RM}, the specified maximal leakage current at V_{RRM}. Since the manufacturer takes into account data scattering and may add additional margins, at a single diode I_R can be significantly lower and V_{BD} will be typically higher; the manufacture extends a warranty, however, only for V_{RRM} and I_{RM}, respectively.

The *I–V* characteristic of a diode is strongly temperature dependent. With increasing temperature

- the leakage current I_R increases, I_R can be several orders of magnitude higher at the typical maximally allowed operation temperature of 150°C than that at room temperature. Both the diffusion component and generation component of the leakage current also increase; see Eq. (3.59) and Fig. 3.13
- the blocking voltage V_{BD} somewhat increases in accordance with the increase in breakdown voltage of avalanche breakdown; see Eq. (3.79) with the temperature-dependent parameters in Eq. (3.87)
- the built-in voltage V_{bi} decreases, because according to Eq. (3.6), the determining parameter for the temperature dependency is the strongly temperature-dependent

n_i^2. This corresponds to a decreased threshold voltage V_S. Also in the derivation of a threshold voltage from Eqs. (3.51) and (3.52), n_i^2 dominates the result.

5.3 Design and Blocking Voltage of the pin-Diode

A dominating parameter for all characteristics of the diode is the width w_B of the low-doped base region. First of all, the base width together with the base-doping concentration determines the blocking voltage. As illustrated in Fig. 5.3, different cases of the field shape can be distinguished.

If w_B is chosen such that the space charge does not reach the n$^+$-layer (triangular field shape), it is called a *non-punch-through (NPT) dimensioning* [Bal87]. If w_B is chosen such that the space charge penetrates the n$^+$-layer, then the field shape is trapezoidal, and the diode is denoted as a *punch-through (PT)* diode. The term "punch-through" is used here in a different meaning as for thyristors. Whereas in thyristors it refers to reaching of the space charge region to a layer of opposite doping, in diodes with PT design the space charge layer is stopped by a highly doped region of the same type of conductivity. The indication of diode design by NPT and PT is widely used. The p$^+$n- and nn$^+$-junctions are assumed to be abrupt.

For NPT design, the equations of Sect. 3.3.2 apply. Particularly the breakdown voltage can be calculated from Eq. (3.79), and the thickness w of the space charge region at breakdown for a doping concentration N_D is given by Eq. (3.81). In the following calculation, the approach (Eq. (3.75)) is used with $n = 7$, so that $\alpha_{eff} = B \cdot |E|^7$ where $B = 2.11 \times 10^{-35}\,\text{cm}^{-35}\,\text{cm}^6/\text{V}^7$ according to Eq. (3.83). According to Eq. (3.85) the condition for NPT is obtained then as

$$w_B \geq \left(\frac{8}{B}\right)^{1/8} \cdot \left(\frac{\varepsilon}{qN_D}\right)^{7/8}$$

For an *ideal* NPT dimensioning, w_B is chosen such that at this point the end w of the space charge is given, $w_B = w$. The condition can be written also as a relationship

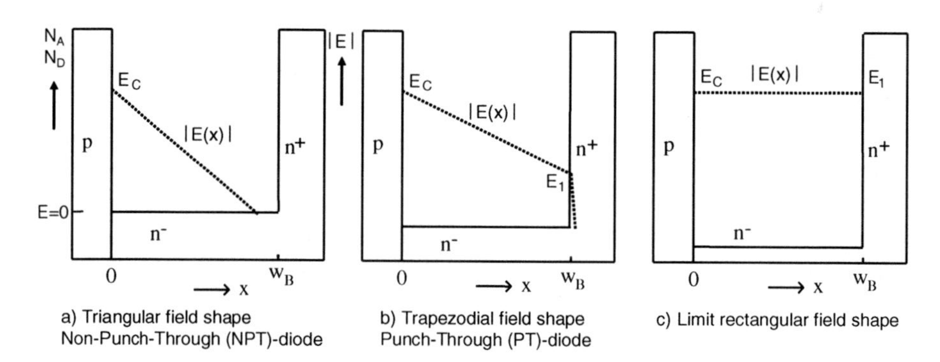

Fig. 5.3 Field distributions at breakdown in pin-diodes with different design

between w_B and the breakdown voltage V_{BD}, which follows by inversion of Eq. (3.80):

$$w_B = 2^{\frac{2}{3}} \cdot B^{\frac{1}{6}} \cdot V_{BD}^{\frac{7}{6}} \tag{5.1}$$

In this condition for NPT design, the doping concentration N_D is not contained, but according to (inverted) Eq. (3.84), it is unambiguously determined by the chosen breakdown voltage V_{BD}.

We consider now diodes with PT dimensioning, where the doping density N_D is lower than for NPT dimensioning. The space charge region penetrates the n^+-layer, where the field drops rapidly to zero as is depicted in Fig. 5.3b. Since the blocking voltage corresponds to the area below the line $|E(x)|$, a higher blocking voltage results from Fig. 5.3b with the same w_B for the now given PT dimensioning.

For PT design, the electric field over the base at breakdown is given by

$$E(x) = -E_c + \frac{q \cdot N_D}{\varepsilon} \cdot x \tag{5.2}$$

where E_c denotes the critical field strength. To calculate E_c, the power approach $\alpha_{eff} = B \cdot |E(x)|^7$ for the effective ionization rate is inserted in the breakdown condition (3.71), which yields

$$\int_0^{w_B} B \cdot \left(E_c - \frac{q \cdot N_D}{\varepsilon} \cdot x \right)^7 dx = 1 \tag{5.3}$$

The integration leads to the equation

$$E_c^8 - \left(E_c - \frac{q \cdot N_D}{\varepsilon} w_B \right)^8 = \frac{8 \cdot q \cdot N_D}{\varepsilon \cdot B} \tag{5.4}$$

Because the absolute value E_1 of the field strength at the nn^+-junction (Fig. 5.3b) is obtained from Eq. (5.2) with $x = w_B$ as

$$E_1 = E_c - \frac{q \cdot N_D}{\varepsilon} \cdot w_B \tag{5.5}$$

Equation (5.4) can be written as

$$E_c = \left(\frac{8 \cdot qN_D}{B\varepsilon} + E_1^8 \right)^{1/8} \tag{5.6}$$

Since $E_1 < E_c$, Eq. (5.6) together with Eq. (5.5) is suitable to calculate E_c by iteration. From E_c one obtains for the breakdown voltage:

$$V_{BD} = \frac{E_1 + E_c}{2} \cdot w_B = \left(E_c - \frac{qN_D}{2\varepsilon} \cdot w_B \right) \cdot w_B \tag{5.7}$$

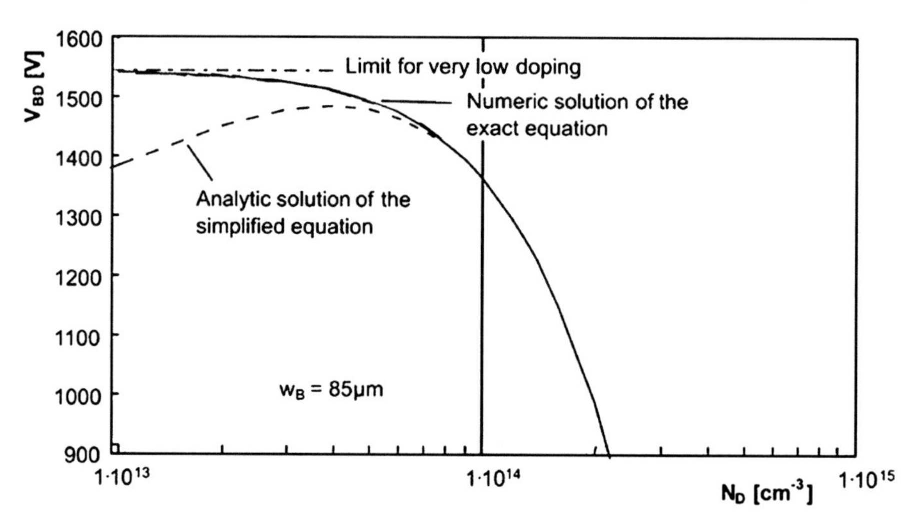

Fig. 5.4 Blocking voltage for a device with given w_B as a function of doping

The breakdown voltage calculated in this way for a given base width (85 μm) is plotted in Fig. 5.4 as a function of N_D (solid line). With decreasing N_D, the breakdown voltage increases monotonously. This holds also for other base widths.

Often one uses a design of moderate or slight punch-through with $E_1 \leq E_c/2$. Then the term for E_1^8 in Eq. (5.6) is negligible and one obtains for the critical field:

$$E_c = \left(\frac{8 \cdot q \cdot N_D}{B \cdot \varepsilon} \right)^{1/8} \tag{5.8}$$

This equation is identical with Eq. (3.78) for $n = 7$. From Eqs. (5.7) and (5.8) the breakdown voltage is obtained as an explicit function of doping density and base width:

$$V_{BD} = \left(\frac{8qN_D}{\varepsilon B} \right)^{\frac{1}{8}} w_B - \frac{1}{2} \frac{qN_D}{\varepsilon} w_B^2 \tag{5.9}$$

This dependency on N_D for $w_B = 85$ is plotted in Fig. 5.4 too (dashed curve). At high N_D down to $8 \times 10^{13}\,\text{cm}^{13}\,\text{cm}^{-3}$, where $E_1/E_c = 0.51$, the curve coincides with the exact curve. At lower N_D, however, the approximate solution becomes inexact and has a maximum at $3.7 \times 10^{13}\,\text{cm}^{-3}$ at the chosen w_B. In some textbooks it is concluded from this approximation that the breakdown voltage of pin-diodes has a maximum at an optimal doping density and from this point downward decreases with decreasing doping. As is shown by the exact curve in Fig. 5.4, this is an error. Neglecting of E_1^8 is only permissible as long as doping is not too low.

The limit for very low doping can be derived immediately from the ionization integral. In this case, the field shape is rectangular, $E_1 = E_C$, and the condition for avalanche breakdown Eq. (3.71) with the polynomial approximation of ionization rates with $n = 7$ is simplified to

$$B \cdot \int_0^{w_B} E_C^7 \mathrm{d}x = 1 \tag{5.10}$$

From this and with $V_{BD} = E_C \cdot w_B$ it follows:

$$V_{BD} = \left(\frac{w_B^6}{B} \right)^{\frac{1}{7}} \tag{5.11}$$

This limiting value for the example $w_B = 85\,\mu\text{m}$ is also depicted in Fig. 5.4. The blocking voltage approaches this limit very quickly already for a doping in the range of $2 \times 10^{13}\,\text{cm}^{-3}$.

For the base width as a function of V_{BD}, Eq. (5.11) yields

$$w_B(\text{PT}, \text{lim}) = B^{\frac{1}{6}} \cdot V_{BD}^{\frac{7}{6}} \tag{5.12}$$

This lowest attainable base width is by a factor $2^{2/3} = 1.59$ smaller than the minimum w_B for NPT design given by Eq. (5.1):

$$w_B(\text{PT,lim}) = 2^{-\frac{2}{3}} w_B(\text{NPT}) \cong 0.63 \cdot w_B(\text{NPT}) \tag{5.13}$$

Although w_B(PT,lim) is not half of w_B(NPT), as would follow for equal critical field strength in both cases, the reduction by the PT design is very significant.

A high field at the nn^+-junction, however, has disadvantages. From a technological viewpoint, an edge termination for this design requires much more effort. Therefore, an only moderate PT dimensioning, e.g. $E_1 \leq 0.5 E_c$, is preferred. If the ratio E_1/E_c is kept independent of the breakdown voltage, the dependency $w_B \sim V_{BD}^{7/6}$ is obtained for each value of E_1/E_c, where however the proportionality factor varies with the field ratio. For $E_1/E_c = 1/2$, the calculation yields

$$w_B(\text{PT}) = 0.70 \cdot w_B(\text{NPT}) \tag{5.14}$$

Figure 5.5 shows the minimum width of the middle region for PT design (5.12) and for NPT design (5.1) as a function of breakdown voltage. The difference between w_B in the two cases results in a significant difference in forward voltage drop. For fast diodes with a rated voltage of 1200 V and with the necessary low carrier lifetime, the difference in V_F amounts up to 0.8 V. For higher voltages the reduction of forward voltage by the PT dimensioning is even larger. This is significant for the conduction losses. Consequently, a PT dimensioning should be applied if possible. Nonetheless, this design also poses challenges, in particular regarding reverse recovery behavior, as will be shown later.

In practice, one has to make further compromises in the dimensioning of diodes. For example, tolerances of background doping N_D, tolerances in the adjustment of

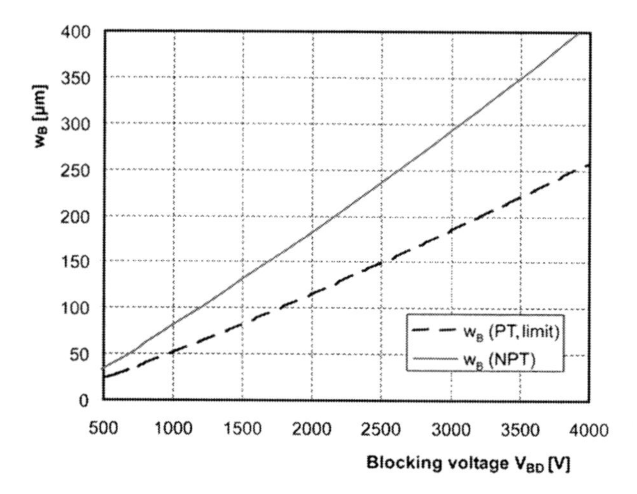

Fig. 5.5 Minimal width w_B of the base for the dimensioning with triangular (NPT) and with rectangular field shape (PT, limit)

w_B, and other aspects have to be considered. Additionally, the junction termination structures do not lead to 100% of the volume breakdown voltage in most cases. Taken from experience, an orientation value for a moderate PT dimensioning can be given by

$$w_B = \chi \cdot V_{BD}^{\frac{7}{6}} \quad \text{with} \quad \chi = 2.3 \times 10^{-6} \text{cm V}^{-\frac{7}{6}} \tag{5.15}$$

It is to be mentioned here that these considerations on dimensioning hold not only for diodes but also for other power devices whose base region contains a higher doped buffer layer to limit the extension of the space charge region. Particularly Schottky diodes, MOSFETs, and modern IGBTs are designed in this way.

A power device has to function over temperature range from about 250 K to more than 400 K. The temperature dependency of the breakdown voltage arises from that of the critical field strength which is described according to Eqs. (3.78) and (3.87) by the temperature-dependent parameters $n(T)$ and $B(T)$. For PT design, we restrict the discussion to the case $E_1/E_c \leq 1/2$. To express V_{BD} as a function of T, one has to use the Eq. (5.9) in a form which contains B and n as variables. This equation is obtained by substituting E_c in Eq. (5.7) by Eq. (3.78):

$$V_{BD} = \left(\frac{(n+1)\, qN_D}{\varepsilon B} \right)^{\frac{1}{n+1}} w_B - \frac{1}{2} \frac{qN_D}{\varepsilon} w_B^2 \tag{5.16}$$

in which the temperature-dependent parameters B and n according to Eq. (3.87) have to be inserted. Since the critical field enters Eq. (5.16) only linearly, while for NPT design $V_{BD}(\equiv V_B)$ is proportional to E_c^2 according to Eq. (3.79), Eq. (5.16) predicts a smaller variation with temperature than Eq. (3.79). Of course, at the transition between PT and NPT, Eqs. (5.16) and (3.79) give identical results. Due to the increase of E_c with T, the width of the space charge region for unlimited extension increases with temperature. Hence a NPT situation existing at 250 K can change into a moderate PT at higher temperatures. Since the blocking capability must be given at the lowest operation temperature, the decrease of V_{BD} from room temperature downward is of particular interest. Often the nn$^+$-junction is not abrupt, but the doping increases gradually. Then the measured temperature dependency is found to be between the predictions of Eqs. (5.16) and (3.79).

Since n and B are arbitrary in Eq. (5.16), the equation can be used also (without making use of the temperature dependency) for PT diodes made of other semiconductors, if the parameters n and B are known for this material. Thus the breakdown voltage of diodes in 4H-SiC-diodes can be calculated using the data for n und B given in Eq. (2.82b).

5.4 Forward Conduction Behavior

5.4.1 Carrier Distribution

At forward bias, the lowly doped base zone of a pin-diode is flooded with carriers injected from the highly doped outer regions. The density of free carriers is increased by a few orders of magnitude above the background doping, and the conductivity in the low-doped zone is strongly enhanced or "modulated." Because in the neutrality condition $n = p + N_D^+$, the doping concentration N_D^+ is negligible, the hole and electron concentrations in the base are approximately equal:

$$n(x) \approx p(x)$$

Figure 5.6 shows the carrier distribution $n = p$ in the middle region calculated for a 1200 V diode at a current density of 160 A/cm^2. Here, the emitter efficiency of

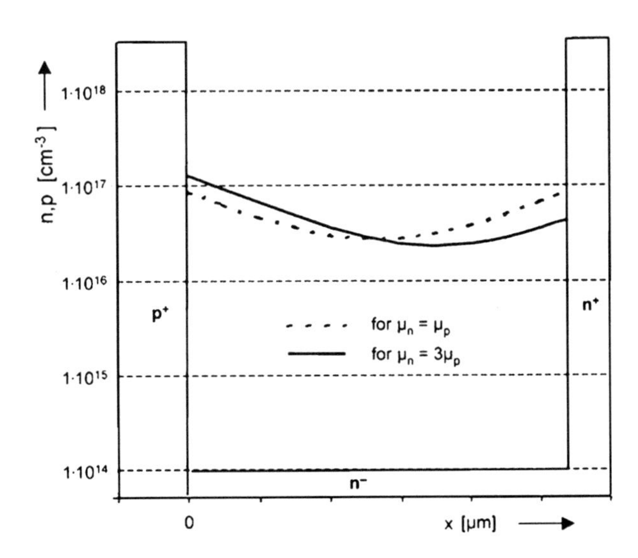

Fig. 5.6 Distribution of charge carriers in the base region at forward conduction. Example of a diode with $\tau_{HL} = 0.48\ \mu s$, $w_B = 108\ \mu m$. The p$^+$- and n$^+$-layers are assumed to be ideal emitters

highly doped regions has been assumed to be 1. n and p exceed the background doping by more than two orders of magnitude. The broken line is obtained if there would be equal mobilities for electrons and holes. Since the electrons in silicon are much more mobile than the holes, an asymmetric distribution, which is increased at the pn-junction, results.

To calculate the carrier distribution, one starts with the transport equations (2.42) and (2.43) which for $n = p$ take the form:

$$j_{\mathrm{p}} = q \cdot \mu_{\mathrm{p}} \cdot p \cdot E - q \cdot D_{\mathrm{p}} \cdot \frac{\mathrm{d}p}{\mathrm{d}x} \tag{5.17}$$

$$j_{\mathrm{n}} = q \cdot \mu_{\mathrm{n}} \cdot p \cdot E + q \cdot D_{\mathrm{n}} \cdot \frac{\mathrm{d}p}{\mathrm{d}x} \tag{5.18}$$

Since in the stationary case, which we consider, the total current density $j = j_{\mathrm{n}} + j_{\mathrm{p}}$ is independent of x according to Eq. (2.87), the electric field strength E is suitably expressed by j. Adding Eqs. (5.17) and (5.18) one obtains

$$j = j_{\mathrm{n}} + j_{\mathrm{p}} = q \cdot (\mu_{\mathrm{n}} + \mu_{\mathrm{p}}) \cdot p \cdot E + q \cdot (D_{\mathrm{n}} - D_{\mathrm{p}}) \cdot \frac{\mathrm{d}p}{\mathrm{d}x} \tag{5.19}$$

Hence, the electric field is

$$E = \frac{\frac{j}{q} - (D_{\mathrm{n}} - D_{\mathrm{p}}) \cdot \frac{\mathrm{d}p}{\mathrm{d}x}}{(\mu_{\mathrm{n}} + \mu_{\mathrm{p}}) \cdot p} \tag{5.20}$$

The term proportional to j is the normal resistive or ohmic field. The second field term determined by the concentration gradient and the difference in the diffusion constants is denoted as "Dember-field". This field compensates the total diffusion current for a given j. Inserting Eq. (5.20) into Eqs. (5.17) and (5.18), one obtains

$$j_{\mathrm{p}} = \frac{\mu_{\mathrm{p}}}{(\mu_{\mathrm{n}} + \mu_{\mathrm{p}})} \cdot j - q \cdot D_{\mathrm{A}} \cdot \frac{\mathrm{d}p}{\mathrm{d}x} \tag{5.21}$$

$$j_{\mathrm{n}} = \frac{\mu_{\mathrm{n}}}{(\mu_{\mathrm{n}} + \mu_{\mathrm{p}})} \cdot j + q \cdot D_{\mathrm{A}} \cdot \frac{\mathrm{d}p}{\mathrm{d}x} \tag{5.22}$$

where D_{A} is the following combination of the single diffusion constants:

$$D_{\mathrm{A}} = \frac{2 \cdot D_{\mathrm{n}} \cdot D_{\mathrm{p}}}{D_{\mathrm{n}} + D_{\mathrm{p}}} \tag{5.23}$$

For this relationship, the Einstein relation $D_{\mathrm{n,p}} = (kT/q)\mu_{\mathrm{n,p}}$ has been used. D_{A} is called "ambipolar diffusion constant" , and the terms $\mp q D_{\mathrm{A}} \cdot \mathrm{d}p/\mathrm{d}x$ are the ambipolar diffusion current densities. These include the current resulting from the Dember field, which makes the ambipolar diffusion current of electrons and holes oppositely equal. In many equations, the ambipolar diffusion constant appears in connection with the high-level lifetime τ_{HL} in the form of the ambipolar diffusion length:

$$L_A = \sqrt{D_A \cdot \tau_{HL}} \tag{5.24}$$

To obtain a differential equation for the carrier concentration, one has to insert Eq. (5.20) or Eq. (5.21) into the relevant continuity equation. According to Eq. (2.84), the continuity equation for holes in the stationary case can be written as

$$\frac{dj_p}{dx} = -q \cdot R = -q \frac{p}{\tau_{HL}} \tag{5.25}$$

where the recombination rate R is expressed according to Eqs. (2.48), (2.50) by p, and the high-level carrier lifetime τ_{HL} by Eq. (2.68), neglecting the equilibrium carrier density. Inserting j_p from Eq. (5.21) the following differential equation for the carrier concentration is obtained:

$$D_A \cdot \frac{d^2 p}{dx^2} = \frac{p}{\tau_{HL}} \tag{5.26}$$

where the mobility ratios and ambipolar diffusion constant are assumed to be constant. In what follows, also the lifetime τ_{HL} is assumed to be constant, which agrees with Eq. (2.68) of the SRH model (for n_r, $p_r << n_0$). Then this differential equation (5.26) has the solutions $p(x) \propto \exp(\pm\, x/L_A)$, where L_A is the ambipolar diffusion length defined in Eq. (5.24). In addition to the differential equation, the wanted carrier distribution has to satisfy conditions at the boundaries of the base region which are given by the emitter efficiency of the p^+- and n^+-regions. In the present section, we assume that these regions are ideal emitters (with emitter efficiency 1), meaning that $j_n = 0$ at the p^+n-junction and $j_p = 0$ at the nn^+-junction. Using these current densities in Eqs. (5.21) and (5.22) one obtains the boundary conditions:

$$q\, D_p \frac{dp}{dx}(0) = -j/2 \quad q\, D_n \frac{dp}{dx}(w_B) = j/2 \tag{5.26a}$$

To the same extent as $D_p < D_n$, the absolute value of the concentration gradient at $x = 0$ (p^+-side of the base region) is higher than at $x = w_B$ (n^+-side). Considering the simplified case $D_n = D_p = D$, the solution of Eq. (5.26) satisfying the conditions (5.26a) is

$$p(x') = \frac{j\, L_A}{2\, q\, D\, \sinh(w_B/(2L_A))} \cosh\left(\frac{x'}{L_A}\right)$$

Here the coordinate x′ has its origin in the middle of the base region: $x' = x - w_B/2$. The symmetrical carrier distribution in Fig. 5.6 (dashed line) was calculated using this equation together with the parameters given in the legend. The obtained cosh-distribution is well known from the chainline. The smaller the ambipolar diffusion length L_A, the more pronounced is the sagging of the carrier distribution.

In the real case of different mobilities of electrons and holes, the carrier distribution can be written as

$$n(x') = p(x') = \frac{j\tau_{\mathrm{HL}}}{2qL_{\mathrm{A}}}\left(\frac{\cosh\frac{x'}{L_{\mathrm{A}}}}{\sinh\frac{w_{\mathrm{B}}}{2L_{\mathrm{A}}}} - \frac{\mu_{\mathrm{n}}-\mu_{\mathrm{p}}}{\mu_{\mathrm{n}}+\mu_{\mathrm{p}}}\cdot\frac{\sinh\frac{x'}{L_{\mathrm{A}}}}{\cosh\frac{w_{\mathrm{B}}}{2L_{\mathrm{A}}}}\right) \tag{5.27}$$

It can be verified that this solution of Eq. (5.26) satisfies the boundary conditions (5.26a). The carrier distribution calculated from Eq. (5.27) for the given parameters is plotted in Fig. 5.6 as solid line. The concentration is at the pn-junction more than a factor 2 higher than that at the nn^+-junction (which later will prove to be a disadvantage for the reverse recovery behavior). The minimum has shifted toward the nn^+-junction. The term with $\sinh(x'/L_{\mathrm{A}})$ in Eq. (5.27) expresses this asymmetry. Since for silicon $\mu_{\mathrm{n}} \approx 3\,\mu_{\mathrm{p}}$, the factor before this term consisting of the mobilities is approximately 0.5. As is noted, integration of Eq. (5.27) yields for the mean carrier concentration across the base:

$$\bar{n} = \bar{p} = \frac{1}{w_{\mathrm{B}}}\int_{-w_{\mathrm{B}}/2}^{w_{\mathrm{b}}/2} p\mathrm{d}x' = \frac{j\cdot\tau_{\mathrm{HL}}}{q\cdot w_{\mathrm{B}}}$$

For real p^+- and n^+-regions with emitter efficiencies < 1, the boundary conditions change and the factors in Eq. (5.27) are no longer valid. The carrier distribution in the base is written in this case suitably in the form:

$$p(x) = \frac{1}{\sinh(w_{\mathrm{B}}/L_{\mathrm{A}})}\left(p_{\mathrm{R}}\sinh\left(\frac{x}{L_{\mathrm{A}}}\right) + p_{\mathrm{L}}\sinh\left(\frac{w_{\mathrm{B}}-x}{L_{\mathrm{A}}}\right)\right) \tag{5.28}$$

$p_{\mathrm{L}}, p_{\mathrm{R}}$ are the concentrations at the left and right edges of the base region and have to be determined from the general boundary conditions (see Sect. 5.4.5).

5.4.2 *Junction Voltages*

In a p^+nn^+-structure one has a space charge region at the p^+n-junction and another at the nn^+-doping step, whereby both are connected with a built-in voltage $V_{\mathrm{bi}}(p^+n)$ and $V_{\mathrm{bi}}(nn^+)$, respectively. If a forward voltage V_{F} is applied, a part of it is used at the junctions to reduce the potential steps there and to raise the injected carrier densities in the base region similarly as for a single pn-junction. Additionally, the forward voltage provides for an ohmic voltage drop V_{drift} over the weakly doped base region needed for the current transport. Hence, if the junction parts are called $V_{\mathrm{j}}(p^+n)$, $V_{\mathrm{j}}(nn^+)$, one obtains

$$V_{\mathrm{F}} = V_{\mathrm{j}}(p^+n) + V_{\mathrm{drift}} + V_{\mathrm{j}}(nn^+) \tag{5.29}$$

The internal voltage steps at the junctions

$$\Delta V(\mathrm{p^+n}) = V_{\mathrm{bi}}(\mathrm{p^+n}) - V_{\mathrm{j}}(\mathrm{p^+n}),$$
$$\Delta V(\mathrm{nn^+}) = V_{\mathrm{bi}}(\mathrm{nn^+}) - V_{\mathrm{j}}(\mathrm{nn^+})$$

are related to the carrier concentrations at the neutral boundaries of the space charge regions via Boltzmann factors. For the hole concentration p_{L} near the $\mathrm{p^+}$n-junction in the neutral base and the electron density n_{R} at the $\mathrm{n^+}$-side of the base (see Fig. 5.6), one obtains

$$\frac{p_{\mathrm{L}}}{\mathrm{p^+}} = \exp\left(-\frac{q \cdot \Delta V(\mathrm{p^+n})}{kT}\right), \quad \frac{n_{\mathrm{R}}}{\mathrm{n^+}} = \exp\left(-\frac{q \cdot \Delta V(\mathrm{nn^+})}{kT}\right) \tag{5.30}$$

where the carrier densities $\mathrm{p^+}$, $\mathrm{n^+}$ of the highly doped regions are given by the doping densities N_{A}, N_{D}. Dividing the relationships (5.30) by the corresponding thermal equilibrium equations (see Chap. 3):

$$\frac{p_{\mathrm{n0}}}{\mathrm{p^+}} = \exp\left(-\frac{q \cdot V_{\mathrm{bi}}(\mathrm{p^+n})}{kT}\right), \quad \frac{N_{\mathrm{D}}}{\mathrm{n^+}} = \exp\left(-\frac{q \cdot V_{\mathrm{bi}}(\mathrm{nn^+})}{kT}\right)$$

one obtains for the external parts to the voltage drop at the junctions:

$$\begin{aligned} V_{\mathrm{j}}(\mathrm{p^+n}) &= \frac{kT}{q} \ln \frac{p_{\mathrm{L}}}{p_{\mathrm{n0}}} = \frac{kT}{q} \ln \frac{p_{\mathrm{L}} \cdot N_{\mathrm{D}}}{n_{\mathrm{i}}^2} \\ V_{\mathrm{j}}(\mathrm{nn^+}) &= \frac{kT}{q} \ln \frac{n_{\mathrm{R}}}{N_{\mathrm{D}}} \end{aligned} \tag{5.31}$$

$V_{\mathrm{j}}(\mathrm{p^+n})$ and $V_{\mathrm{j}}(\mathrm{nn^+})$ differ significantly from one another depending on the doping density N_{D} of the base region. By leveling out the strong difference of built-in voltages, they make the internal voltage steps $\Delta V(\mathrm{p^+n})$ and $\Delta V(\mathrm{nn^+})$ more similar. The sum of both, the total external junction voltage V_{j}, does not depend on N_{D}:

$$V_{\mathrm{j}} \equiv V_{\mathrm{j}}(\mathrm{p^+n}) + V_{\mathrm{j}}(\mathrm{nn^+}) = \frac{kT}{q} \ln \frac{p_{\mathrm{L}} \cdot n_{\mathrm{R}}}{n_{\mathrm{i}}^2} \tag{5.32}$$

These equations hold independently of the injection level, but they will be used below for high-injection conditions where $p_{\mathrm{L}} = n_{\mathrm{L}}$ and $n_{\mathrm{R}} = p_{\mathrm{R}}$.

For the example of Fig. 5.6 with $p^+ = 2 \times 10^{18}\,\mathrm{cm^{-3}}$, $N_{\mathrm{D}} = 7 \times 10^{13}\,\mathrm{cm^{-3}}$ as doping of the middle layer, and $n^+ = 1 \times 10^{19}\,\mathrm{cm^{-3}}$, one obtains $V_{\mathrm{bi}}(\mathrm{p^+n}) = 0.721\,\mathrm{V}$, $V_{\mathrm{bi}}(\mathrm{nn^+}) = 0.307\,\mathrm{V}$. With p_{L} and p_{R} from the asymmetrical carrier distribution in Fig. 5.6, one attains $V_{\mathrm{j}}(\mathrm{p^+n}) = 0.654\,\mathrm{V}$, $V_{\mathrm{j}}(\mathrm{nn^+}) = 0.161\,\mathrm{V}$, $V_{\mathrm{j}} = 0.815\,\mathrm{V}$. For the internal voltage steps, the results are $\Delta V(\mathrm{p^+n}) = 0.067\,\mathrm{V}$, $\Delta V(\mathrm{nn^+}) = 0.136\,\mathrm{V}$.

5.4.3 Voltage Drop Across the Middle Region

Now the voltage V_{drift} that drops across the middle region has to be calculated. V_{drift} results by integration of the electric field which is given in Eq. (5.20). It contains in the nominator as second term the Dember field, which is proportional to the gradient of the carrier density. This term leads to the Dember voltage V_{Dem}. Using the Einstein relation (2.44), it follows from Eq. (5.20) that

$$V_{\mathrm{Dem}} = \frac{kT}{q} \cdot \frac{\mu_{\mathrm{n}} - \mu_{\mathrm{p}}}{\mu_{\mathrm{n}} + \mu_{\mathrm{p}}} \cdot \ln \frac{p_{\mathrm{L}}}{p_{\mathrm{R}}} \tag{5.33}$$

For a symmetric carrier distribution this voltage term vanishes since $p_{\mathrm{L}} = p_{\mathrm{R}}$. But also for the actual asymmetrical distributions V_{Dem} is very small. For the example in Fig. 5.6, $V_{\mathrm{Dem}} = 14.3\,\mathrm{mV}$ results, which can be neglected.

From the term in Eq. (5.20) which is proportional to the current density, the voltage over the middle region is obtained as

$$V_{\mathrm{drift}} = \frac{j}{q \cdot (\mu_{\mathrm{n}} + \mu_{\mathrm{p}})} \int_{0}^{w_{\mathrm{B}}} \frac{\mathrm{d}x}{p(x)} \tag{5.34}$$

For a homogeneous distribution $p(x) = \mathrm{const}$ the integral would be equal to w_{B}/p. For a not too strongly inhomogeneous distribution, this holds approximately if p is replaced by the average value $\bar{p}$. This leads to

$$V_{\mathrm{drift}} = \frac{j \cdot w_{\mathrm{B}}}{q \cdot (\mu_{\mathrm{n}} + \mu_{\mathrm{p}}) \cdot \bar{p}} \tag{5.35}$$

As can be verified by carrying out the integration in Eq. (5.34) with the exact carrier distribution (5.27) (see below), this is a good approach for $w_{\mathrm{B}}/L_{\mathrm{A}} \leq 3$, but can be used as a rough approximation up to $w_{\mathrm{B}}/L_{\mathrm{A}} = 4$. For $w_{\mathrm{B}}/L_{\mathrm{A}} = 3$ the error made with Eq. (5.35) is only 7%, although the inhomogeneity is already considerable ($\cosh(1.5) = 2.35$). The density of carriers represents a stored charge

$$Q_{\mathrm{F}} = q \cdot A \cdot w_{\mathrm{B}} \cdot \bar{p} \tag{5.36}$$

and, with this, it follows that

$$V_{\mathrm{drift}} = \frac{I_{\mathrm{F}} \cdot w_{\mathrm{B}}^2}{(\mu_{\mathrm{n}} + \mu_{\mathrm{p}}) \cdot Q_{\mathrm{F}}} \tag{5.37}$$

where $I_{\mathrm{F}} = A \cdot j$ denotes the forward current (A is the area of the device).

5.4.4 Voltage Drop in the Hall Approximation

For the further calculation, we assume now as in Sect. 5.4.1 that recombination in the p^+- and n^+-regions is negligible (emitter efficiency 1). This is suggested because the excess charge in the base is orders of magnitude higher than in the end regions (see Fig. 5.7). In the theory of diode characteristic this case is called Hall approximation [Hal52]. Since the minority carrier currents in the end regions are neglected, integration of the continuity equation (5.25) yields for the current density:

$$j = \frac{q \cdot w_B \cdot \bar{p}}{\tau_{HL}} \tag{5.38}$$

Multiplying Eq. (5.38) with $A\tau_{HL}$, one obtains for the stored charge:

$$Q_F = I_F \cdot \tau_{HL} \tag{5.38a}$$

Hence from Eq. (5.37) it follows that

$$V_{drift} = \frac{w_B^2}{(\mu_n + \mu_p) \cdot \tau_{HL}} \quad \text{for} \quad w_B/L_A < 3 \tag{5.39}$$

Expressing τ_{HL} by the ambipolar diffusion length L_A defined in Eq. (5.24) together with Eq. (5.23), Eq. (5.39) can be written in the form

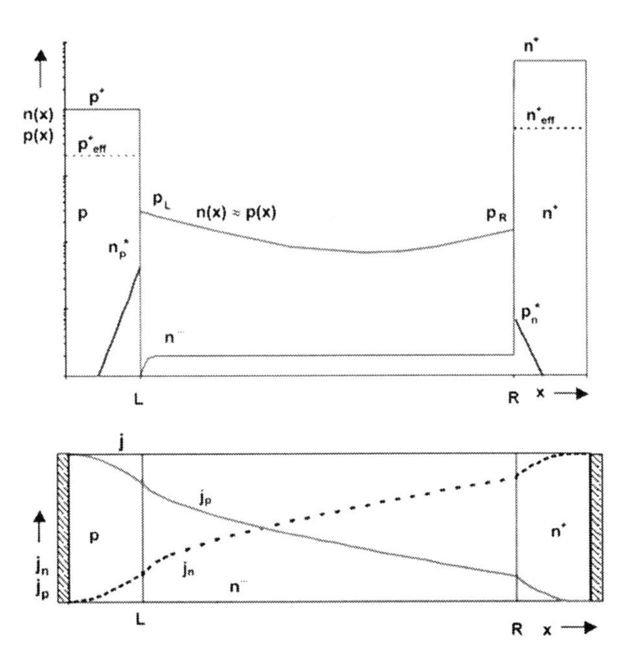

Fig. 5.7 pin-Diode with consideration of recombination in the border regions

$$V_{\text{drift}} = f(b) \cdot \frac{kT}{q} \left(\frac{w_B}{L_A} \right)^2 \quad \text{with} \quad f(b) = \frac{2\,b}{(1+b)^2} \tag{5.40}$$

where $b = \mu_n/\mu_p$. Equation (5.40) can be found in many textbooks on device physics. The factor $f(b)$ is always $\leq 1/2$. For silicon, the mobility ratio is $b \approx 3$ which yields $f(b) \approx 3/8$. Like Eq. (5.37), Eqs. (5.39) and (5.40) apply up to about $w_B/L_A = 4$. Even for $w_B = 4L_A$, V_{drift} amounts only to 0.16 V according to Eq. (5.40).

For a more pronounced sagging of the carrier distribution, the voltage drop is mainly produced by the region of lowest concentration, hence V_{drift} is underestimated by Eqs. (5.39) and (5.40). The accurate formula for V_{drift} is obtained by carrying out the integral in Eq. (5.34) with the exact distribution (5.27). With $d = w_B/2$ the result can be written as

$$V_{\text{drift}} = 4f(b) \cdot \frac{kT}{q} \cdot \sinh\left(\frac{d}{L_A}\right) \cdot \cosh\left(\frac{\Delta}{L_A}\right) \cdot \operatorname{arctg}\left(\frac{\sinh(d/L_A)}{\cosh(\Delta/L_A)}\right) \tag{5.41}$$

where Δ denotes the distance of the minimum of $p(x)$ from the middle of the base. The asymmetry enhances the voltage drop a little. In the limit $d/L_A >> 1$, Eq. (5.40) turns into

$$V_{\text{drift}} = \pi f(b) \cdot \frac{k\,T}{q} \cdot \cosh\left(\frac{\Delta}{L_A}\right) \exp\left(\frac{d}{L_A}\right) \quad \text{for } w_B >> 2 \cdot L_A \tag{5.41a}$$

According to Eqs. (5.39) to (5.41), the voltage drop across the middle region does not depend (explicitly) on the current. The increase of current is neutralized by a proportional increase of the carrier concentration and so the ratio $j/\bar{p}$ in (5.35) is constant. With mobilities and carrier lifetime in the base assumed constant, this implies a current-independent voltage V_{drift}. Actually, the decrease of μ_n and μ_p causes an increase of V_{drift} with current. According to (5.38), the concentration $\bar{p}$ reaches very high values even at normal forward current densities. At such high concentrations, also the *lifetime* is significantly reduced because of Auger recombination. These effects together with the slight increase of the junction voltage according to Eq. (5.32) lead to an appreciable increase of the forward voltage with current. In the range where the Auger recombination predominates strongly and hence the lifetime in the base is very inhomogeneous, however, the analytical approach above becomes insufficient.

However, the application range of the equations is anyway restricted to much smaller concentrations and current densities. As mentioned, they are based on the condition of negligible recombination in the emitter regions, and this becomes significant at a decade smaller $\bar{p}$ than the Auger recombination. Typically the Hall approximation applies up to a current density of about 5 to 30 A/cm^2, depending on the lifetime and width of the base region. Above this limit the experimental forward voltage (see Fig. 5.2) is noticeably higher and increases stronger with current than calculated. This is supported by the observation that the stored charge increases essentially slower with current in the relevant range than according to the linear dependency (5.38a). In the next section, the theory will be expanded taking into account the emitter recombination, which resolves these discrepancies.

That the cuductivity of the base region is enhanced very strongly by the high injection remains valid. Due to this conductivity modulation, even high-voltage devices with their wide, weakly doped base region can be operated with high forward current densities without causing a very high forward voltage.

5.4.5 *Emitter Recombination, Effective Carrier Lifetime, and Forward Characteristic*

The influence of emitter recombination on the injection efficiency of pn-junctions has been described in detail in Section (3.4). Using (3.93) for the junctions of a forward-biased pin-diode, a minority hole current $j_p(n^+) = q \cdot h_n \cdot p_R^2$ flows into the n^+-region, and an electron current $j_n(p^+) = qh_p \cdot p_L^2$ is flowing in the p^+-region, where h_p and h_n are the constants of the respective emitter region given by Eq. (3.101) (if recombination at the contact is neglected). In the case of Fig. 5.6, these currents amount to $j_n(p^+) \approx 32\,A/cm^2$ and $j_p(n^+) \approx 8\,A/cm^2$, if h_p and h_n have a typical value of $2 \times 10^{-14} cm^4/s$ (see Fig. 3.21). Hence, the current due to recombination in the end regions amounts to 25% of the base recombination current in this case where the lifetime in the base is very small (0.48 μs). For a higher τ_{HL}, higher current density or intentionally reduced emitter efficiency, the emitter recombination current can predominate.

To calculate the influence of emitter recombination on the forward characteristics of a pin-diode, we introduce an effective carrier lifetime τ_{eff} by [Sco69]

$$\frac{\int_{-\infty}^{\infty} \Delta p \quad dx}{\tau_{eff}} = \int_{-\infty}^{\infty} \frac{\Delta p}{\tau_p} dx \tag{5.42}$$

By this definition, τ_{eff} is a mean carrier lifetime of the structure including the emitter recombination. The integration extends from a point deep in the p^+-region ($x = -\infty$) over the base to a point deep in the n^+-layer ($x = \infty$). In the base where the injection level is high, the excess hole concentration Δp is equal to $p = n$, and in the p^+-region the recombination rate $\Delta p/\tau_p$ can be equated to the minority recombination rate $\Delta n/\tau_n(n^+) \approx n/\tau_p$. To realize the importance of the effective lifetime for device characteristics, we use the continuity equation (2.84) which in one-dimensional form can be written as

$$-\frac{\partial j_p}{\partial x} = q \cdot \frac{\Delta p}{\tau_p} + q \cdot \frac{\partial \Delta p}{\partial t} \tag{5.43}$$

Since the hole current deep in the p^+-region equals the total current ($j_p(-\infty) = j$), and deep in the n^+-region is zero ($j_p(\infty) = 0$), the integration of Eq. (5.43) yields

$$j = q \cdot \int_{-\infty}^{\infty} \frac{\Delta p}{\tau_p} \mathrm{d}x + q \cdot \frac{\mathrm{d}}{\mathrm{d}t} \int_{-\infty}^{\infty} \Delta p \, \mathrm{d}x \tag{5.44}$$

Inserting Eq. (5.42) and multiplying with the area, one obtains

$$I = \frac{Q}{\tau_{\mathrm{eff}}} + \frac{\mathrm{d}Q}{\mathrm{d}t} \tag{5.45}$$

where I denotes the current and Q the stored charge of excess carriers: $Q \equiv qA \int \Delta p \mathrm{d}x$. Equation (5.45) is a generally valid equation of charge dynamics. For a stationary forward current I_F, it takes the form:

$$Q_F = I_F \cdot \tau_{\mathrm{eff}} \tag{5.46}$$

where Q_F is the stored charge for this special case. According to Eq. (5.46), the effective lifetime can be directly determined by measuring Q_F for a given forward current I_F. Compared with Eq. (5.38a), the lifetime τ_{HL} in the base is replaced by the effective lifetime of the structure in Eq. (5.46). When the current is interrupted, τ_{eff} represents the decay time of the stored charge, since for $I_F = 0$ Eq. (5.45) yields $\mathrm{d}Q/\mathrm{d}t = -Q/\tau_{\mathrm{eff}}$.

Immediately important for the *I–V* characteristic is that within the widely applicable approximations (5.35) and (5.37), the effective lifetime determines the voltage drop across the middle region. By insertion of Eq. (5.46) in Eq. (5.37) one obtains

$$V_{\mathrm{drift}} = \frac{w_B^2}{(\mu_n + \mu_p) \cdot \tau_{\mathrm{eff}}} \tag{5.47}$$

This equation is the generalization of Eq. (5.39) for the real case that the emitter efficiency of the junctions is below unity. Equation (5.47) is applicable if $w_B < 4 \cdot \sqrt{D_A \cdot \tau_{HL}}$ (see the discussion to Eq. (5.35)).

We evaluate the effective lifetime now in dependence on device parameters and on the stored charge or the mean concentration $\bar{p}$ in the base region. By splitting up the integration interval on the right-hand side of Eq. (5.42) into the three neutral regions with constant lifetime one obtains (see Fig. 5.7)

$$\frac{1}{\tau_{\mathrm{eff}}} \int_{-\infty}^{\infty} \Delta p \, \mathrm{d}x = \frac{1}{\tau_n} \int_{-\infty}^{x_p} n \, \mathrm{d}x + \frac{1}{\tau_{HL}} \int_{L}^{R} p \, \mathrm{d}x + \frac{1}{\tau_p} \int_{x_n}^{\infty} p \, \mathrm{d}x \tag{5.48}$$

The equilibrium minority carrier concentrations and likewise the contributions of the space charge layers from x_p to L and from R to x_n are neglected on the right-hand side. Since the integrals are proportional to the respective stored charges, Eq. (5.48) can be written as

$$\frac{Q}{\tau_{\mathrm{eff}}} = \frac{Q_n(\mathrm{p}^+)}{\tau_n(\mathrm{p}^+)} + \frac{Q_B}{\tau_{HL}} + \frac{Q_p(\mathrm{n}^+)}{\tau_p(\mathrm{n}^+)} \tag{5.49}$$

where Q_B denotes the stored charge in the base, $Q_n(p^+), Q_p(n^+)$ are the stored charges of minority carriers in the p$^+$- and n$^+$-regions, respectively, and $Q = Q_B + Q_n(p^+) + Q_p(n^+)$ denotes the total stored charge. Because of the low injection in the end regions and the relative small minority carrier diffusion length, the stored charges $Q_n(p^+), Q_p(n^+)$ are small compared with the stored charge $Q_B = q \cdot w_B \cdot \bar{p}$, if the base width is not too small and the injection level not extremely high. Equation (5.49) shows that in spite of the small stored charges $Q_n(p^+), Q_p(n^+)$, the *recombination* in the end regions can be significant if the lifetimes $\tau_n(p^+)$, $\tau_p(n^+)$ are correspondingly smaller than τ_{HL}. The latter can be caused by Auger recombination, a high density of recombination centers in the outer layers (see the discussion to Eq. (3.108)) or by a design of the end regions leading to high surface recombination.

Insertion of the exponential minority carrier distribution in the first and third integrals on the right side of Eq. (5.48) yields the connection with the emitter parameters introduced in Sect. 3.4. For the n$^+$-layer, we obtain with Eqs. (3.43), (3.100) and (3.101) and neglecting the equilibrium density p_{n0}:

$$\frac{1}{\tau_p}\int_{x_n}^{\infty} p\,\mathrm{d}x = \frac{L_p}{\tau_p}(n^+)\cdot p_n^* = \frac{L_p}{\tau_p}\cdot\frac{p_R^2}{n^+}e^{\Delta E_g/kT} = h_n \cdot p_R^2 \tag{5.50}$$

The bandgap narrowing ΔE_g results in an enhancement of the minority carrier concentration $p_n{}^*$ and hence of the emitter parameter h_n. The analogous equation holds for the recombination integral over the p$^+$-region (first term on the right-hand side of Eq. (5.48)):

$$\frac{1}{\tau_n}\int_{-\infty}^{L} n\mathrm{d}x = \frac{L_n(p)}{\tau_n(p)}\cdot n_p^* = \frac{L_n}{\tau_n}\cdot\frac{p_L^2}{p^+}e^{\Delta E_g/kT} = h_p \cdot p_L^2 \tag{5.51}$$

If the integral on the left side in Eq. (5.48) is approximated by $w_B \cdot \bar{p}$, neglecting the stored minority carrier charges in the end regions, one obtains from Eq. (5.48)

$$\frac{1}{\tau_{eff}} = \frac{1}{\tau_{HL}} + h_p \cdot \frac{p_L^2}{w_B \cdot \bar{p}} + h_n \cdot \frac{p_R^2}{w_B \cdot \bar{p}} \tag{5.52}$$

To correlate the mean concentration $\bar{p}$ with the concentrations p_L and p_R at the boundaries, the carrier distribution is used in the form (5.28). One obtains

$$\bar{p} = \frac{1}{w_B}\int_{L}^{R} p\,\mathrm{d}x = \frac{L_A}{w_B}\cdot\tanh\left(\frac{d}{L_A}\right)\cdot(p_L + p_R) \tag{5.53}$$

For simpler writing we use again the letter d for $w_B/2$. Using Eq. (5.53), Eq. (5.52) can be written as

$$\frac{1}{\tau_{eff}} = \frac{1}{\tau_{HL}} + \frac{H}{d}\cdot\left(\frac{d/L_A)}{\tanh\left(d/(L_A)\right)}\right)^2\cdot\bar{p} \tag{5.54}$$

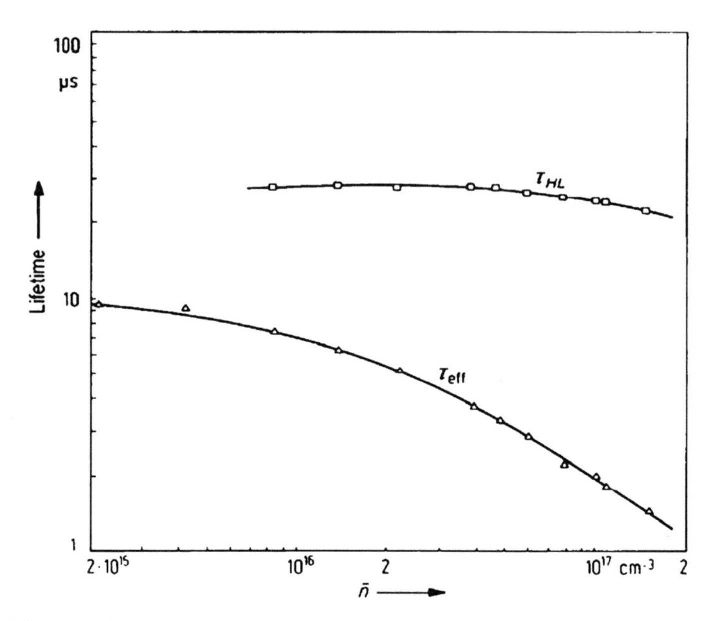

Fig. 5.8 Effective lifetime τ_{eff} of a forward-biased pin-diode and lifetime τ_{HL} in the base region as functions of the mean carrier concentration in the base. See the text. From [Sco82]

where

$$H = 2 \cdot \frac{\eta^2 h_\text{p} + h_\text{n}}{(\eta + 1)^2} \tag{5.55}$$

with $\eta = p_\text{L}/p_\text{R}$. The quantity H corresponds to a first approximation often independent of $\bar{p}$. For a symmetrical distribution ($\eta = 1$), H reduces to $(h_\text{n} + h_\text{p})/2$. With a typical H-value of $2 \times 10^{-14}\,\text{cm}^4/\text{s}$, one obtains from Eq. (5.54) for a base width $w_\text{B} = 200\,\mu\text{m}$ at $\bar{n} = \bar{p} = 1 \times 10^{17}\text{cm}^{-3}$:

$$\frac{1}{\tau_\text{eff}} = \frac{1}{\tau_\text{HL}} + \frac{1}{5.0\,\mu\text{s}} \left(\frac{d/L_\text{A}}{\tanh(d/L_\text{A})} \right)^2 \quad \text{at}\, \bar{p} = 1 \times 10^{17}\,\text{cm}^{-3}$$

Even if the carrier lifetime τ_HL is very high, the effective carrier lifetime remains below 5 μs in this case, the value given by the recombination in the border regions for $w_\text{B}/L_\text{A} \to 0$.

Measurements of the effective lifetime τ_eff and the high-level lifetime τ_HL in the base are plotted in Fig. 5.8 as functions of the mean concentration $\bar{n} = \bar{p}$ in the base. $\tau_\text{eff} = Q_\text{F}/I_\text{F}$ was determined by measuring the stored charge Q_F, which simultaneously delivers the concentration $\bar{n}$. The base lifetime τ_HL was determined from the carrier distribution measured via the recombination radiation profile. While τ_HL is found to be nearly independent of the carrier concentration, τ_eff decreases by an order of magnitude in the shown range. This is in agreement with the above theory; particularly the variation of τ_eff with $\bar{n}$ can be described by Eqs. (5.52) and

(5.54). A point not in agreement with these equations is that τ_{eff} does not tend to the base lifetime τ_{HL} for small $\bar{n}$, but remains considerably smaller. We will come back to this effect at the end of this paragraph.

The boundary concentrations p_L, p_R depend strongly on the parameters h_p, h_n. In some modern fast power diodes, the injection efficiency of the p-emitter is made small while the n^+-region keeps the usual form with a small parameter h_n. This leads to an inversion of the carrier distribution compared with Fig. 5.6 and by means of this to an improved reverse recovery behavior. The influence of the h-values on the carrier distribution follows from the continuity of the electron current at the p^+n-junction and the hole current at the nn^+-junction. Using Eqs. (5.21) and (5.22) together with Eq. (5.28) the boundary conditions can be written as

$$\frac{D_n}{D_n + D_p} j - q \frac{D_A}{L_A} \left(\frac{p_L}{\tanh(w_B/L_A)} - \frac{p_R}{\sinh(w_B/L_A)} \right) = q\, h_p p_L^2 \tag{5.56a}$$

$$\frac{D_p}{D_n + D_p} j + q \frac{D_A}{L_A} \left(\frac{p_L}{\sinh(w_B/L_A)} - \frac{p_R}{\tanh(w_B/L_A)} \right) = q\, h_n\, p_R^2 \tag{5.56b}$$

Without solving these equations explicitly for p_L and p_R, one can see that p_L becomes smaller if h_p is enhanced. In the limit of very large emitter recombination (right-hand side) compared with the recombination in the base, the ratio $\eta = p_L/p_R$ tends to $(D_n h_n/(D_p h_p))^{1/2}$, as follows by division of Eqs. (5.56a) and (5.56b). A calculation of η as a function of $\bar{p}$ and the device parameters h_p, h_n, w_B, and L_A is given in [Sco69]. The quantity H, which according to Eq. (5.54) determines the effective lifetime and hence the voltage drop V_{drift} varies less than h_p because η decreases with increasing h_p. A measured carrier profile for a 1200 V diode using the mentioned design principle is shown later in Fig. 5.33 in Sect. 5.7.4. In this diode the p-emitter layer had a doping $N_A = 5\times10^{16}\,\text{cm}^{-3}$ and a thickness $w_p = 2\,\mu\text{m}$, which according to Eq. (3.106) results in $h_p = 2.6\times10^{-12}\,\text{cm}^4/\text{s}$. The n^+-region had a doping concentration of about $10^{19}\,\text{cm}^{-3}$ and an estimated h_n - value of $2\times10^{-14}\,\text{cm}^4/\text{s}$. As shown by the figure, the boundary concentration p_L is about a factor 4 smaller than p_R. However, the recombination term $h_p p_L^2$ in Eq. (5.52) is still seven times higher than $h_n p_R^2$.

From Eqs. (5.46) and (5.54), the current density can be expressed as a function of the concentration $\bar{p}$:

$$j = \frac{Q_F/A}{\tau_{eff}} = \frac{q \cdot w_B \cdot \bar{p}}{\tau_{eff}} = q \cdot \bar{p} \cdot \left(\frac{w_B}{\tau_{HL}} + 2 \cdot H \cdot \left(\frac{d/L_A}{\tanh(d/L_A)} \right)^2 \cdot \bar{p} \right) \tag{5.57}$$

The voltage drop across the base follows by insertion of Eq. (5.54) into Eq. (5.47)

$$V_{drift} = \frac{w_B}{\mu_n + \mu_p} \cdot \left(\frac{w_B}{\tau_{HL}} + 2H \cdot \left(\frac{d/L_A}{\tanh(d/L_A)} \right)^2 \cdot \bar{p} \right) \tag{5.58}$$

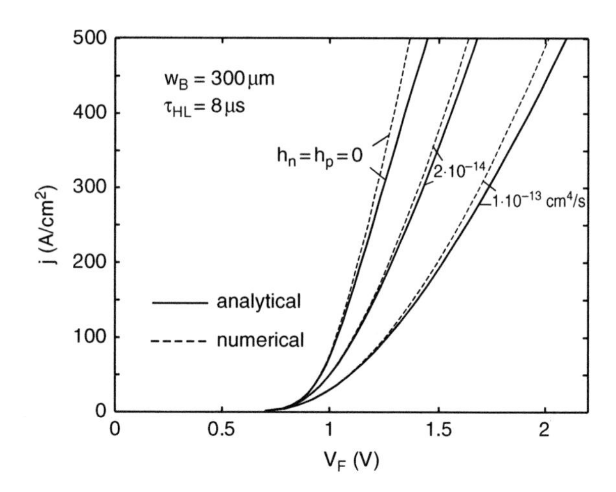

Fig. 5.9 Current–voltage characteristics calculated with different emitter parameters

The junction voltage Eq. (5.32) can be written using Eq. (5.53):

$$V_{\mathrm{j}} = 2\,\frac{kT}{q}\,\ln\left(\frac{2\sqrt{\eta}}{1+\eta}\cdot\frac{d/L_{\mathrm{A}}}{\tanh(d/L_{\mathrm{A}})}\cdot\frac{\bar{p}}{n_{\mathrm{i}}}\right) \tag{5.59}$$

where as before $\eta = p_{\mathrm{L}}/p_{\mathrm{R}}$. With Eqs. (5.57), (5.58), and (5.59) the current–voltage characteristic is given in a parameter representation $j(\bar{p})$, $V_{\mathrm{F}}(\bar{p}) = V_{\mathrm{j}}(\bar{p}) + V_{\mathrm{drift}}(\bar{p})$. This form of the characteristic has the advantage that also the mobilities and ambipolar diffusion constant as well as the lifetime in L_{A} can be inserted as functions of $\bar{p}$ to consider carrier–carrier scattering and, if necessary, Auger recombination.

Forward characteristics calculated with these equations for different values of h_{p}, h_{n} are shown in Fig. 5.9 (solid curves). The dimensions of the base region are suitable for a 3 kV-diode. Carrier-carrier scattering is taken into account as described in chapter 2. The neglect in the step from (5.34) to (5.35) is abandoned using the inhomogeneity factor given by the later equation (5.63). In the case $h_{\mathrm{n}} = h_{\mathrm{p}} = 2\cdot 10^{-14}$ cm^4/s a realistic characteristic for diodes with these dimensions is obtained. For $h_{\mathrm{n}} = h_{\mathrm{p}} = 0$ (Hall case), the concentration $\bar{p}$ runs so high at high current densities that Auger recombination in the base becomes very significant. Within the above model, Auger recombination is considered replacing $1/\tau_{\mathrm{HL}}$ by $1/\tau_{HL} + (c_{A,n} + c_{A,p})\cdot \bar{p}^2$ (see Eq. (2.56)). To test the accuracy of the model, also characteristics calculated numerically are plotted in Fig. 5.9 (dotted lines). They were obtained by determining the carrier distribution for each current density from the differential equation (5.25) using a lifetime including Auger recombination and variable mobilities in (5.21); the mobility sum in (5.34) was pulled under the integral. As is seen, the analytical approach is fairly accurate. Above the current range of the figure and in the Hall case, the analytical model is no longer suited. However, the Hall case is anyway hypothetical.

Sometimes the direct relationship between forward voltage and current density is desirable. To derive it, Eq. (5.57) is resolved first for $\bar{p}$, which yields

$$\frac{1}{\bar{p}} = \frac{q\, w_B}{2\, j\, \tau_{HL}} \left(1 + \sqrt{1 + \frac{2\, \tau_{HL}\, H}{q\, D_A\, \tanh^2(d/L_A)} \cdot j} \right) \tag{5.60}$$

Inserting this into Eq. (5.35) one obtains for the voltage drop across the middle region:

$$V_{drift} = \frac{w_B^2}{2\,(\mu_n + \mu_p)\, \tau_{HL}} \left(1 + \sqrt{1 + \frac{2\, \tau_{HL}\, H}{q\, D_A\, \tanh^2(d/L_A)} \cdot j} \right) \tag{5.61}$$

The junction voltage as a function of current density is obtained by insertion of Eq. (5.60) into Eq. (5.59). Hence the forward voltage $V_F = V_j + V_{drift}$ is given now as a function of the current density j.

Contrary to the Hall approximation (5.39), the voltage drop V_{drift} depends now explicitly on the current density. For a sufficiently small emitter quantity H or small current density j, Eq. (5.61) reduces to Eq. (5.39). Since H cannot be made smaller than about $1 \times 10^{-14}\,\mathrm{cm^4/s}$, however, the current-dependent term in Eq. (5.61) becomes soon significant. The reference current density

$$j_0 \equiv \frac{q\, D_A\, \tanh^2(d/L_A)}{2\, \tau_{HL}\, H}$$

around which the current density j comes into play is fairly small. If for example $H = 2 \times 10^{-14}\,\mathrm{cm^4/s}$, $d = 100\,\mu\mathrm{m}$, $\tau_{HL} = 4\,\mu\mathrm{s}$, $D_A = 15\,\mathrm{cm^2/s}$, j_0 amounts only to 11.1 A/cm^2. If H is made large for above-mentioned reasons, j_0 will be still smaller. For $j >> j_0$, Eq. (5.61) turns into

$$V_{drift} = \frac{w_B^2}{(\mu_n + \mu_p)\, L_A\, \tanh(d/L_A)\, \sqrt{2}} \cdot \sqrt{H \cdot j/q} \tag{5.62}$$

V_{drift} is here proportional to the square root of the current density, as far as the variation of the mobilities can be neglected. $\bar{p}$ is proportional to $\sqrt{j}$ in this case according to Eq. (5.60); hence, $j/\bar{p}$ in Eq. (5.35) is also proportional to $\sqrt{j}$. This

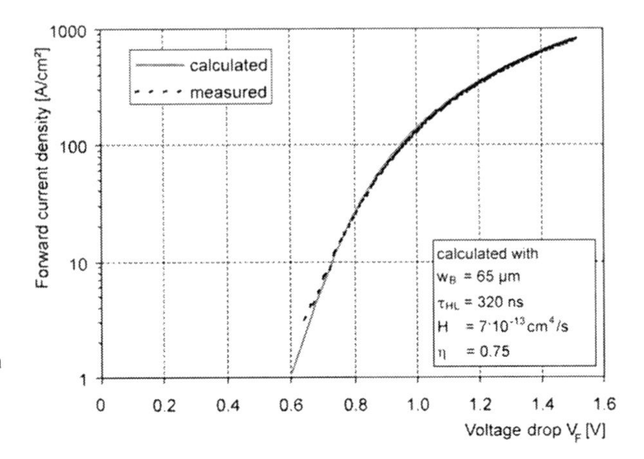

Fig. 5.10 Measured forward characteristics of a fast 600 V diode compared to calculation results with consideration of recombination in the emitter regions

relationship is in accordance with measurements, which often yield a dependency $V_F - V_j$ proportional to $\sqrt{j}$ for power diodes.

Figure 5.10 shows the measurement of the forward characteristics of a fast 600 V diode and the result of a calculation with Eqs. (5.59), (5.60), and (5.61). For the dependency of μ_n, μ_p, D_A on the carrier concentration, the equations given in Appendix A are used. The calculation agrees with the measurement within a wide range of the current density.

If the carrier distribution is very inhomogeneous, i.e. very unsymmetrical or if it is characterized by a $d/L_A > 2$, the approximations (5.35) or (5.47), underlying our calculation, underestimates the voltage drop V_{drift} considerably. Carrying out the integral in Eq. (5.34) with the exact carrier distribution $p(x') = \cosh((x' - \Delta)/L_A$ results in a voltage drop V_{drift} which has been given for the Hall case by Eq. (5.41). According to this equation, V_{drift} is by the factor

$$f = \left(\frac{L_A}{d}\right)^2 \cdot \sinh\left(\frac{d}{L_A}\right) \cdot \cosh\left(\frac{\Delta}{L_A}\right) \cdot \mathrm{arctg}\left(\frac{\sinh(d/L_A)}{\cosh(\Delta/L_A)}\right) \tag{5.63}$$

higher than given by the approximation (5.40) (Δ denotes the distance of the minimum of $p(x)$ from the middle of the base). Since the carrier distribution has the same general cosh-form also if emitter recombination is present, the inhomogeneity factor (5.63) applies generally and has to be added correctly to the V_{drift} expressions (5.47), (5.58), (5.61), and (5.62). In Fig. 5.9 the factor has been taken into account.

The above equations can be used also for IGBTs to describe the *I–V* characteristics in the saturated current range. The conduction behavior of several types of IGBTs is mainly determined by the p-emitter, since very shallow and low-doped emitter regions are implemented. An IGBT characteristic will be shown later in Fig. 10.2. In the range above the rated current one finds often a nearly linear or resistive characteristic.

In Fig. 5.8 it is observed that τ_{eff} for small $\bar{n}$, in the range of 10^{15} cm^{-3}, is nearly constant like the lifetime τ_{HL} in the base, but is essentially smaller than τ_{HL}. This result has been found for all samples till now. It has led to the conclusion that there is a linear recombination part entering the stored charge, which is not located in the volume of the base but must be located in the emitters or thin boundary layers of the base not resolved by the radiation profiles [Sco79, Coo83]. This recombination part called "linear emitter recombination" has been mentioned already in Sect. 3.4 and is expressed by Eq. (3.108). In Fig. 5.33 in Sect. 5.7.4, the phenomenon appears again: The lifetime in the volume of the base is higher than given by Q_F/I_F at small $\bar{p}$ where the quadratic emitter recombination is negligible. The sagging of the concentration toward the middle of the base is therefore smaller. For the incorporation of this effect into the theory of the *I–V* characteristic we refer to the literature [Sco79].

Above calculations can be used to implement the parabolic shape of the forward characteristics of modern fast diodes in a circuit simulator. Such simulators need simplified models for the forward characteristics of diodes, and this description by a parabola is in most cases much closer to reality than the description with the Hall approximation, or the often used oversimplified description with a straight line.

5.4.6 Temperature Dependency of the Forward Characteristics

For low current, the forward voltage decreases with temperature, because the part of the voltage that drops at the pn-junction, i.e. the term $V_j(p^+n)$ given in Eq. (5.31), decreases with increasing n_i^2. The diode at low-current condition can thus be used as a temperature sensor (see Chap. 11, Fig. 11.19). At increased current density, the temperature dependency of V_{drift} predominates and here, opposing effects have to be considered:

- The mobilities decrease with temperature (see Fig. 2.13 and Appendix A1). According to Eq. (5.47), this leads to an increase in V_{drift}.
- The carrier lifetime increases with increasing temperature which, in turn, leads to a decrease in V_{drift}.

Since both effects are opposing, the resulting behavior depends on the specific technology, especially of the temperature dependency of the effect of the used recombination centers.

By using radiation-induced recombination centers, a curve as is depicted in Fig. 5.11 on the right-hand side is measured. The lines for the characteristics at 25 and 150°C intersect at 150–200 A/cm^2, a current density which is typical for rated current of a 1200 V fast diode.

Even though an intersection point is likewise found for diodes without recombination centers (rectifier diodes for grid frequency) and for diodes using gold as recombination center, this intersection occurs for those devices typically at a three times higher current density.

By using platinum combined with a not very weakly doped p-emitter, no intersection point is found in the relevant current range. An example is given in Fig. 5.11 (left). In this case, the forward voltage strongly decreases with temperature. This

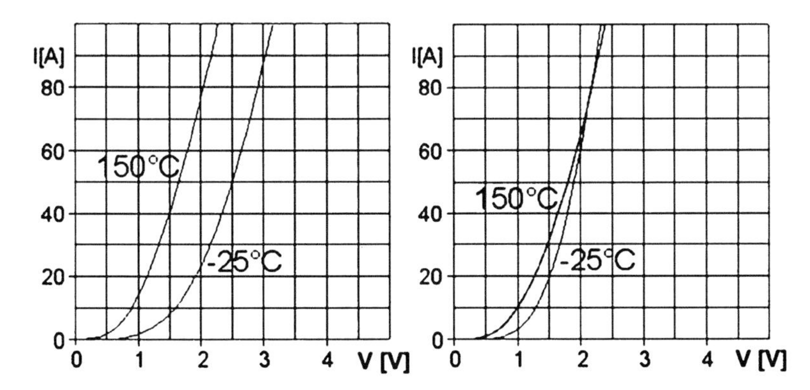

Fig. 5.11 Forward characteristics of fast 1200 V diodes and its temperature dependency. *Left*: platinum-diffused diode. *Right*: diode with radiation-induced recombination centers (CAL diode). Active area is 0.32 cm^2

is advantageous in terms of conduction losses, but this temperature behavior is very unfavorable for parallel connection of diodes. As the result of tolerances in the manufacturing process, there is always some variation of the forward voltage for different samples. In case of parallel connection, the diode with the lower voltage drop will attract more current. Consequently, it will generate higher conduction losses, its temperature will increase. That will result in a further reduction of the forward voltage, so that this diode will attract a further increased fraction of the current, and so on. A pronounced negative temperature dependency with a V_F decrease of more than 2 mV/K endangers thermal instability in parallel connection. A diode like that in Fig. 5.11 (right) is, on the other hand, well suited for parallel connection.

Diodes connected in parallel are coupled thermally:

- via the substrate for paralleling of several diodes in a module,
- via the heat sink with paralleling of modules.

In the case of a weakly negative temperature coefficient, this coupling is usually sufficient to avoid a thermal runaway of the diode with the lowest forward voltage. In the case of diodes with a negative temperature coefficient of < -2 mV/K, it is recommended to decrease the current load in parallel connection to a lower value as would result from the sum of the current ratings of the single diodes. This measure is known as "derating".

5.5 Relation Between Stored Charge and Forward Voltage

Especially with fast diodes, a trade-off between the demands for fast switching – low stored charge, etc. – and for low forward voltage has to be made. If the carrier lifetime is adjusted to be low, then the forward voltage increases according to Eq. (5.61). According to Eq. (5.37), the voltage V_{drift} and the stored charge Q_F are related by the equation:

$$Q_F = \frac{w_B^2 \cdot I_F}{V_{drift}\left(\mu_n + \mu_p\right)} \tag{5.64}$$

The forward voltage has been partitioned into the parts V_{drift} in the base region and the voltage drops $V_j(p^+n)$ at the pn-junction and $V_j(nn^+)$ at the nn$^+$-junction (see Eq. (5.29)). These terms are combined in Eq. (5.32) to the junction voltage V_j, $V_F = V_j + V_{drift}$ Replacing V_{drift}, Eq. (5.64) results in

$$Q_F = \frac{w_B^2 \cdot I_F}{\left(V_F - V_j\right)\left(\mu_n + \mu_p\right)} \tag{5.65}$$

Equation (5.65), like Eq. (5.35), is a good approximation for $w_B/L_A < 4$. This hyperbolic relation is shown in Fig. 5.11. In this figure, Q_F according to Eq. (5.65) is drawn for a fast 600 V diode with $w_B = 65\,\mu$m. For comparison, the experimental results for Q_{RR} of a fast diode with this dimensioning are also shown. Q_{RR}

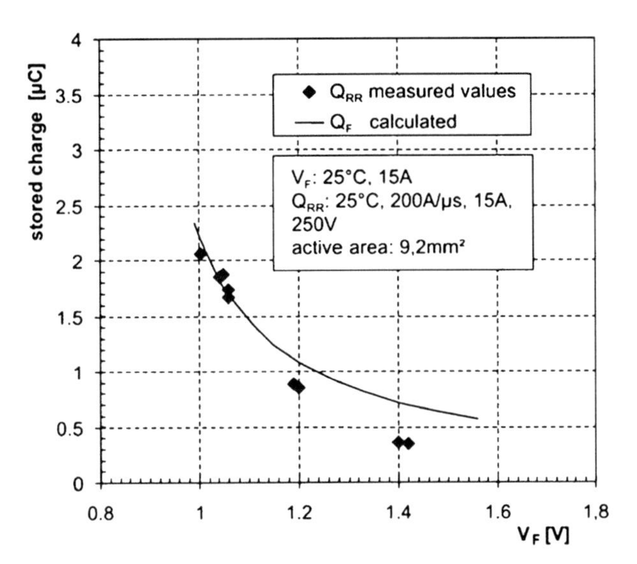

Fig. 5.12 Relation of stored charge to forward voltage for a fast 600 V diode

distinguishes from Q_F by the amount of charge recombining during the measurement duration.

A hyperbolic relation as shown in Fig. 5.12 results for every technology; how far it can be shifted to lower values, however, is an evaluation criterion for the specific design. The base width w_B contributes squared to Q_F as well as to V_F. Therefore, w_B must be kept as low as possible by taking into account all requirements to the diode.

Nevertheless, up to now nothing has been stated about the time-dependent waveform in which the stored charge emerges during the reverse recovery procedure. This, however, is most significant and is to be discussed in the following paragraphs regarding the turn-off behavior. However, first the turn-on behavior will be investigated.

5.6 Turn-On Behavior of Power Diodes

At the transition of the diode into the conducting state, the voltage first increases to the turn-on voltage peak V_{FRM} (forward recovery maximum) before it drops down to the forward voltage. Figure 5.13 shows the definition of V_{FRM} and the turn-on time t_{fr}, in which t_{fr} is defined as the time interval between the instant of 10% forward voltage and the instant at which the voltage has dropped down again to the 1.1-fold value of the steady-state forward voltage.

This old definition comes from a time in which thyristors were the dominating devices in power electronics; low current slopes were usual and V_{FRM} amounted to several volts. This definition does not pertain to freewheeling diodes and snubber diodes used in circuits with IGBTs as switching elements, because with them such steep slopes di/dt of the current occur that V_{FRM} may reach – for example in a poorly

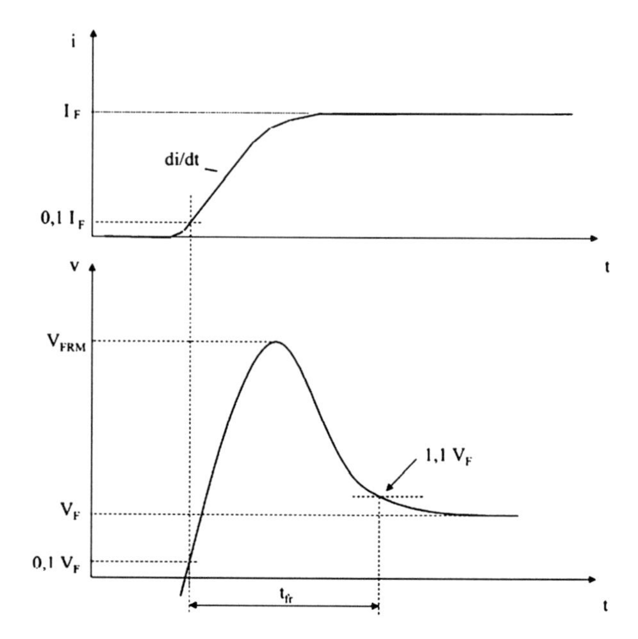

Fig. 5.13 Characteristic parameters of the turn-on behavior of power diodes

designed 1700 V diode – a value of 200 – 300 V which is more than 100 times the value of V_F. Reading out the time at the value of $1.1 \times V_F$ is no longer possible in the measurement.

A low V_{FRM} is one of the most important requirements for diodes in snubber- and clamping circuits, since these circuits work just after the diode has turned on.

The forward recovery voltage peak is an important feature also for freewheeling diodes designed for a blocking voltage of 1200 V and more. At turn-off of the IGBT, the freewheeling diode turns on; the di/dt at the IGBT turn-off creates at the parasitic inductance a voltage peak on which V_{FRM} is superimposed. The sum of both components can lead to a critical voltage peak.

The measurement of this behavior is not trivial, since the inductive component and V_{FRM} cannot be distinguished in an application-conform chopper circuit. The measurement is possible only at an open setup directly at the bond wires of the diode. Such a measurement is depicted in Fig. 5.14. Here, the turn-on of two diodes is shown, one of which (standard diode) is designed with a very wide w_B in order to achieve a soft recovery behavior at turn-off. For the CAL diode presented in comparison, w_B is kept as low as possible. Under the same measurement conditions, i.e. the same parameters for controlling the IGBT turn-off, a V_{FRM} of 84 and 224 V results for the CAL diode and the standard diode, respectively. It is possible to analyze the worst-case scenario. At a step function current form ($\mathrm{d}i/\mathrm{d}t = \infty$), the maximally occurring voltage corresponds to the resistance of the base without carrier injection, multiplied by the current density:

$$V_{FRM} = \frac{w_B \cdot j}{q \cdot \mu_n \cdot N_D} \tag{5.66}$$

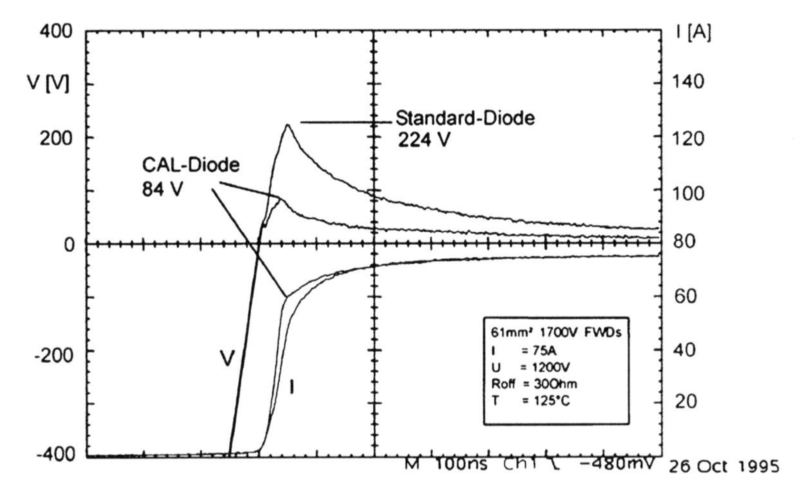

Fig. 5.14 Turn-on of two diodes with different width w_B of the low-doped layer

This equation can be used as long as $j < qN_D v_{sat}$. The design of a diode for higher blocking voltage requires a lower doping N_D and a wider base width w_B, which increases the resulting voltage peak strongly. Figure 5.15 shows the results according to Eq. (5.66), whereby N_D and w_B were selected for fast diodes of different voltage ranges, and the mobility μ_n for $T = 400\,K$ from Fig. 2.12 was used.

For a diode designed for 600 V, a voltage peak amounting to only some 10 V can occur. For a diode designed for 1700 V, however, this peak may amount to more than 200 V, and for a diode designed for the voltage range of > 3000 V, more than 1000 V are possible. Moreover, the effects of recombination centers have not yet

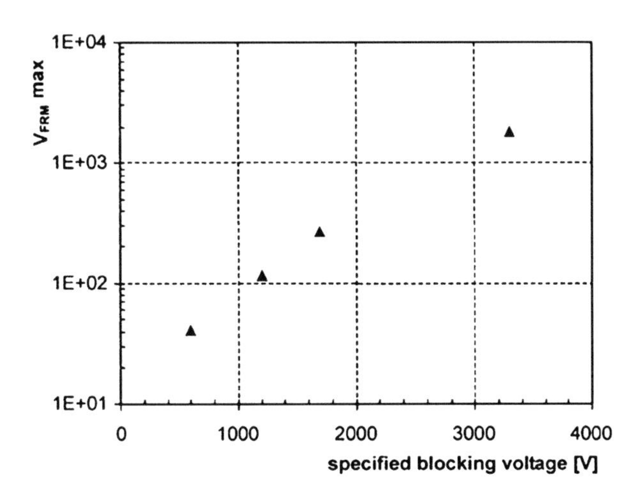

Fig. 5.15 Calculated worst-case voltage peak at turn-on of diodes at $T = 125°C$. Compensation effects are not considered

been considered in this calculation. It is known that the recombination center gold has a compensating effect. The level at $E_C - 0.54$ eV is of acceptor type. In a low-doped n-region, the density of these acceptors compensates a part of the background doping density. The resulting decreased effective doping has to be applied in Eq. (5.66), and the voltage peak can then become significantly higher. In practice, there are no current slopes occurring with the form of a step function; however, current slopes with di/dt in the range of > 2000 A/μs are to be expected in IGBT applications as shown in Fig. 5.14.

The importance of the turn-on behavior of freewheeling diodes was underestimated for a long time. Just after high-voltage IGBTs of > 3000 V were introduced and failures in the application occurred, the anti-parallel diode to the IGBT was found as a reason: if it creates a high forward recovery peak V_{FRM}, this voltage peak is applied to the IGBT in the reverse direction. The reverse blocking capability is not specified for common IGBTs, since the collector-side pn-junction has no defined junction termination. To counteract this problem, more attention was paid to the turn-on of the freewheeling diodes.

Regarding switching losses, the turn-on behavior of the diode is not significant. Even if high voltage peaks occur, the turn-on process is very fast, and turn-on losses amount to only some percent of the turn-off losses or of the conduction losses of the diode. For thermal calculations, turn-on losses can be neglected in verymost cases.

5.7 Reverse Recovery of Power Diodes

5.7.1 Definitions

With the transition from the conducting to the blocking state, the charge stored in a diode has to be removed. This charge causes a current flow in the reverse direction of the diode. The reverse recovery behavior signifies the time-dependent waveforms of this current and the corresponding voltage.

The simplest circuit to measure this effect is the circuit according to Fig. 5.16, whereby S represents an ideal switch, I_F an ideal current source, V_{bat} an ideal voltage source, L an inductor, and D the diode being considered. After closing the switch S, the progression of current and voltage as shown in Fig. 5.17 occurs at a the diode. Figure 5.17 exemplifies a diode with soft recovery behavior, whereas Fig. 5.18 depicts two examples of the current waveform of diodes with a snappy reverse recovery behavior.

First, the definitions will be explained using the circuit in Fig. 5.16 and the waveform in Fig. 5.17. In the circuit of Fig. 5.16, after closing the switch S it holds

$$L \cdot \frac{\mathrm{d}i}{\mathrm{d}t} + v(t) = -V_{\mathrm{bat}} \tag{5.67}$$

where $v(t)$ is the time-dependent voltage at the diode. First, during the current decay, the voltage at the diode is in the range of the forward voltage V_F which is in the range

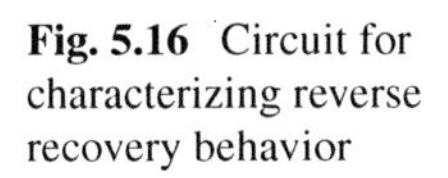

Fig. 5.16 Circuit for characterizing reverse recovery behavior

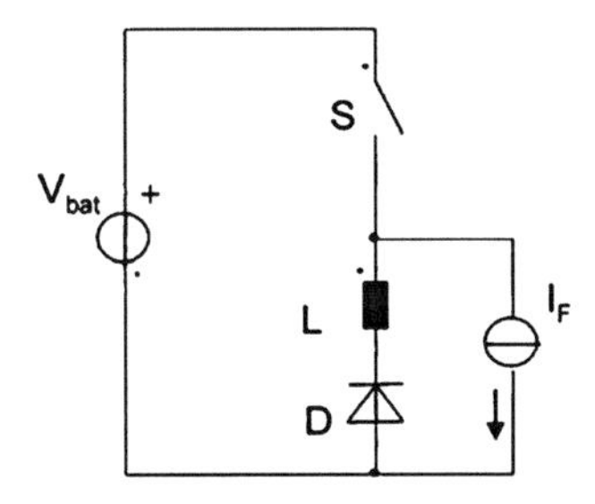

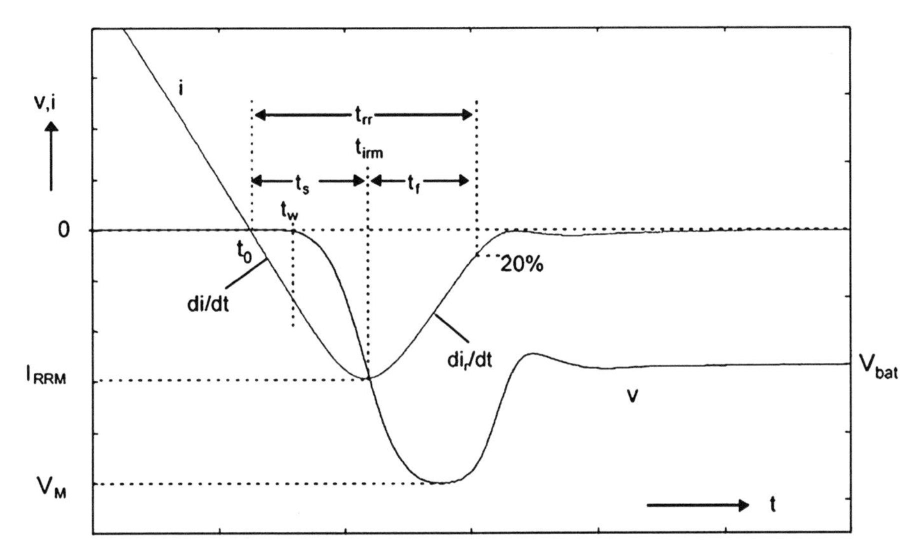

Fig. 5.17 Waveforms of current and voltage for a soft recovery diode during the reverse recovery process in a circuit according to Fig. 5.16 and definitions of some characteristic values of the recovery behavior

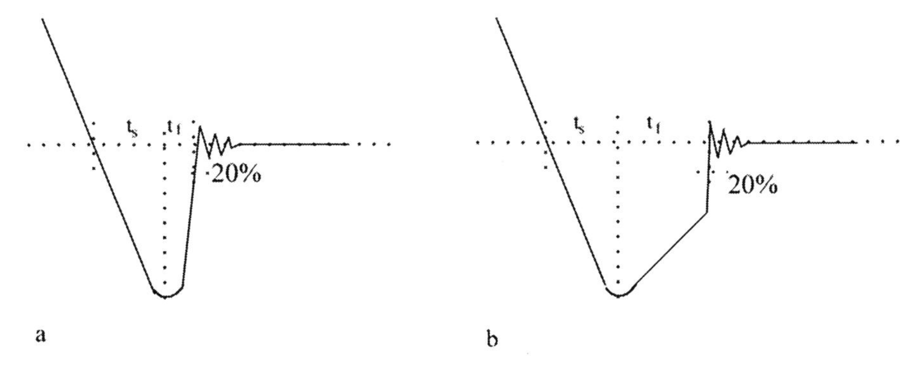

Fig. 5.18 Current waveform for two different possibilities of snappy reverse recovery behavior

of 1 – 2 V, and $v(t)$ can be neglected. The current slope at commutation is determined by the voltage and inductance:

$$-\frac{\mathrm{d}i}{\mathrm{d}t} = \frac{V_{\mathrm{bat}}}{L} \tag{5.68}$$

The zero-crossing point of the current occurs at t_0. At t_{w} the diode starts to take over the voltage; at this instant the pn-junction of the diode is free of charge carriers. At the same point in time, the current deviates from the linear slope. At t_{irm} the reverse current attains its maximum I_{RRM}. At t_{irm}, it holds that $\mathrm{d}i/\mathrm{d}t = 0$ and from Eq. (5.67), $v(t) = -V_{\mathrm{bat}}$ results.

After t_{irm} the reverse current decays down to the level of the static leakage current. The shape during this interval solely depends on the diode. If this decay is steep, a snappy reverse recovery behavior is given. If this decay occurs slowly, however, a soft recovery behavior is indicated. This slope $\mathrm{d}i_{\mathrm{r}}/\mathrm{d}t$, which is often not linear, leads to an induced voltage $L \cdot \mathrm{d}i_{\mathrm{r}}/\mathrm{d}t$ that adds on the battery voltage.

The switching time t_{rr} is defined as the time between t_0 and the point in time at which the current has decayed down to the value of 20% of I_{RRM}. With the subdivision of t_{rr} in t_{f} and t_{s} as is shown in Fig. 5.17, formerly the following "soft factor" s was defined as the quantitative parameter of the reverse recovery behavior:

$$s = \frac{t_{\mathrm{f}}}{t_{\mathrm{s}}} \tag{5.69}$$

whereby, e.g. $s > 0.8$ signified that a diode can be called to be "soft."

This definition, however, is very insufficient. According to it, a current shape as in Fig. 5.18a would be snappy, but a current shape as in Fig. 5.18b would be accepted as soft. While in Fig. 5.18b $t_{\mathrm{f}} > t_{\mathrm{s}}$ is given and $s > 1$ according to Eq. (5.69), a very steep slope, a reverse current snap-off, occurs in a part of the reverse recovery waveform.

The following definition of the soft factor is better:

$$s = \left| \frac{-\left.\frac{\mathrm{d}i}{\mathrm{d}t}\right|_{i=0}}{\left(\frac{\mathrm{d}i_{\mathrm{r}}}{\mathrm{d}t}\right)_{\max}} \right| \tag{5.70}$$

The applied current slope must be measured at the zero-crossing point, and the $\mathrm{d}i_{\mathrm{r}}/\mathrm{d}t$ caused by the diode is measured at its maximal value. The measurement must be executed at less than 10% and at 200% of the specified rated current. For soft recovery the value of $s > 0.8$ is again required. With this definition, also a behavior like that shown in Fig. 5.18b is considered to be snappy. Additionally, this definition includes the observation that small currents are especially critical for reverse recovery behavior.

The term $\mathrm{d}i_{\mathrm{r}}/\mathrm{d}t$ determines the occurring voltage peak, and $v(t)$ in Eq. (5.67) has the maximal amplitude at the maximal slope:

$$V_{\mathrm{M}} = -V_{\mathrm{bat}} - L \cdot \left(\frac{\mathrm{d}i_{\mathrm{r}}}{\mathrm{d}t}\right)_{\max} \tag{5.71}$$

Thus, the voltage peak V_M occurring under special conditions or the induced voltage $V_{ind} = V_m - V_{bat}$ can be used as a quantitative definition for the reverse recovery behavior. As conditions, V_{bat} and the applied di/dt must be indicated.

This definition, however, is also insufficient, because even more parameters have an influence on the reverse recovery behavior:

1. The temperature: In most cases, high temperatures are more critical for the reverse recovery behavior. For some fast diodes, however, room temperature or a lower temperature is a more critical condition for possible occurrence of snappy recovery behavior.
2. The applied voltage V_{bat}: Higher voltage leads to worse recovery behavior.
3. The value of the inductor L: According to Eq. (5.71), with increased L the voltage at the diode is increased; this makes the conditions for the diode harder.
4. The commutation velocity di/dt: A rise in di/dt leads to greater danger of oscillations and snap-off of the current. The reverse recovery behavior has an increased tendency toward snappy behavior.

All these different influences cannot be covered by a simple quantitative definition. The circuit according to Fig. 5.16 and the definitions according to Eqs. (5.69) and (5.70) can only be used to show the effect of different design parameters. In fact, the reverse recovery behavior has to be evaluated by using the waveforms of current and voltage, which are measured under application-conform conditions.

The application-conform double-pulse measurement circuit is shown in Fig. 5.19. Compared to the circuit in Fig. 5.16, the ideal switch is replaced by a real switch, e.g. an IGBT. The ideal current source is replaced by an ohmic-inductive load consisting of R and L. The commutation velocity is given by the transistor; for an IGBT it is adjustable by the resistor R_{on} in the gate circuit as described later in Chap. 10. V_{bat} is the battery voltage, which is assisted by a capacitor C. The wiring between capacitor, IGBT, and the diode together form a parasitic inductance.

In Fig. 5.20 the drive signals for the IGBT, the current in the IGBT, and the current in the diode are shown for the double-pulse mode. By turning off the IGBT, the load current is transferred to the freewheeling diode. At the next turn-on of the IGBT, the diode is commutated – the point in time when the characteristic reverse

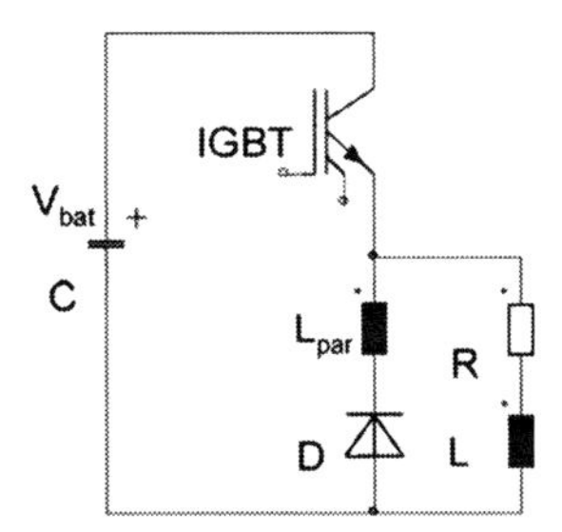

Fig. 5.19 Application-conform double-pulse circuit for measuring the reverse recovery behavior

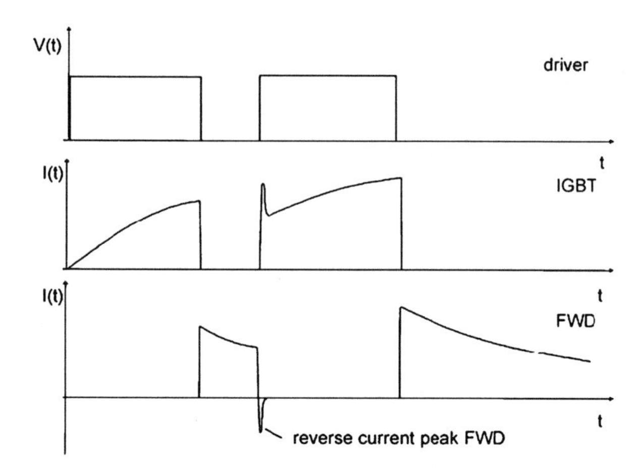

Fig. 5.20 Signal of the driver, current in the IGBT, and current in the freewheeling diode (FWD) at double-pulse measurement with the circuit according to Fig. 5.19

recovery of the diode occurs. At turn-on, the IGBT has to conduct additionally the reverse current of the freewheeling diode.

This reverse recovery event is shown at higher time resolution in Fig. 5.21 for a soft recovery diode. Figure 5.21a presents the current and voltage waveforms in the IGBT and the resulting power loss $v(t) \cdot i(t)$ at the turn-on. In addition, Fig. 5.21b displays the current and voltage waveforms for the freewheeling diode as well as the power losses in the diode.

While the IGBT has to conduct the reverse recovery current maximum I_{RRM} of the freewheeling diode additionally to the load current, the voltage at the IGBT is still in the range of the battery voltage V_{bat} (1200 V in Fig. 5.21a). In this instant the maximum of the turn-on loss in the IGBT occurs.

The reverse recovery current waveform of the diode can be divided into two phases:

1. The waveform until I_{RRM} and the subsequent decay of the reverse current with di_r/dt. In a soft recovery diode, $|di_r/dt|$ is in the range of $|di/dt|$. The reverse current peak I_{RRM} causes the most stress for the switching device.
2. The tail current phase during which the reverse current slowly phases out. A useful definition of the switching time t_{rr} is hardly possible for such a waveform. The tail current phase causes the main losses in the diode, since now there is a high voltage across the diode. Even though a snappy diode without tail current would feature less switching losses in the diode, it is detrimental for the application due to generated voltage peaks and oscillations. Slow and soft waveforms are desired. For the IGBT the tail current phase of the diode causes less stress, since in this phase the voltage at the IGBT has already decayed down to a low value.

The diode switching losses in Fig. 5.21b are represented on the same scale as those for the IGBT in Fig. 5.21a; the diode losses in the application are small

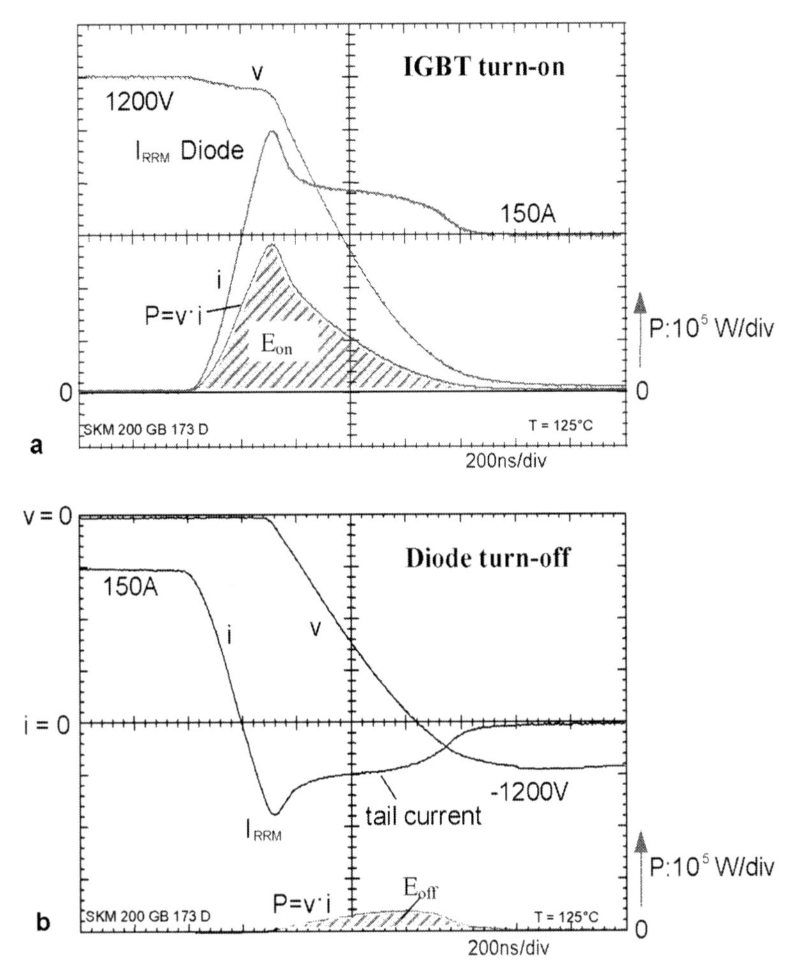

Fig. 5.21 Current waveform, voltage waveform, and power losses at turn-on of the IGBT (**a**) and simultaneous turn-off of the diode (**b**) during the measurement of the diode recovery behavior in a double-pulse circuit according to Fig. 5.19

compared to the losses in the IGBT. With regard to the total losses in the interaction of both devices, it is therefore important to keep the reverse recovery current peak I_{RRM} low, and to take care that the main part of the stored charge of the diode is extracted in the tail phase. As the tail current causes the main part of the switching losses in the diode, it must also be limited. Typically, the switching losses in the diode are lower than those in the transistor (compare Fig. 5.21a and b). Regarding its contribution to the total losses, the most important characteristic for the diode is the reverse recovery current peak I_{RRM} that must be as low as possible.

For a typical application in a current range of 100 A, in which the semiconductor devices of the chopper circuit are packaged inside of a single module, the parasitic inductance L_{par} is in the range of 40 nH or below. This leads to no significant overvoltage. Since there is no longer an ideal switch but rather a real transistor, the transistor still takes a part of the voltage during the reverse recovery phase of the diode and lowers the applied battery voltage by $v_C(t)$. After turn-on of the transistor, Eq. (5.67) is now valid in a modified form:

$$L_{par} \cdot \frac{di}{dt} + v(t) = -V_{bat} + v_C(t) \tag{5.72}$$

The voltage occurring at the diode after I_{RRM} is now

$$v(t) = -V_{bat} - L_{par} \cdot \frac{di_r}{dt} + v_C(t) \tag{5.73}$$

where $v_C(t)$ is the voltage drop across the transistor in this phase. For soft recovery diodes it is typical that at moderate commutation velocities di/dt of up to 1500 A/μs and minimized parasitic inductance, the absolute value of the voltage at the diode $v(t)$ is smaller than V_{bat} and no voltage peak occurs.

As long as the circuit according to Fig. 5.19 is applied and the parasitic inductance is kept low, one can use the following definition:

A diode exhibits a soft recovery behavior if, under all relevant application conditions in an application conform circuit, no overvoltage is caused at the diode by a reverse recovery current snap-off.

The relevant conditions cover the whole current range, all commutation velocities which can occur in the application and the temperature range from −50°C up to 150°C.

This definition is valid provided that there are not excessively high commutation velocities (> 6 kA/μs) or high parasitic inductances (>50 nH) in the circuit. If L_{par} is increased and the switching characteristics of the IGBT approach the characteristics of an ideal switch, which means that $v_C(t)$ approaches zero, then the circuit in Fig. 5.19 approaches the circuit in Fig. 5.16. In this case, voltages peaks are unavoidable also for soft recovery diodes.

In high-power modules with a rated current of 1200 A and more, 24 or more IGBT chips are connected in parallel. Such modules exhibit a large volume and it is very difficult to achieve a low parasitic inductance when connecting these modules to a three-phase inverter. In this range it makes sense to investigate the diode with an increased parasitic inductance. Thus, it should be investigated whether there are conditions at which a reverse current snap-off occurs. In such a reverse current snap-off, the waveform usually is similar to the waveform shown in Fig. 5.18b. The reverse current snap-off may even occur relatively late in the reverse recovery waveform, at the end of the tail current.

5.7.2 Reverse Recovery Related Power Losses

The turn-off energy of the diode per switching event is generally (see Fig. 5.21b):

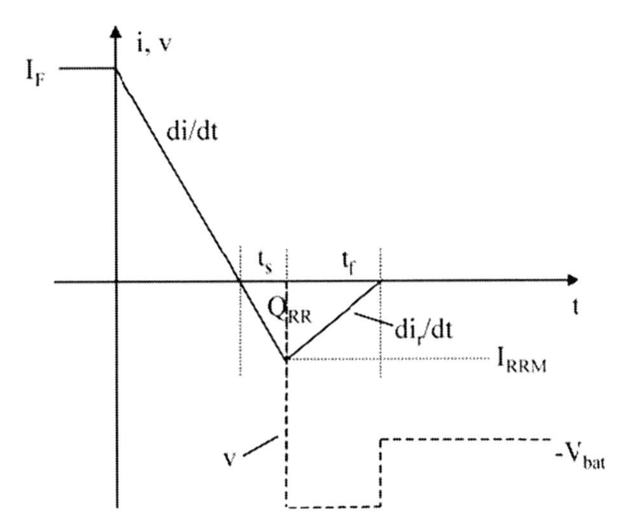

Fig. 5.22 Simplified waveform of current and voltage at the diode during turn-off in the circuit according to Fig. 5.16

$$E_{\mathrm{off}} = \int_{t_s+t_f} v(t) \cdot i(t)\mathrm{d}t \tag{5.74}$$

A simplified estimation can be given for two cases. The first is the case of the circuit according to Fig. 5.16 and the waveform in Fig. 5.17. This waveform is drawn in a simplified way in Fig. 5.22.

During the time until current zero crossing and during the time t_s, the voltage is simply assumed to be $v = 0$. For $t > t_s$ the diode takes over the voltage. If a linear decay of i_r during t_f is assumed, it holds during t_f that

$$i_r(t) = -I_{\mathrm{RRM}} + \frac{I_{\mathrm{RRM}}}{t_f} \cdot t \tag{5.75}$$

$$v = -V_{\mathrm{bat}} - L \cdot \frac{\mathrm{d}i_r}{\mathrm{d}t} = -V_{\mathrm{bat}} - L \cdot \frac{I_{\mathrm{RRM}}}{t_f} = \text{const.} \tag{5.76}$$

Thus, it follows:

$$E_{\mathrm{off}} = \frac{1}{2} \cdot L \cdot I_{\mathrm{RRM}}^2 + \frac{1}{2} \cdot V_{\mathrm{bat}} \cdot I_{\mathrm{RRM}} \cdot t_f \tag{5.77}$$

The first term on the right-hand side of Eq. (5.77) can be modified using Eq. (5.67). Accordingly, it holds that $L = -\frac{V_{\mathrm{bat}}}{\mathrm{d}i/\mathrm{d}t} = \frac{V_{\mathrm{bat}}}{I_{\mathrm{RRM}}/t_s}$. With this, one obtains from Eq. (5.77):

$$E_{\mathrm{off}} = \frac{1}{2} \cdot V_{\mathrm{bat}} \cdot I_{\mathrm{RRM}} \cdot t_s + \frac{1}{2} \cdot V_{\mathrm{bat}} \cdot I_{\mathrm{RRM}} \cdot t_f = \frac{1}{2} \cdot I_{\mathrm{RRM}} \cdot t_{rr} \cdot V_{\mathrm{bat}} = Q_{\mathrm{RR}} \cdot V_{\mathrm{bat}} \tag{5.78}$$

The diode turn-off losses are thus directly proportional to Q_{RR}.

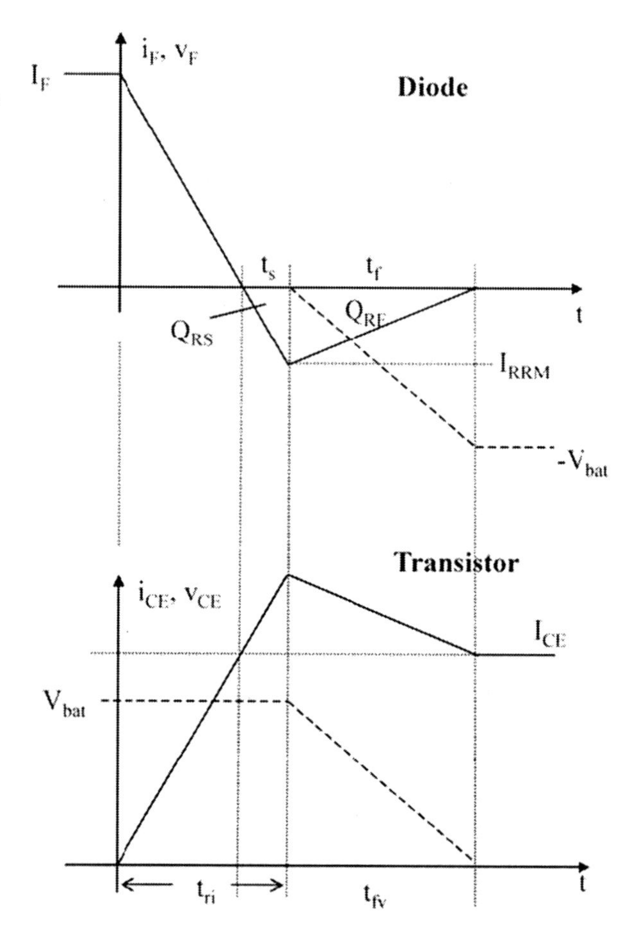

Fig. 5.23 Simplified waveforms of current and voltage at the diode and in the transistor for the circuit according to Fig. 5.19

This simplified consideration is valid only for the circuit in Fig. 5.16 in which main effects are determined by the inductor L. Also for the second case, the application-conform circuit according to Fig. 5.19 with the waveforms according to Fig. 5.21, a simplified estimation of the switching losses is possible. The waveforms from Fig. 5.21 are illustrated in a simplified way in Fig. 5.23. Here, the parasitic inductance L_{par} is neglected, and the voltage slope at the diode is only determined by the voltage slope at the transistor $v_C(t)$. The voltage fall time of the transistor t_{fv} is assumed to be equal to the reverse current fall time t_f of the diode. The current and voltage waveforms are again idealized by straight lines.

The reverse recovery charge Q_{RR} of the diode is subdivided into the charge Q_{RS} occurring during the storage time t_s and the charge Q_{RF} occurring during the reverse current fall time t_f. It holds that $Q_{RR} = Q_{RS} + Q_{RF}$. The turn-off energy loss in the diode per switching event is then

$$E_{off} = \frac{1}{3} Q_{RF} \cdot V_{bat} \tag{5.79}$$

The parameters Q_{RR} and I_{RRM} at specified conditions $\mathrm{d}i/\mathrm{d}t$ and V_{bat} are given in the data sheets of well-specified, modern freewheeling diodes. From I_{RRM} and $\mathrm{d}i_F/\mathrm{d}t$, one can calculate Q_{RS} as

$$Q_{RS} = \frac{1}{2} t_s \cdot I_{RRM} = \frac{1}{2} \cdot \frac{I_{RRM}}{\mathrm{d}i_F/\mathrm{d}t} \cdot I_{RRM} = \frac{1}{2} \cdot \frac{I_{RRM}^2}{\mathrm{d}i/\mathrm{d}t} \tag{5.80}$$

and because of $Q_{RF} = Q_{RR} - Q_{RS}$, it follows:

$$E_{off} = \frac{1}{2} V_{batt} \cdot \left(Q_{RR} - \frac{1}{2} \cdot \frac{I_{RRM}^2}{\mathrm{d}i/\mathrm{d}t} \right) \tag{5.81}$$

If this result is compared with the result (5.78), the case of the inductor-determined turn-off process, then one can see the turn-off energy loss is less than half of the estimation in Eq. (5.78). The transistor has relieved the stress for the diode by its $v_C(t)$ which drops across the transistor during the diode turn-off event and which slowly decays during the turn-on of the transistor.

Nevertheless, for this reduction of losses in the diode, the switching transistor has to pay the price during its turn-on. From the same simplified consideration in Fig. 5.23, it can be derived that assuming a freewheeling diode without reverse recovery current maximum – and thereby without a stored charge – the idealized turn-on energy loss in the transistor would be

$$E_{on}(\mathrm{tr,id}) = \frac{1}{2} \cdot (t_{ri} - t_s) \cdot I_{CE} \cdot V_{bat} + \frac{1}{2} \cdot t_{fv} \cdot I_{CE} \cdot V_{bat} \tag{5.82}$$

By the freewheeling diode, the following terms are generated additionally:

the dissipated energy because the current-increase time t_{ri} is prolonged by t_s; in this interval, the current as well as the voltage are high
the power losses caused by Q_{RS}
the power losses caused by Q_{RF}

This leads to additional losses ΔE_{on} – in the order of the enumeration:

$$\Delta E_{on} = t_s \cdot I_F \cdot V_{bat} + Q_{RS} \cdot V_{bat} + \frac{2}{3} Q_{RF} \cdot V_{bat} \tag{5.83}$$

Comparing this with Eq. (5.79), one can state that the losses in the transistor, generated by the diode, are higher than the losses in the diode itself by the first terms in Eq. (5.83) and additional twice of the diode losses according to Eq. (5.79) occurs as last term in Eq. (5.83).

The turn-on losses in the transistor amount to $E_{on}(\mathrm{tr}) = E_{on}(\mathrm{tr,id}) + \Delta E_{on}$. The sum of the losses caused by the diode in the transistor and in the diode is

$$E_{off} + \Delta E_{on} = t_s I_F V_{bat} + Q_{RR} V_{bat} \tag{5.84}$$

This is significantly higher than the estimation in Eq. (5.78); in summary, an excessive price was paid for relieving the stress in the diode.

These estimations show that the requirement for diodes is a low reverse recovery peak I_{RRM}, and thereby a t_s which is as low as possible.

In real circuits, a parasitic inductance that causes additional losses in the diode has to be considered. If the parasitic inductance dominates and if the relief of the diode by the voltage decay of the transistor can be neglected, then the losses in the diode again approach the situation that was described with Eq. (5.78). Comparatively high parasitic inductances have to be taken into account, for example, in some traction applications.

In the application, neither an ideal switch nor a circuit without any inductance can be presumed. Thus, Eq. (5.74) must be used for an exact determination of the switching losses. The waveforms of i and v are recorded with an oscilloscope, multiplied, and integrated over the total switching time. The simplified estimation given here comply with measured values in a low-inductive circuit with an accuracy of $\pm 20\%$.

5.7.3 Reverse Recovery: Charge Dynamic in the Diode

Figure 5.6 has shown the flooding of the base of the diode with free carriers for the forward conduction state. The reverse recovery behavior is determined by the internal behavior of the stored plasma and the time-dependent shape during its removal. First, this shall be investigated qualitatively. Figure 5.24 depicts the simulation of the stored carriers in a snappy diode, whereas Fig. 5.25 shows the same for a soft recovery diode.

At forward conduction, the n^--base of the diode is flooded with free electrons and holes in a range of 10^{16} cm^{-3}; they build up a neutral plasma in which the density of electrons n and of holes p are approximately equal. After commutation a neutral plasma zone with $n \approx p$ is still present in the diode until the time t_4. The removal of free carriers takes place toward the cathode by the electron current and toward the anode by the hole current, the plasma removal process occurs in the outer circuit as reverse current. In the case of the snappy diode (Fig. 5.24), both edges of the plasma meet shortly after t_4, and suddenly the source for feeding the reverse current vanishes. The reverse current is interrupted abruptly, the reverse recovery behavior is snappy.

The time-dependent shape of the remaining plasma for a soft recovery diode is shown in Fig. 5.25. The doping profile is the same as in Fig. 5.24 in this example, but the on-state carrier concentration ($t = t_0$) is changed resulting from an inhomogeneous carrier lifetime, i.e. a low lifetime at the pn-junction and a higher lifetime at the nn$^+$-junction. During the whole reverse recovery process, a neutral plasma remains inside the diode and feeds the reverse current. At the instant t_5 the diode has taken over the applied voltage. A plasma decay as shown in Fig. 5.25 leads to a tail current as depicted in Fig. 5.21b.

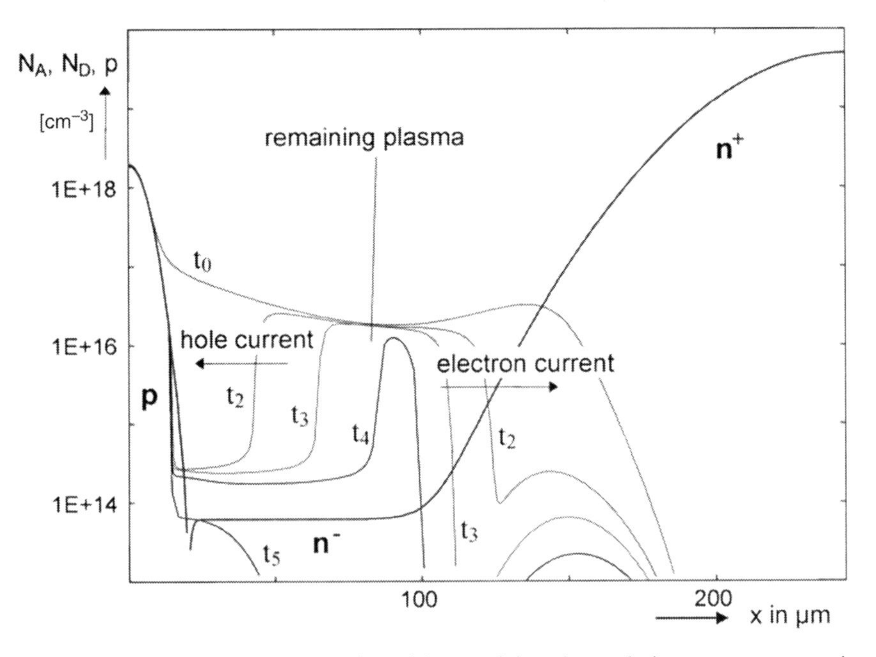

Fig 5.24 Doping profile and concentration of the remaining plasma during reverse recovery in a snappy diode (ADIOS simulation)

Whether soft recovery behavior is achieved depends on the time-dependent decay of the plasma and how this process is controlled suitably. It took a comparatively long time until the reverse recovery behavior was mastered.

For analyzing the dynamic behavior of the plasma, two cases will be investigated in more detail in the following. The first case refers to the effects during the increase in voltage and is based on the model of Benda and Spenke [Ben67]. With this model is investigated under which conditions a reverse current snap-off or a soft recovery behavior is to be expected. Thereafter, a further analysis is conducted for the case that the device has already taken the voltage with a soft recovery shape of the current and that despite this, a snap-off does occur at the end of the tail current.

For investigating the first case based on the model of Benda and Spenke, the following simplifications are assumed:

- The pn- and nn^+-junctions are assumed to be abrupt, and the charge within the highly doped regions is neglected.
- The plasma edges are treated as abrupt.
- Since the amount of charge carriers removed by the current is much higher than the recombination of charge carriers, recombination in this time interval is neglected.
- Additionally, the hyperbolic asymmetric shape of the initial plasma (Fig. 5.6 respectively Eq. (5.27)) is, as a first step, simplified by a constant plasma density, whereby $n = p = \bar{n}$ is assumed.

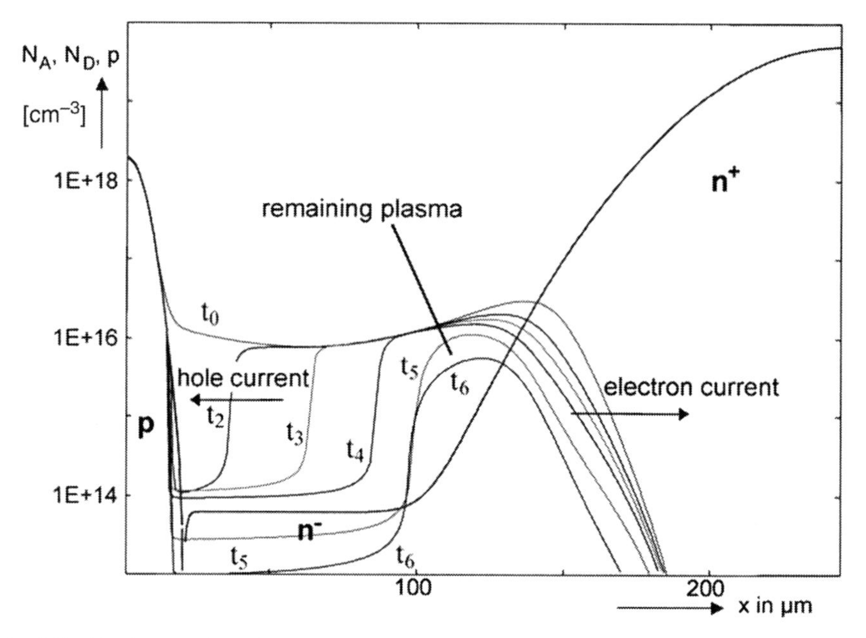

Fig. 5.25 Doping profile and concentration of the remaining plasma in a soft recovery diode (ADIOS simulation). Compared to Fig. 5.24 the difference is caused by an inhomogeneous carrier lifetime

Figure 5.26 illustrates this simplified model. In the plasma the current consists of electron current and hole current; it holds that $j = j_n + j_p$ whereby

$$
\begin{aligned}
j_p &= \frac{\mu_p}{\mu_n + \mu_p} j \\
j_n &= \frac{\mu_n}{\mu_n + \mu_p} j
\end{aligned}
\tag{5.85}
$$

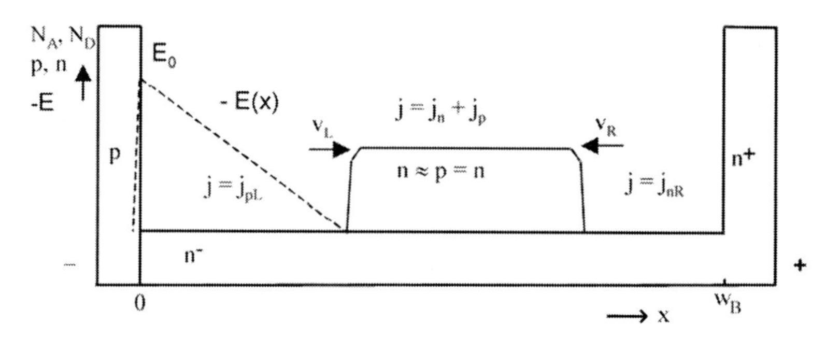

Fig. 5.26 Simplified drawing of the removal of the internal plasma in a diode

On the left side of the plasma zone, toward the pn-junction, a part of the n⁻-base has become free of carriers and a space charge is building up. The current flows here as a hole current $j = j_{pL}$. From the continuity condition for the current at the border of plasma and space charge, it follows:

$$j_{pL} = j_n + j_p \tag{5.86}$$

For the difference of the hole current, it is therefore valid at this position that

$$j_{pL} - j_p = \Delta j_p = j_n \tag{5.87}$$

Using Eq. (5.85) it follows:

$$\Delta j_p = \frac{\mu_n}{\mu_n + \mu_p} j \tag{5.88}$$

Both sides of Eq. (5.85) multiplied by the time interval $\mathrm{d}t$ lead to a differential charge

$$\Delta j_p \cdot \mathrm{d}t = \frac{\mu_n}{\mu_n + \mu_p} j \cdot \mathrm{d}t \tag{5.89}$$

This differential charge corresponds to a charge $q \cdot \bar{n} \cdot \mathrm{d}x$ stored in a volume element $\mathrm{d}x$ while the plasma zone is shortened by $\mathrm{d}x$. Consequently, Eq. (5.89) becomes

$$q \cdot \bar{n} \cdot \mathrm{d}x = \frac{\mu_n}{\mu_n + \mu_p} j \cdot \mathrm{d}t \tag{5.90}$$

and for the velocity of the movement of the left-hand side border of the plasma to the right-hand side, one obtains

$$|v_L| = \frac{\mathrm{d}x}{\mathrm{d}t} = \frac{\mu_n}{\mu_n + \mu_p} \cdot \frac{j}{q\bar{n}} \tag{5.91}$$

Analogously, this can be analyzed for the right side of the plasma zone. From the continuity condition for the current at the border between the plasma zone and the plasma-free zone at the right-hand side, it follows:

$$j_{nR} = j_n + j_p \tag{5.92}$$

and this leads to

$$|v_R| = \frac{\mu_p}{\mu_n + \mu_p} \cdot \frac{j}{q \cdot \bar{n}} \tag{5.93}$$

For silicon, with $\mu_n \approx 3\mu_p$, one obtains $v_L \approx 3v_R$: The plasma zone is removed from the side of the pn-junction thrice as fast as from the side of the nn^+-junction. With the assumed constant density of free carriers in the plasma, both plasma fronts will finally meet at

$$w_x = \frac{v_L}{v_L + v_R} \cdot w_B = \frac{3}{4} w_B \tag{5.94}$$

The electric field is built up by the space charge in the part of the n^--layer between the pn-junction and border of the plasma. The electric field cannot penetrate the plasma zone, since this is neutral. Hence, the electric field will have a triangular shape. The voltage across the device corresponds to the area below the line $-E(x)$ in Fig. 5.26. The plasma-free zone at the right side does not contribute to the voltage in this first approximation.

Therefore, the device can take over as much voltage as possible with a width w_x before a reverse current snap-off occurs. This voltage shall be designated as threshold voltage for snappy behavior V_{sn}. Since electrons are not present in the space charge region, one obtains

$$V_{sn} = \frac{1}{2}\frac{q \cdot N_D}{\varepsilon} w_x^2 \tag{5.95}$$

The current j in this region flows as a hole current, and the holes have the same polarity as the positively charged donor ions of the background doping. Equation (5.95) is valid as long as the density of holes p, which are traveling through the space charge, is low, $p << N_D$, and can be neglected. With increasing p, N_D has to be replaced by $N_{eff} = N_D + p$. Therefore, the gradient of the electric field and thereby the area below the line $-E(x)$ is increased by the current. The device can thus take over more voltage. Maximally, this voltage can take the value at which avalanche breakdown sets in. Rewriting Eq. (5.1), one obtains the maximal voltage V_{sn} as a function of w_x to be

$$V_{sn} = \left(\frac{w_x^6}{2^4 \cdot B}\right)^{\frac{1}{7}} \tag{5.96}$$

This is the maximal voltage for triangular electric field shape. Equation (5.96) predicts a higher value for V_{sn} than Eq. (5.95). In most cases, the prediction from Eq. (5.95) is too low, especially for switching with high di/dt which is typical in applications with IGBTs as switching devices; at these conditions the hole concentration p cannot be neglected. Equation (5.96) is closer to the experimental observation. It should be noted here that V_{sn} is always significantly lower than the breakdown voltage V_{BD} of the device. Since with a wide w_B also increases w_x, this shifts the snap-off of the reverse current to a higher voltage.

Earlier, some suggestions to achieve soft recovery behavior in fast diodes have been based on this movement of the plasma fronts and proposed a wide w_B [Mou88]. Even some recent solutions are applying this approach. Here, the diode must be designed for triangular field shape (NPT), and w_x is determined using Eq. (5.1). Moreover, the doping is chosen as high as possible for the respective voltage according to Eq. (3.84), w_x results from Eq. (3.85) or Eq. (5.1) and finally, according to Eq. (5.94), $1/3 \cdot w_x$ is added to determine w_B.

As was already mentioned in Sect. 5.3 in the investigations on dimensioning of fast diodes, the minimal width of the n^--layer of a NPT diode is the $2^{2/3}$-fold of the minimal width of the n^--layer of a PT diode – the diode with the smallest width of the n^--region for the required voltage. Compared to the minimal width $w_{B\,min} = w_{B(PT,lim)}$ of Eq. (5.12), the suggestions above lead to

$$w_B = \frac{1}{0.63} \cdot \frac{4}{3} w_{Bmin} \cong 2,1 \cdot w_{Bmin} \tag{5.97}$$

This results in a significant increase in the forward voltage drop, to which w_B contributes with the power of 2 or even exponentially. To avoid this high forward voltage, the charge-carrier lifetime can be increased, but this contradicts the requirement that the diode must be fast. Hence, further measures are necessary to achieve soft recovery behavior, especially under condition of fast switching events in IGBT circuits. With the behavior of the internal plasma as in Fig. 5.25, a soft recovery is obtained without making w_B as thick.

Up to now, the plasma profile was assumed to be homogeneous and constant over the base of the diode, which contradicts reality. However, the simplified approach allows useful conclusions to be drawn also for an inhomogeneous distribution. According to Fig. 5.5 or Eq. (5.27), the plasma density is increased at the pn-junction because of the different mobilities of electrons and holes. This is now taken into account by the densities $\bar{n}_L$ at the side of the pn-junction and $\bar{n}_R$ at the side of the nn$^+$-junction. The term $\bar{n}_L$ stands for the average density of carriers close to the pn-junction. Division of Eq. (5.93) with Eq. (5.91) leads to

$$\frac{v_R}{v_L} = \frac{\mu_p}{\mu_n} \cdot \frac{\bar{n}_L}{\bar{n}_R} \tag{5.98}$$

Furthermore, the quantity $\eta = p_L/p_R$ introduced in Sect. 5.4.5 in Eq. (5.55) shall be used, and we approximate that this proportionality is valid within some distance to the respectively junction, e.g. $\eta = \bar{n}_L/\bar{n}_R = \bar{p}_L/\bar{p}_R$. Inserted into Eq. (5.94), this results in the following equation:

$$w_x = \frac{1}{1 + \frac{\mu_p}{\mu_n} \cdot \eta} \cdot w_B \tag{5.99}$$

If the profile in Fig. 5.6 is simply expressed with $\eta = 2$, then Eq. (5.99) yields

$$w_x = \frac{3}{5} \cdot w_B \tag{5.100}$$

The diode would snap-off even at a lower voltage than estimated with Eq. (5.94), or a diode with abrupt highly doped border regions and homogeneous lifetime in the base would have to be made even thicker than according to Eq. (5.97).

On the other hand, if the profile in Fig. 5.6 can be inverted to maintain a higher density of carriers at the nn$^+$-junction compared to the pn-junction, this will be advantageous [Sco89]. If, for example $\eta = 1/3$, then Eq. (5.99) leads to $w_x = 0.9 w_B$, and w_B would have to be widened by a much smaller amount to obtain a sufficient w_x. Such a distribution can be achieved by special structures of the p-emitter, or by using a p-region which is much lower doped than the n$^+$-region, or by an inhomogeneous carrier lifetime which, at the pn-junction, is much lower than deep in the base. Such measures are used in modern fast diodes (see Sect. 5.7.4).

The above analytical approach using a simplified model gives us a good understanding of the physics of the recovery process. Considering the results of the numerical simulation shown in Fig. 5.24, one notices the following deviations from the simple model:

- The p- and n$^+$-regions usually have diffusion profiles; hence, the pn- and nn$^+$-junctions are not abrupt.
- The very low gradient of the doping concentration at the nn$^+$-junction in the figure results in a later start of the carrier removal at the nn$^+$-junction than at the pn-junction. Therefore, the low doping gradient at the nn$^+$-junction is advantageous for the recovery behavior. The doping gradient at the pn-junction, on the other hand, should be as abrupt as possible, because this involves an early rise in the reverse voltage and a reduced peak reverse current.
- Of course, the gradients at the edges of the plasma cannot be infinite. However, the non-abrupt transition to the depletion regions does not greatly affect the velocity at the edges, because after a time interval *dt* the form of the carrier concentrations at the boundaries is approximately unchanged and only the thickness of the uniform plasma region is reduced.

Another aspect not included in the above model is avalanche generation which often takes place during fast switching, since free carriers, still present in the space charge region, enhance the electric field. This "dynamic avalanche" will be discussed later in Chap. 12.

Now the second case will be considered: the device has successfully sustained the interval of the voltage increase, whereby the reverse recovery in this interval was soft. The device has taken over the applied voltage, but the plasma was not completely removed. Resulting from the remaining plasma, a tail current still flows. This tail current flows through the space charge as hole current, there holds $j = j_p$. Under the given condition of a high field, the holes flow with the drift velocity $v_{d(p)}$, which approximates the saturation drift velocity v_{sat}. Accordingly, the hole density in the space charge is

$$p = \frac{j}{q \cdot v_{sat}} \tag{5.101}$$

This influences the gradient of the electric field in the space charge region as

$$\frac{dE}{dx} = \frac{q}{\varepsilon}(N_D + p) \tag{5.102}$$

This situation is illustrated in Fig. 5.27 for a diode with low base-doping N_D. The voltage can be assumed to be constant in the investigated time interval: consequently, the area under $-E(x)$ is constant. The hole density is one factor in dE/dx, and p is determined by the hole current extracted from the remaining plasma. This hole current leads to a removal of the remaining plasma. By applying Eq. (5.101), Eq. (5.91) is changed to:

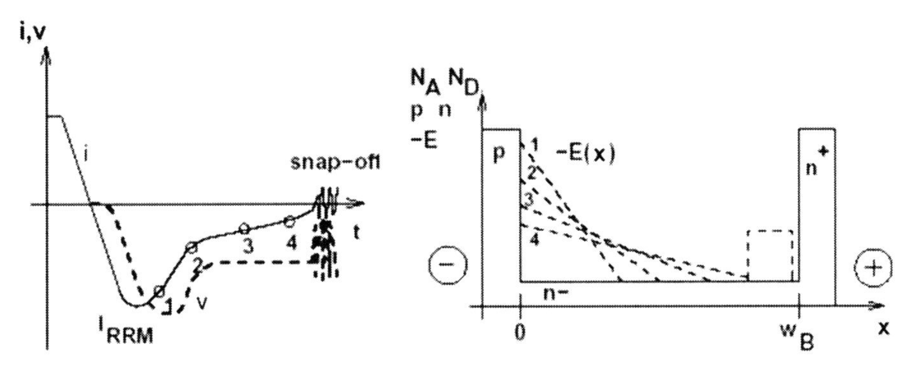

Fig. 5.27 Reverse current snap-off at the end of the tail current. *Left*: Waveform of current and voltage. *Right*: Electric field at different points in time

$$|v_l| = \frac{\mathrm{d}x}{\mathrm{d}t} = \frac{\mu_n}{\mu_n + \mu_p} \cdot \frac{p \cdot v_{sat}}{\bar{n}} \tag{5.103}$$

As p and j decrease, $\mathrm{d}E/dx$ becomes lower and the space charge layer widens. However, if the space charge reaches the end of the base while still a significant current is flowing, the source of the current will suddenly vanish and the current will then snap-off. The electrical field springs from a triangular to a trapezoidal shape.

To avoid this effect, the device must be capable of taking the space charge at a given voltage without a punch of the space charge to the n^+-layer. The voltage limit at which the space charge reaches the n^+-layer at a given background doping N_D and a base width w_B, is

$$V_{sn} = \frac{1}{2}\frac{q \cdot N_D}{\varepsilon} w_B^2 \tag{5.104}$$

For this case which is described in more detail in [Fel04], w_x can be set to w_B. As long as the battery voltage V_{bat} stays below V_{sn}, no reverse current snap-off will occur. In the static case or upon occurrence of voltage peaks, the diode can bear much more voltage, since then a trapezoidal space charge can be built up, and the breakdown voltage V_{BD} is much higher than V_{sn}.

The simplified description up to now cannot give an answer how a diode transits from the plasma in Fig. 5.26 to that in Fig. 5.27, since only the field-induced drift components of the current were considered. Taking into account the diffusion components, the movement of the plasma layer backward to the cathode can be described. The diffusion components on the right side of the plasma become dominating in the tail current phase, and the remaining plasma can move from the position in Fig. 5.26 to that in Fig. 5.27. This is explained in [Bab08].

The effect of reverse current snap-off in the tail phase can especially occur with diodes designed for higher voltages (> 2000 V). In such applications, often a low $\mathrm{d}i/\mathrm{d}t$ is applied, because the circuit contains significant parasitic inductances. This is the case in some applications for very high power control. Usually in such applications, the battery voltage is limited to 66% of the breakdown voltage V_{BD} for which the device is designed. In order to keep V_{sn} high, the doping N_D is not allowed to be too low. This, however, contradicts the demands of cosmic ray stability, see Sect. 12.7. Hence, an optimal trade-off has to be found.

5.7.4 Fast Diodes with Optimized Reverse Recovery Behavior

Recombination centers are implemented in all fast silicon pin-diodes. The characteristics of the employed recombination centers gold, platinum and radiation-induced centers have been described in Sects. 2.7 and 4.9 . With the density of recombination centers, the charge carrier lifetime and thereby the stored charge Q_{RR} is decreased. However, there is no direct relationship between the concentration of recombination centers – as long as their axial distribution is kept constant – and the form of the reverse current decay. Therefore, it cannot be determined by the recombination center density whether the behavior will be soft or snappy.

It was already shown that the recovery behavior will become soft, if only the width w_B of the lowly doped base region is wide enough. Yet this leads to very high forward conduction losses and/or switching losses, which can not be accepted in most cases. The modern concepts aim to adjust soft recovery behavior without using a strongly enhanced base width.

5.7.4.1 Diodes with a Doping Step in the Low-Doped Layer

To avoid an excessively wide w_B and to minimize the disadvantages resulting from it, in 1981 an n$^-$-layer with a step in the doping density was suggested by Wolley und Bevaqua [Wol81]. The doping profile of this diode is shown in Fig. 5.28. Approximately in the middle of the base, the doping is augmented by a factor of 5 – 10. Such layers are manufactured using a two-step epitaxy process.

When the space charge builds up and the field penetrates the higher doped layer n_2, it decreases there with a steeper gradient. At turn off, the remaining plasma zone

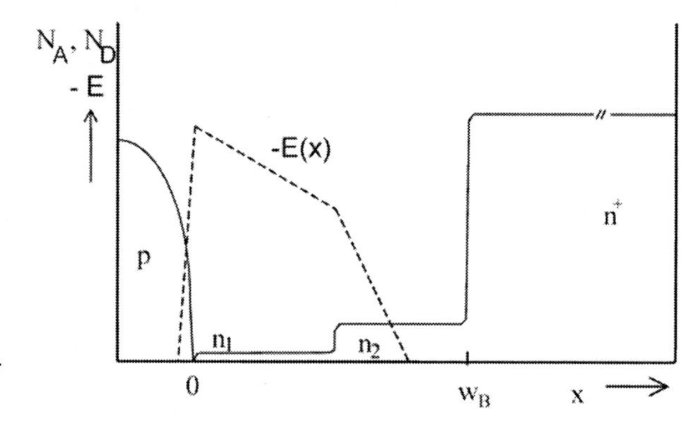

Fig. 5.28 Diodes with a step in the doping concentration of the low-doped layer

is located in the layer n_2. The voltage which the device can take corresponds to the area below the line $-E(x)$. This area is larger than that for a triangular field shape. The threshold voltage for possible snappy reverse recovery behavior is shifted to a higher value. This measure is nowadays often used for diodes manufactured in epitaxial technology in the voltage range of up to 600 V. For diodes of higher voltage, especially above 1200 V, the fabrication of an epitaxial layer with the necessary thickness is too laborious. For soft recovery behavior under all conditions relevant for the application, usually one of the following concepts is used.

5.7.4.2 Diodes with Anode Structures for Improving the Recovery Behavior

It was already shown that the carrier concentration in the flooded base of the pin-diode with highly doped border regions is higher at the pn-junction than at the nn^+-junction (see Fig. 5.6). This is disadvantageous for the reverse recovery behavior. Therefore, concepts were developed to invert this distribution: the concentration shall be higher at the nn^+-junction than at the pn-junction. This principle was explained already in [Sco89], and there were several approaches to realize this by using structures in the p-anode layer.

For example, Schottky junctions cannot inject holes. Thus, it is obvious that implementation of Schottky junctions on a part of the area will lead to the desired distribution, considering the average concentration over the area. This measure is often discussed in the literature.

The "*m*erged *p*in/*S*chottky" (MPS) diode consists of sequenced p-layers and Schottky regions [Bal98] (Fig. 5.29a). The distance between the p-layers is chosen to be so small that, in case of blocking voltage, the Schottky junction is shielded from the electric field, and only low field-strength occurs at its position. Consequently, the high leakage current of a Schottky junction is avoided. Figure 5.30 presents the forward characteristics of a MPS diode similar to that of [Bal98] calculated with the device simulator TCAD DESSIS [Syn07]. The structure with $w_B = 65\ \mu m$ is designed for a blocking voltage of 600 V. Compared here are the forward characteristics of the pin region, of the Schottky region, and of the parallel connection of both in a MPS diode. At low current density the MPS diode approaches the characteristics of a Schottky diode. In the range of 200 A/cm^2, the typical current density of fast 600 V diodes at the rated current, the advantage caused by the Schottky regions is only small. At increased current density, the MPS diode has a higher forward voltage because of the loss of area of p-emitter regions.

If the MPS diode is designed for a blocking voltage of 1000 V and greater, then the range of lower forward voltage drop shifts to lower current density, because then the voltage drop in the low-doped base region predominates. But the effect that the area of the p-region is reduced and thereby the injection of carriers at the anode side of the device is reduced, maintains its function. An inverted distribution of free carriers is formed (see Fig. 5.31, right-hand side).

Since the p-layers must be arranged closely to shield the Schottky regions from the electric field, the possibility is limited to make their share of the surface area small. A further idea to improve the MPS diode is the "*t*rench *o*xide *p*in *S*chottky"

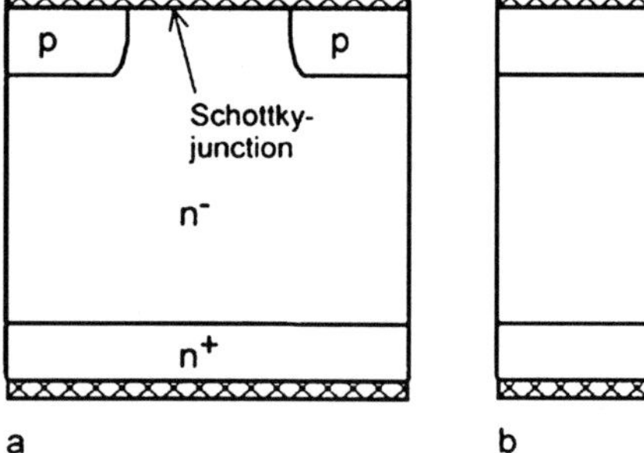

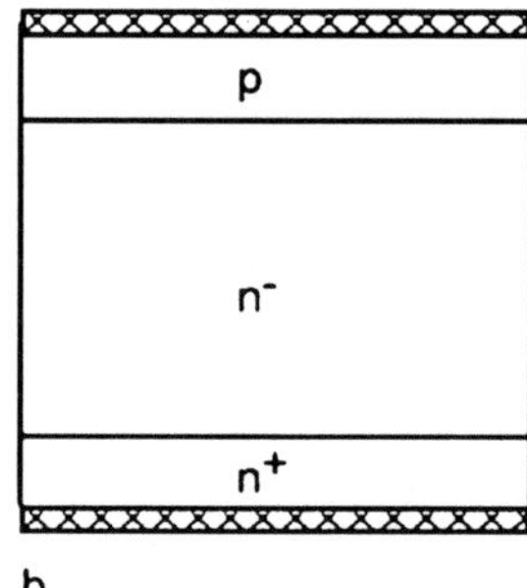

Fig. 5.29 p-emitter for improving the reverse recovery behavior: (**a**) emitter structures of the merged pin/Schottky diode, (**b**) uniformly reduced p-doping, low emitter efficiency

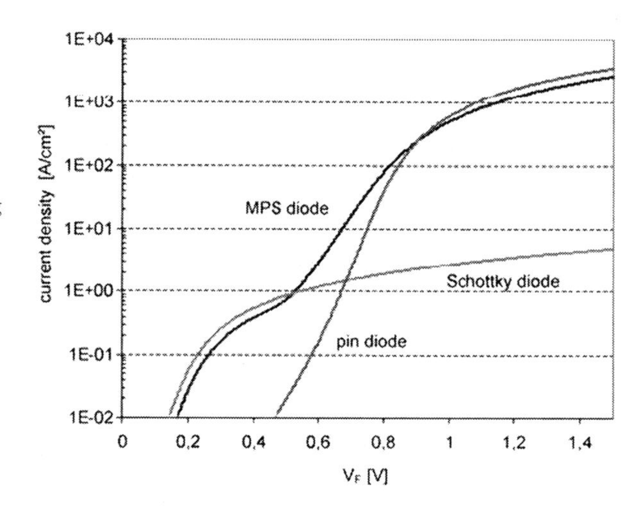

Fig. 5.30 Forward characteristics of a MPS diode with 50% Schlottky area. The characteristics of the pin-region and the Schottky region are shown in comparison. Simulation using TCAD DESSIS

(TOPS) diode [Nem01] which was presented by Fuji. The structure is shown in Fig. 5.31. The p-anode regions are placed at the bottom of the trench cells. A Schottky contact is formed at the semiconductor surface. Hole injection takes place only by the p-layers that are connected via the resistor of the polysilicon layer in the trench. Thus, the integral hole injection over the area is strongly reduced. This results in a profile of free carriers as is shown in Fig. 5.31 on the right side, which exhibits a stronger inversion of the internal plasma.

A soft recovery behavior can be expected from this plasma profile. The p-regions at the bottom of the trenches effectively shield the surface against the electric field; therefore, the leakage current is kept low with a narrow arrangement of the trenches.

There are several further concepts of structures at the anode side, among them, also structures with diffused p^+- and n^+-regions. What they all have in common is the reduction of the area of layers that are injecting holes and, thus, the reduction of the density of free carriers at the pn-junction.

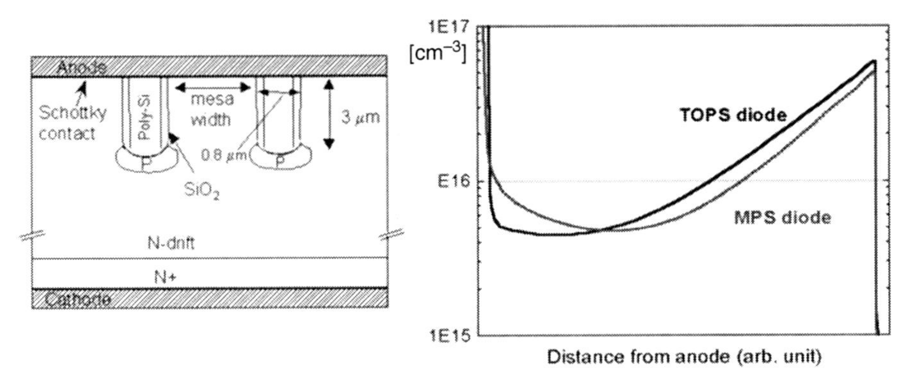

Fig. 5.31 Trench oxide pin Schottky (TOPS) diode. Structure (*left*) and vertical distribution of holes in the forward conduction state (*right*). Figure taken from [Nem01] © 2001 IEEE

5.7.4.3 The EMCON Diode

Instead of reducing the emitter area by emitter structures, an homogeneous p-layer with high emitter recombination, as is shown in Fig. 5.29b, can also lead to the desired inverted distribution of the plasma. This concept is applied in the "*em*itter-*con*trolled" (EMCON) diode [Las00]. It needs much less effort in production compared to the MPS diode or the TOPS diode.

The EMCON diode utilizes a p-emitter of low emitter efficiency. The emitter parameter h_p, introduced in Eq. (3.94), can be expressed for a p-emitter which is not too highly doped as follows:

$$h_p = \frac{D_n}{p^+ \cdot L_n} \tag{5.105}$$

To reduce the emitter efficiency γ according to Eq. (3.96), h_p must have a high value. According to Eq. (5.105), this can be done if the doping density of the emitter p^+ is chosen low and also if the effective diffusion length L_n is adjusted to a low value. Both measures are used in the EMCON diode. p^+ must be sufficiently high to avoid a punch-through of the electric field to the semiconductor surface. For a thin p-layer, L_n is approximately the same as the penetration depth of the p-layer w_p, and also this depth is small in an EMCON diode. Under this condition it holds that

$$h_p = \frac{D_n}{p^+ \cdot w_p} = \frac{D_n}{G_n} \tag{5.106}$$

where $G_n = p^+ w_p$ is the Gummel number of the emitter under the condition of an abrupt emitter [Sze81]. The term G_n represents the number of doping atoms per area. For the diffused emitter of the EMCON diode, it is more precisely expressed as

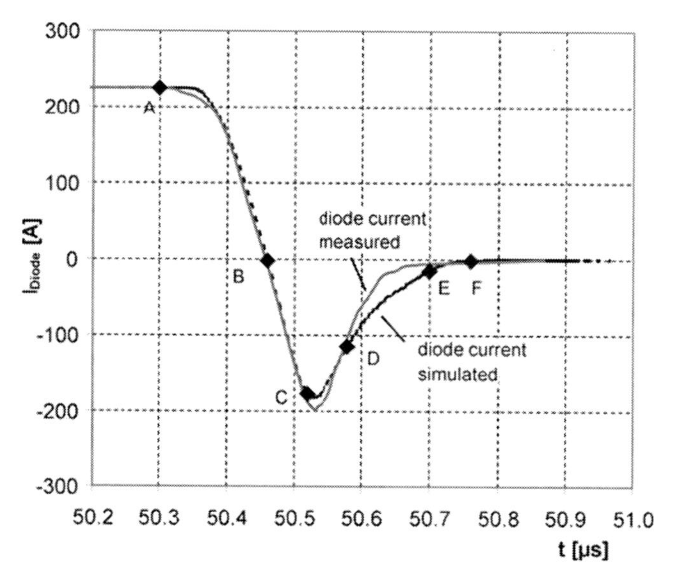

Fig. 5.32 Turn-off behavior of a 1200 V EMCON-HE diode, measured and simulated at 600 V, 25°C, 225 A/cm^2

$$G_{\mathrm{n}} = \int_{0}^{w_p} p(x)\mathrm{d}x \tag{5.107}$$

Using the Gummel number according to Eq. (5.107) increases h_{p} compared to the abrupt emitter. High h_{p} and therewith low γ leads to a decreased p_{L}, as is necessary to create the inverted distribution. According to Eq. (5.52), h_{p} is the dominating factor for the emitter recombination. High h_{p} means that a significant share of the total recombination takes place inside the p-emitter or at the surface.

An emitter of high doping and high efficiency is applied at the cathode side of the EMCON diode; hence, the plasma density at this side is high.

Figure 5.32 depicts the turn-off waveform of an EMCON-HE diode. The measured turn-off behavior is compared to the numerical device simulation. Here, the simulated waveform agrees sufficiently well with the measured characteristic. The numerical simulation enables visualization of the effects inside the device. The simulated distribution of the density of free carriers is shown in Fig. 5.33 for the indicated time steps in Fig. 5.32.

At forward conduction the diode is flooded with free carriers. In Fig. 5.33 the rhombic dots show the measured carrier distribution in an EMCON diode at forward conduction as is obtained by the internal laser deflection method [Deb96]. The density of free holes calculated with the device simulator (line A in Fig. 5.33 for instant A in Fig. 5.32) is in good agreement with the measured density. The hole density represents the plasma density; $n \approx p$ applies for the regions of high flooding below the lines A, B, C, etc. For the initial distribution at instant A, it holds

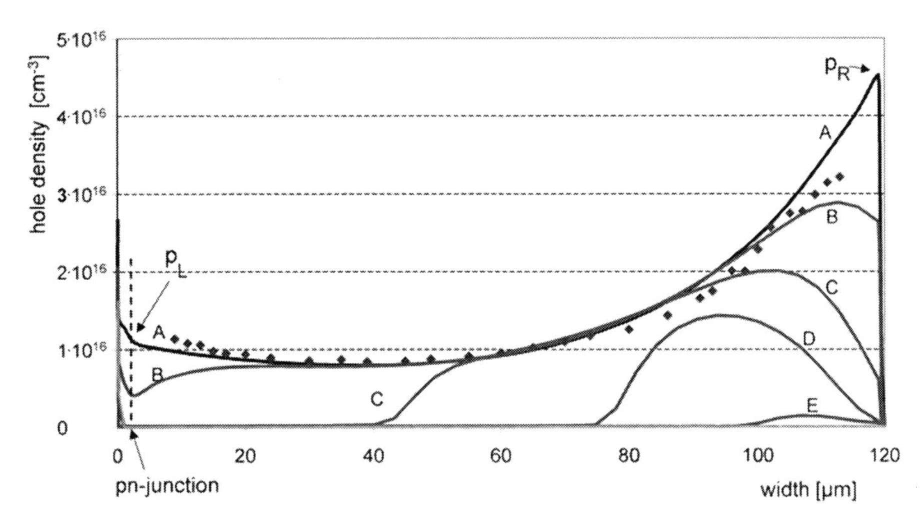

Fig. 5.33 Carrier distribution in an EMCON diode for forward conduction and during commutation. Measured plasma distribution in the forward conduction state (*rhombuses*), simulated hole distribution for the forward conduction state (*line A*), removal of the internal stored free-carriers (on example of holes) during commutation (*line B–E*)

that $\eta = p_L/p_R \approx 0.25$; this is attained by a strong emitter recombination in the anode.

During commutation and change of the polarity of the voltage (C, D, E in Figs. 5.32 and 5.33), the hole current flows to the negatively poled anode on the left side, and the electron current flows to the positively poled cathode. As is illustrated in Fig. 5.33, the removal of the stored charge occurs for the instants B, C, D, E. At the instant C, the diode has reached the maximal reverse current. After that, a current can still flow resulting from the removal of the still existing plasma which ensures a soft recovery behavior.

The low p-emitter efficiency of the EMCON diode leads to the disadvantage of a reduced surge current capability. A structure proposed in [Sco89], the so-called SPEED-diode containing high p-doped regions, could improve the surge current capability.

5.7.4.4 The CAL Diode

For the carrier lifetime adjustment in the diodes, often a platinum diffusion is used, whereby the vertical profile of the recombination centers (Fig. 4.23) results from the diffusion process and cannot be modified. The advantage of an adjusted profile of recombination centers, created by implantation of light ions, was already recognized early [Sil85, Won87]. But at that time this technology, which requires particle accelerators of energy of up to 10 MeV, was only available for research issues. The situation changed in the early 1990s of the last century. Then, the interest of basic researchers in high-energy physics migrated to the GeV region, and particle accelerators in the medium energy range became available for semiconductor production.

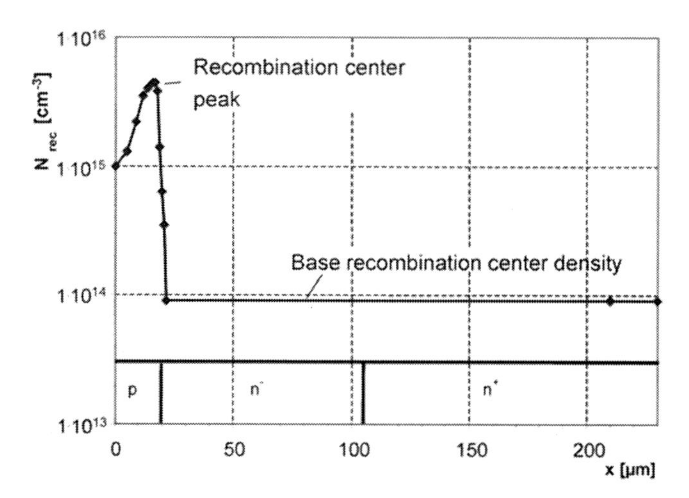

Fig. 5.34 Profile of recombination centers in the CAL diode (scheme)

The first diode that reached the maturity of a series product with this technology was the so-called *c*ontrolled *a*xial *l*ifetime (CAL) diode [Lut94], designed for a blocking voltage of 1200 V in 1994. Meanwhile several manufacturers have applied this concept, whereby devices of up to 9 kV have been realized and are commercially available.

Figure 5.34 illustrates the profile of recombination centers in the CAL diode. The recombination center peak is created by implanting He^{2+}-ions. From this implantation, a profile of lattice defects similar to those in Fig. 4.13 results. Its depth can be adjusted by the energy of the helium implantation and its peak density by the dose. In combination with this implantation, the base recombination center density is adjusted as homogeneous across the base, preferably by electron irradiation. Three degrees of freedom – the recombination center peak depth, its height, and the base recombination center density – are available for adjusting the reverse recovery behavior.

It is most suitable to locate the recombination center peak close to the pn-junction. The main requirement is a low reverse recovery current peak I_{RRM}. For this the pn-junction must be free of carriers at an early moment of the reverse recovery period. The relation between forward voltage V_F and I_{RRM} is better, the closer the recombination center peak is located to the pn-junction. In the CAL diode, the peak of radiation-induced recombination centers is located in the p-layer close to the pn-junction as is shown in Fig. 5.34. Consequently, the main part of multi-vacancies, which act as generation centers (see Sect. 4.9), is outside of the space charge, and the arrangement leads to a low leakage current.

An inverted plasma distribution in the on-state is achieved by this arrangement of the recombination center peak. The on-state plasma distribution shown in Fig. 5.25 is calculated with a recombination center profile according to Fig. 5.34.

The reverse recovery behavior of the CAL diode was already displayed in Fig. 5.21b. The reverse recovery current peak, which can be adjusted by the height of the recombination center peak, is reduced, and the main part of the stored charge is extracted during the tail current phase. The tail current can be adjusted by the base recombination center density. An increased base recombination center density shortens the tail current time, but this has the drawback of increased forward voltage. With the given degrees of freedom, the reverse recovery behavior can be adjusted in a wide range. Thus, a diode can be designed that exhibits soft switching behavior under all conditions relevant in the application, especially also at low current.

5.7.4.5 The Hybrid Diode

Modern fast diodes have been considerably optimized with the described concepts. Thus, there is only a small difference between the CAL diode and the EMCON-HE diode in the voltage range of 1200 V with respect to the relationship of forward voltage drop V_F and reverse recovery charge Q_{RR} [Lut02]. Moreover, there are indications that the limits for possible optimization of fast 1200 V rated diodes based on silicon have almost been approached. However, by parallel and series connection of different diode designs, these limits given for a single device can be exceeded to some degree.

The hybrid diode [Lut00] is shown in Fig. 5.35a. It consists of a parallel connection of two diodes with contrary switching behavior, whereby on the one hand, a snappy diode D_E is used whose low-doped middle layer is designed as thin as

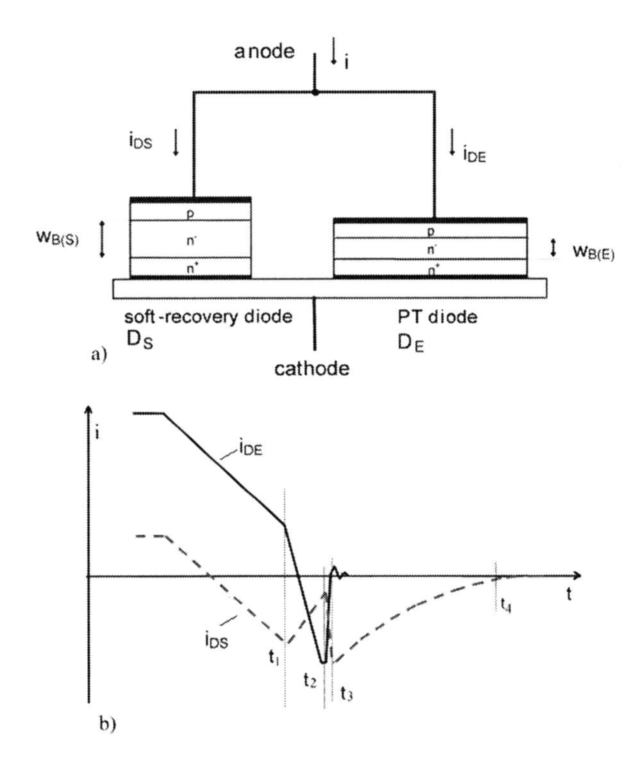

Fig. 5.35 Hybrid diode: (**a**) structure; (**b**) current waveform in both partial diodes during reverse recovery

possible; this design is also known as a "punch-through" (PT) diode. On the other hand, a soft recovery diode D_S is used. With the correct adjustment of the characteristics of both diodes in parallel connection, the low forward voltage drop of a PT diode combined with soft recovery behavior of the soft diode can be obtained simultaneously.

The function principle is presented in Fig. 5.35b. The snappy diode D_E carries the main part of the forward current, whereas the soft diode D_S carries a smaller part. At commutation, the current i_{DS} through D_S crosses zero first and reaches at the instant t_1 its turning point in the reverse current. At this moment, the diode D_E is still carrying forward current. At t_1, the pn-junction of D_S is free of carriers. Then, the diode D_E is commutated with an increased slope di/dt. The total current is still impressed by the outer circuit.

At t_2 the pn-junction of D_E is free and similarly the reverse recovery current maximum of D_E is reached. Between t_2 and t_3 the hard snap-back of the reverse current i_{DE} in D_E occurs. The reverse current i_{DS} in D_S, which contains still stored plasma, increases with the same slope. The total current as the sum of i_{DS} and i_{DE} shows no reverse current snap-off. Therefore, no high voltage peak is induced. The remaining plasma in the soft recovery diode D_S is removed between t_3 and t_4. The behavior of the combined arrangement is soft.

For an effective function of the hybrid diode, the diode D_S has to deliver sufficient charge after the reverse current snap-off in D_E. Thus, the soft diode D_S must be flooded with sufficient plasma, and for this, it must carry between 10 and 25% of the total forward current. The forward voltage drop of both diodes must match to achieve this. In the practical realization, a very fast epitaxial diode with low width w_B of the middle layer is used as diode D_E. Its forward voltage at rated current is in the range of 1.1 V. As diode D_S, a CAL diode with $^1/_6$ of the area of D_E is connected in parallel. This special CAL diode has the double width w_B compared to the diode D_E, but by implementing a low recombination center density in the base, its forward voltage drop is adjusted to achieve the desired current distribution between both diodes. D_S features a large tail current, but it is capable of surely overtaking the reverse current snap-off of the epitaxial diode with a sixfold area.

The hybrid diode is proven in application in drives for fork lifts and other electrical vehicles. In these vehicles the battery voltage is typically in the range of 80 V. The applied step-down converters are built up using MOSFETs. To achieve low conduction losses, the MOSFETs must be designed for a voltage not higher than 160 – 200 V. For the induced voltage peak according to Eq. (5.71):

$$V_{ind} = -L_{par} \cdot \left(\frac{di_r}{dt} \right)_{max} \tag{5.108}$$

Only 80 V margin are available for V_M if the rated voltage of 160 V should not be exceeded for a battery voltage of 80 V. The controlled current is typically high in these applications, i.e. in the range of 200–700 A. High currents lead to a large footprint of the modules, and a significant parasitic inductance L_{par} can hardly be avoided. Therefore in Eq. (5.108), the term di_r/dt, the slope of the current after

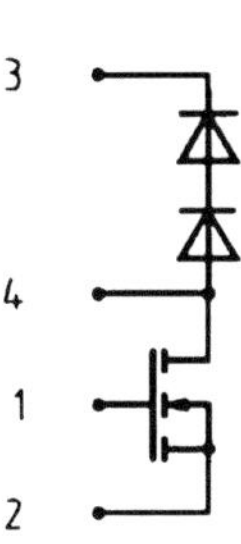

Fig. 5.36 Configuration of a tandem diode and a MOSFET intended for the application as step-up converter

the reverse current maximum, has to be low to keep the induced voltage in the range that is allowed in such applications. The hybrid diode has proven its capability under these hard requirements.

5.7.4.6 The Tandem Diode

The tandem diode consists of a series connection of two fast diodes in a common housing. An example [IXY00] is shown in Fig. 5.36. The configuration is intended for the application in a step-up converter for power factor correction. The concept of the tandem diode aims to offer diodes with a reverse recovery charge Q_{RR} as low as possible for the application at very high switching frequencies.

Equation (5.37) describes an approximate relationship between the stored charge and the base width w_B of a diode. The minimal base width, however, is determined by the required blocking voltage. To express the base width as a function of the desired voltage, we assume the case of a moderate PT dimensioning, as it was given in Eq. (5.15):

$$w_B = \chi \cdot V_{BD}^{\frac{7}{6}} \text{with} \chi = 2,3 \times 10^{-6} \text{cm V}^{-\frac{7}{6}}$$

Inserting this into Eq. (5.37), one obtains

$$Q_F = I_F \frac{\chi^2}{V_{drift}\left(\mu_n + \mu_p\right)} \cdot V_{BD}^{\frac{7}{3}} \tag{5.109}$$

Now the recovery charge of the diodes connected in series, measured as reverse current integral, is equal to the stored charge of a single diode in the series, hence $Q_{RR}^{(series)} = Q_{RR}$. Furthermore, $Q_{RR} = Q_F$ can be assumed as long as the recombination during the reverse recovery is low. However, the total forward voltage of a series of n diodes is

$$V_F = n \cdot \left(V_{drift} + V_j\right) \tag{5.110}$$

Expressing the drift voltage of a single diode V_{drift} by the total forward voltage V_F, Eq. (5.109) yields

$$Q_F = I_F \cdot \frac{\chi^2}{\left(\mu_n + \mu_p\right)} \cdot \frac{\left(\frac{V_{BD}}{n}\right)^{\frac{7}{3}}}{\left(\frac{V_F}{n} - V_j\right)} \tag{5.111}$$

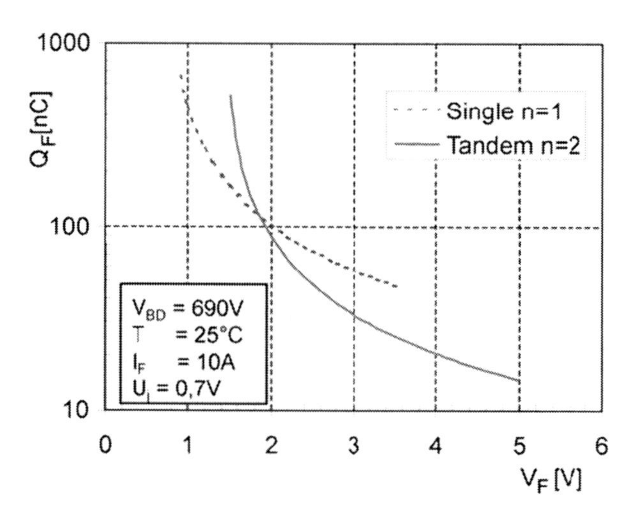

Fig. 5.37 Relation between stored charge and forward voltage for a single diode and for a tandem diode

This relation between stored charge and forward voltage is presented in Fig. 5.37 for a single diode ($n = 1$) and for a tandem diode with $n = 2$. For a forward voltage higher than approximately 1.9 V, the tandem diode shows a better trade-off between V_F and Q_F. The tandem diode recommends itself for applications where very high frequencies and low switching losses are required as the primary feature, but where the current is low and conduction losses are less important. If a forward voltage of 3 V is allowed, a significant reduction of the stored charge can be achieved with the tandem diode. The conduction losses are shared between two devices, which is advantageous regarding the thermal management.

Since a PT dimensioning was assumed for the partial diodes in the above equations, no soft recovery behavior can be expected under these conditions. But in the intended applications with voltages of around 600 V, the converters are usually build up very compactly, so that low parasitic inductances can be realized. In combination with the typically applied low currents, the soft recovery requirements can be reduced, also considering that soft recovery leads to switching losses during current decay. In the considered voltage range, no RC network for symmetrical voltage sharing is necessary using modern devices. Competing with the tandem diode are Schottky diodes made from GaAs or SiC that are also used for very high frequencies.

5.7.4.7 MOS-Controlled Diodes

The idea underlying the MOS-controlled diode (MCD) is to improve diode properties by introducing a third electrode, a MOS gate. Here, we have to anticipate some concepts of Chap. 9 about the MOSFET – the *m*etal *o*xide *s*emiconductor *f*ield *e*ffect *t*ransistor. The basic form of the MCD has the same structure as a power MOSFET. As is shown in Fig. 5.38, the MOSFET contains a pn^-n^+-diode, which consists of the n^+-drain region, the weakly doped n-region and the p-well which is connected with the source metallization, the anode of the diode [Hua94]. The diode

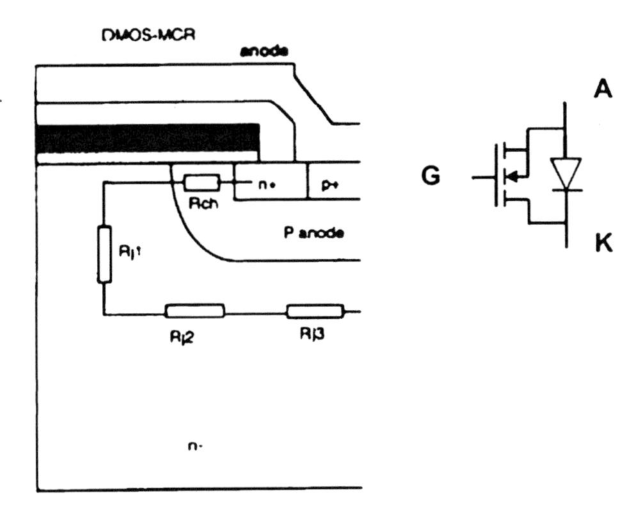

Fig. 5.38 Basic structure of the MOS-controlled diode (MCD) and equivalent circuit. Figure taken from [Hua94]

lies parallel to the MOS channel and conducts if the voltage changes its polarity; it is an "anti-parallel" diode. The voltage at the drain electrode is negative in the diode conducting state. The corresponding equivalent circuit is shown in Fig. 5.38 at the right-hand side.

By opening the channel of the MOSFET with a positive gate voltage, a current path parallel to the pn-junction of the diode is created. If the voltage drop along the channel is smaller than the set-in junction voltage of the diode ($\approx$ 0.7 V at room temperature), almost the whole current is flowing through the channel, and the MCD operates in a unipolar mode without carrier injection. Therefore, during commutation no stored charge of injected carriers is to be extracted. Operating in this mode, the structure was called a "synchronous rectifier" [Shm82]. At a channel voltage higher than the set-in junction voltage, charge carriers are injected by the pn-junction. This range of operation, however, is usually excluded when speaking of a synchronous rectifier.

Normally, the MCD is used in a different manner: since the channel is closed during most of the time of forward conduction, the MCD works like a pin-diode. The channel is opened a short time before commutation, so that the pn-junction is nearly shortened via the n-channel. Therefore, the injection by the anode emitter is strongly decreased. This can be expressed also using an effective emitter efficiency of the p-well *and* the channel. Analogously to Eq. (3.96), this can be written as

$$\gamma = 1 - \frac{j_{\mathrm{n}}}{j} \tag{5.112}$$

where j_{n} now denotes the electron current through the channel. If j_{n} is increased by opening the channel shortly before commutation, γ is strongly decreased and hence also the carrier density is decreased at the anode side of the diode structure. In this way, a drastic reduction of the recovery charge can be achieved. In addition, the inverted carrier distribution leads to a soft recovery behavior.

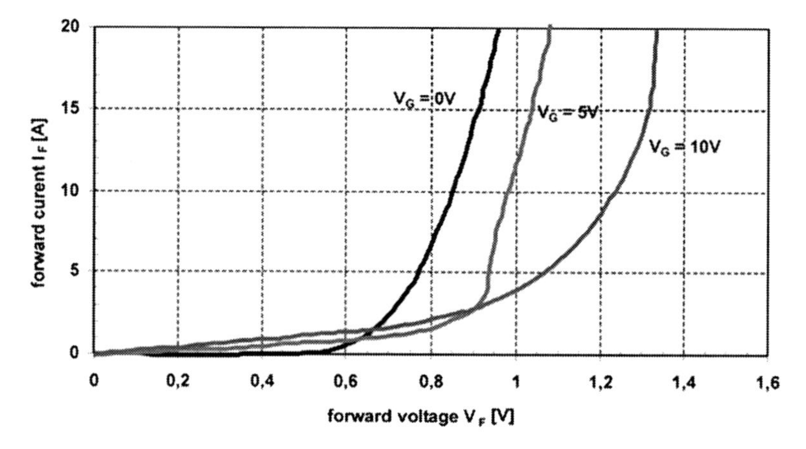

Fig. 5.39 Characteristics of the 1000 V MOSFET IXFX21N100Q in the third quadrant

The parallel operation of the pin-diode and the MOS channel is illustrated in Fig. 5.39. A conventional 1000 V MOSFET was found which shows the effect, i.e. the IXFX21N100Q of the manufacturer IXYS. With the channel closed ($V_G = 0$ V), the current is very small below the set-in junction voltage of approxssimately 0.6 – 0.7 V. With opened channel ($V_G = 5$ V and $V_G = 10$ V), significant current flows already at smaller voltages where a resistive characteristic is observed. This is caused by the channel resistance, R_{ch}, and resistance components arising from a vertical (R_{j1}) and lateral current paths (R_{j2}, R_{j3}) through the weakly doped n-region. On account of R_{ch} the total resistance decreases with increasing gate voltage. This is the operation range of the synchronous rectifier. In the range above 0.7 V, the forward voltage drop at open channel is significantly higher than that in the case of closed channel, as can be seen especially for the higher gate voltage. During the operation cycle of the MCD, the diode works only shortly before turn-off in this part of the characteristics.

In order that opening of the channel results in effective shunting of the current, the voltage drop caused by the external current in the channel and the other resistance components must be smaller than the set-in junction voltage [Hua94]. This holds only up to a critical current I_{crit} given by (at room temperature):

$$I_{crit} = \frac{0.7\ \mathrm{V}}{R_{CH} + R_{j1} + R_{j2} + R_{j3}} \tag{5.113}$$

I_{crit} should be as high as possible and, hence, the resistances R_{Ch} and R_{j1}, R_{j2}, R_{j3} must be as small as possible. An MCD structure optimized for this requirement is shown in Fig. 5.40 [Hua95]. By using a trench structure, the resistances R_{j1} and R_{j2} are removed completely. If additionally the n-region directly under the p-layer is provided with a higher doping, depicted in Fig. 5.40 as n-buffer, also the resistance

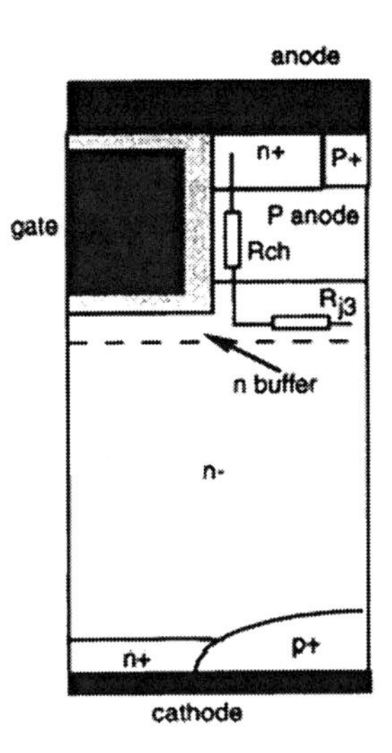

Fig. 5.40 Trench-MCD cell which uses an additional buffer layer at the anode side. Figure from [Hua94]

R_{j3} should be small. But the allowed buffer doping at this position is very closely limited, because it reduces the blocking capability.

The MCD must be switched from the state of fully flooded base, at which no voltage is applied at the gate, to the state of a lower plasma density in the base by applying a voltage to the gate. In this process the main part of the stored charge must be removed. If this is removed by recombination, one has to consider a time of some 10 μs, which is too long for practical application. An n-buffer additionally makes the extraction of holes more difficult. Therefore, an additional p-layer is implemented at the cathode side, which facilitates the extraction of holes to the negatively poled cathode easier.

With these measures, the stored charge before commutation can be reduced significantly; in device simulation even a reduction by a factor of 20–40 was shown [Hua94]. Nevertheless, the stored charge cannot be completely eliminated.

The fundamental disadvantage of all the previously shown variants is that before a blocking voltage is applied to the device, the channel must be closed again. Figure 5.41 depicts an MCD as replacement for a diode in a commutation circuit with an IGBT as switch. The current flows in the freewheeling path, before the diode is commutated from the conducting to the blocking mode at the next turn-on of the IGBT. If the channel of the MCD is open while voltage is applied, the current will not flow into the load but via the channel; a short in the circuit occurs in the commutation branch. Therefore, the channel must be closed beforehand; this point in time must be met with an accuracy of approximately 100 ns. It is hard to ensure this with a driver circuit, however. This disadvantage – that when a positive gate voltage is applied, no blocking capability is given – is the main factor that impedes the implementation of MCD and trench-MCD in practical applications.

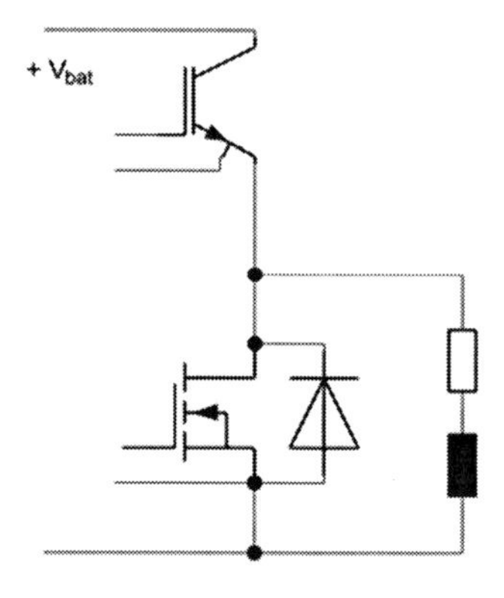

Fig. 5.41 MCD in a commutation branch with IGBT as switching device

An example for a modified solution without this fundamental disadvantage is the emitter-controlled diode (ECD), which is shown in Fig. 5.42 [Dru01]. A low-doped p-zone is connected to the p^+-zone or to the p-well. The highly doped n^+-zone is arranged in the p-well; the path to the low-doped p-zone is controlled by a MOS channel. The channel can be prolonged by a very shallow n^+-layer above the low-doped p-layer. Geometry and doping of the respective layers are chosen such that if a positive voltage is applied to the gate and the channel is open, the current takes the path via the low-doped p-layer. For this, the sum of all voltage drops across this path, i.e. the junction voltage at the pn^--junction, the voltage drop in the channel and further resistive parts in the path, must be lower than the junction voltage of the p^+n^--junction.

Without positive voltage at the gate, a $p^+n^-n^+$-structure is given in which a plasma distribution with increasing density to the pn-junction arises, similar to Fig. 5.6. In the case of a channel opened by a positive voltage at the gate, an inverted plasma distribution is given similar to Fig. 5.33 or Fig. 5.31 (right side). The turn-off process is executed only from the open-channel mode; in this mode a turn-off with soft recovery behavior can be expected. In Fig. 5.43 the internal plasma distribution is compared for both modes. The special progress of the MCD is that, also with open channel, a structure with a blocking capability is given.

The ECD is explained in detail in [Dru01, Dru03] and has not been realized in practice as yet. Nevertheless, it is possible that this idea or similar ideas are used as a foundation for future optimization of diodes in the voltage range of >3 kV. Even though the necessary effort seems to be high at first glance, one has to consider that, nowadays in applications with IGBTs in this voltage range, the possible switching frequencies are limited especially by the reverse recovery behavior of the diode. A progress in diodes can lead to an advantage on the system level that might justify the necessary effort.

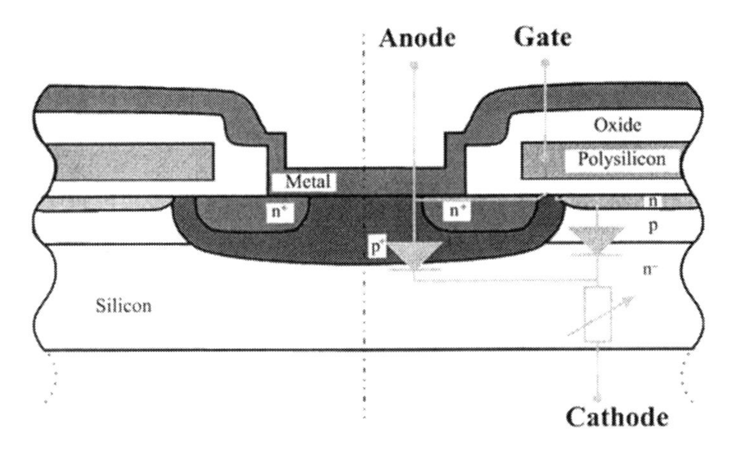

Fig. 5.42 Emitter-controlled diode (ECD). Figure according to [Dru03]

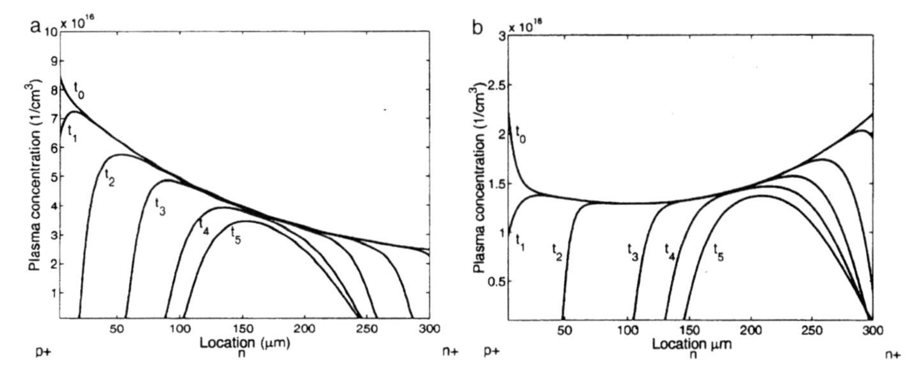

Fig. 5.43 Plasma distribution in the ECD during the reverse recovery process: (**a**) $V_G = 0$, closed channel, (**b**) positive gate voltage, open channel. Figure taken from [Dru01] © 2003 IEEE

5.7.4.8 Diodes with Cathode Side Hole Injection

While all measures up to now modified the anode side, it was found that also the cathode side, the nn^+-junction, can be used to improve the reverse recovery behavior. Structures which inject at reverse recovery additional holes from the cathode side are the field charge extraction (FCE) structure [Kop05] and the controlled injection of backside holes (CIBH) structure [Chm06]. Since there are additional p-layers at the cathode side, no electric field can build up between the plasma and said layers. It can be achieved that the plasma does not detach from the cathode. Instead of Eq.

(5.93), $|v_r| = 0$ applies and from Eq. (5.94) results $w_x = w_B$. This extends the voltage range of soft recovery. Moreover, it was shown for the CIBH diode that, if at the end of the reverse recovery process the space charge punches to the n^+-layer (see Eq. (5.104), the backside p-layers inject additional holes and damps possible oscillations [Fel08].

Since these structures were developed with the intention to increase the capability of the diode to withstand dynamic avalanche, they will be described in more detail in Sect. 12.4.

5.8 Outlook

For the soft recovery behavior of fast diodes, appropriate solutions have been found in the voltage range smaller than 2000 V. Even though further optimization is possible, there are indications that the design is already approaching the limits of what is possible for pin-diodes in silicon. Still there is a potential of hybrid structures for improving diodes.

For the voltage range of 3000 V and more, considerable work still has to be done to realize diodes with satisfactory reverse recovery behavior in applications with very high power. These applications combine switching slopes, which are much steeper than formerly occurring using thyristors and GTOs, with a significant parasitic inductance in the commutation circuit. For these applications the devices have to be optimized to avoid voltage peaks and oscillations even under such hard conditions. Furthermore, dynamic ruggedness is very important; this will be explained in Chap. 12.

Regarding applications with high switching frequencies, Schottky diodes are the better choice. Schottky diodes from GaAs are available as single diodes for 300 V, and they may be developed to reach the voltage range of 600 V. Although Schottky diodes in SiC have already been established for the range of 600–1200 V (see Chap. 6), they can also be designed for a higher voltage range. Because of problems involving material quality and defects, SiC devices are still available only with comparatively small area. For applications in the wide field of motor drives, they should become available in sufficient area (of up to 1 cm^2) to carry enough current per device and at costs that are competitive to silicon devices.

Moreover, there are encouraging results in research regarding pin-diodes in SiC; in particular, here blocking voltages far above 10 kV are possible with a single diode. First results seem to show that the physics of reverse recovery, which was previously explained, can be used in a similar way for analysis and optimization of SiC pin-diodes [Bar07]. However, applications in the high-power range require high currents and, thus, large device areas too.

Therefore, diodes from silicon will probably dominate the market for a long time and still further work for optimizing them is necessary. In the field of fast diodes, Si and SiC will most likely exist in parallel for some time.

References

[Bab08] Baburske R, Heinze B, Lutz J, Niedernostheide FJ: "Charge-carrier Plasma Dynamics during the Reverse-recovery Period in p^+-n^--n^+ diodes, IEEE Trans. Electron Devices ED-55 No 8, pp. 2164–2172 (2008)

[Bal87] Baliga BJ, Modern Power Devices, John Wiley & Sons, New York 1987

[Bal98] Baliga BJ: "Power Devices" in S.M. Sze: Modern Semiconductor Device Physics, John Wiley & Sons, New York 1998

[Bar07] Bartsch W, Thomas B, Mitlehner H, Bloecher B, Gediga S: "SiC-Powerdiodes: Design and performance" Proceedings European Conference on Power Electronics and Applications EPE, (2007)

[Ben67] Benda HJ, Spenke E: "Reverse Recovery Process in Silicon Power Rectifiers", Proceedings of the IEEE, Vol 55 No 8 (1967)

[Chm06] Chen M, Lutz J, Domeij M, Felsl HP, Schulze, HJ: "A novel diode structure with Controlled Injection of Backside Holes (CIBH)". Proceedings of the ISPSD, Neaples pp. 9–12 (2006)

[Coo83] Cooper RN: "An investigation of recombination in Gold-doped pin rectifiers", Solid-St. Electron. 26, 217–226 (1983)

[Deb96] Deboy G et al: "Absolute measurement of carrier concentration and temperature gradients in power semiconductor devices by internal IR-Laser deflection", Microelectronic Engineering 31, 299–307 (1996)

[Dru01] Drücke D, Silber D: "Power Diodes with Active Control of Emitter Efficiency", Proceedings of the ISPSD, Osaka, pp. 231–234 (2001)

[Dru03] Drücke D: Neue Emitterkonzepte für Hochspannungsschalter und deren Anwendung in der Leistungselektronik, Dissertation, Bremen 2003

[Fel04] Felsl HP, Falck E, Pfaffenlehner M, Lutz J: "The Influence of Bulk Parameters on the Switching Behavior of FWDs for Traction Application", Proceedings Miel 2004, Niš/Serbia & Montenegro, 2004

[Fel08] Felsl HP, Pfaffenlehner M, Schulze H, Biermann J, Gutt T,. Schulze HJ, Chen M, Lutz J: "The CIBH Diode – Great Improvement for Ruggedness and Softness of High Voltage Diodes" ISPSD 2008, Orlando, Florida, pp. 173–176 (2008)

[Hal52] Hall RN, "Power rectifiers and transistors", Proc IRE 40, 1512–1518 (1952)

[Hua94] Huang Q: "MOS-Controlled Diode - A New Class of Fast Switching Low Loss Power Diode" VPEC, pp. 97–105 (1994)

[Hua95] Huang Q, Amaratunga GAJ: "MOS Controlled Diodes - A new Power Diode" Solid-St. Electron. 38 No 5, 977–980 (1995)

[IXY00] IXYS data sheet FMD 21-05QC (2000)

[Kop05] Kopta A, Rahimo M: "The Field Charge Extraction (FCE) Diode – A Novel Technology for Soft Recovery High Voltage Diodes" Proc. ISPSD Santa Barbara, pp. 83–86 (2005)

[Las00] Laska T, Lorenz L, Mauder A: "The Field Stop IGBT Concept with an Optimized Diode", Proceedings of the 41th PCIM, Nürnberg (2000)

[Lut94] Lutz J, Scheuermann U: "Advantages of the New Controlled Axial Lifetime Diode", Proceedings oft the 28th PCIM, Nuremberg (1994)

[Lut00] Lutz J, Wintrich A: "The Hybrid Diode - Mode of Operation and Application", European Power Electronics and Drives Journal Vol. 10 No. 2 (2000)

[Lut02] Lutz J, Mauder A: "Aktuelle Entwicklungen bei Silizium-Leistungs-dioden", ETG-Fachbericht 88, VDE-Verlag Berlin (2002)

[Mou88] Mourick P, Das Abschaltverhalten von Leistungsdioden, Dissertation, Berlin 1988

[Nem01] Nemoto M et al: "Great Improvement in IGBT Turn-On Characteristics with Trench Oxide PiN Schottky Diode", Proceedings of the ISPSD, Osaka (2001)

[Sco69] Schlangenotto H, Gerlach W: "On the effective carrier lifetime in psn-rectifiers at high injection levels", Solid-St. Electron. 12, pp. 267–275 (1969)

[Sco79] Schlangenotto H, Maeder H: "Spatial Composition and Injection Dependence of Recombination in Silicon Power Device Structures", IEEE Trans. Electron Devices Ed-26, No 3, pp. 191–200 (1979)

[Sco82] Schlangenotto H, Silber D, Zeyfang R: "Halbleiter-Leistungsbauelemente - Untersuchungen zur Physik und Technologie", Wiss. Ber. AEG-Telefunken 55 Nr. 1–2 (1982)

[Sco89] Schlangenotto H et al, "Improved Recovery of Fast Power Diodes with Self-Adjusting p Emitter Efficiency", IEEE Electron Dev. Letters Vol. 10. pp. 322 – 324 (1989)

[Shm82] Shimada Y, Kato K, Ikeda S, Yoshida H: "Low input capacitance and low loss VD-MOSFET rectifier element", IEEE Trans. Electron Devices, Volume 29, Issue 8, pp. 1332–1334 (1982)

[Sil85] Silber D, Novak WD, Wondrak W, Thomas B, Berg H: "Improved Dynamic Properties of GTO-Thyristors and Diodes by Proton Implantation", IEDM, Washington (1985)

[Syn07] Advanced tcad manual. Synopsys Inc. Mountain View, CA. Available: http://www.synopsys.com (2007)

[Sze81] Sze SM, Physics of Semiconductor Devices. John Wiley & Sons, New York 1981

[Wol81] Wolley ED, Bevaqua SF: "High Speed, Soft Recovery Epitaxial Diodes for Power Inverter Circuits", IEEE IAS Meeting Digest (1981)

[Won87] Wondrak W, Boos A, "Helium Implantation for Lifetime Control in Silicon Power Devices," Proc. of ESSDERC 87, Bologna, pp. 649–652, (1987)

Chapter 6
Schottky Diodes

Schottky diodes are unipolar devices, which means that only one type of carrier is available for the current transport. If they are designed for large blocking voltages, the resistance of the base will increase strongly due to the lack of charge carrier modulation, as will be shown in the following. Schottky power diodes have been used for a long time, but in the last years they have gained an increased importance in the medium power range:

- Si Schottky diodes in the voltage range up to approximately 100 V to be used as freewheeling diodes for MOSFETs. Their advantage is the low junction voltage and the absence of stored charge. At turn-off from conducting to blocking mode, only the capacitive recharge of the junction capacitance needs to be considered. This makes them useful for very high switching frequencies.
- Schottky diodes from semiconductor materials with wide bandgap. With these materials, much higher blocking voltages are possible due to the higher critical fields. However, since the increased bandgap leads to a comparatively high junction voltage in a bipolar device (see Fig. 3.11), a Schottky junction with a much smaller junction voltage has become an attractive alternative.

6.1 Aspects of the Physics of the Metal–Semiconductor Junction

Figure 6.1 shows schematically the band structure for a metal-to-semiconductor interface. To understand the charge carrier transport across this junction we need to define some quantities relating the electron energies in the metal and semiconductor to the energy of a free electron, the so-called vacuum level.

$$\Phi_M\text{: work function for the metal}$$

This is the energy, which must be added to an electron to allow it to escape from the metal, and this is equivalent to the difference between Fermi level, E_F, which is located in the conduction band of a metal and the vacuum level. In the semiconductor,

J. Lutz et al., *Semiconductor Power Devices*, DOI 10.1007/978-3-642-11125-9_6,

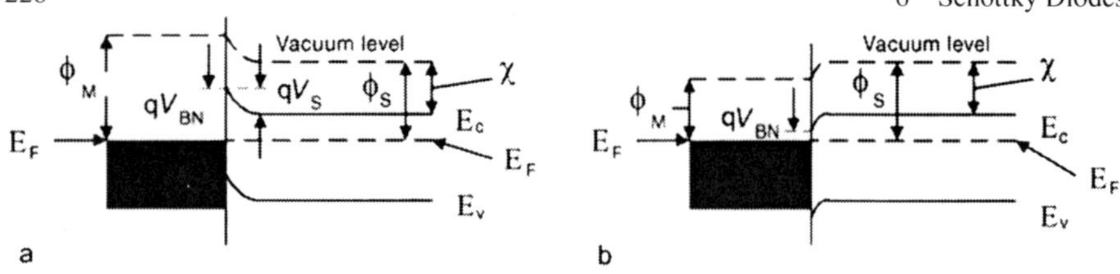

Fig. 6.1 Metal–semiconductor junction for an n-type semiconductor. (**a**) Schottky junction, (**b**) ohmic junction. Figure taken from [Ben99] reprint with permission of John Wiley & Sons, Inc.

Φ_S: work function for the semiconductor

The work function for a semiconductor is also defined as the distance between the Fermi level and the vacuum level. However, the Fermi level of a non-degenerate semiconductor is positioned between the valence band and the conduction band and no electron is allowed to have this energy. Therefore additionally we need to define:

χ: electron affinity for the semiconductor

This is the energy needed to remove an electron located at the bottom of the conduction band, E_C, where most of the conduction electrons reside, to the vacuum level outside the semiconductor.

If a semiconductor and a metal, for which $\Phi_M > \Phi_S$, are put into contact, the electrons close to the metal will leave the semiconductor until again the condition of thermal equilibrium is fulfilled, i.e. the Fermi level is constant throughout the whole structure (Fig. 6.1a). In the semiconductor this leads to a layer depleted of electrons and a space charge region will be formed by positively charged donors, setting up a barrier for further electron transition across the junction. From the figure it can be seen that the height of this barrier for electrons moving from the metal to the semiconductor is given by

$$eV_{BN} = \Phi_M - \chi \tag{6.1}$$

For electrons moving from the semiconductor to the metal the barrier is accordingly

$$eV_K = \Phi_M - (\chi + (E_C - E_F)) \tag{6.2}$$

where V_K is the contact voltage. This junction has a rectifying characteristic: A negative voltage at the metal in relation to the semiconductor will broaden the space charge region and increase the barrier. The metal–semiconductor junction is now in the blocking mode. On the other hand, a positive voltage at the metal with respect to the semiconductor is narrowing the space charge and decreases the barrier. The metal–semiconductor junction is now in forward conduction mode and electrons can flow from the semiconductor to the metal.

In the case of Fig. 6.1b, where $\Phi_M > \Phi_S$ is given, no depletion layer is formed. Instead an enhancement layer of electrons is built up at the semiconductor surface

Table 6.1 Richardson constant A^* for different semiconductor materials

Si	120 A/(cm^2 K^2)	[Sze81]
GaAs	8 A/(cm^2 K^2)	[Sze81]
SiC	400 A/(cm^2 K^2)	[Tre01]

and the barrier is lowered. For $\Phi_M = \chi + (E_C - E_F)$ the barrier vanishes, and for a Φ_M which is small enough a free flow of electrons through the junction is possible. Such a junction is called ohmic junction. For real contacts a large quantity of surface states are always formed by the interruption of the crystal lattice and this induces an additional barrier. In order to produce a good ohmic contact, usually a highly doped semiconductor with a doping > 10^{18} cm^{-3} is necessary. In this case the barrier is so thin that charge carriers can easily tunnel through.

6.2 Current–Voltage Characteristics of the Schottky Junction

The rectifying characteristics of the Schottky junction can be described by an equation for the *I–V* characteristics, in analogy to related equation for the *I–V* characteristics of the pn-junction (3.50)

$$j = j_s \cdot \left(e^{qV/kT} - 1\right) \tag{6.3}$$

with

$$j_s = A^* \cdot T^2 \cdot e^{-qV_{BN}/kT} \tag{6.4}$$

This saturation blocking current j_s is different from that of the pn-junction. In the equation for j_s the term A^* is a material-specific constant, the so-called Richardson constant.

Equation (6.4) is only a first approximation and valid for small voltages and small currents. One important correction term is the image force, $\Delta\Phi$, which must be introduced for larger biasing. It corresponds to a decrease of the work function at a metal surface in the presence of a high electric field [Pau76]. With the image force Eq. (6.4) becomes

$$j_s = A^* \cdot T^2 \cdot e^{-q(V_{BN} - \Delta\Phi)/kT} \tag{6.5}$$

with

$$\Delta\Phi = \sqrt{\frac{q \cdot E}{4 \cdot \pi \cdot \varepsilon}} \tag{6.6}$$

The image force $\Delta\Phi$ is nearly independent of the applied voltage. It is the main reason for the "soft" characteristics of Schottky diodes under reverse bias, giving an increased blocking current close to avalanche breakdown. The effect of the image force is observed especially for Schottky diodes with low barrier heights, but is of less importance for large barrier heights. Under forward biasing the image force leads to an increased barrier height, resulting in an increased forward voltage drop. This effect is important at high current densities of 200–300 A/cm^2. Typical values of the image force potential are in the range of 15–30 mV.

In addition to the image force, there are several other effects that must be considered for a detailed treatment of the Schottky interface [Pau76]. The effect of various correction terms is sometimes summarized by an extended Richardson constant, A^{**}, in the expression for the saturation blocking current j_S. Typically the value for the extended Richardson constant is about 10–20% lower than A^*. Also an ideality factor n is sometimes introduced in the exponent of Eq. (6.3) to obtain a better agreement with experimentally measured values.

$$j = j_S \cdot \left(e^{q \cdot V / n \cdot k \cdot T} - 1\right) \tag{6.7}$$

The value for n is between 1 and 2, but for good Schottky diodes a value of 1.02–1.06 can be obtained. The main reason for a non-ideal behavior, i.e. $n \neq 1$, are generation and recombination centers in the space charge region and at the interface.

Figure 6.2 shows the ideal I–V characteristics of a Schottky diode for very small current densities calculated with Eq. (6.3). Compared to the result with the equation for the pn-junction, the Schottky barrier shows a several orders of magnitude higher saturation leakage current j_S, which in this case is as high as 6.2 μA/cm^2. Furthermore, the total leakage current of a real device will often be significantly larger than the ideal current shown below, as is always the case in

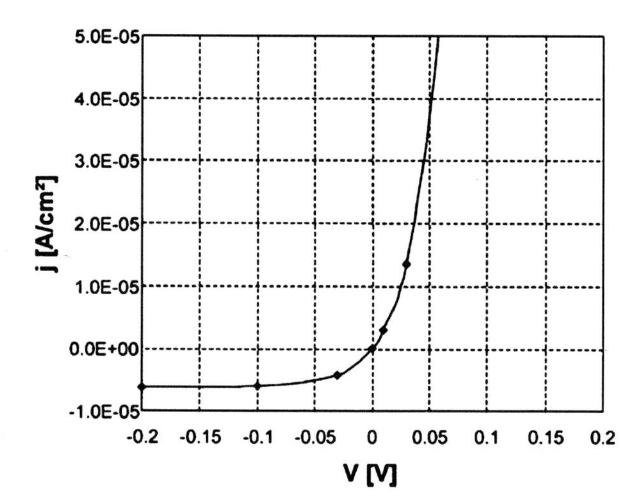

Fig. 6.2 Calculated I–V characteristic of a Schottky diode. The example shows Pd_2Si on Si with a barrier height of 0.73 eV at a temperature of 300 K

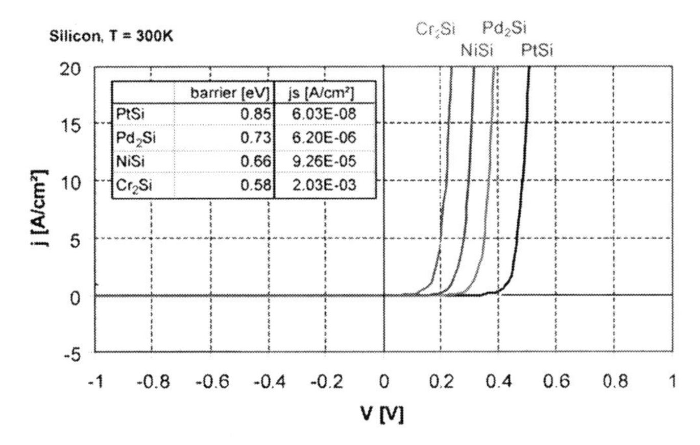

Fig. 6.3 Calculated *I–V* characteristics for Si Schottky diodes for different contact materials according to data from G. Berndes, IXYS Semiconductor GmbH [Ber97]

power devices. This is caused by effects of technology, processing, and junction termination.

Potential barriers for different contact materials are given in inserted table in Fig. 6.3. The calculated *I–V* characteristics using these barriers are shown in Fig. 6.3 on a larger current scale, showing the different threshold voltages for different materials.

For Cr_2Si, for instance, the threshold voltage is only 0.22 V (at 10 A/cm^2), but the leakage current will be about 2 mA/cm^2 for this type of barrier. Moreover, the leakage current will increase with temperature according to Eq. (6.4). The Cr_2Si barrier is therefore only suited for a low blocking voltage, e.g., for diodes in switched-mode power supplies for low voltages. For 100 V Schottky diodes, typically PtSi is used as contact material. For tuning and optimization of the barrier and the contact properties, also composite silicides are used. The *I–V* characteristic is strongly temperature dependent ($j_S \sim T^2$), so the threshold voltage is decreasing with temperature. It should also be pointed out that the series resistance is neglected in the idealized equation (6.3).

In the forward characteristics of a Schottky diode a voltage drop across the low-doped middle layer is added, which is in the given case of a unipolar device a voltage drop across an ohmic resistance. This leads to the total forward voltage drop V_F:

$$V_F = V_S + R_{diff} \cdot I_F = V_S + R_\Omega \cdot I_F \tag{6.8}$$

Here V_S is the voltage drop across the Schottky junction. The middle layer is not modulated since it is a unipolar device, and R_Ω will be significantly larger compared to a pin diode with the same w_B (Note: For a pin-diode, R_{diff} is only a technical definition to simplify the *I–V* characteristics and it is not derived from semiconductor physics).

6.3 Structure of Schottky Diodes

Figure 6.4 shows schematically the structure of a Schottky diode. For power diodes only n-doped silicon is used in practice, because of the higher mobility of electrons. Epitaxial silicon is used for the low-doped n^--layer which is grown on a substrate with high n^+-doping.

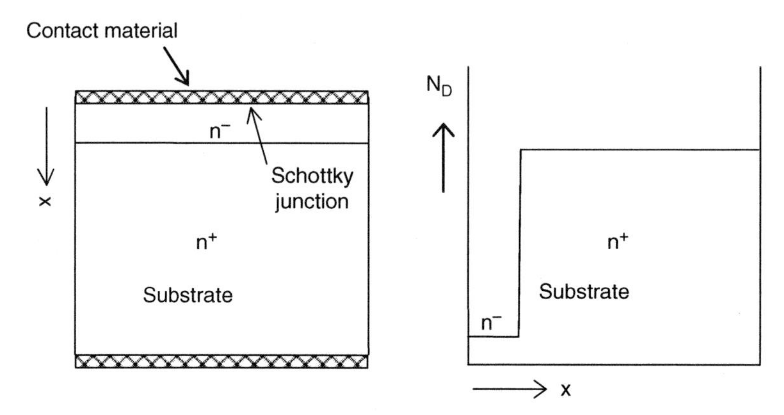

Fig. 6.4 Schematic structure of a Schottky diode

Real Schottky diodes need additional junction termination structures, but this is not shown in Fig. 6.4. Commonly used terminations are field plates, the diffusion of potential rings, or a JTE structure. Sometimes combinations of field plate and potential ring or JTE structures are also implemented. For details of these junction terminations, see Chap. 4.

The low-doped layer must sustain the reverse voltage. Since a Schottky diode is a unipolar device, the width of this layer w_B is directly proportional to the ohmic voltage drop in the forward direction.

6.4 Ohmic Voltage Drop of a Unipolar Device

The resistance of the low-doped middle layer in a unipolar device is calculated by the following expression:

$$R_\Omega = \frac{w_B}{q \cdot \mu_n \cdot N_D \cdot A} \tag{6.9}$$

where A is the active area of the device. For a defined area, only the width w_B, the doping (which is equal to the density of free electrons), and the mobility of electrons determine the resistance.

To sustain the required blocking voltage V_{BD}, the middle layer must have a width and a doping as discussed in Chap. 3. First, a triangular electric field is considered.

For silicon this results in a maximal possible doping N_D for a required breakdown voltage V_{BD} according to inversion of Eq. (3.84):

$$N_D = 2^{-\frac{1}{3}} \cdot B^{-\frac{1}{3}} \cdot \frac{\varepsilon}{q} \cdot V_{BD}^{-\frac{4}{3}} \tag{6.10}$$

The width w_B must be chosen minimally equal to w required for the space charge. By insertion of Eq. (6.10) into Eq. (3.81), one obtains

$$w_B = 2^{\frac{2}{3}} \cdot B^{\frac{1}{6}} \cdot V_{BD}^{\frac{7}{6}} \tag{6.11}$$

Equations (6.10) and (6.11) inserted into Eq. (6.9) results in

$$R_\Omega = \frac{2 \cdot B^{\frac{1}{2}} \cdot V_{BD}^{\frac{5}{2}}}{\mu_n \cdot \varepsilon \cdot A} \tag{6.12}$$

Equation (6.12) gives a first approximation for the resistance of the middle layer based on the assumption of a triangular field or a NPT-design. For a PT-design, or a trapezoidal shape of the electric field, the doping N_D is lowered and the width w_B is decreased, in the same way as described in Fig. 5.3 and Eqs. (5.2), (5.3), (5.4), (5.5), (5.6), (5.7), (5.8), and Eqs. (5.9). A decrease of both N_D and w_B will counteract each other in Eq. (6.9). There will be a minimum in this expression for the doping, which occurs for a moderate PT-design [Dah01].

This effect can be estimated if Eq. (5.9), applicable for a moderate PT dimensioning, is written in the form of a quadratic equation for w_B:

$$\frac{1}{2}\frac{qN_D}{\varepsilon} w_B^2 - \left(\frac{8qN_D}{\varepsilon B}\right)^{\frac{1}{8}} w_B + V_{BD} = 0 \tag{6.13}$$

The solution of this equation is

$$w_B = \frac{\varepsilon}{qN_D}\left(\left(\frac{8qN_D}{\varepsilon B}\right)^{\frac{1}{8}} \pm \sqrt{\left(\frac{8qN_D}{\varepsilon B}\right)^{\frac{1}{4}} - 2 \cdot \frac{qN_D}{\varepsilon} \cdot V_{BD}}\right) \tag{6.14}$$

where only the negative sign of the square root is of technical relevance. It must be kept in mind that Eq. (6.14) is no longer valid if the design changes to a NPT-design or to an extreme PT structure producing a nearly rectangular shape of the field. If now Eq. (6.14) is inserted into Eq. (6.9), the relation between R_Ω and N_D shown in Fig. 6.5 is obtained, exemplifying a device which is designed for a blocking voltage of 240 V.

According to Fig. 6.5 the lowest on-state resistance for a 240 V design will occur for a doping between 1.1 and 1.3 × 10^{15} cm^{-3}. The thickness of the middle layer w_B will be in the range of 11–13 μm. This is smaller than w_B for a device with

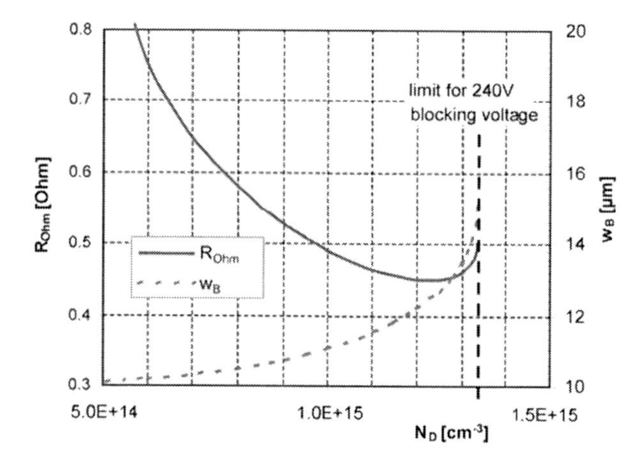

Fig. 6.5 Calculated base width w_B and resistance R_Ω for a 240-V silicon Schottky diode in dependency on the background doping. Device active area is 1 mm^2.

NPT-design, which would require a w_B of minimally 15 μm. The shape of the electric field is significantly trapezoidal, almost as shown in Fig. 5.3b. The minimum resistance R_Ω in the figure is 0.45 Ω; however, for the NPT-design the result from Eq. (6.12) is 0.5 Ω.

This value for a NPT-design, which according to Eq. (6.10) results in a doping level of $N_D = 1.34 \times 10^{15}$ cm^{-3}, can also be seen in Fig. 6.5. Estimation for other blocking voltage will lead to similar results. Equation (6.12) can now be written in a more useful form:

$$R_{\Omega,\min} = 0.9 \cdot \frac{2 \cdot B^{\frac{1}{2}} \cdot V_{BD}^{\frac{5}{2}}}{\mu_n \cdot \varepsilon \cdot A} \tag{6.15}$$

Figure 6.6 shows the width of the middle layer w_B for a trapezoidal field shape as a function of the breakdown voltage V_{BD}, and the corresponding resistance R_Ω of the middle layer of an unipolar device according to Eq. (6.15). For the mobility μ_n

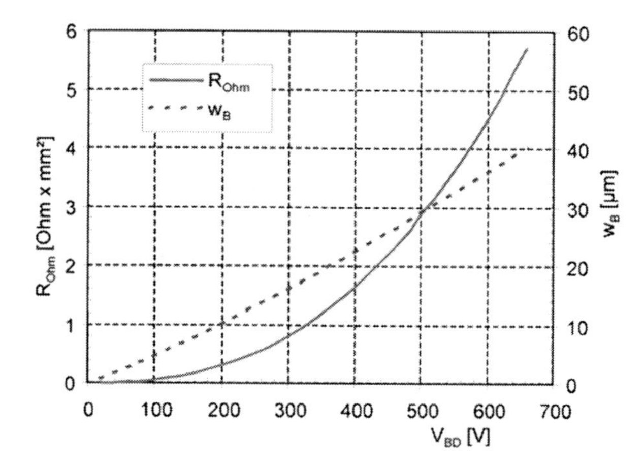

Fig. 6.6 Width of the middle layer and resistivity of the middle layer in dependence on breakdown voltage

the parameters of Appendix A have been used. From Eq. (6.15) one can see that the resistance is growing more than a power of two with an increased breakdown voltage. This is caused by the reduced doping concentration which is necessary for enhanced blocking voltage additionally to the increase of the base width w_B. As a result, unipolar devices for high blocking voltage will have a very high on-state resistance.

In practice, however, a deviation from this series resistance behavior is often observed for silicon Schottky diodes. To achieve sufficient blocking capability of a Schottky diode, a p-doped potential ring is often implemented at the edge of the active area, similar to Fig. 4.19. In this case a pn-junction is connected in parallel to the Schottky junction, and at a forward voltage in the range of the diffusion voltage of the pn-junction, this junction starts to inject charge carriers. This injection of minority carriers will give a significantly lower resistance in the middle layer compared to the resistance calculated with Eq. (6.15). Additionally, the resistance should increase with temperature due to the reduction of the mobility with temperature according to Eq. (6.15). However, this is often not found in measurements of Si Schottky diodes. For Schottky diodes based on SiC the resistance shows the temperature dependence as expected for a decreasing μ_n in Eq. (6.9). Details on this issue will be discussed below, see Fig. 6.10

Example: Design of a Silicon Schottky Diode for a Rated Voltage of 200 V

Because some spread of data is to be expected from tolerances in used materials and fabrication processing steps (roughly 10% as a role of thumb) and to have some safety for tolerances in the measurement technique (up to 10%), the calculation is done for a breakdown voltage of 240 V. From Eq. (6.15) a value for $R_\Omega A =$ 0.45 $\Omega{\cdot}mm^2$ is obtained. At a current density of 1.5 A/mm^2 – a typical current density at rated current – results a voltage drop of 0.68 V. As Schottky contact material PtSi can be used. This material leads to a threshold voltage of 0.5 V, and therefore a forward voltage $V_F = V_S + R_\Omega A j = 1.18$ V is to be expected. In practice, however, less than 0.9 V is measured. This discrepancy can be explained by the bipolar effect, as discussed above.

In comparison, a fast pin epitaxial diode designed for voltages of 200 V can be fabricated with such a low stored charge that it is comparable to the capacitive charge of a Schottky diode. Even though the junction voltage of a Si pin-diode is in the range 0.7 –0.8 V, such fast pin epitaxial diodes does not reach a higher voltage drop than 1.0 V at the current density assumed in this example.

However, if the Schottky diode is designed for a rated voltage of 100 V, the same consideration leads to $R_\Omega A = 0.082$ $\Omega{\cdot}mm^2$. If we take the same PtSi barrier, the resulting on-state voltage is $V_F = V_S + R_\Omega A{\cdot}j = 0.62$ V. Such a low value cannot be reached with a pin-diode. For still lower voltages, the advantage of the Schottky diode compared to a pin diode will be even higher.

6.5 Schottky Diodes Based on SiC

The mostly used SiC polytype is 4H which has a bandgap of 3.26 eV. For a SiC pn-junction this leads to a diffusion barrier or built-in potential of ≈ 2.8 V, as can be seen in Fig. 3.11. The high forward voltage drop resulting from this barrier is a clear disadvantage for SiC device technology. However, this disadvantage can be avoided by using a Schottky junction instead of a pn-junction.
Equation (6.3) leads to the threshold voltages compared in the *I–V* characteristics in Fig. 6.7 with values of $V_{\mathrm{BN}} = 0.85$ eV for PtSi on Si, and with $V_{\mathrm{BN}} = 1.27$ eV for Ti on SiC. The different Richardson constants for SiC and Si are given in Table 6.1. At 10 A/cm^2 one finds for this idealized case that the SiC Schottky diode has a threshold voltage of approximately 0.9 V, and this value can be accepted.

One of the main advantages of SiC compared to silicon is a ten times higher critical electric field strength for avalanche breakdown. In analogy to Si, the doping and width of the middle layer can be calculated for SiC. If a general approach for the ionization rates as power law is used, $\alpha_{\mathrm{eff}} = B \cdot |E|^n$, as already introduced in Chap. 2 in Eq. (2.76), with $B = C/E_0^n$, we obtain a general solution for the relation between avalanche breakdown and doping as it was given in Eq. (3.79). Equation (3.79) solved for N_{D} results in

$$N_{\mathrm{D}} = \frac{\varepsilon}{q} \cdot 2^{\frac{n+1}{1-n}} \left(\frac{B}{n+1}\right)^{\frac{2}{1-n}} V_{\mathrm{BD}}^{\frac{n+1}{1-n}} \tag{6.16}$$

For w_{B}, a more general expression than (6.11) can be derived too. The procedure similar to Chap. 3, Eqs. (3.77), (3.78), Eq. (3.79), (3.80), and (3.81) is used and Eq. (6.16) is inserted into the expression (3.81) for w. It leads for the investigated NPT-dimensioning to

$$w_{\mathrm{B}} = 2^{\frac{n}{n-1}} \left(\frac{B}{n+1}\right)^{\frac{1}{n-1}} V_{\mathrm{BD}}^{\frac{n}{n-1}} \tag{6.17}$$

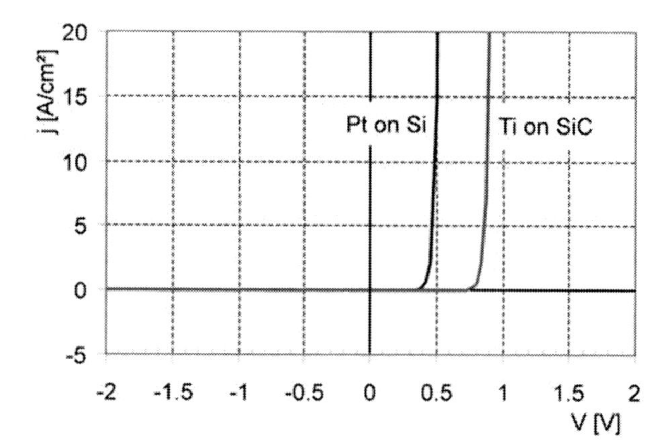

Fig. 6.7 Threshold voltage for Schottky diodes based on Si and SiC

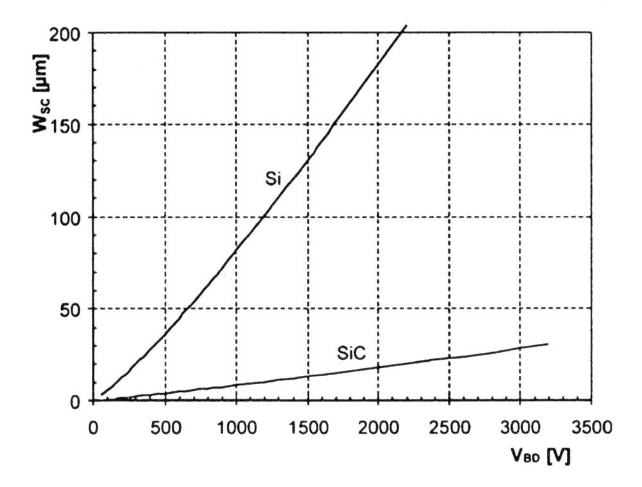

Fig. 6.8 Width of the space charge at V_{BD} for Si and SiC, assuming a triangular field shape

For details on this estimation, see Sect. 3.3.

The width of the space charge for Si and SiC in relation to the blocking voltage under the assumption of a triangular electric field is shown in Fig. 6.8. For Si, $B = 2.107 \times 10^{-35}\,\text{cm}^6\,\text{V}^{-7}$ and $n = 7$ is used as before, for SiC $B = 2.18 \times 10^{-48}\,\text{cm}^{7.03}\,\text{V}^{-8.03}$ and $n = 8.03$ is used, see Eq. (2.82b) based on the proposal from [Bar09]. The figure shows that for SiC the device can be made a factor of 10 thinner, and additionally a much higher doping concentration N_D is possible, compare Fig. 3.19. According to Eq. (6.9) this will lead to a much lower on-state resistance R_Ω.

For a more general expression of the on-state resistance R_Ω, Eqs. (6.16) and (6.17) are inserted into Eq. (6.9) resulting in

$$R_\Omega = \frac{2^{\frac{2n+1}{n-1}} \left(\frac{B}{n+1}\right)^{\frac{3}{n-1}} V_{BD}^{\frac{2n+1}{n-1}}}{\varepsilon \mu_n A} \tag{6.18}$$

With the same procedure as for the Si Schottky diode, one can estimate that a moderate PT-design is most suited for achieving a resistance R_Ω as low as possible for SiC. For the calculation of R_Ω also the mobility of electrons and its dependency on the doping in SiC must be known. The situation is more complicated in SiC since the mobility is anisotropic, i.e. parallel to the crystallographic c-axis it has a higher value than in the orthogonal direction. However, this asymmetry is neglected in most cases, often the so-called analytical model is used as given in Appendix A. The mobility of electrons in n-doped SiC at 25 °C is well described by [Ser94]

$$\mu_n = \frac{947}{1 + \left(\frac{N_D}{1.94 \times 10^{17}\,\text{cm}^{-3}}\right)^{0.61}} \frac{\text{cm}^2}{\text{V s}} \tag{6.19}$$

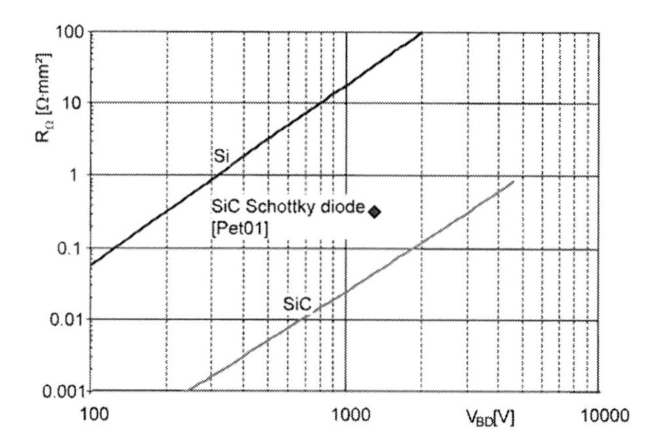

Fig. 6.9 Unipolar resistance $R_{\Omega,\mathrm{min}}$ for Si and SiC as a function of the breakdown voltage

The "effective" electron mobility μ_n in SiC is lower than in Si and it also decreases rapidly for increasing doping. This doping dependency has the consequence that in a device with PT-design, and therefore also lower background doping, the mobility can be higher. The same procedure as used for the derivation of Eq. (6.15) leads to the following expression:

$$R_{\Omega,\mathrm{min}} = 0.88 \cdot \frac{2^{2.43}\left(B/9.03\right)^{0.43} V_{\mathrm{BD}}^{2.43}}{\varepsilon \mu_{\mathrm{n}} A} \tag{6.20}$$

Because the influence of the doping on the mobilities is more pronounced in SiC for a 300–600 V device, decreased doping of the PT-design is of more effect as in Si. This effect is expressed by the pre-factor of 0.88. Finally, for SiC the relative dielectric constant $\varepsilon_{\mathrm{r}} = 9.66$ is used. The resistance $R_{\Omega,\mathrm{min}}$ derived in this way is shown in Fig. 6.9. The curves are approximately linear on this double-logarithmic scale and they are sometimes referred to as the silicon or silicon carbide "unipolar limits." The specific on-resistance resistance for SiC is approximately a factor of 500 lower than for Si at the same rated voltage, or for the same R_{Ω}, a one decade higher blocking voltage can be achieved.

The graphs for the "unipolar limit" vary in the literature by some small amount. One reason for this is that some authors do not consider that the critical electric field strength depends on the shape of the electric field. This field shape dependency is taken into account here by using the respective approach for the ionization rates for Si and SiC.

When dimensioning real devices some tolerances must always be considered and one can only approximate these limits. For SiC the crystal quality is still insufficient and this can have a large influence on the critical field; therefore the possible high field strength before avalanche breakdown cannot be fully used in the design of practical devices. Often the doping is chosen even lower and the width of the middle layer is chosen still higher than given by the expressions to limit the electric field at rated voltages to values below 1.5 MV/cm.

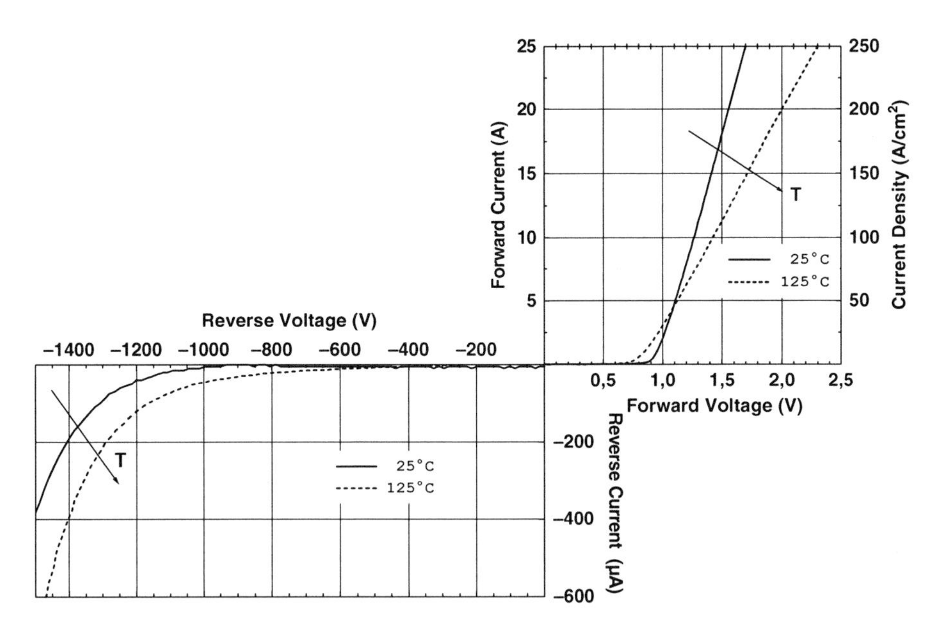

Fig. 6.10 *I*–*V*-characteristic of a 1200-V SiC Schottky diode with 10 mm^2 active area at 25 and at 125°C. Figure taken from [Pet01]

In Fig. 6.10 the *I*–*V* characteristic of a 1200-V SiC Schottky diode is shown. The specific resistance of this device at room temperature, which can be extracted from this characteristic, is also indicated in Fig. 6.9. The "silicon limit" is exceeded by nearly two decades, nevertheless, there is still a high potential left for optimization of SiC Schottky diodes.

The first SiC Schottky diodes were developed by the German company SiCED and introduced on the market by Infineon for rated voltages of 300 , 600, and 1200 V. The charge which occurs at commutation of a Schottky diode into the reverse direction is caused by the junction capacitance and this quantity is not temperature dependent. SiC Schottky diodes are therefore superior to solutions using bipolar Si diodes in all applications that require a high switching frequency (switched mode power supplies, power factor correction, etc.) [Zve01]. The higher price of a SiC diode compared to Si pn-diodes can be compensated by reduction of inductive or capacitive components in the circuit, which is possible because of the allowed higher switching frequency.

The Ti Schottky barrier can theoretically be used for devices up to 3000 V blocking capability. Figure 6.10, taken from [Pet01], shows the *I*–*V* characteristics of a SiCED Schottky diode rated for 1200 V with 10 mm^2 active area. The characteristic is shown for 25 and 125°C. At 25°C one can recognize the threshold voltage in agreement with Fig. 6.7. At higher temperature, the differential resistance is increasing because of the decrease of electron mobility with temperature.

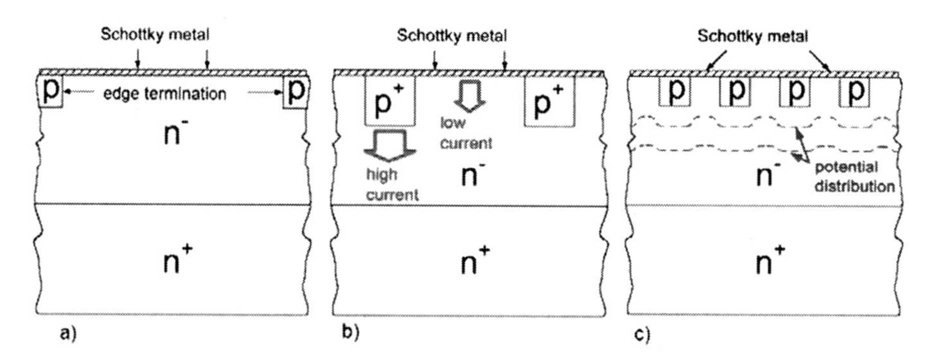

Fig. 6.11 Structure of (**a**) usual SiC Schottky-diode, (**b**) MPS diode structure with p^+-regions to improve the surge current capability [Bjo06], and (**c**) MPS diode structure with p-regions to shield the Schottky junction from high electric fields [Sin00]. Figure similar to [Hei08], Copyright 2008 IEEE

In reverse direction a leakage current increasing with temperature can be observed. In theory the breakdown voltage should increase with the temperature due to the decreasing avalanche coefficients and increasing critical field strength. The decrease of blocking voltage with temperature for real devices and the "soft" shape of the blocking characteristics is probably caused by lattice defects and/or surface defects.

To improve the SiC Schottky diode, merged pin-Schottky structures (MPS-diodes) have been introduced [Bjo06, Sin00]. Most Schottky diodes contain already a p-zone at the edge of the active area for field termination as shown in Fig. 6.11a. In Fig. 6.11b additional high p-doped areas are implemented, which act as p-emitters and inject carriers if the forward voltage drop is above the junction voltage of the SiC pin-diode. This is the case in the range of 2.8 V, as discussed above. An additional solution is the implementation of p-layers in analogy with silicon merged pin-Schottky (MPS) diodes as shown in Fig. 5.28. Such a structure is shown in Fig. 6.11c. These p-layers are separated by a relatively small distance of n^--material with the aim to completely shield the Schottky contact from the high electric field in reverse bias.

The surge current behavior of these structures was investigated in [Hei08]. The *I–V* characteristic of diodes with the three structures in Fig. 6.11 is shown in Fig. 6.12. The structure with p^+-regions (b) shows the lowest forward voltage drop for high current. The effect of the pin-diode is clearly noticeable for currents above 20 A, when the p-layers inject current. This structure shows a high surge current capability. A similar effect, although not as strong, can also be seen for structure (a), where the pin-contribution caused by the edge region injects carriers. Structure (c) shows no pin-behavior. Instead, the *I–V* characteristic deviates from the ohmic behavior at currents above 20 A. At high forward voltage, a significant electric field arises across the middle layer, which has a thickness of only some micrometers. The increased field will then cause the electron mobility to decrease and results in an increased forward voltage drop. A further significant contribution is the temperature increase during measurement. Therefore, the surge current capability is limited for

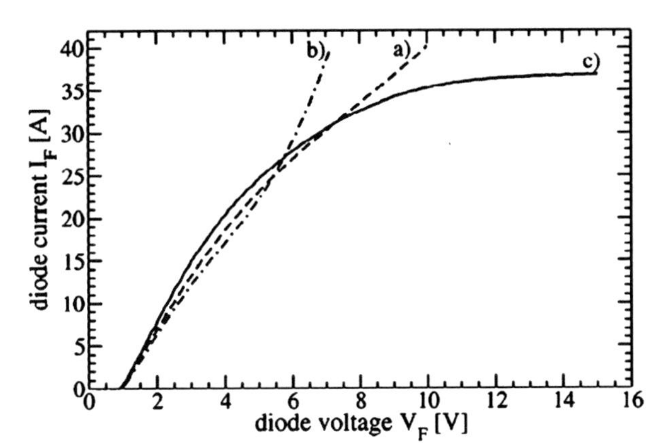

Fig. 6.12 Forward I–V-characteristic for SiC diodes for the three structures is shown in Fig. 6.11. Specified voltage 600 V, rated current 4 A for all three diodes. Figure similar to [Hei08], Copyright 2008 IEEE

this structure. However, a structure like Fig. 6.11c is appropriate for high–voltage SiC Schottky diodes in the voltage range of 2 –3 kV or even higher, because it avoids the reverse bias leakage current of the Schottky junction.

SiC Schottky diodes have today found a market and they are preferred for several applications that involve high switching frequencies. SiC devices are still young and when we compare to Si devices, we must remember that the silicon technology has been improved for several decades to reach the device performance achieved today. For silicon carbide there is still a high potential for optimization of crystal quality as well as for processing technology and device design.

References

[Bar09] Bartsch W, Schoerner R, Dohnke KO: "Optimization of Bipolar SiC-Diodes by Analysis of Avalanche Breakdown Performance", Proceedings of the ICSCRM 2009, paper Mo-P-56 (2009)

[Ben99] Benda V, Govar J, Grant DA, Power Semiconductor Devices, John Wiley & Sons, New York (1999)

[Ber97] Berndes G, Strauch G, Mößner S (IXYS Semiconductor GmbH): "Die Schottky-Diode - ein wiederentdecktes Bauelement für die Leistungshalbleiter-Hersteller", Kolloquium Halbleiter-Leistungsbauelemente, Freiburg (1997)

[Bj006] Bjoerk F, Hancock J, Treu M, Rupp R, Reimann T, "2nd Generation 600 V SiC Schottky Diodes Use Merged pn/Schottky Structure for Surge Overload Protection", Proceedings of the APEC 2006

[Dah01] Dahlquist F, Lendenmann H, Östling M: "A high performance JBS rectifier - design considerations", Material Science Forum Vols. 353–356 p 683 (2001)

[Hei08] Heinze B, Lutz J, Neumeister M, Rupp R: "Surge Current Ruggedness of Silicon Carbide Schottky- and Merged-PiN-Schottky Diodes" ISPSD 2008, Orlando, Florida, USA (2008)

[Pau76] Paul R, Halbleiterdioden, VEB Verlag Berlin 1976

[Pet01] Peters D, Dohnke KO, Hecht C, Stephani D: "1700 V SiC Schottky Diodes scaled up to 25A", Materials Science Forum Vols. 353–356 pp. 675–678 (2001)

[Ser94] Schaffer WJ, Negley GH, Irvine KG, Palmour JW, "Conductivity anisotropy in epitaxial 6H and 4H SiC" Materials Research Society Symposium Proceedings, Vol 339, p. 595–600 (1994)

[Sin00] Singh R et al, "1500 V 4 Amp 4H-SiC JBS Diodes", Proceedings of the ISPSD, Toulouse (2000)

[Sze81] Sze SM, Physics of Semiconductor Devices. John Wiley & Sons, New York 1981

[Tre01] Treu M, Rupp M, Kapels H, Bartsch W, Material Science Forum Vols. 353–356 pp. 679–682 (2001)

[Zve01] Zverev I. et al: "SiC Schottky Rectifiers: Performance, Reliability and key application", Proceedings of the 9th EPE, Graz (2001)

Chapter 7
Bipolar Transistors

The transistor was invented in 1947 first in the form of a point contact transistor, whose emitter and collector were formed by sharp metal wires on a germanium block as base [Bar49, Sho49]. Soon it was clear that the metal semiconductor junctions at the point contacts can be replaced by two closely coupled pn-junctions. The first paper on a bipolar transistor of silicon with diffused emitter and base was published in 1956 [Tan56]. For a bipolar transistor as power switch, the emitter and base are fine interdigitated structures whose distance in the range of 30 μm have to be mastered; the technology for this was available in the 1970s. For a certain time, the bipolar transistor was the most important switching device in power electronics. But already at the end of the 1980s, the IGBT was introduced (see Chap. 10), and the IGBT began to replace the bipolar power transistor. Nowadays power converters are no longer equipped with bipolar transistors. In niche markets, e.g., as line deflection transistors in TVs, they have survived. However, recently, activities to develop bipolar SiC transistors were started.

7.1 Function of the Bipolar Transistor

The bipolar transistor has an npn-structure or pnp-structure. Therefore, it comprises two subsequent pn-junctions. The power transistor, with the exception of the voltage range below 200 V, always has an npn-structure as schematically shown in Fig. 7.1.

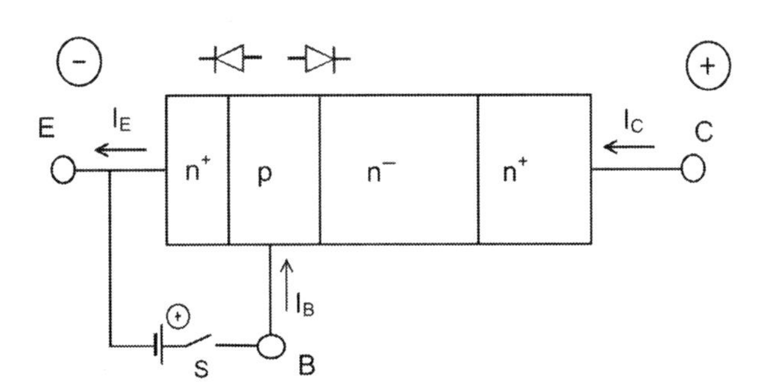

Fig. 7.1 npn-power transistor shown schematically

J. Lutz et al., *Semiconductor Power Devices*, DOI 10.1007/978-3-642-11125-9_7,

When a positive voltage is applied to the collector C, the pn-junction between base B and collector is biased in blocking direction and the pn-junction between base and emitter E is biased in forward direction. With open base, the electron density is low in the base. The base p-doping is in the range between 10^{16} and 10^{17} cm^{-3}; a density $n_{p0} = n_i^2/p$ in the range of 10^4 cm^{-3} can be achieved from the relation (2.6). In spite of a high voltage at the collector, the transistor carries only a very small current; it is in the blocking state.

If the switch S is closed and a positive current I_B is injected at the base contact, the n^+p-junction is biased in forward direction, and so the base is flooded with electrons. But with the base current I_B, not only the emitter current is increased, the electrons in the p-base now have a high gradient of carrier density in the direction of the blocking base–collector junction. They diffuse in this direction and into the low-doped n^--layer. If an electric field is build up, they are accelerated by the field toward the collector.

The current gain α in a common-base circuit is defined by

$$I_C = \alpha I_E + I_{CB0} \tag{7.1}$$

with the leakage current I_{CB0}, measured between base and collector with an open emitter. Furthermore, a current gain β in a common-emitter circuit is defined by

$$I_C = \beta I_B + I_{CE0} \tag{7.2}$$

with the leakage current I_{CE0}, measured between emitter and collector with an open base. According to Fig. 7.1, I_C is the load current which is to be controlled, therefore β is the current gain of the load current in relation to the controlling base current.[1]

If the relation

$$I_E = I_C + I_B \tag{7.3}$$

resulting from Fig. 7.1 is used and the leakage currents I_{CE0} and I_{BE0} are neglected, Eq. (7.2) is solved for β and finally inserted into Eq. (7.3); it follows that

$$\beta = \frac{I_C}{I_B} = \frac{I_C}{I_E - I_C} = \frac{I_C/I_E}{1 - I_C/I_E} = \frac{\alpha}{1 - \alpha} \tag{7.4}$$

The same result is also obtained if the exact definitions (7.1) and (7.2) are used, then the leakage currents must be converted. Equation (7.4) can be solved for α which results in

$$\alpha = \frac{\beta}{\beta + 1} \tag{7.5}$$

The closer α approaches 1 the higher is the current gain β of the collector current. For example $\alpha = 0.95$ results in $\beta = 19$.

[1] Strictly speaking it must be distinguished between the DC current gain $A = I_C/I_E$ and the small-signal current gains $\alpha = \Delta I_C/\Delta I_E$, the same holds for β. This is neglected in the following simplified treatment.

7.2 Structure of the Bipolar Power Transistor

Figure 7.2 shows the structure of a power transistor. The emitter regions are mostly arranged in stripes, the width of the emitter fingers of power transistors is typically in the range of 200 μm. Base and emitter fingers are arranged in an alternating sequence, interleaved like the teeth of two combs.

The collector region is divided into a low-doped n^--layer to take over the electric field and an adjacent higher doped n^+-layer. The diffusion profile of the bipolar transistor along the vertical line A–B through an emitter region in Fig. 7.2 is shown in Fig. 7.3. A diffusion profile of this type features the "triple-diffused" bipolar power transistor. The term "triple-diffused" stands for the subsequent diffusion of the deep collector layer, followed by the p-base layer and finally by the n^+-emitter diffusion. In this case, the n^+-layer is a deep-diffused layer with a Gauss-shaped

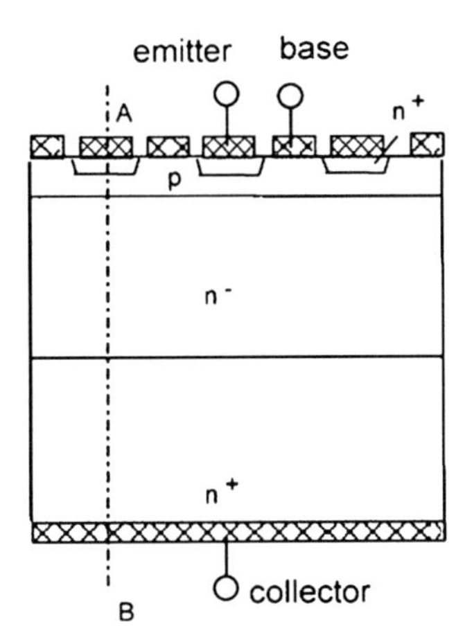

Fig. 7.2 Structure of a power transistor

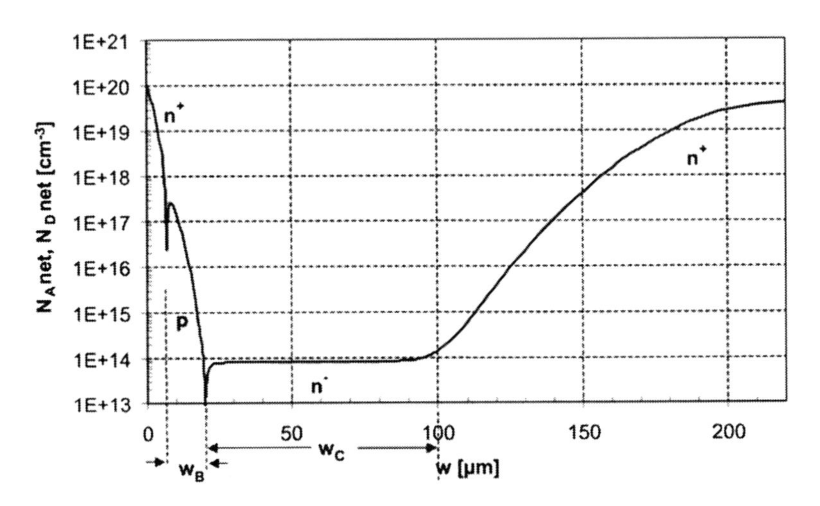

Fig. 7.3 Doping profile of a 1200 V bipolar power transistor along the line A–B in Fig. 7.2

profile of the doping atoms. This deep-diffused layer can also be replaced by an epitaxial layer; then a transistor fabricated from epitaxial wafers is given. But the abrupt junction from the n^+-layer to the n^--layer in epitaxial transistors leads to some disadvantages; see later paragraphs on second breakdown.

7.3 *I–V* Characteristic of the Power Transistor

In Fig. 7.4, a measurement of the forward *I–V* characteristic of a power transistor is shown. Already at a low collector voltage, e.g., at 0.4 V, a comparatively high current density is reached. This is not possible for a device where a pn-junction is forward biased, because of the junction voltage in the range of 0.7 V. In a bipolar transistor in this operation mode, both pn-junctions are forward biased, and the voltage at the pn^--junction is opposed to that at the n^+p-junction. This region of the forward characteristics, which features very low voltage drop, is denoted as saturation region.

The quasi-saturation region, in which the current slightly increases with increasing voltage, follows next to the saturation region. For a higher voltage – not shown in Fig. 7.4 – the bipolar transistor enters the active region. There, the collector current is nearly constant for a given base current and independent of the increasing collector voltage.

From this shape of the forward characteristic, the short-circuit capability of the transistor results. The current is limited, even if there is a short circuit at the load. If in a basic circuit according to Fig. 5.18 a short circuit occurs shorting the load consisting of R and L, then the voltage across the transistor increases until the applied external voltage V_{bat} drops across the transistor. The value of the short-circuit current which occurs in short-circuit mode is given by the transistor *I–V* characteristic and the applied base current. Very high losses are generated in this operation mode; however, if the short circuit is detected within some microseconds by appropriate supervising functions in the driver circuit and is turned-off by the driver, the device will survive this event.

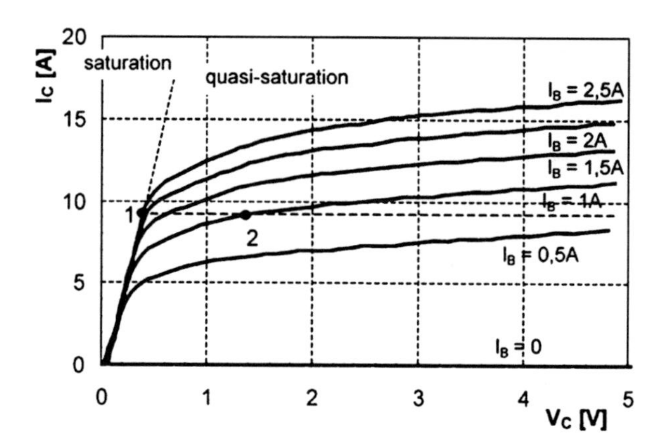

Fig. 7.4 Forward *I–V* characteristic of a bipolar transistor type BUX 48A

7.4 Blocking Behavior of the Bipolar Power Transistor

In Eq. (7.1), I_{CB0} is the leakage current measured between base and collector. To determine the leakage current between collector and emitter at open base, Eq. (7.1) can be used. At open base, $I_C = I_E = I_{CE0}$ holds, and from Eq. (7.1) results

$$I_{CE0} = \alpha \cdot I_{CE0} + I_{CB0} \tag{7.6}$$

solved for I_{CE0}:

$$I_{CE0} = \frac{I_{CB0}}{1 - \alpha} \tag{7.7}$$

The leakage current between collector and emitter is therefore always higher than the leakage current between collector and base. With $\alpha = 0.9$, I_{CE0} is 10 times larger than I_{CB0}.

Also the onset value of avalanche breakdown for a voltage between collector and emitter is lower at open base compared to a voltage between collector and base. To calculate this, Eq. (7.1) must be extended by the effect of avalanche multiplication. The current $\alpha\ I_E$ enters the space charge from the collector side and is enhanced by the electron multiplication factor M_n. The leakage current I_{CB0}, mainly created by generation in the space charge, is enhanced by its multiplication factor M_{SC} (compare Sect. 3.3 on avalanche breakdown, especially Eq. (3.68)). It holds that

$$I_C = M_n \alpha I_E + M_{SC} I_{CB0} \tag{7.8}$$

At open base, $I_C = I_E$ is given, which for the collector current results in

$$I_C = \frac{M_{SC} I_{CB0}}{1 - M_n \alpha} \tag{7.9}$$

Therefore, the collector current grows to infinity for $M_n \alpha = 1$ or it holds that

$$M_n = \frac{1}{\alpha} \tag{7.10}$$

for avalanche breakdown between collector and emitter, whereas the avalanche breakdown between collector and base arises only if M_n grows to infinity. For high current gain(α close to 1), the blocking capability at open base is strongly reduced. For $\alpha = 0.9$, it is sufficient as the condition for avalanche breakdown that the multiplication factor M_n grows to $1/0.9 = 1.11$.

Because the ionization factors for electrons are much higher than that for holes, $\alpha_n > \alpha_p, M_n$ grows much faster with the electric field (see Fig. 3.13). M_n determines the avalanche breakdown and the approximation by an effective multiplication factor M or an effective ionization rate α_{eff} is no longer valid. The breakdown voltage V_{CEO} is considerably smaller than V_{CB0}. Figure 7.5 shows the difference between

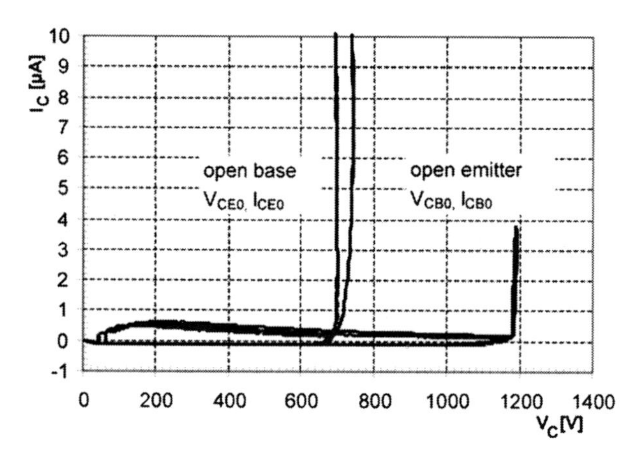

Fig. 7.5 Blocking characteristic of the transistor BUX 48A with open base and with open emitter

V_{CB0} and V_{CE0} for a commercially available bipolar transistor. For this example, V_{CE0} amounts about 60% of V_{CB0}.

For a pnp-transistor, we obtain in analogy

$$M_p = \frac{1}{\alpha} \tag{7.11}$$

Because of $M_p < M_n$, the difference from V_{CEO} to V_{CB0} will be smaller for a pnp-transistor.

In Sect. 3.3, with Eq. (3.69) an often used approximation for the multiplication factor at a voltage $V < V_{CE0}$ was given [Mil57]

$$M = \frac{1}{1 - (V/V_{CB0})^m} \tag{7.12}$$

where V is an applied voltage smaller than V_{CE0}. With Eq. (7.10), it leads for $V = V_{CE0}$ to

$$V_{CE0} = (1 - \alpha)^{\frac{1}{m}} \, V_{CB0} \tag{7.13}$$

where a value of 5 for m is given as typical for a bipolar npn-transistor [Ben99]. The value $m = 5$ leads to agreement with measurements, if for α the value at typical forward operation conditions is used where both β and α are high. However, static avalanche breakdown effects appear at low current where the current amplification is low, since a high share of the current recombines in the base layer. If this is considered, $m = 2.2$ as suggested in Sect. 3.3 is a better choice.

An approximation for a transistor with typical structure can also be found for V_{CE0} [Ben99]

$$V_{CE0} = K_1 w_C \tag{7.15}$$

where w_C is the width of the low-doped collector zone, as defined in Fig. 7.3, and K_1 is given as 10^5 V/cm.

The difference between reverse blocking voltage and reverse current of a transistor structure with open base and the reverse blocking voltage of a pure pn-junction is very important for devices with several pn-junctions. It is also very fundamental in the practical application of the bipolar transistor, since for a voltage applied at open base, the device goes into breakdown at a much lower voltage and it is potentially destroyed. On the other hand, if a negative voltage is applied to the base with respect to the emitter, both pn-junctions of the transistor are biased in reverse direction. The leakage current of both junctions is extracted as base current, and no more interaction of both pn-junctions occurs. For this case, the blocking capability between collector and emitter is approximately the same as between collector and base. A similar effect occurs if base and collector terminal are connected by a short circuit. In practical application, a small negative voltage is applied at the base of the transistor if a high reverse voltage emerges at the collector. Usually a negative voltage at the base is applied for turning off the collector current, and it is maintained continuously during the blocking mode.

7.5 Current Gain of the Bipolar Transistor

According to definition (7.1), it holds that

$$\alpha = \frac{j_C - j_{CB0}}{j_E} \tag{7.16}$$

Equation (7.16) is multiplied in nominator and denominator by the electron current j_{nB} which is injected by the emitter into the base

$$\alpha = \frac{j_{nB}}{j_E}\frac{j_C - j_{CB0}}{j_{nB}} = \gamma\alpha_T \tag{7.17}$$

The first term on the right-hand side of Eq. (7.17) corresponds to the emitter efficiency γ, which was already introduced in Sect. 3.4 with Eq. (3.95) and which is given for an n-emitter by

$$\gamma = \frac{j_{nB}}{j_{nE} + j_{pE}} \tag{7.18}$$

For an n-emitter, this definition describes the share of the electron current j_{nB}, injected into the base, related to the total emitter current.

The second term in Eq. (7.17) is denoted as transport factor α_T

$$\alpha_T = \frac{j_C - j_{CB0}}{j_{nB}} \tag{7.19}$$

For an npn-transistor, this corresponds to the share of the electron current injected by the emitter which arrives at the collector. For $j_C = j_{CB0}$ and therefore $\alpha_T = 0$, only the leakage current arrives at the emitter. For an npn-transistor with high current gain, γ as well as α_T shall be close to unity, so that α exhibits a value close to 1.

Now the emitter efficiency γ shall be investigated in more detail. Recombination at the pn-junction between emitter and base shall be neglected; this is feasible at current densities higher than 1 mA/cm^2. The electron currents on both sides of this pn-junction are therefore assumed to be equal, $j_{nE} = j_{nB}$. Then it follows that

$$\gamma = \frac{j_{nB}}{j_{nB} + j_{pE}} = \frac{1}{1 + \frac{j_{pE}}{j_{nB}}} \tag{7.20}$$

The minority carrier current j_{pE} penetrating into the emitter can be expressed by

$$j_{pE} = q \cdot \frac{D_p}{L_p \cdot N_E} \tag{7.21}$$

and for the electron current j_{nB} penetrating into the base at the condition of low injection it holds

$$j_{nB} = q \cdot \frac{D_n}{L_n \cdot N_B} \tag{7.22}$$

Equations (7.21) and (7.22) are inserted into Eq. (7.20). For the case of low injection, that means that the flooding with free carriers is smaller than the doping of the base N_B, the emitter efficiency γ can now be expressed by

$$\gamma = \frac{1}{1 + \frac{D_p}{D_n} \cdot \frac{N_B}{N_E} \cdot \frac{w_B}{L_p}} \tag{7.23}$$

In this equation L_n, the diffusion length of the electrons in the base, was replaced by the width of the base w_B, since w_B is always smaller than L_n for the high carrier lifetime typical in bipolar transistors. Further L_p in the emitter will be small, after all w_B and L_p will be in the same order or magnitude. Therefore, the dominating term in Eq. (7.23) is the quotient N_B/N_E. To achieve γ close to 1, the doping of the emitter N_E must be much higher than the doping of the base N_B. Equation (7.23) gives a first approximation for these relations, which are very important for the design of a transistor.

Equation (7.23) is valid for the case of low injection. Furthermore, bandgap narrowing was not considered in Eq. (7.23); that means a not too high doping of the n$^+$-emitter was presumed. Additionally, the description of the emitter efficiency in Eq. (7.23) does not contain the current dependency of the emitter efficiency. Based on Sect. 3.4, the emitter efficiency can be investigated in more detail. The n-emitter is characterized by the emitter parameter h_n. Auger recombination and bandgap narrowing determine an n-emitter with high doping; in analogy to Eq. (3.104) it holds that

$$h_{\rm n} = e^{\Delta E_{\rm g}/kT} \sqrt{D_{\rm p} \cdot c_{\rm A,n}} \tag{7.23b}$$

With the Auger coefficient $C_{\rm A,p} = 2.8 \times 10^{-31}$ cm^6/s and with the mobility $\mu_{\rm p}$ of 79 cm^2/V s (estimated for a doping of 1×10^{19} cm^{-3}) and using bandgap narrowing according to Slotboom and DeGraaf as described in Eq. (2.25), one obtains

$$h_n \approx 2 \times 10^{-14}\,{\rm cm}^4/{\rm s}$$

The experience of the authors with design and fabrication of bipolar transistors has shown that the parameter $h_{\rm n}$ is in the range between 1×10^{-14} cm^4/s and 2×10^{-14} cm^4/s up to a doping of 5×10^{19} cm^{-3} and it only increases at higher doping resulting from Eq. (7.23b) with bandgap narrowing according to Eq. (2.25). A high value for $h_{\rm n}$ indicates a reduced emitter efficiency.

The emitter efficiency can be expressed in analogy to Eq. (3.96) as

$$\gamma = 1 - q h_{\rm n} \frac{p_{\rm L}^2}{j} \tag{7.24}$$

To estimate γ, the density of the free carriers $p_{\rm L}$ at the junction between emitter and p-base must be known. In a power transistor, shown again schematically in Fig. 7.6 (top), a low-doped collector layer with a thickness $w_{\rm C}$ follows adjacent to the p-base. This layer is responsible for taking over the space charge in the forward blocking mode of the transistor. It must be flooded with free carriers in on-state to achieve a low voltage drop in the conduction mode. The effective base width increases from $w_{\rm B}$ to $w_{\rm B}+w_{\rm C}$ (Fig. 7.6), referred as Kirk-effect [Kir62].

The transistor is in the mode of saturation at high base current and low voltage $V_{\rm C}$. This point is marked in the I–V characteristic in Fig. 7.4 with the number 1. The holes injected in the base diffuse also into the low-doped collector region. A conductivity-modulated zone is built up in which $n \approx p$ holds. The shape of the carrier density at point 1 in the I–V characteristic is shown in Fig. 7.6 with the

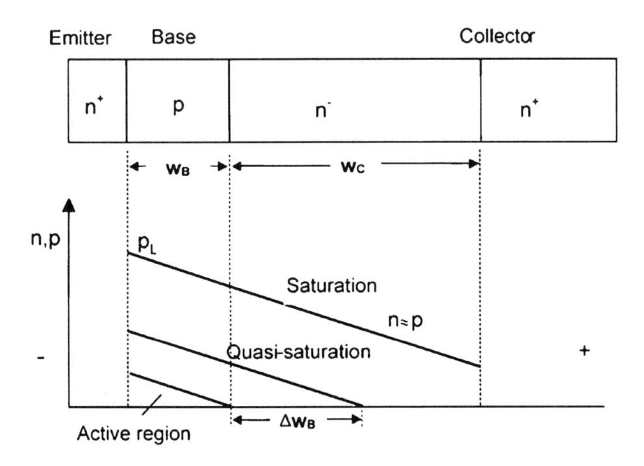

Fig. 7.6 Density of the plasma of free carriers in a bipolar transistor for varied base current at constant collector current

graph for the mode of saturation. The p-base as well as the low-doped collector layer are flooded with free charge carriers. The current transport from the collector to the emitter can be considered as solely carried by electrons, since the diffusion of holes is oriented in the direction against the electric field, whereas the diffusion of electrons is supported by the electric field.

The density of free carriers decreases from the emitter towards the collector. For the distribution across base and low-doped collector layer, Eq. (5.26) still applies. This equation leads to a sagging carrier distribution. The grade of deviation from a linear shape is determined by the charge carrier lifetime. It turns out to be more pronounced for a lower carrier lifetime. In high-quality bipolar transistors, the carrier lifetime is high and the loss by recombination is low; therefore, this sagging is neglected in Fig. 7.6. This is a good approximation for $L_n > 2(w_B + w_C)$.

Therefore, with the assumed simplified relation

$$\frac{dp}{dx} = -\frac{p_L}{w_B + w_C} \tag{7.25}$$

which applies for the point of transition from the saturation mode to the quasi-saturation mode, see Fig. 7.6, and with $j_p = 0$ and $j = j_C$, the result obtained for the collector current is based on Eq. (5.21)

$$j_C = \frac{\mu_n + \mu_p}{\mu_p} q D_A \frac{p_L}{w_B + w_C} \tag{7.26}$$

Using Eq. (5.23) and the Einstein relations (2.44)

$$j_C = 2qD_n \frac{p_L}{w_B + w_C} \tag{7.27}$$

is obtained. Solved for the density of free carriers at the emitter-base junction

$$p_L = \frac{j_C (w_B + w_C)}{2D_n q} \tag{7.28}$$

follows. Equation (7.28) inserted into Eq. (7.24) now allows the estimation of the emitter efficiency. For a transistor with $w_C = 50\,\mu\text{m}$, for example, and $w_B = 10\,\mu\text{m}$, a value for p_L of $2\times10^{16}\,\text{cm}^{-3}$ results for a current density of 30 A/cm^2. The emitter efficiency is $\gamma = 0.96$. For 10 A/cm^2, a value of $\gamma = 0.99$ is obtained. The emitter efficiency depends strongly on the current density. With increasing collector current, the emitter efficiency and thereby the current gain decreases. This is observed for all power transistors.

The transport factor α_T was the second factor for the current gain according to Eq. (7.17), it can be written for $L_n > 2w_{eff}$ [Sze81, Ben99]

$$\alpha_T = 1 - \frac{w_{eff}^2}{2L_n^2} \tag{7.29}$$

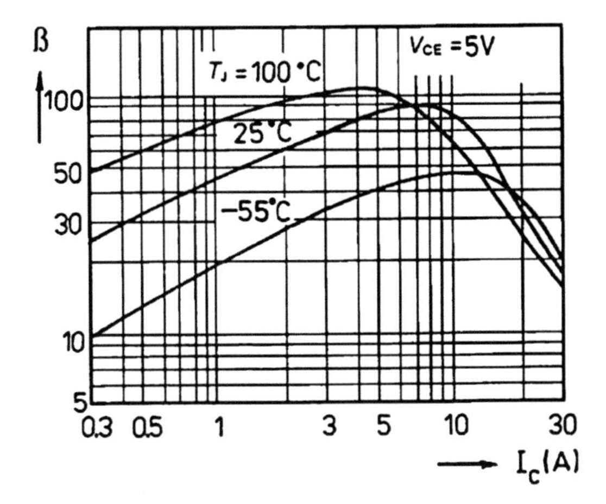

Fig. 7.7 Current gain β as function of collector current and temperature. Data by M. Otsuka, Development of High Power Transistors for Power Use, Toshiba 1975. Reprint from [Ben99] with permission of John Wiley & Sons, Inc.

where L_n represents the diffusion length in the base, which is combined with the carrier lifetime according to Eq. (3.48). w_{eff} is the effective width of the base, which may be shortened compared to w_B at increased voltage, see Fig. 7.8. Equation (7.29) is usually derived for the low-injection condition, for details see [Ben99]. To attain α_T close to 1, w_B must be chosen small and L_n must be the maximum possible. A high current gain requires a short base and a carrier lifetime in the base as high as possible.

The current gain of the bipolar transistor depends on the current and the temperature. For the current gain β, this is shown in Fig. 7.7.

For very small current, β is small; the current injected into the base recombines to a large extent in the base layer. For increasing current, β reaches a maximum; for further increasing current, at the condition of high injection, it decreases again. This is attributed to the emitter efficiency, which decreases as described by Eqs. (7.24) and (7.28). Additionally, the temperature dependency must be taken into consideration. β increases with temperature for small and medium current, because the carrier lifetime increases with temperature. For high current, the decrease of β with temperature is shifted to lower currents. The rated current of a bipolar transistor is typically in the range where β und α are already declining.

7.6 Base Widening, Field Redistribution, and Second Breakdown

If the base current is reduced while the collector current remains constant, the transistor transits into the mode of quasi-saturation; in the I–V characteristic in Fig. 7.4, this is the shift from point 1 to point 2. Now the low-doped collector layer is flooded with plasma of free carriers only up to Δw_B in Fig. 7.6. Only electrons carry the current in the region $w_C - \Delta w_B$ which is free of plasma, and because of the low doping, a significant resistive voltage drop is created. It can be described in analogy to Eq. (6.9) by

$$\Delta V_{CE} = \frac{j_C (w_C - \Delta w_B)}{q\mu_n N_D} \tag{7.30}$$

If the collector current is kept constant and the base current is decreased, then Δw_B decreases as shown in Fig. 7.6. For $\Delta w_B = 0$, only the base w_B is flooded with free carriers; the transistor has now reached the active region. At this point, $\Delta V_{CE} = 13.4$ V results for the example of a 600 V transistor with $w_C = 60\ \mu$m, background doping $N_D = 1 \times 10^{14}$ cm^{-3}, $j_C = 50$ A/cm^2, and $\mu_n = 1400$ cm^2/Vs.

Now the pn-junction between base and collector has become free of carriers. If now the voltage is increased, a space charge will build up. For a moderate voltage this is shown in Fig. 7.8 with line 1, for increased voltage line 2. The flooded zone is pushed back further into the base by the amount that the space charge penetrates the higher doped base layer. The base width is shortened to the effective base width w_{eff}, which leads to a slight increase in the current gain. In literature, this is described as Early effect (named after its discoverer James M. Early, 1922–2004) [Ear52]. Power transistors are usually not operated in the active region. However, they pass through the active region at switching instants.

For a voltage in the active region close to the voltage V_{CE0}, the electric field is shown in Fig. 7.8, line 2. The space charge has build up over nearly the whole layer w_C for this case.

The collector current now flows through the space charge as electron current. The condition of a high electric field applies, and the electrons travel with the drift velocity $v_d \approx v_{sat}$ over almost the whole distance. For the electron density holds

$$n = \frac{j}{q v_{sat}} \tag{7.31}$$

The shape of line 2 in Fig. 7.8 is given only as long as the density of electrons travelling through the n$^-$-layer is small compared to the background doping N_D. The negatively charged electrons compensate the background doping and according to Poisson's equation it is

$$\frac{dE}{dx} = \frac{q}{\varepsilon}(N_D - n) \tag{7.32}$$

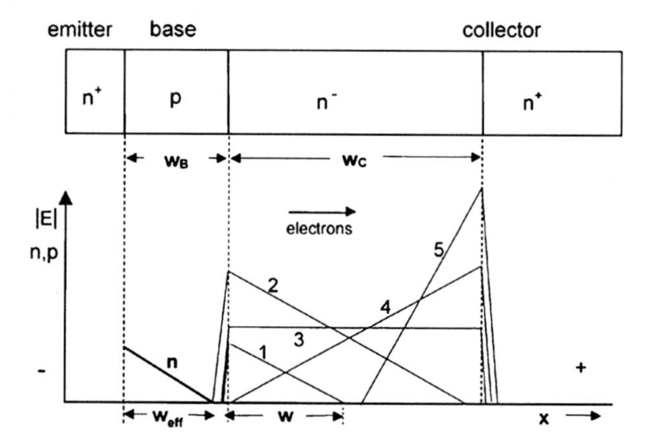

Fig. 7.8 Electric field in the active mode of the bipolar transistor. 12: increasing voltage, V_{CE}; 23, 4, 5: constant voltage; increasing current I_C

If the collector current increases by increased base current, the condition will be reached that the density of electrons flowing through the space charge is equal to the background doping:

$$\frac{j}{qv_{\text{sat}}} = N_{\text{D}} \tag{7.33}$$

For this case, $\mathrm{d}E/\mathrm{d}x = 0$ applies and an almost rectangular electric field with the shape of line 3 in Fig. 7.8 exists. For a bipolar transistor with the background doping $N_{\text{D}} = 1 \times 10^{14}$ cm^{-3}, Eq. (7.33) is fulfilled at a current density of approximately 160 A/cm^2, using $v_{\text{sat}} = 10^7$ cm/s.

If the collector current is further increased by an increased base current, then $n > N_{\text{D}}$ holds, and the gradient of the electric field changes its sign, see line 4 in Fig. 7.8 [Hwa86]. The electric field is redistributed and the field maximum shifts from the pn-junction to the nn$^+$-junction. A further increase in the current leads to a further increas in field strength (line 5) and finally avalanche breakdown at the nn$^+$-junction will occur.

This is the onset of second breakdown. This effect was explained first by Phil Hower [How70]. Second breakdown is destructive: holes, generated by avalanche at the nn$^+$-junction, are accelerated in the layer w_{C}. The front side of the transistor is in the active mode, the arriving holes act as an additional base current, and even more electrons are generated by the emitter, etc. A positive-feedback loop is established. Such a mechanism is usually destructive.

To avoid this critical condition, a safe operating area (SOA) is defined for the transistor for the case of positively biased base – FBSOA, forward-biased SOA – and for the case of negatively biased base – RBSOA, reverse-biased SOA. The turn-off is a particularly critical condition in this respect. During turn-off with inductive load, the voltage must increase first before the current can decrease. The transistor passes through the active region of the I–V characteristic. The specification of an RBSOA sets a limit for the voltage against which the transistor can be turned off. Figure 7.9 shows an example.

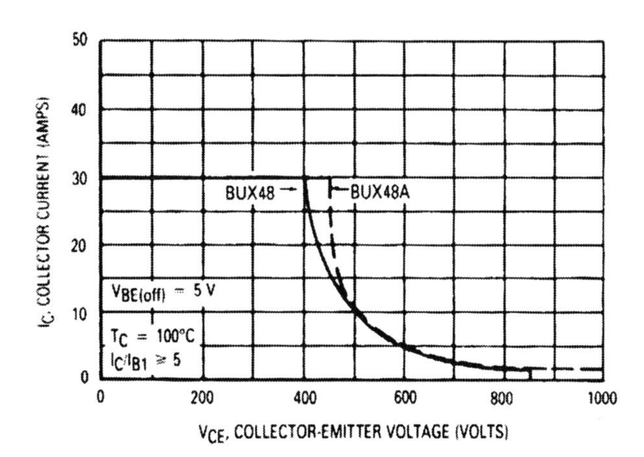

Fig. 7.9 RBSOA of the Motorola transistor BUX 48

The turn-off process is also critical because of the following reason: the current below an emitter finger is extracted from both edges toward the center. At turn-off with an inductive load, a small area in the center of the emitter finger finally remains, which carries the total current. Thus, the current density is increased and the mechanism according to Eqs. (7.31) to (7.33) is triggered at lower currents. Some design measures are possible to increase the SOA. The width of the emitter fingers can be reduced and the pitch of the structure can be diminished. Also special emitter structures were tested, which maintain the current flow at the edge of the emitter, e.g., the ring emitter structure [Mil85]. These structures showed an extended RBSOA and a considerably improved stability against second breakdown.

Finally, a shallow gradient of the diffusion profile at the nn^+-junction, as shown in Fig. 7.3, is a countermeasure against the formation of a field peak at the nn^+-junction. For a slightly increasing n-doping toward the collector, the electric field can penetrate significantly into the n-collector layer and a higher voltage can be sustained before the conditions for avalanche breakdown are encountered. Usable transistors in the range of 1000–1400 V could be manufactured with a diffusion profile similar to Fig. 7.3.

7.7 Limits of the Silicon Bipolar Transistor

If the transistor is designed for higher voltage, the low-doped collector layer w_C must be widened. Since the function of the bipolar transistor is based on diffusion of holes into this low-doped collector region, the current gain decreases with increasing w_C. The Motorola transistor in Fig. 7.4 exhibits $\beta = 10$ only up to a collector current of 10 A. The high base current requires a high effort and generates significant losses in the base drive units.

The required base current could be reduced down to acceptable values by the introduction of double- and triple-stepped Darlington transistors [Whe76]. Darlington transistors with blocking voltages of 1200–1400 V became available with up to 100 A of controllable current per single die. With Darlington transistors, high-switching frequencies are no longer possible. But the requirements for variable speed motor drives with switching frequencies in the range of 5 kHz could be fulfilled by Darlington transistors.

An extension to higher voltages is not possible with silicon. Meanwhile, a field-controlled device, the IGBT, was found for the motor drive applications. IGBTs are much easier to control and cause lower losses in the driver. Therefore, bipolar power transistors have been widely taken from the market of power devices and replaced by IGBTs. But the knowledge of the effects in bipolar transistors is essential for a deeper understanding of the effects in more complex power devices.

7.8 SiC Bipolar Transistor

With SiC, a bipolar transistor with a much thinner collector layer and therefore a drastically reduced w_C is possible. Figure 6.8, which determines the necessary width of the low-doped region, is also valid for dimensioning a SiC bipolar transistor. According to the small w_C, an acceptable current gain can be achieved even for transistors with blocking voltages above 1000 V. To obtain a high current gain, state-of-the-art SiC epitaxial growth and surface passivation are important. SiC bipolar transistors can be fabricated with very low on-state voltage drop, if ohmic contacts with low contact resistivity are formed [Lee07].

Figure 7.10 shows the measurement of a SiC "large area" bipolar transistor, fabricated by TranSiC AB [Dom09]. A value for β of 35 was achieved for a BJT with open base breakdown voltage V_{CE0} of 2.3 kV. It should be noted that the characteristic is nearly ohmic in the saturation mode, and an "on-resistance" of 0.03 Ω can be observed, which is about 0.45 Ω·mm^2 for this device with 15 mm^2 active area.

The SiC transistor establishes the opportunity to operate a high-voltage device with a voltage drop clearly below 1 V at room temperature. Further, SiC has the advantage of a possible higher doping of the region w_C. Therefore, the effect of second breakdown according to Eqs. (7.31), (7.32), and (7.33) is to be expected only at a very high current density which is outside the range of possible operation. Additionally, high operation temperatures are possible with SiC, but then a reduced current gain and an increased on-resistance must be taken into account. Nevertheless, the progress in SiC technology might revitalize the interest in bipolar transistors again.

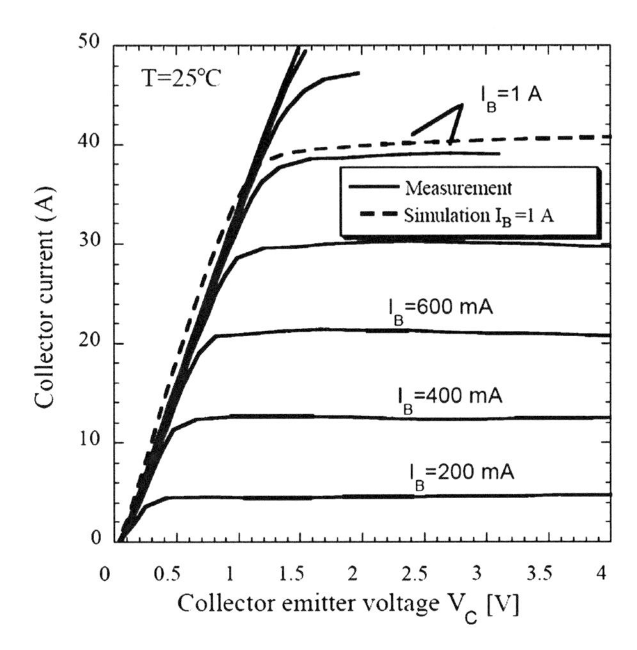

Fig. 7.10 *I–V* characteristic of a SiC bipolar transistor. Active area 15 mm^2, breakdown voltage $V_{CE0} = 2.3$ kV

References

[Bar49] Bardeen J, Brattain WH, "Physical Principles involved in transistor action", Phys. Rev. 75, 1208–1225, 1949

[Ben99] Benda V, Govar J, Grant DA, Power Semiconductor Devices, John Wiley & Sons, New York 1999

[Dom09] Domeij M, Zaring C, Konstantinov AO, Nawaz M, Svedberg JO, Gumaelius K, Keri I, Lindgren A, Hammarlund B, Östling M, Reimark M, : "2.2 kV SiC BJTs with low V_{CESAT} fast switching and short-circuit capability" to be published in Proceedings of the 13th International Conference on Silicon Carbide and Related Materials, Nuremberg (2009)

[Ear52] Early JM: "Effects of Space-Charge Layer Widening in Junction Transistors" Proceedings of the IRE Voll 40, Issue 11, pp. 1401–1406 (1952)

[How70] Hower PL, Reddi K: "Avalanche injection and second breakdown in transistors" IEEE Trans. Electron Devices, ED-17, p. 320 (1970)

[Hwa86] Hwang K, Navon DH, Ting-Wei Tang, Hower PL: "Second breakdown prediction by two-dimensional numerical analysis of BJT turnoff", IEEE Trans. Electron Devices, 33, Issue 7, pp. 1067–1072 (1986)

[Kir62] Kirk CT: "A theory of transistor cut-off frequency (fT) fall-off at high current density", IEEE Trans. Electron Devices, ED-23, p. 164 (1962)

[Lee07] Lee HS, Domeij M, Zetterling CM, Östling M, Heinze B, Lutz J: "Influence of the base contact on the electrical characteristics of SiC BJTs" ISPSD 2007, Jeju, Korea

[Mil57] Miller SL: "Ionization rates for holes and electrons in silicon", Phys. Rev., vol 105, pp.1246–1249 (1957)

[Mil85] Miller G, Porst A, Strack H: "An advanced high voltage bipolar power transistor with extended RBSOA using 5 μm small emitter structures", 1985 International Electron Devices Meeting, Vol. 31, pp. 142–145 (1985)

[Sho49] Shockley W: "The theory of p-n junctions in semiconductors and p-n junction transistors", Bell System Techn. J. 28, S. 435–489 (1949)

[Sze81] Sze SM, Physics of Semiconductor Devices. John Wiley & Sons, New York 1981

[Tan56] M. Tanenbaum and D. E. Thomas, "Diffused emitter and base silicon transistor", Bell Syst. Tech. J. 35, pp. 1–22, 1956

[Whe76] Wheatley CF, Einthoven WG, "On the proportioning of chip area for multistage darlington power transistors," IEEE Trans. Electron Devices, ED-23, pp. 870–878, 1976

Chapter 8
Thyristors

The thyristor was the dominating switching device in power electronics for a long time. It was described already in 1956 [Mol56] and introduced to the market in the early 1960s [Gen64]. The acronym SCR (Silicon Controlled Rectifier) was primarily used for a thyristor in early publications and is still occasionally in use today. In its basic structure, a thyristor can be fabricated without very fine structures and with low-cost photolithography equipment. The thyristor is still widely used in applications with low switching frequencies, such as controlled input rectifiers which are applied at the grid frequency of 50 or 60 Hz. A further actual application field of the thyristor is the power range that cannot be reached with other power devices – the range of very high blocking voltages and very high currents. For high-voltage DC power transmission, thyristors with 8 kV blocking voltage and more than 5.6 kA rated current have been introduced in 2008 as a single device in the size of a 6-inch wafer [Prz09].

8.1 Structure and Mode of Function

Figure 8.1 shows the structure of a thyristor in a simplified drawing. The device consists of four layers forming three pn-junctions. The p-doped anode layer is located at the bottom, followed by the n-base, the p-base, and finally the n^+-doped cathode layer.

The three pn-junctions formed by the four alternately doped layers are marked by diode symbols J_1, J_2, and J_3 in Fig. 8.1. If a voltage is applied in the forward blocking direction, the junctions J_1 and J_3 are biased in forward direction and the junction J_2 is biased in reverse direction as long as the device is in the forward blocking state. Thus, across J_2 a space charge region with a high electric field will build up (Fig. 8.1c). It penetrates widely into the weakly doped n^--layer.

If a voltage is applied in reverse blocking direction of the thyristor, the junction J_2 is forward biased; J_1 and J_3 are biased in reverse direction. Because of the high doping on both sides of the junction J_3, the avalanche breakdown voltage of it is usually relatively low ($\approx$ 20 V). The main part of the applied voltage is taken by the junction J_1, the shape of the electric field is shown in Fig. 8.1d. Since the same weakly doped n^--layer takes the electric field and since the upper and the lower p-layers are usually fabricated simultaneously from both sides in a single diffusion

J. Lutz et al., *Semiconductor Power Devices*, DOI 10.1007/978-3-642-11125-9_8,

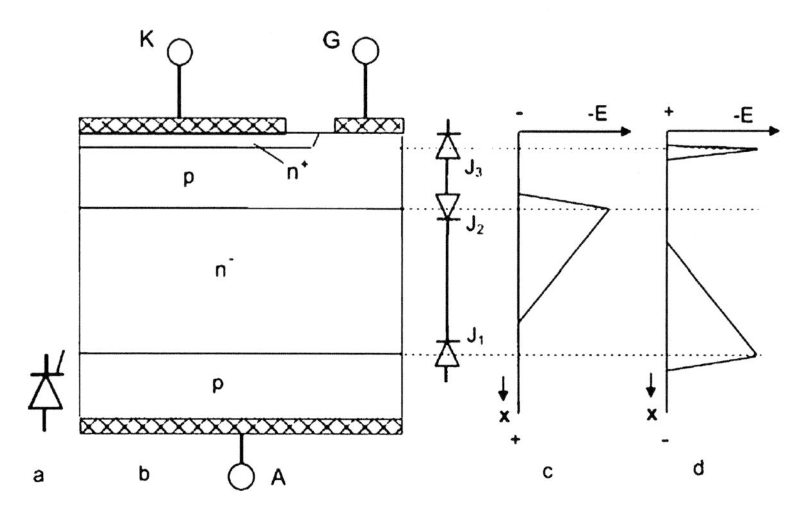

Fig. 8.1 Thyristor. (**a**) Symbol, (**b**) pn-structure, (**c**) shape of the electric field in the forward blocking mode, and (**d**) shape of the electric field in the reverse blocking mode

step, the blocking capability of the thyristor is nearly the same for both directions (if the npn-transistor is shorted, see Sect. 8.4). The thyristor is a symmetrically blocking device.

The thyristor can be divided into two partial transistors, a pnp-transistor and an npn-transistor, with the common base circuit current gains α_1 and α_2, respectively (Fig. 8.2).

Then we obtain for the collector current I_{C1} of the pnp-partial transistor according to Eq. (7.1)

$$I_{C1} = \alpha_1 \cdot I_{E1} + I_{p0} = \alpha_1 \cdot I_A + I_{p0} \tag{8.1}$$

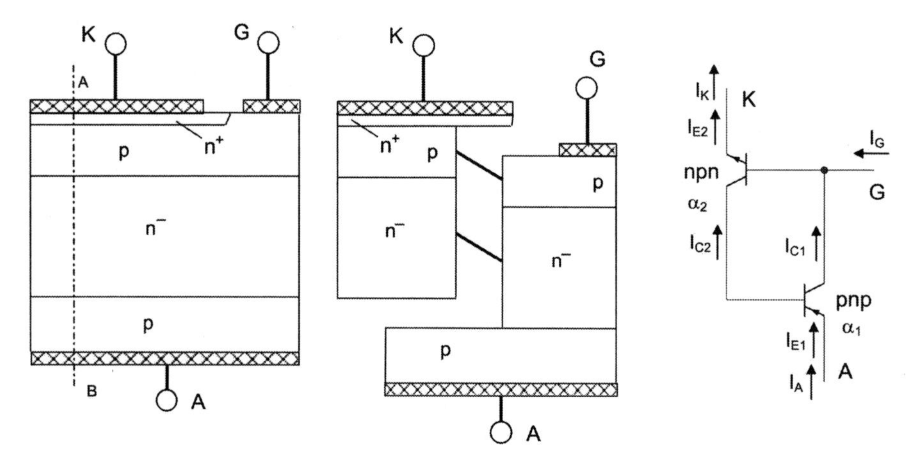

Fig. 8.2 Partition of the thyristor in two partial transistors and equivalent circuit

where I_{p0} is the diffusion leakage current from the middle weakly doped n^--layer. In the same way we obtain for the npn-partial transistor

$$I_{C2} = \alpha_2 \cdot I_{E2} + I_{n0} = \alpha_2 \cdot I_K + I_{n0} \tag{8.2}$$

with I_{n0} as the diffusion leakage current in the p-base. The anode current I_A is the sum of both partial currents I_{C1} and I_{C2}:

$$I_A = I_{C1} + I_{C2} = \alpha_1 \cdot I_A + \alpha_2 \cdot I_K + I_{p0} + I_{n0} \tag{8.3}$$

From the balance of currents flowing into the device and out of the device it holds additionally that

$$I_K = I_A + I_G \tag{8.4}$$

Equation (8.4) inserted into Eq. (8.3) leads to

$$I_A = \alpha_1 \cdot I_A + \alpha_2 \cdot I_A + \alpha_2 \cdot I_G + I_{p0} + I_{n0} \tag{8.5}$$

Equation (8.5) resolved for I_A results in an expression for the anode current

$$I_A = \frac{\alpha_2 \cdot I_G + I_{p0} + I_{n0}}{1 - (\alpha_1 + \alpha_2)} \tag{8.6}$$

which can be used as long as avalanche multiplication can be neglected. From Eq. (8.6) one can see that I_A rises to infinity when the denominator in Eq. (8.6) approaches zero. The current gains α_1 and α_2 are in turn dependent on the current. For very low currents they are close to zero and they increase with current as shown for the bipolar transistor in Fig. 7.7. The trigger condition of the thyristor is therefore

$$\alpha_1 + \alpha_2 \geq 1 \tag{8.7}$$

If the triggering condition is fulfilled, the anode current has the tendency to increase infinitely, this is valid even if $I_G = 0$ in Eq. (8.6). The thyristor is in the forward conduction mode. In this mode, there is an internal positive feedback loop that is established by the current amplification of the two partial transistors. Both transistors are in the saturation mode, this leads to a low forward voltage drop that is comparable to the voltage drop across a forward biased diode.

Equivalent to Eq. (8.7) is the condition $\beta_1\beta_2 \geq 1$. At low current, both α and β grow with increasing current, see Fig. 7.7. Especially, said conditions are also fulfilled if the small-signal current gains $\alpha' = \Delta I_C/\Delta I_E$ or $\beta' = \Delta I_C/I\Delta_B$ are used [Ger79], which are larger than α and β at low current. In nowadays fabricated power thyristors, α_2 and β_2 are determined by the cathode emitter shorts and are zero at low gate current, and therefore the trigger function is mainly adjusted by the cathode shorts. For more details see Sect. 8.4.

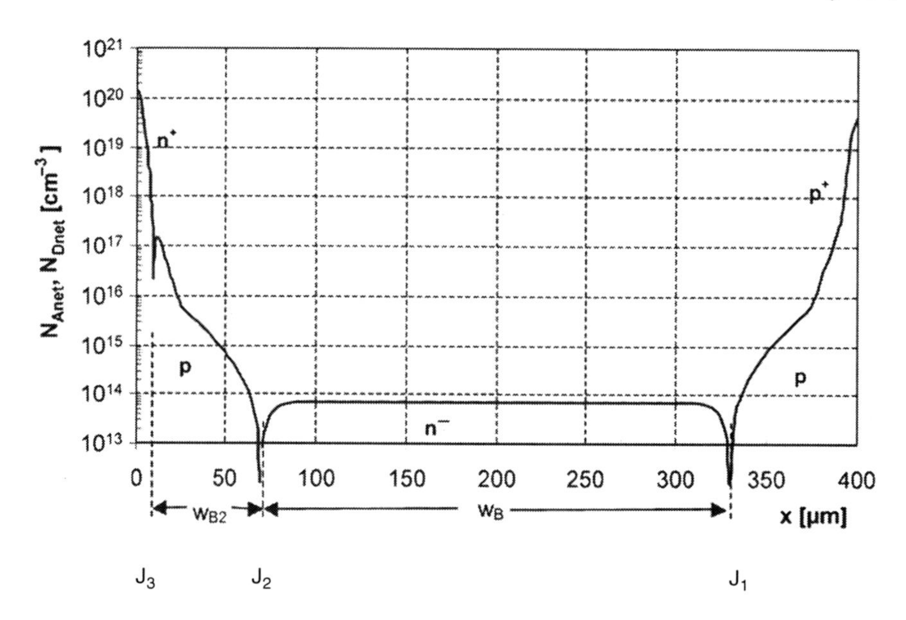

Fig. 8.3 Diffusion profile of a thyristor designed for 1600 V

An exemplary diffusion profile of a thyristor along the line A–B in Fig. 8.2 is shown in Fig. 8.3. The fabrication of a thyristor starts with a weakly doped n^--wafer. Usually, both p-layers are created simultaneously by diffusion: predeposition with an acceptor dopant, e.g., aluminum, at both wafer surfaces and a subsequent high-temperature drive-in step. To create the deep pn-junctions J_1 and J_3, which are typical for high-voltage thyristors, aluminum is a suitable dopant because of its relatively fast diffusion in silicon. To adjust the required doping concentrations at the n^+p-junction J_3 and near the anode contact, additional p-diffusions are applied. Thus, the final doping profile in the p-layers can be approximated by superposition of several Gauss-type profiles. The junctions J_1 and J_2 both exhibit a very shallow gradient of the diffusion profile at the p-side: This facilitates the fabrication of a junction termination structure with beveled edges as shown in Figs. 4.16 and 4.17. A further structure, the edge diffusion structure, is shown in Fig. 4.22. In all cases, the base of the pnp-transistor with thickness w_B and doping concentration N_D determines the blocking capability of the thyristor in both directions.

8.2 *I–V* Characteristic of the Thyristor

In forward direction two branches of the *I–V* characteristic exist: the forward blocking mode and the forward conduction mode. A simplified drawing of the *I–V* characteristic is shown in Fig. 8.4. In the forward blocking mode, the maximally allowed voltage V_{DRM} is defined at a leakage current $I_{DD,max}$. In reverse direction, the maximal voltage V_{RRM} is specified at a maximal allowed current $I_{RD,max}$.

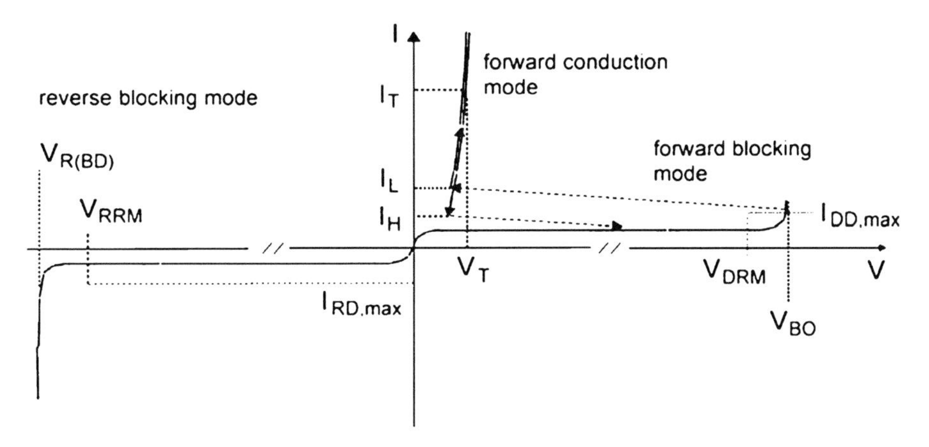

Fig. 8.4 Simplified *I–V* characteristic of the thyristor and some important thyristor parameters

The data sheet values for V_{DRM}, V_{RRM} may differ considerably from the values measured for a real device, as mentioned already for the *I–V* characteristic of diodes. In reverse direction the blocking capability is limited by $V_{R(BD)}$. In forward direction the blocking capability is denoted by the breakover voltage V_{BO}. At a voltage higher than V_{BO} the device is triggered and switches to the forward conduction mode. This mode of triggering, breakover triggering, is usually avoided in power thyristors. Usually the thyristor is fired by the gate. At breakover triggering, especially in large-area thyristors, the device may be locally overstressed by uncontrolled local current concentration, and destruction is possible.

In forward conduction mode, the voltage drop V_T is defined at a specified current I_T. The branch of the *I–V* characteristic for high current is similar to the forward characteristic of a power diode. The middle layer is flooded with free carriers, and similar current densities as in a power diode are possible. Again, as already discussed for the *I–V* characteristic of power diodes, the data sheet value V_{Tmax}, the maximally allowed forward voltage drop, is higher than the real value V_T of a device, since there is some unavoidable variation in the electrical characteristics of devices. Therefore, the manufacturer usually specifies values with a certain safety margin.

Dedicated parameters of the forward characteristic are

The latching current I_L: The minimal anode current which must flow at the end of a 10 μs trigger pulse to safely switch the thyristor into the conduction mode and to maintain the conduction mode safely when the gate signal returns to zero.

The holding current I_H: The minimal anode current necessary to maintain the thyristor in conduction mode without gate current and which ensures that the conducting thyristor will not extinguish. A decrease of current below I_H can lead to a turn-off of the thyristor.

Since the device is not completely flooded with carriers at the initial phase of the turn-on process, $I_L > I_H$ always applies. The latching current is typically twice as large as the holding current.

8.3 Blocking Behavior of the Thyristor

Avalanche breakdown as limit for the blocking capability is already known from the paragraphs on power diodes and transistors. For a thyristor, there is a second limit of the blocking capability, the punch through effect: The space charge region, which spreads across the n^--layer with increasing applied voltage, may arrive at the adjacent layer of opposite doping.[1] Holes will be accelerated in the electric field and the blocking capability is no longer given.

For simplification a triangular electric-field shape across the n^--layer (Fig. 8.1c or d) is assumed in the following discussion. Furthermore, the penetration of the space charge region into the p-layer of the blocking pn-junction shall be neglected. The avalanche breakdown voltage and its dependency on the background doping were already calculated in Sect. 3.3. It is given by Eq. (3.84) for the triangular electric field shape. This dependency is drawn in Fig. 8.5 as line (1). It is equal to the dependency in Fig. 3.17, lower part. Additionally, the width of the space charge region is given by Eq. (3.58). Solving Eq. (3.58) for the voltage and neglecting the small V_{bi} leads to

$$V_{PT} = \frac{1}{2}\frac{qN_D}{\varepsilon}w_B^2 \tag{8.8}$$

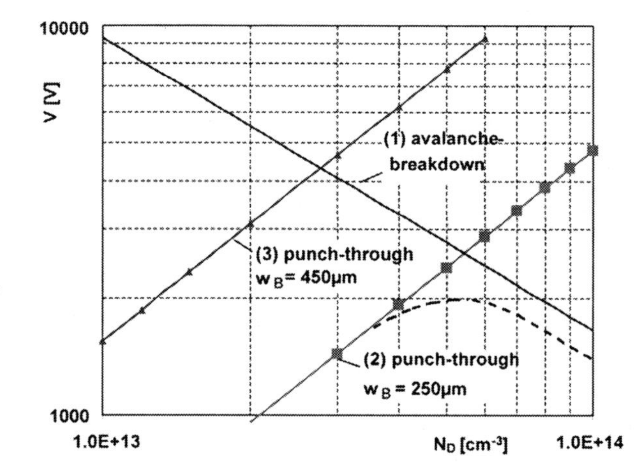

Fig. 8.5 Blocking capability of a thyristor: Avalanche breakdown voltage as a function of N_D, punch-through voltage for two different widths of the n^--layer

[1] Already for diodes, the term "punch through" was used, however, in a different meaning of the word. When in diodes the space charge region extends through the entire n^--layer and penetrates into an n^+-layer, the blocking capability can further increase. See Section 5.3.

In this equation, the voltage was set to $V_r = V_{PT}$, the voltage at which the space charge region reaches the region of opposite doping at the position $w = w_B$. For $w_B = 250$ and 450 μm, the calculated values of V_{PT} are drawn in Fig. 8.5 as line (2) and (3), respectively.

The optimal design parameters for the base width w_B and its doping concentration N_D for a thyristor with a blocking voltage somewhat above 1600 V can be estimated by considering the intersection point of lines (1) and (2) in Fig. 8.5. If the doping concentration is reduced below the concentration at the intersection point, the avalanche breakdown voltage will increase, but the space charge region will reach the opposite p-layer at a voltage lower than the avalanche breakdown voltage and punch through will limit the blocking capability.

In the following it shall be investigated how close one can approach the limits given by avalanche breakdown and punch through. In reverse direction, we neglect the small voltage at J_3 (which is usually shorted, see Sect. 8.4). The behavior at the blocking junction J_1 is equivalent to that of a bipolar pnp-transistor in open base configuration [Her65]. According to Eq. (7.11), the blocking capability of this junction is lower than for a pn-diode. Avalanche breakdown sets on already if $M_p \alpha_1 = 1$ holds:

$$M_p = \frac{1}{\alpha_1} \tag{8.9}$$

Only for $\alpha_1 = 0$, the avalanche breakdown voltage of the pn-junction reaches the value derived for diodes. Since $M_p << M_n$ applies (see Fig. 3.15), the effect is not as strong as in an npn-transistor. The onset of avalanche breakdown is reduced to lower voltage, as shown in Fig. 8.5 by the dotted line, the reduction of the breakdown voltage is most pronounced close to the intersection point of lines (1) and (2).

In forward direction the blocking junction is J_2. Its blocking capability is denoted by the breakover voltage V_{BO}. We can use the trigger condition (8.6), set $I_G = 0$ and take into account the multiplication factors for the hole current in the pnp-transistor and for the electrons in the npn-transistor. The breakover voltage will be reached for

$$I_A = \frac{M_{SC} \cdot I_{SC} + M_p I_{p0} + M_n I_{n0}}{1 - \left(M_p \cdot \alpha_1 + M_n \cdot \alpha_2\right)} \tag{8.10}$$

For α_1 and α_2 in Eq. (8.10), the small signal current gains must be used. Thebreakover voltage will be reached for $M_p \alpha_1 + M_n \alpha_2 = 1$. Since $M_n >> M_p$ holds, the forward breakover voltage will be very sensitive to α_2. Only for $\alpha_2 = 0$ the blocking capability will be the same as in reverse direction.

Both current gains are dependent on temperature and increase with temperature for low currents. To ensure a blocking capability of the thyristor at higher temperatures, α_2 must be reduced for low currents; since symmetrical blocking capability is required it must be zero for low current. This can be achieved by the implementation of emitter shorts.

8.4 The Function of Emitter Shorts

The current gain of a transistor depends not only on current but also on temperature. At low temperature the current gain is low; it increases with temperature (Chap. 7, Fig. 7.7). This has the consequence that the breakdown in open-base configuration, Eq. (7.10), will be reached at a lower voltage when the temperature is increased. For the thyristor, the breakover voltage V_{BO} will decrease strongly. This behavior is shown in Fig. 8.6 by the dotted line for a thyristor withoutemitter shorts.

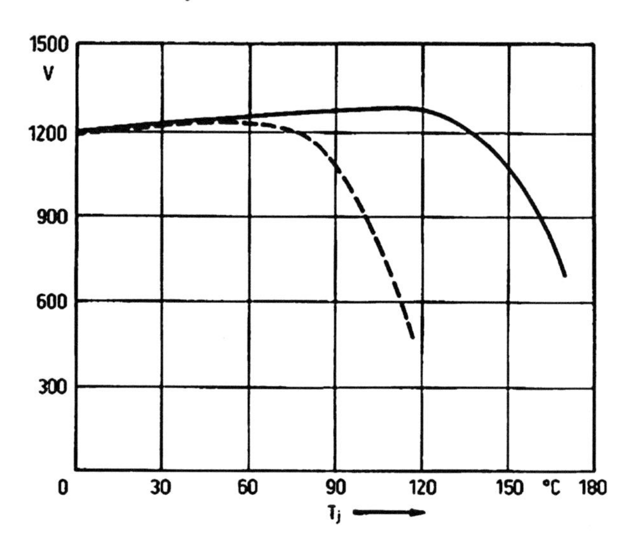

Fig. 8.6 Temperature dependency of the breakover voltage V_{BO}. *Dotted line*: Without emitter shorts. *Full line*: With emitter shorts. Figure taken from [Ger79] © 1979 Springer

By the implementation of emitter shorts on the cathode side (Fig. 8.7), a shunt parallel to the pn-junction between the base and the emitter of the npn-partial transistor is created [Chu70, Ger79, Rad71]. The current coming from the pnp-transistor flowing into the base of the npn-transistor is conducted via this shunt to the cathode contact. The shunt resistance is determined by the lateral distances between the shunts and the doping concentration of the p-base. If the current is high enough, the voltage

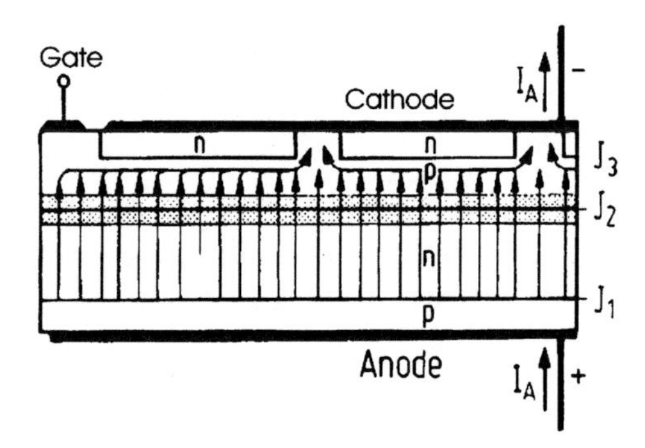

Fig. 8.7 Arrangement of emitter shorts at the cathode side of a thyristor. Figure according to [Ger79] © 1979 Springer

drop across the short gets sufficiently high and a perceptible current gain of the npn-partial transistor arises.

The cathode emitter shorts determine the effective α_2 and thus widely the dynamic and static characteristics of the thyristor. The dependency on the forward blocking capability on temperature for a thyristor with emitter shorts is drawn in Fig. 8.6 with the full line. With appropriate design of the emitter shorts, the thyristor will have the same blocking capability in forward and in reverse direction even at increased temperature.

Even if a thyristor is provided with emitter shorts, its blocking capability remains sensitive to temperature variations because of the temperature dependency of the current gain of the partial transistors. In most thyristors the maximal allowed operating temperature is restricted to 125°C, in some special thyristors a little bit above. At increased temperature, thyristors show a significantly increased leakage current compared to diodes.

8.5 Modes to Trigger a Thyristor

A thyristor can be triggered

1. By a *gate current* I_G. This is the most common mode to trigger a thyristor. For technical applications, the following characteristics are given:
   I_{GT}, V_{GT}: Minimal current and minimal voltage that must be provided by a gate unit for secure triggering of the thyristor.
   I_{GD}, V_{GD}: Maximal current and maximal voltage at the gate, at which a thyristor will surely not trigger. To avoid unwanted triggering by disturbing signals, which could occur, for example, by electromagnetic cross talk between cables and drive units, these thyristor parameters are very important.
2. By *exceeding the breakover voltage.* In usual power thyristors, this trigger mode is strictly avoided. However, special modifications of thyristor structures, e.g. SIDACs or SIDACtors [SID97], use breakover triggering to act as protection devices against too high voltages. They are connected in parallel to a device or an integrated circuit for protection. They trigger at a voltage higher than V_{BO} and protect other parts of the circuit from overvoltage. Their voltage range is limited to small and medium voltages in the range of some 10 V up to some 100 V. Further, pnpn-thyristor structures are used as electrostatic discharge (ESD) protection structures in integrated circuits, where the same principle of breakover triggering is applied.
3. By a *voltage pulse with a slope* d*v*/d*t* *above the critical* d*v*/dt_{cr} in forward direction. If such a voltage pulse occurs, the junction capacity of the pn-junction J_2 is charged. If the slope dv/dt is high enough, the generated displacement current may be sufficient to trigger the thyristor. dv/dt triggering is an unwanted triggering event. For the application of a thyristor, a maximal allowed dv/dt_{cr} is defined.

4. By a *light pulse* with photons that penetrate into the space charge region across the junction J_2 [Sil75, Sil76, Chk05]. If the energy of the arriving photons is high enough, electrons transit from the valence band up to the conduction band. The generated electron–hole pairs are separated in the electric field immediately, the electrons flow to the anode, the holes to the cathode. The generated current has the same effect as a current supplied by the gate. If the light power is high enough, the triggering condition (8.7) can be fulfilled. Light triggering is preferably used in case of series connection of thyristors. This is the case, for example, in applications of high-voltage direct current transmission (HVDC), where total voltages up to several 100 kV have to be controlled. The possibility to trigger a thyristor via a glass fiber cable is of high advantage because of the given electrical insulation with this type of signal cable.

The cathode emitter shorts (Sect. 8.4) determine widely the triggering of a thyristor [Sil75]. For activating the npn-transistor, the voltage drop below the emitter to the next emitter short $V = R_{\text{p-base}} \cdot I_G$ must approach the built-in voltage V_{bi} of the n^+p-junction between emitter and p-base, which is in the order of 0.7 V at 300 K. Since V_{bi} will decrease with temperature and $R_{\text{p-base}}$ will increase due to the reduced mobility of holes at elevated temperature, the trigger condition will be much earlier fulfilled at a high operation temperature of 125°C.

Emitter shorts reduce the trigger sensitivity, increase the trigger current, and increase the critical voltage slope dv/dt_{cr} [Sil75]. While the thyristor is most sensitive to dv/dt_{cr} at 125°C, it must also be ensured that the necessary trigger current I_{GT} at low temperature – room temperature or down to −40°C – becomes not too high. The necessary compromise is especially difficult for a light-triggered thyristor, where a low trigger power is required in the face of losses, e.g., in the glass fiber cables, and still a sufficiently high dv/dt_{cr} must be maintained [Sil76].

8.6 Trigger Front Spreading

With injection of a gate current into a thyristor only the region close to the gate is switched into the conduction mode. The width of the conducting region at the first instant of the turn-on phase is only of the order of some fractions of millimeters. The situation immediately after triggering is illustrated in Fig. 8.8.

The triggered region spreads over the cathode area with a velocity v_z in the range of 50 – 100 μm/μs or 50 – 100 m/s, which is very small for electronic effects. Since there will be only a small lateral voltage drop in lateral direction, the spreading of the initial region of high carrier density is slow. It will take a time of typically 100 – 200 μs until the front of the primary carrier plasma in a thyristor spreads across a length of 1 cm. This slow spreading of the triggered region is a severe limitation for some thyristor applications. It limits the allowed current slope di/dt for the anode current.

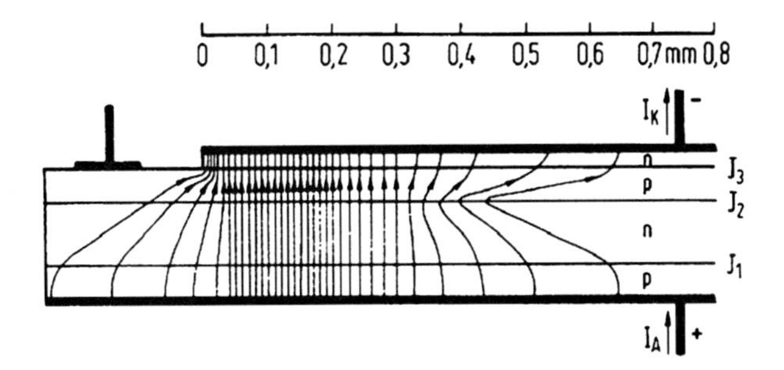

Fig. 8.8 Current distribution in the thyristor immediately after triggering. Figure taken from [Ger79] © 1979 Springer

The trigger front spreading velocity is approximately proportional to the square root of the current density:

$$v_Z \sim \sqrt{j} \tag{8.11}$$

Furthermore, v_Z is reduced by emitter shorts: in the region of shorts it decreases strongly. Emitter shorts reduce v_Z to approx. 30 μm/μs for a current density of 100 A/cm^2. Additionally, the carrier lifetime is of influence. If the carrier lifetime is reduced, for example, by a gold diffusion, and if, as is necessary for a fast thyristor, strong emitter shorts are implemented, the trigger front spreading velocity v_Z may be as low as 10 μm/μs equivalent to 10 m/s – a sprinter at the Olympic Games is running faster!

Thyristors with a higher blocking capability require a larger thickness w_B of the n$^-$-base, v_Z decreases with increasing w_B of the thyristor. For a 4.5-kV thyristor with a high carrier lifetime, v_Z will be in the range of 20 μm/μs. The dependence v_Z on w_B is given by

$$v_Z \sim \frac{L_A}{w_B} \tag{8.12}$$

where L_A is the ambipolar diffusion length. A low spreading velocity increases the danger of overstress of the thyristor by an increased local current density in the emitter region close to the gate contact. Therefore, measures to increase the di/dt capability are necessary.

8.7 Follow-Up Triggering and Amplifying Gate

Thyristors for applications at line frequency are characterized by a high carrier lifetime and a high ambipolar diffusion length L_A. They are manufactured according to the basic structure described above. Their rated current and di/dt capability are of the order of 100 A and 150 A/μs, respectively. These properties are sufficient for applications at line frequency. For applications requiring a higher current capability, it would be possible to increase the gate area, this results in an increase of the initially triggered region. However, this strategy requires a higher gate trigger current I_{GT} and thus a higher effort for the gate drive unit. To avoid this drawback, follow-up triggering was introduced.

First, a pilot thyristor is triggered, which in turn triggers the main thyristor. Thus, the main part of the power necessary for triggering is not provided by the gate drive circuit, but by the main current.

The principle of operation of a thyristor structure consisting of a pilot thyristor and a main thyristor is illustrated in Fig. 8.9a. Between the gate and the main cathode K, a small auxiliary cathode K′ is arranged. A gate current will trigger the thyristor K′. Its cathode current flows via the resistor R and generates a voltage drop at the resistor. Thus, between K′ and K a positive voltage builds up, it generates an electric field in the p-base and a lateral hole current to the n-emitter of the main cathode K, which triggers the main thyristor.

In an advanced configuration the resistor R is integrated into the thyristor (Fig. 8.9b). It is realized by the extension of the n-emitter layer and consists

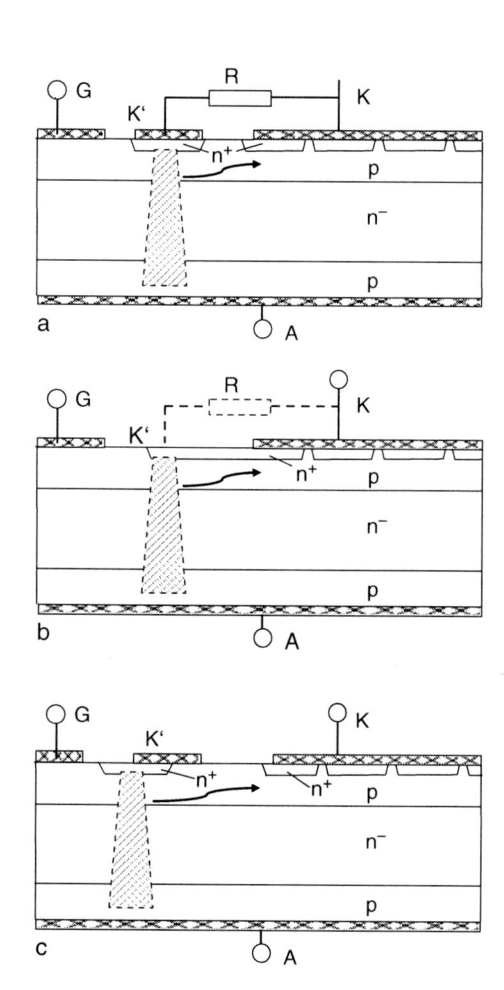

Fig. 8.9 Follow-up triggering of a thyristor. (**a**) Principle, (**b**) lateral-field emitter, integration of the resistor in the n⁺-emitter layer, (**c**) amplifying gate

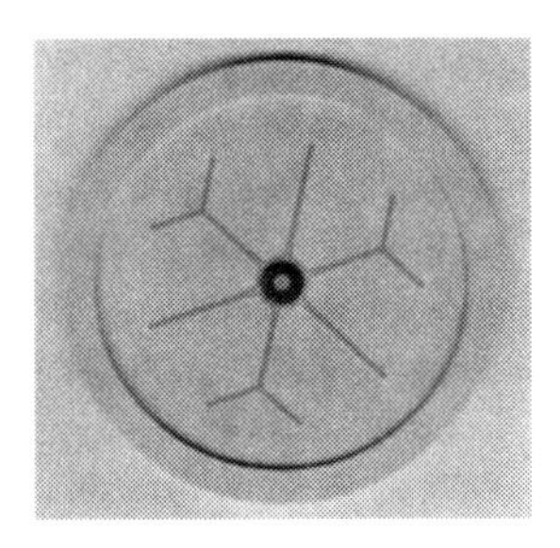

Fig. 8.10 Gate structure of a light-triggered thyristor, diameter 119 mm. Manufacturer Infineon

essentially of its lateral resistance in the region reaching from the pilot thyristor to the left border of cathode contact K. This structure was introduced as lateral-field emitter [Ger65] and allows much higher d*i*/d*t* slopes at turn on.

In a thyristor with amplifying gate (AG) [Gen68], shown in Fig. 8.9c, the cathode metal of the auxiliary thyristor K′ is connected to the p-base of the main thyristor. After turn-on of the auxiliary thyristor, the current flows across the lateral resistance of the p-base to the main cathode K. The total current of the cathode K′ acts as gate current for the main thyristor K. In addition, the overlapping of the cathode contact K′ and the p-base forms a cathode short and improves the d*i*/d*t* capability.

The amplifying gate can be formed in miscellaneous geometrical shapes, for example, stripes, or other distributed structures spreading across the area of the main cathode. A structure of a large area thyristor is shown in Fig. 8.10. This thyristor is designed for applications of high-voltage direct current transmission (HVDC) and has a blocking capability of 8000 V, a rated current of 3570 A ($T_{\text{case}} = 60°\text{C}$), and a d*i*/d*t* immunity of 300 A/μs.

The main thyristor area is triggered by a four-stage amplifying gate structure located in the center of the device. The three inner AGs are ring shaped whereas the fourth AG is distributed about the main cathode. This amplifying gate design ensures triggering of the main thyristor by a wide initial trigger front.

The central structure of such a high-power thyristor can be very complex; especially protection functions can be integrated. Figure 8.11 shows details in the center of the structure in Fig. 8.10 [Scu01, Nie01]. The left edge in Fig 8.11 corresponds

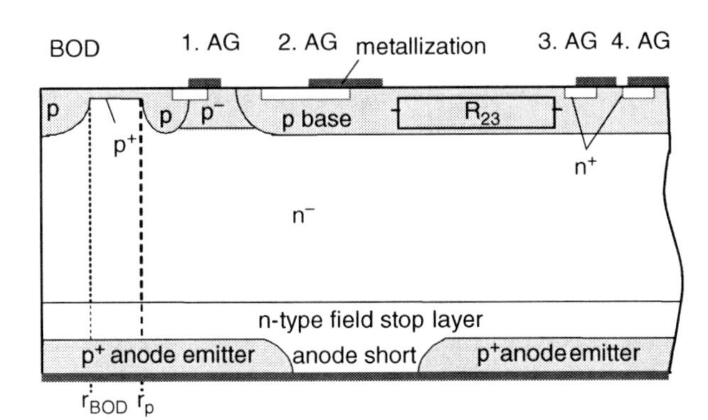

Fig. 8.11 Cross-section of the central amplifying gate structure of a modern high-power thyristor, consisting of four amplifying gates and several protection structures. Figure from [Nie07]

to the chip center in Fig. 8.10. The fourth AG is distributed over the main cathode area to increase the region in which triggering of the main cathode area starts. Direct light triggering of the thyristor is possible by irradiating the photosensitive area in the center of the device. The typical light power to turn on the thyristor is of the order of 40 mW.

An overvoltage protection is integrated by a breakover diode (BOD) in the center of the structure. Its avalanche current triggers the thyristor via the AG structure when an overvoltage is applied to the device. The voltage level V_{BOD} at which the overvoltage protection function is activated can be adjusted by the distance between the central p-region with radius r_{BOD} and the concentric p-ring with an inner radius r_p (Fig. 8.11).

The shunt resistance of the weakly doped p^- region below the n^+-emitter of the innermost AG is adjusted such that the dv/dt capability of this AG is lower than that of the other AGs and the main cathode area. By this way a reliable dv/dt protection function is integrated into the thyristor, because the device is turned on safely by the innermost AG when the anode-to-cathode voltage rises at a rate higher than the rated maximum dv/dt value. The resistor R_{23} between the second and the third AGs protects the two innermost AGs from being destroyed when the thyristor is turned on with a high current rate di/dt.

A thyristor with a structure in Fig. 8.11 needs high effort in design and fabrication. With the integrated self-protection functions, it includes already elements that support a reliable operation. Integration of these self-protection functions therefore reduces the number of electronic components of the total power electronic system.

8.8 Thyristor Turn-Off and Recovery Time

Only special configurations of the thyristor can be turned off via the gate, these are the Gate Turn-Off (GTO) thyristors which will be treated later. The usual turn-off of a thyristor happens by zero crossing of the anode current, which is given for applications operating in an AC circuit. In case of forward conduction, the base of the thyristor is flooded with free carriers, similar to the internal plasma at forward conduction of a diode. At commutation into the reverse direction, a reverse current will occur, caused by the stored charge. This process is similar to the turn-off process in a pin-diode. The stored charge must be removed down to a very small rest charge, before a voltage in forward direction can be re-applied to the thyristor. The time which is necessary for this charge removal, and which must be minimally set as hold-off interval to avoid unwanted triggering of the thyristor, is denoted as the recovery time t_q. A thyristor is not able to withstand a forward voltage pulse with the rated blocking voltage or the rated maximum dv/dt_{cr} value until the charge-carrier plasma is almost completely removed from the n^--base.

Figure 8.12 shows the definition of the recovery time. The current slope di_T/dt of the anode current is determined by the external circuit. Equation (5.68) is valid similar as in diodes. The anode current crosses zero, and a reverse current peak occurs due to the removal of the stored charge.

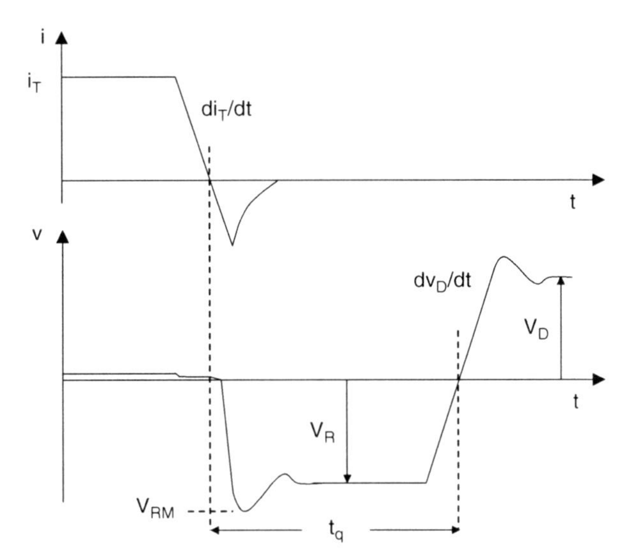

Fig. 8.12 Definition of the recovery time t_q of a thyristor

In a typical thyristor first the junction J_3 is depleted of charge carriers when the current is turned off. However, the blocking capability of this junction is usually only between 10 and 20 V, because of the relatively high doping concentration of the p-base in the range of 10^{17} cm^{-3}, and additionally it is provided with cathode shorts. Thus, the reverse current is increasing until the junction J_1 is depleted of carriers. Then, the thyristor starts to take over the reverse voltage V_R and a short instant later the reverse recovery current reaches its maximum value.

After having reached the maximum, the reverse current decreases and a voltage peak V_{RM} is generated, similar to the behavior of a diode during the turn-off period. For thyristors, this period is less critical than for fast diodes. One reason for this is the large width w_B of the n$^-$-layer typically used in thyristors. In most cases, the reverse current decays slowly and a tail current is observed. During this period, charge carriers are still stored in the region close to the junction J_2.

While the thyristor is in the reverse blocking mode, the polarity of the applied voltage changes. A voltage V_D with a defined slope dv_D/dt is applied in forward direction, the thyristor remains in the blocking mode (forward blocking mode) and does not switch into the on-state mode, if $t > t_q$ applies (Fig. 8.12). Otherwise, the control of the circuit is lost. The thyristor must be able to withstand the applied forward voltage, thus the forward voltage V_D is allowed to be applied only after a certain time interval. This minimum time interval between zero crossing of the current i_T and zero crossing of an applied forward voltage V_D is denoted as turn-off time t_q.

The recovery time of a thyristor is much higher than the switching time of a diode. For the case of charge removal without applying a voltage V_R in reverse direction, and neglecting also V_D, one can estimate according to [Ger79]

$$t_q \approx 10 \cdot \tau \tag{8.13}$$

where τ is the charge carrier lifetime in the n-base. However, t_q is specified at high operation temperature, whereas τ is typically measured at room temperature. In modern thyristors, t_q depends strongly on the cathode shorts. For turn-off with an applied reverse voltage V_R, Eq. (8.13) can only be used for an estimation of the upper limit value of t_q [Ger79]. If a voltage in reverse direction is applied, only in that part of the base, in which the space charge region has build up, the stored carriers are removed efficiently by the electric field. The recovery time t_q depends on the application conditions:

- the forward current I_T: t_q increases with increasing forward current;
- the temperature: t_q increases with temperature because τ is increasing with temperature;
- the voltage slope dv/dt. This voltage slope must be smaller than the critical voltage slope dv/dt_{cr} in any case. The closer dv/dt approaches dv/dt_{cr} , the less rest charge is allowed and the higher is t_q

With a 100 A 1600 V thyristor, which is in its basic form triggered via a gate in the center, t_q is in the range of 200 μs. With a high-power 8-kV thyristor, t_q is in the range of 550 μs. In large-area thyristors, a forward voltage pulse appearing before t_q may turn on the thyristor somewhere in the main cathode area in an uncontrolled way and – in the worst case – may lead to a destruction of the thyristor. Special structures have been introduced to integrate a t_q protection function into the thyristor [Nie07].

The recovery time limits the maximal frequency range allowed in thyristor applications. With diffusion of gold t_q can be reduced and also with a higher density of cathode shorts. Fast gold-diffused thyristors reached a t_q down to 10 – 20 μs. Since modern power devices with turn-off capability are meanwhile available, the interest in fast thyristors has vanished. For the high-voltage range > 3 kV, the development of fast thyristors was never successful.

8.9 The Triac

In a triac (triode AC switch) two thyristors are integrated in an anti-parallel configuration in a single device. The triac was introduced early [Gen65]. Figure 8.13 shows the structure.

For a triac, one can no longer distinguish between anode and cathode, therefore the notations “Main Terminal 1” and “Main Terminal 2,” MT_1 and MT_2 are conventionally used.

The triac can be triggered via a common gate in both directions. Its *I–V* characteristic presents in both the first and the third quadrant a conducting branch and a blocking branch. The triac may replace two thyristors in an AC converter, but only within some limits.

The triac is described in more detail in [Ger79]. For the application, the main restriction is due to the fact that at zero crossing of the current the triac must block

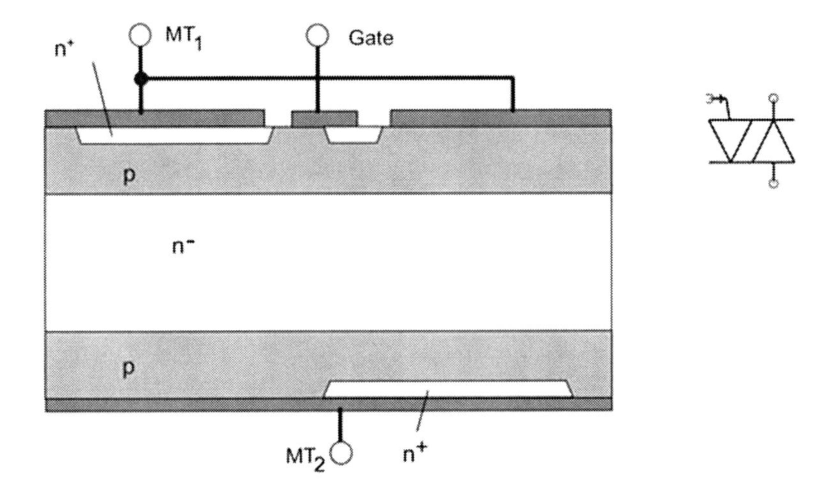

Fig. 8.13 Triac structure and symbol

the voltage in reverse direction. However, in conduction mode, the device is flooded with free charge carriers. If the commutation is executed with a too high d*i*/d*t*, still a part of the stored charge remains after zero crossing of the current. If now a voltage with a too high d*v*/d*t* is applied, an unwanted re-triggering occurs. The device will not transit to the blocking block mode and the possibility to control the converter is lost.

Therefore the allowed current slopes d*i*/d*t* and voltage slopes d*v*/d*t* are limited drastically for a triac – d*i*/d*t* to some 10 A/μs and d*v*/d*t* to the order of 100 V/μs. This allows the use of triacs only for applications with comparatively small current and moderate voltage. In fact, in such applications they are often used. An example for triac application is the AC converter for controlling of a medium-power heater.

If currents higher than 50 A have to be controlled, two thyristors in anti-parallel configuration are commonly used instead of a triac.

8.10 The Gate Turn-Off Thyristor (GTO)

To provide a thyristor with an active turn-off capability, several special measures are necessary. The Gate Turn-Off (GTO) thyristor was introduced in the 1980s [Bec80]. In the voltage range above 1400 V it was superior to the bipolar power transistor, the competing device at that time. But with the introduction of the IGBT (see Chap. 10) and the capability to design the IGBT for a high voltage, the GTO thyristor was superseded by the IGBT, since a GTO thyristor requires a high negative gate current for turn-off and the effort for the drive unit is high. The GTO thyristor is nowadays used in the power range that is not reached by the IGBT. Up to 6 kA 6 kV GTO thyristors fabricated from one single 150-mm wafer are available [Nak95]. From the GTO structure, the new device Gate Commutated Thyristor (GCT) was derived. It exhibits an improved ruggedness and an extended safe operation area.

Equation (8.6) was derived for the trigger condition of a thyristor and describes the dependency of the trigger condition on the current gains of both partial transistors. From this equation, also a turn-off condition can be derived. If in Eq. (8.6) the leakage currents of the partial transistors are neglected, we obtain

$$I_{\mathrm{A}} = \frac{-\alpha_2 \cdot I_{\mathrm{G}}}{(\alpha_1 + \alpha_2) - 1} \tag{8.14}$$

For turn-off, a negative gate current $-I_{\mathrm{G}}$ is necessary. Similar to the current gain β for the bipolar transistor, a turn-off gain β_{off} for the turn-off process of a GTO thyristor can be defined:

$$\beta_{\mathrm{off}} = \frac{I_{\mathrm{A}}}{-I_{\mathrm{G}}} \tag{8.15}$$

With Eq. (8.14) it follows that for the turn-off gain

$$\beta_{\mathrm{off}} = \frac{\alpha_2}{(\alpha_1 + \alpha_2) - 1} \tag{8.16}$$

A high turn-off gain requires on the one hand a high current gain α_2 of the npn-partial transistor, on the other hand the denominator $(\alpha_1 + \alpha_2 - 1)$ must be small and ideally approaches zero. In other words, the sum of the current-gain factors, $\alpha_I + \alpha_2$, should be only slightly larger than 1. However, this leads to an increased trigger current I_{GT}, an increased latching current I_{L}, and finally to an increased forward voltage drop V_{T} of the thyristor. The demand for a high current gain is therefore in contradiction to the requirement for low conduction losses. GTO thyristors show typically a turn-off gain β_{off} between 3 and 5. Thus, to turn off a 3000-A GTO thyristor, for example, the drive unit must supply a current of 1000 A.

In fact, Eq. (8.16) is of low value for the design of a GTO thyristor with high turn-off capability. Most important is the homogeneous operation of the large number of segments in a high-current GTO thyristor [Shi99].

Further, Eqs. (8.14), (8.15), and (8.16) hold only if the lateral voltage drop of the turn-off current below the emitter can be neglected [Wol66]. The observance of conditons (8.14), (8.15), and (8.16) alone is not sufficient to achieve a gate turn-off cabability of a thyristor. The GTO is thyristor distinguished from the conventional thyristor by its emitter structure, which consists of separated emitter fingers (Fig. 8.14). The width b of the fingers must be small, since the charge carriers below the emitter fingers must be extracted via the gate contact at turn-off. The width b is typically between 100 and 300 μm in modern GTO thyristors. The GTO thyristors consist of a large amount of emitter fingers. They are typically used to control very high currents, therefore a large device area is necessary and it is usual to fabricate a single GTO thyristor from a wafer.

Figure 8.15 shows such a GTO thyristor, fabricated from a 100-mm wafer. The gate contact is formed as a ring located between four inner and four outer rings with emitter fingers. The main reason for this arrangement is to ensure that the distance

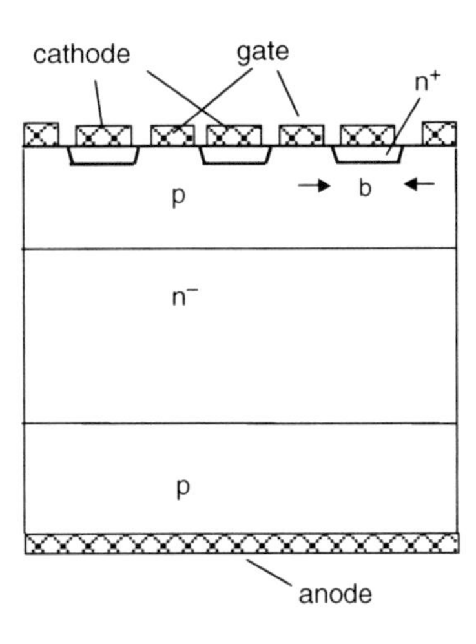

Fig. 8.14 Gate turn-off (GTO) thyristor

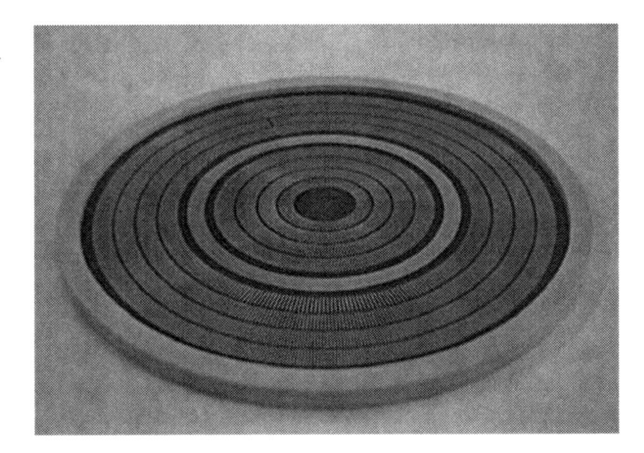

Fig. 8.15 Arrangement of the emitter fingers of a 4.5 kV GTO thyristor fabricated from a 100 mm silicon wafer; final device diameter 82 mm. Figure from Infineon

of the gate contact from the most distant fingers is small enough to allow an efficient extraction of the charge carriers and that the voltage drop in the gate metallization is not too high.

During turn-off, the holes in the p-base are removed by the negative gate voltage and flow toward the gate contact. The charge carrier plasma which transports the anode current is first removed from the edge of the emitter finger, the residue of the plasma is located in the center of the finger (Fig. 8.16). The hole current must flow laterally underneath the emitter finger. Before the anode current finally is interrupted, a small region in the middle of the emitter fingers, or even of only very few single emitter fingers, takes the total anode current. This is the weak point of the GTO thyristor. To achieve a turn-off capability for high current, it is essential that the resistance of the p-base below the emitter finger is not too high.

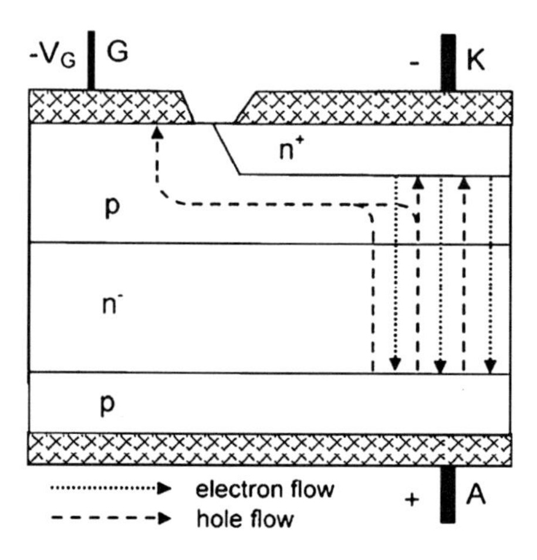

Fig. 8.16 Current flow in a single finger of the GTO thyristor at turn-off

The maximal current I_{Amax} that can be turned off by a GTO thyristor is determined by the breakdown voltage $V_{\mathrm{GK(BD)}}$ of the n$^+$p-junction between the cathode and the gate and by the lateral resistance R_{p} of the p-base below an emitter finger:

$$I_{A\,\max} = \beta_{\mathrm{off}} \cdot \frac{V_{\mathrm{GK(BD)}}}{R_{\mathrm{p}}} \tag{8.17}$$

where

$$R_{\mathrm{p}} \sim \rho \cdot b \tag{8.18}$$

and ρ is the specific resistivity of the p-base below the emitter finger. In a GTO thyristor with fingers of a width $b = 300\,\mu\mathrm{m}$, the resistivity ρ below the emitter must be four times smaller than in a conventional thyristor. This requires a sufficiently high p-base doping concentration N_{A}. Simultaneously, a sufficient blocking voltage $V_{\mathrm{GK(BD)}}$ of the n$^+$p-junction between cathode and gate is required. This blocking voltage is given by Eq. (3.84). However, the doping concentration that determines the breakdown voltage of the emitter-gate junction in the GTO thyristor is the doping concentration N_{A} of the p-base. Therefore, N_{A} must not be too high. Typical doping concentrations are of the order of $10^{17}\,\mathrm{cm}^{-3}$, resulting in breakdown voltages $V_{\mathrm{GK(BD)}}$ of about 20 – 22 V. The applied gate voltage at turn-off is usually −15 V.

Values > 4 for β_{off} do not increase the turn-off capability significantly. Decisive for the GTO design is the second term in Eq. (8.17).

With these measures the plasma can be efficiently extracted from the p-base of the GTO thyristor. However, the plasma of charge carriers still remains in the wide n$^-$-layer. Thus, additional measures are necessary to remove the charge carriers in this region. The first GTO thyristor generation utilized a gold diffusion to adjust a low carrier lifetime in the n$^-$-layer. However, the gold diffusion is very difficult to control with sufficient accuracy (see Sect. 4.9).

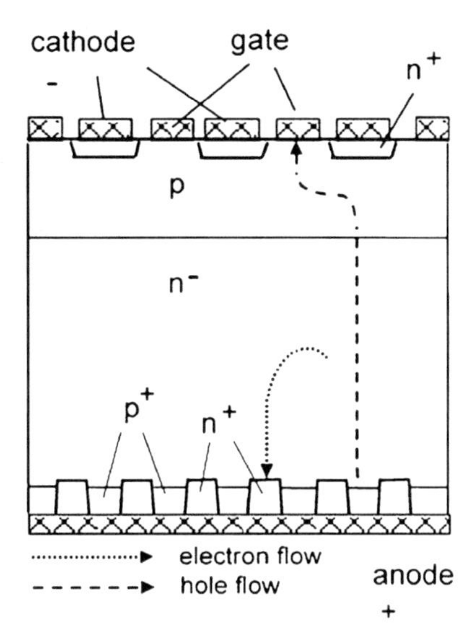

Fig. 8.17 GTO thyristor with emitter shorts at the anode side

An effective improvement was the implementation of shorts at the anode side. The structure of a GTO thyristor with anode shorts is shown in Fig. 8.17. The hole current is extracted via the gate and the injection of electrons from the n^+-emitter is stopped. The electrons in the n-base are removed via the anode shorts by the high positive voltage at the anode side. The injection of the anode emitter is interrupted and the charge carriers are removed effectively.

A GTO thyristor with anode shorts has no blocking capability in reverse direction. In most applications this is no disadvantage, since an inverse freewheeling diode is connected in parallel to the GTO thyristor in the power circuit. In modern GTO thyristors, anode shorts and charge carrier lifetime control are combined. An implantation of proton or helium nuclei is preferably used for adjustment of the carrier lifetime. The region with high density of recombination centers is located close to the p^+-anode layer, since at this position their effect on the stored charge is most effective.

Despite all these measures, the slope dv/dt of the voltage applied to a GTO thyristor at turn-off must be strongly restricted. This is done by an RCD circuit, often denoted as "snubber" (Fig. 8.18). The slope dv/dt of the increasing voltage is limited by the capacitor C.

Figure 8.19 finally shows the turn-off process of the GTO thyristor. The negative gate current increases up to the value I_{GRM}, just then the anode current begins to fall. The turn-off delay time t_{gs} is defined as the time interval between the moments when the gate current I_G crosses zero and the anode current is dropped to 90% of starting anode current I_{T0}. The anode current then falls steeply within the fall time t_{gf}. During this period, a voltage peak V_{pk} occurs in the waveform of the anode voltage. The value of V_{pk} is determined by the parasitic inductance in the snubber circuit and by the forward recovery voltage peak V_{FRM} of the snubber diode D, the

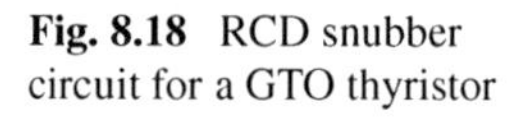
Fig. 8.18 RCD snubber circuit for a GTO thyristor

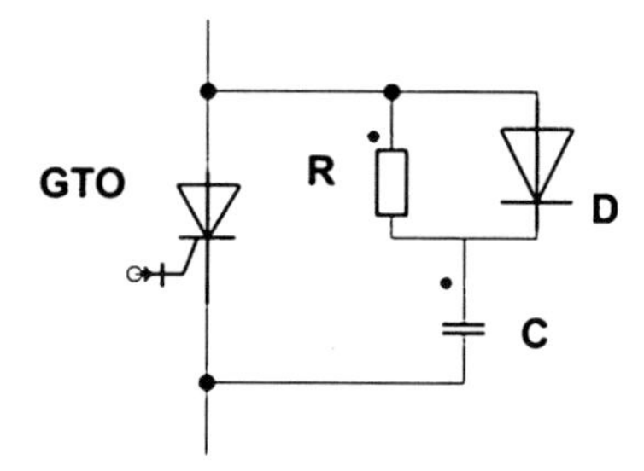

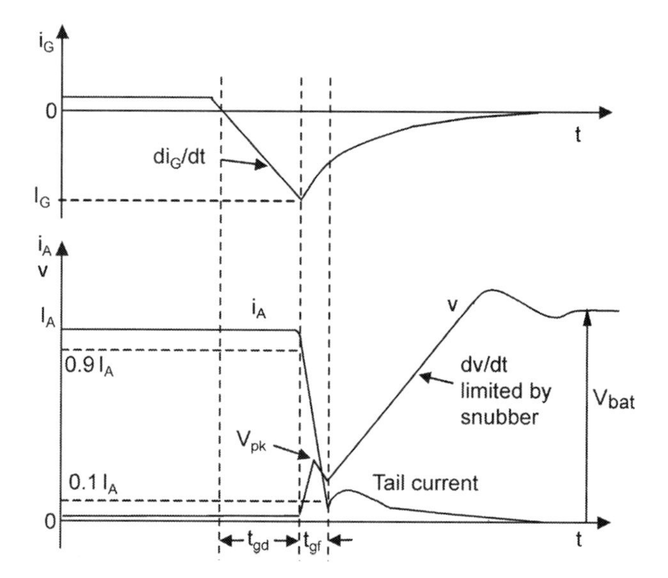

Fig. 8.19 Turn-off characteristic of a GTO thyristor

last term typically dominates. Just after V_{pk} the effect of the snubber starts. The slope dv/dt of the voltage is then limited by the capacitor C.

In a GTO thyristor, a tail current follows after the interval t_{gf}. This tail current is generated by the extraction of the stored charge in the part of the n-base close to the anode junction. Its duration is of the order of several microseconds, and it generates the main part of the switching losses during the turn-off phase. The implementation of effective anode shorts and the adjustment of the charge carrier lifetime reduce the tail current.

Even if the gate control unit is properly designed, two further factors are of disadvantage for the application of GTO thyristors:

1. The requirement for an RCD snubber. For a high voltage >3 kV, the capacitor is very voluminous and expensive, especially under the additional constraint of low internal inductance.

2. As already shown, the charge removal below the emitter fingers starts at the finger edges. Before the anode current is decreasing, a narrow region in the center of the finger remains which carries the total anode current. The larger the device, the more difficult it is to achieve a homogeneous operation of all fingers, and it may happen at the end of the turn-off period that a few of them or even a single finger has to carry the total anode current. This is the weak point of the GTO thyristor, because the respective finger may be destructed under such conditions.

8.11 The Gate-Commutated Thyristor (GCT)

The operation principle of the GCT [Gru96] is to work with a drive unit that has the capability to transfer the total anode current within a very short time into the gate unit. The GCT is turned off with a turn-off gain $\beta_{\mathrm{off}} = 1$.

The GCT consists of the semiconductor device, a gate connection with very low inductance realized by a printed circuit board PCB in coplanar design, and the drive unit assembled with low-inductive capacitors and low-resistive MOSFETs (Fig. 8.20). The drive unit must be capable to deliver the gate current in the range of the anode current within 1 μs. This is a severe challenge; especially the resistance R_{G} and the parasitic inductance L_{G} in the gate drive circuit must be extremely low. In the gate drive circuit, the differential equation holds [Lin06]:

$$L_{\mathrm{G}} \cdot \frac{\mathrm{d}i_{\mathrm{G}}}{\mathrm{d}t} + R_{\mathrm{G}} \cdot i_{\mathrm{G}} = V_{\mathrm{G}} \tag{8.19}$$

The solution of this equation is

$$i_{\mathrm{G}}(t) = \frac{V_{\mathrm{G}}}{R}\left(1 - \exp\left(-\frac{R_{\mathrm{G}}}{L_{\mathrm{G}}} \cdot t\right)\right) \tag{8.20}$$

Fig. 8.20 GCT with gate drive unit. The gate wiring consists of wide metal layers on thePCB and is connected in the form of a ring to the device

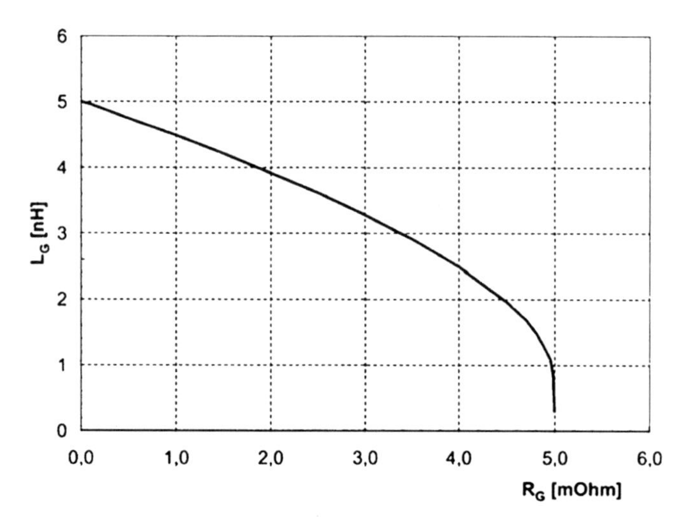

Fig. 8.21 Maximum allowed parasitic inductance L_G in the gate circuit of a 4000 A GCT. Figure according to [Lin06]

The voltage V_G is limited by the avalanche breakdown voltage of the pn-junction between gate and cathode. The requirement is that the gate current must increase within a time t_{gs} up to the anode current I_A. Thus, the maximally allowed inductance L_G is given by

$$L_G = -\frac{R_G t_{gs}}{\ln\left(1 - \frac{R_G}{U_G} \cdot I_A\right)} \tag{8.21}$$

The maximally allowed inductance L_G is shown in Fig. 8.21 as a function of the gate resistance R_G for the conditions $V_G = 20\,\text{V}$, anode current $I_A = 4000\,\text{A}$, and allowed $t_{gs} = 1\,\mu\text{s}$. The resistance R_G and the parasitic inductance L_G must be kept extremely small to allow an operation as GCT. For $R_G > 5\,\text{m}\Omega$, no solution exists for the specified values of V_G, I_A, and t_{gs}.

It must be considered that L_G as well as R_G consists not only of the external wiring. The resistance R_{on} of the used MOSFETs, the wiring, the resistance of the connections within the semiconductor housing, and the resistance of the gate metallization of the semiconductor die contribute to R_G. The MOSFETs must have a blocking capability slightly higher than 20 V. In this voltage range very low-resistive MOSFETs are available today. In this operation mode the npn-partial transistor is turned off abruptly. The positive feedback loop in the thyristor is interrupted. The extraction of the charge carriers below the emitter fingers still starts at the edges of the fingers, but the problem of narrow filaments during turn-off is solved to a great extent. The GCT can be operated without RCD snubber.

There is no requirement for a reverse blocking capability of a GCT. Therefore, it is possible to implement an n-doped buffer layer in front of the anode region and to design the device with a moderate trapezoidal shape of the electric field (PT dimensioning). With this additional buffer layer, the thickness w_B of the n^--layer can be reduced. Consequently, conduction losses and turn-off losses are diminished. A drive unit which must supply a current as high as the current to be controlled requires high effort without any doubt. But the power, which must be supplied by the driver unit, is not higher than the driver power for a GTO thyristor with the same current turn-off capability. In contrast, it is reduced. In a GCT the negative gate current increases rapidly and the high current flows only during a short time interval. The total charge which is extracted via the gate is even smaller than in a GTO thyristor, since in a GTO still new carriers are injected by the emitter junction J_3 during the increase of the negative gate current in the interval t_{gs} (Fig. 8.19), i.e. the n^+-emitter replenishes part of the charge that must be extracted by the gate driver. In a GCT, only the charge stored before turn-off must be removed. The GCT can be operated with even half of the drive power of a GTO [Lin06].

Although only few modifications of the silicon die are made in a GCT compared to a GTO, the CCT constitutes a significant progress. The main weaknesses of the GTO are solved to a large extent.

References

[Bec80] Becke HW, Misra RP: "Investigations of gate turn-off structures" International Electron Devices Meeting, Volume 26, pp. 649–653 (1980)

[Ben99] Benda V, Govar J, Grant DA, Power Semiconductor Devices, John Wiley & Sons, New York 1999

[Chu07] Chu, CK: "Geometry of thyristor cathode shunts" IEEE Trans. Electron Devices, Volume 17, Issue 9, pp. 687–690 (1970)

[Chk05] Chukaluri EK, Silber D, Kellner-Werdehausen U, Schneider C, Niedernostheide FJ, Schulze HJ: "Recent Developments of High-Voltage Light-Triggered Thyristors". Proceedings 36th Power Electronics Specialists Conference PESC ′05 pp. 2049–2052 (2005)

[Gen64] Gentry FE, Gutzwiller FW, Holonyak N, Von Zastrow EE: Semiconductor Controlled Rectifiers: Principles and Applications of p-n-p-n Devices, Principle-Hall Inc, New York (1964)

[Gen65] Gentry FE, Scace RI, Flowers JK: "Bidirectional triode pnpn-switches", Proc. IEEE 53 pp. 355–369 (1965)

[Gen68] Gentry FE, Moyson J: "The amplifying gate thyristor", IEEE International Electron Devices Meeting, Volume 14, p 110 (1968)

[Ger65] Gerlach W: "Thyristor mit Querfeldemitter" Z. angew. Phys 17 pp. 396–400 (1965)

[Ger79] Gerlach W, Thyristoren, Springer Verlag Berlin 1979

[Gru96] Gruening H, Odegard B, Ress J, Weber A, Carroll E, Eicher S: "High-Power Hard-Driven GTO Module for 4.5kV/3kA Snubberless Operation" Proceedings of the PCIM, pp. 169–183 (1996)

[Her65] Herlet A: "The maximum blocking capability of silicon thyristors", Solid-St. Electron., Vol. 8, Issue 8, pp. 655–671 (1965)

[Lin06] Linder S, Power Semiconductors, EPFL Press, Lausanne, Switzerland, 2006

[Mol56] Moll JL, Tannenbaum M, Goldey M, Holoniak N: "p-n-p-n Transistor Switches" Proceedings IRE 44 pp. 1174–1182 (1956)

[Nak95] Nakagawa T, Tokunoh, F, Yamamoto M, Koga S: "A new high power low loss GTO", Proceedings of the 7th ISPSD, pp. 84–88 (1995)

[Nie01] Niedernostheide FJ, Schulze HJ, Kellner-Werdehausen U: "Self Protected High Power Thyristors", Proceedings of the PCIM, Nuremberg, 51–56 (2001)

[Nie07] Niedernostheide FJ, Schulze HJ, Felsl HP, Laska T, Kellner-Werdehausen U, Lutz J: "Thyristors and IGBTs with integrated self-protection functions" IET Journal Circuits, Devices & Systems, Volume 1, Issue 5, Page(s):315–320 October 2007

[Prz09] Przybilla J, Dorn J, Barthelmess R, Kellner-Werdehausen U, Schulze HJ, Niedernostheide FJ: "Diodes and thyristor — Past, presence and future", Proceedings 13th European Conference on Power Electronics and Applications, EPE ′09 (2009)

[Rad71] Raderecht, PS: "A review of the ′shorted emitter′ principle as applied to p-n-p-n silicon controlled rectifiers", International Journal of Electronics, Vol. 31, Issue 6, p541 (1971)

[Scu01] Schulze HJ, Niedernostheide FJ, Kellner-Werdehausen U, "Thyristor with Integrated Forward Recovery Protection", Proceedings of the ISPSD, Osaka, 199–202 (2001)

[Shi99] Shimizu Y, Kimura S, Kozaka H, Matsuura N, Tanaka T, Monma N: "A study on maximum turn-off current of a high-power GTO", IEEE Trans. Electron Devices, Vol 46, Issue 2, pp. 413–419 (1999)

[Sid97] SIDACtor General Information http://www.ryston.cz/pdf/teccor/specelec.pdf

[Sil75] Silber D, Füllmann M: "Improved gate concept for light activated power thyristors", 1975 International Electron Devices Meeting, Volume 21, Page(s): 371–374 (1975)

[Sil76] Silber D, Winter W, Füllmann M: "Progress in light activated power thyristors", IEEE Trans. Electron Devices, Volume 23, Issue 8, Aug pp. 899–904 (1976)

[Wol66] Wolley ED: "Gate turn-off in pnpn devices" IEEE Trans. Electron Devices ED-13, pp. 590–597 (1966)

Chapter 9
MOS Transistors

9.1 Function Principle of the MOSFET

The MOSFET basic structure was investigated early [Hof63]. For the comprehension of the function of a MOSFET (metal oxide semiconductor field effect transistor), the surface of the semiconductor may be examined at first. The surface of a semiconductor is always a disturbance of the ideal lattice due to the lack of the neighboring atom. Therefore, a thin oxide will always be built up on the surface or other atoms and molecules are adsorbed. Thus, these surface layers are normally electrically charged.

A p-type semiconductor may be given as an example. Assumed is a positive charge on the surface (Fig. 9.1).

For a small positive charge we obtain

$$|qV_S| < E_i - E_F$$

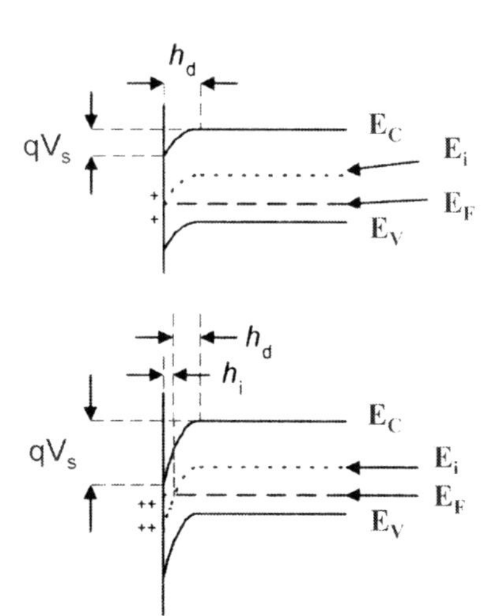

Fig. 9.1 Semiconductor surface, p-type semiconductor, positive charge on the surface

J. Lutz et al., *Semiconductor Power Devices*, DOI 10.1007/978-3-642-11125-9_9,

The hole-concentration/density on the surface is being reduced. The conduction band and the valence band are bent downward. A depletion zone of the thickness h_{d} is built up.

An increased positive charge yields

$$|qV_{\mathrm{S}}| > E_i - E_{\mathrm{F}}$$

Conduction and valence band are bent even stronger, for a small region at the surface the Fermi level is now closer to the conduction band than to the valence band. An inversion layer of the thickness h_{i} is being formed, in which electrons are the majority charge carriers.

Next to this is the depletion layer with the thickness h_{d}, which separates the inversion layer from the p-type area.

Having a negative charge on the p-type semiconductor, an accumulation layer of holes will be formed. The n-type semiconductor behaves similarly: With a positive charge at the surface, the accumulation layer is formed, with a negative surface charge a depletion zone is formed, with an increased negative charge the inversion layer.

Next, we assume a thin oxide film on the p-type semiconductor and apply a metallization on it. On this metal film a positive voltage is applied. Furthermore, two n^+-areas are added and bonded as source and drain region. We have the simplest case of a lateral MOS field effect transistor shown in Fig. 9.2 [Hof63].

A positive voltage has the same effect as the positive surface charge: When there is a sufficient positive voltage at the gate, both n-areas are connected by the inversion layer. Due to the gate voltage $V_{\mathrm{G}} > V_{\mathrm{T}}$ a current can flow between the drain and the source.

Gate-Source Threshold Voltage V_{T} (n-channel-MOSFET):
The threshold voltage is the gate voltage, at which the generated electron concentration equals the concentration of the acceptors.

It can be distinguished as

n-channel MOSFET: An n-type channel is formed in a p-area.
p-channel MOSFET: A p-type channel is formed in an n-area

On closer inspection, it has to be considered that the oxide contains positive charges on to the boundary surface to the semiconductor. These charges are in the

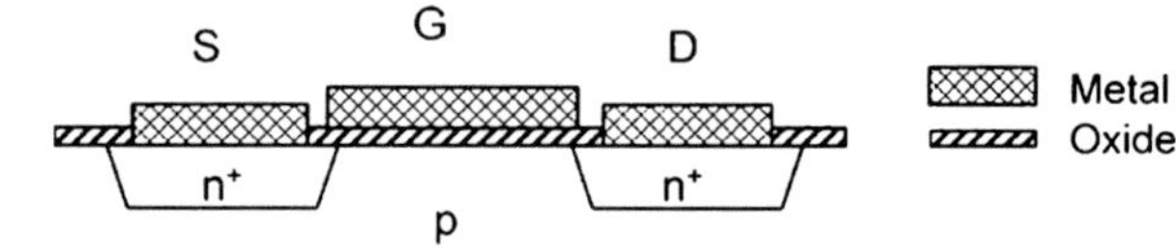

Fig. 9.2 Lateral n-channel MOSFET. Description: S, source; G, gate; D, drain

order of 5×10^9 to 1×10^{11} cm^{-2}. Moreover, the gate area of power MOSFETs consists of a heavily n-doped polysilicon layer (see Figs. 9.4 and 9.5) and a potential difference already exists between gate and semiconductor due to the differing positions of the Fermi level in the n$^+$-doped polysilicon and in the p-type semiconductor (concerning the n-channel MOSFET). Both effects function in the same way as an external positive gate voltage and result in a reduction of the threshold voltage V_T. In case of a low doping of the p-area and a high oxide charge of the n-channel MOSFET, V_T is negative, even without gate voltage a channel exists. The definition of the threshold voltage mentioned above remains valid.

It can be distinguished as

Depletion type: $V_T < 0$. The device is normally on and does not block before a negative gate source voltage $V_G < V_T$ is applied.
Enhancement type: $V_T > 0$. An n-channel only develops when $V_G > V_T$ (normally off device).

Usually, MOSFETs of the enhancement type are used in power electronics, because of the normally-off feature. Almost always n-channel MOSFETs are used, being more advantageous since the mobility of the electrons is much higher than that of the holes (see Chap. 2). Typically, the threshold voltage of modern devices is adjusted between 2 and 4 V.

9.2 Structure of Power MOSFETs

The configuration shown in Fig. 9.2 will sustain little drain-source voltage. Thus, a structure such as in Fig. 9.3, which is named DMOS (D = double diffused), is used from 10 V upward. In front of the drain is a n$^-$-area, the drain extension area, which takes over the blocking voltage.

Lateral DMOS transistors are frequently used in power ICs and in monolithic integrated power semiconductor circuits ("smart power"). But they have the disadvantage of having a low current–load capacity, because the n$^-$-area demands a large part of the surface of the semiconductor. If real "power" has to be controlled, a vertical MOSFET is realized by arranging the area for the electric field vertically

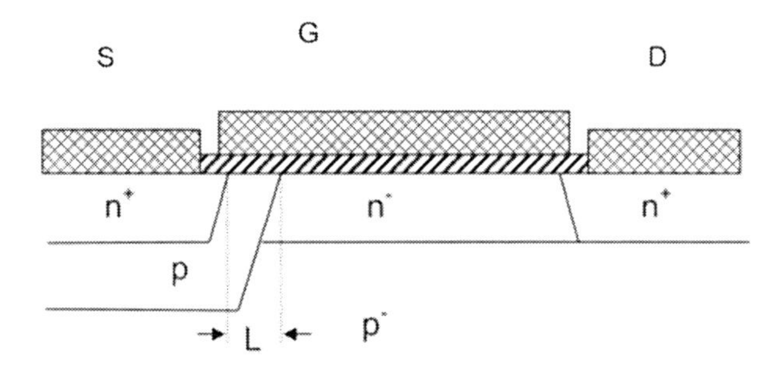

Fig. 9.3 Lateral DMOS

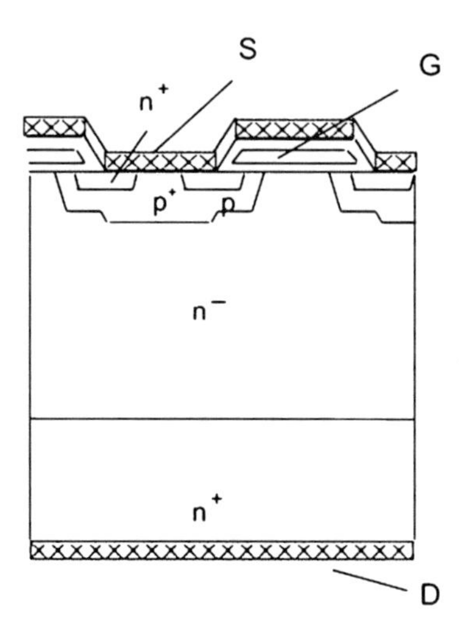

Fig. 9.4 Vertical DMOS transistor. The gate electrode is polysilicon

(Fig. 9.4). Consequently, the volume of the semiconductor is utilized and the surface can be used for the formation of the cells.

On the surface of the semiconductor the individual cells are formed, which consist of p-wells and diffused n^+-source areas. A cross section of a cell can be seen in Fig. 9.4. The p-well is connected to the source metallization, so that the parasitic npn-transistor is shorted. In order that the short has very little resistance, the doping is increased at this spot by an additional p^+-implantation, followed by a diffusion step. At the edges of the well is the channel, which is covered with the thin gate oxide. Above the oxide the gate electrode is applied, usually consisting of a heavily

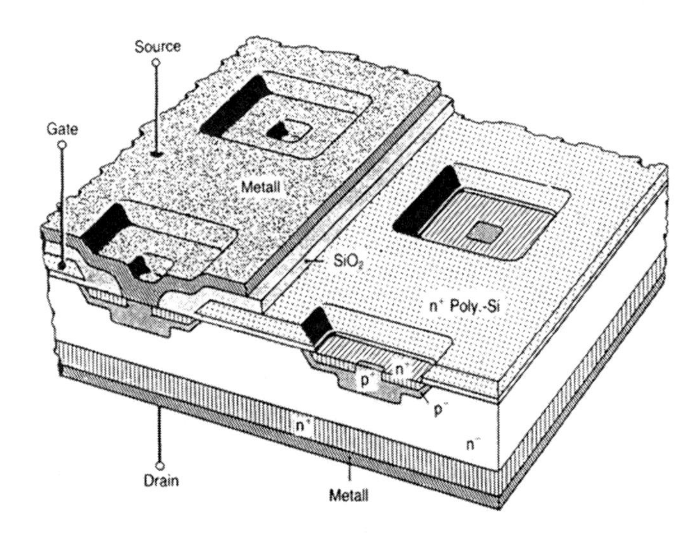

Fig. 9.5 Cell structure of a vertical DMOS. From [Ste92]

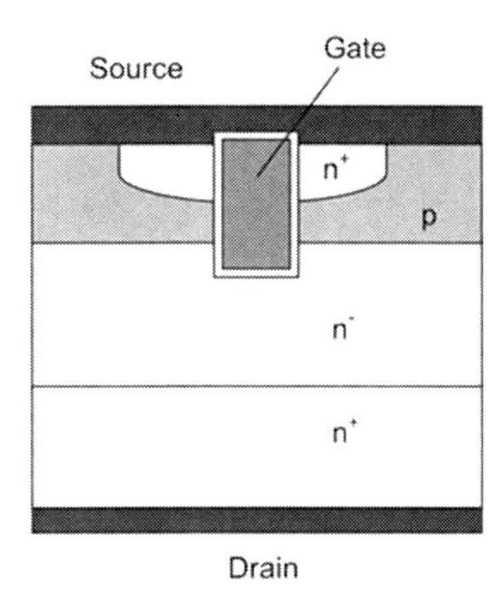

Fig. 9.6 Vertical trench MOSFET

doped n$^+$ polysilicon. At one spot, mostly in the center of the chip, the gate electrode is conducted to the surface and there bonded by a bond wire.

Because the current has to flow through the inversion channel, many single cells are formed to gain a larger width of the channel. An example is given in Fig. 9.5. Here, the cells are of square form and arranged in a square pattern. The semiconductor surface area is used even more effectively with a hexagonal pattern, where the single cells are hexagonal, the so-called HEXFET structure.

The vertical DMOS (also called VDMOS) transistors are used in a wide field of applications. Since the second half of the 1990s, with the introduction of the trench MOS [Sod99] a further improvement has been made, in which the channel area is vertically arranged too (Fig. 9.6). Due to this, a much smaller on-state resistance can be gained especially in the lower voltage range <100 V.

9.3 Current–Voltage Characteristics of MOS Transistors

Figure 9.7 shows the current–voltage characteristics of the MOSFET. The device is in the blocking state as long as a positive voltage V_D between drain and source is given and V_G is smaller than the threshold voltage V_T. The blocking voltage of the MOSFET is limited by the avalanche breakdown. Because the npn-transistor is suspended by a low resistance short, the blocking voltage of the MOSFET corresponds to the blocking voltage of the diode, which is formed by the p-well, the low doped base region and the n$^+$-layer.

A current carrying channel is built for $V_G > V_T$, resulting in the given current–voltage characteristics. Similar to the current gain of bipolar transistors a transconductance is defined here. For low voltages V_D, the current–voltage characteristics have the form of a straight line. For a defined gate voltage V_G the resistance $R_{DS(on)}$ is indicated.

The transition between the ohmic region and the pinch-off region is called the quasi saturation. This region is described by a parabolic curve.

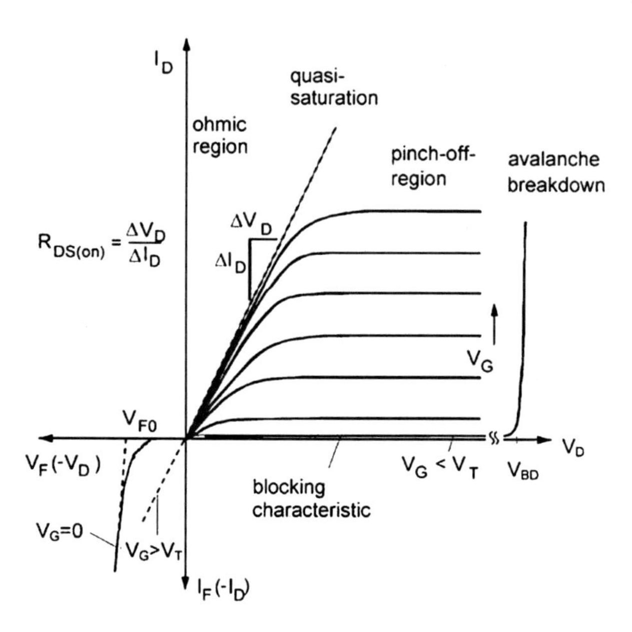

Fig. 9.7 Current–voltage characteristic of the MOSFET

In the reverse direction of the MOSFET a forward-biased diode is present. As for a power diode, this diode forward characteristic is often approximated by a threshold voltage V_{F0} and a differential resistance.

9.4 Characteristics of the MOSFET Channel

Due to the oxide layer of the gate and the gate electrode a capacitor is built above the channel. Its area specific capacity can be described with

$$C_{\mathrm{ox}} = \frac{\varepsilon_0 \cdot \varepsilon_{\mathrm{r}}}{d_{\mathrm{ox}}} \tag{9.1}$$

The oxide has the thickness d_{ox} (i.e. <100 nm). The relative permittivity of the oxide is $\varepsilon_{\mathrm{r}} = 3.9$ (for SiO_2). Having the gate voltage V_{G} higher than the threshold voltage V_{T} an inversion channel is created, as it is shown in Fig. 9.8a. As long as the voltage drop caused by the current in the channel can be neglected, the charge of the inversion channel yields

$$Q_{\mathrm{s}} = C_{\mathrm{ox}} \cdot (V_{\mathrm{G}} - V_{\mathrm{T}}) \tag{9.2}$$

The carriers, forming this charge, are available for the current transport in the inversion channel. As long as the pinch-off can be neglected in the channel, the resistance of the channel is

$$R_{\mathrm{ch}} = \frac{L}{W \cdot \mu_{\mathrm{n}} \cdot Q_{\mathrm{s}}} = \frac{L}{W \cdot \mu_{\mathrm{n}} \cdot C_{\mathrm{ox}} \cdot (V_{\mathrm{G}} - V_{\mathrm{T}})} = \frac{1}{\kappa \cdot (V_{\mathrm{G}} - V_{\mathrm{T}})} \tag{9.3}$$

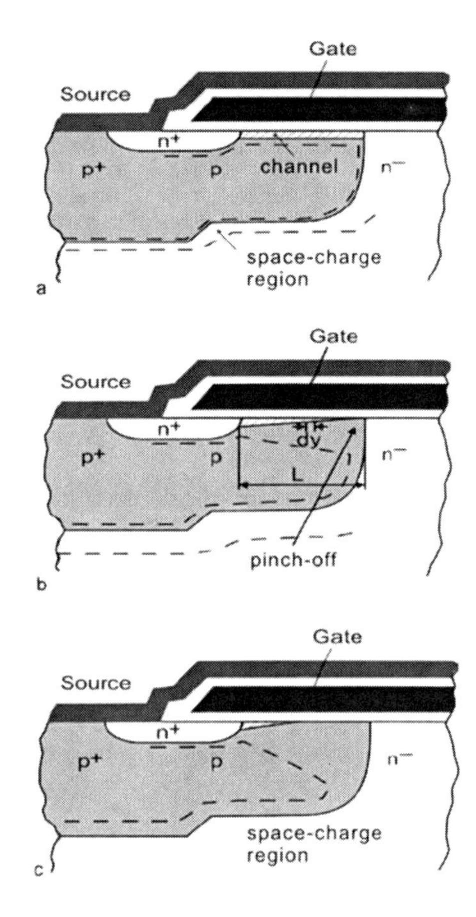

Fig. 9.8 MOSFET channel. (**a**) Ohmic region, $V_D << V_G - V_T$, (**b**) pinch-off, $V_D = V_G - V_T$, (**c**) channel length shortening $V_D >> V_G - V_T$

where L is the length of the channel (for example, 2 μm, see Fig. 9.3) and W is the entire width of it. In Fig. 9.3, W is vertical to the plane of projection and corresponds to the circumference of the single cell multiplied by the number of cells (see Fig. 9.5). Having a high cell density, a large W is achieved, and thus a low channel resistance. W can amount to some 100 m per cm^2 of a chip surface in modern semiconductor devices. The parameters dependent on geometry can be summarized in

$$\kappa = \frac{W \cdot \mu_n \cdot C_{ox}}{L} \tag{9.4}$$

Equation (9.3) is valid for the ohmic region in Fig. 9.7, that is, for the area in which the voltage drop across the channel can be neglected with regard to its influence on Q_S.

As can be seen in Eq. (9.3), the channel resistance is affected by the carrier mobility. In Sect. 2.6 it has been shown that the mobility μ_p only amounts to approximately one third of μ_n. It is for this reason that whenever it is possible, the n-channel MOSFET is used in power electronics.

With increasing current a voltage drop $V(y)$ develops across the channel. The channel narrows, see Fig. 9.8b. Along the length of the channel y a charge $Q(y)$ will exist. For an element dR of the resistance R_{CH}, with Eq. (9.3) one obtains

$$\mathrm{d}R = \frac{\mathrm{d}y}{W \cdot \mu_n \cdot Q(y)} \tag{9.5}$$

with

$$Q(y) = C_{ox} \cdot (V_G - V_T - V(y)) \tag{9.6}$$

In a segment dR the voltage drop is

$$\mathrm{d}V = I_D \cdot \mathrm{d}R \tag{9.7}$$

Inserting Eqs. (9.6) and (9.5) into Eq. (9.7) yields

$$I_D = W \cdot \mu_n \cdot C_{ox} \cdot (V_G - V_T - V(y)) \cdot \frac{\mathrm{d}V}{\mathrm{d}y} \tag{9.8}$$

The voltage V_D drops between the boundaries $y = 0$ and $y = L$:

$$\int_0^L I_D \cdot \mathrm{d}y = W \cdot \mu_n \cdot C_{ox} \cdot \int_0^{V_D} (V_G - V_T - V(y)) \cdot \mathrm{d}V \tag{9.9}$$

Integration leads to the following $I_D(V_G, V_D)$ characteristic:

$$I_D = \kappa \cdot \left((V_G - V_T) \cdot V_D - \frac{1}{2} V_D^2 \right) \tag{9.10}$$

for $V_D \leq V_G - V_T$. The characteristic corresponds to the parabolic section (quasi saturation) in Fig. 9.7. For small V_D it verges into

$$I_D = \kappa \cdot (V_G - V_T) \cdot V_D \tag{9.11}$$

and corresponds to the ohmic region, as already indicated in Eq. (9.3). The passing into the pinch-off region results from Eq. (9.10) for $\mathrm{d}I_D/\mathrm{d}V_D = 0$. Afterward the channel is pinched off for

$$V_D = V_G - V_T \tag{9.12}$$

For a larger V_D, inserting Eq. (9.12) into Eq. (9.10) yields the characteristics in the pinch-off region. In this region, the current remains almost constant even for increased voltage V_D

$$I_{Dsat} = \frac{\kappa}{2} (V_G - V_T)^2 \tag{9.13}$$

The transconductance is defined by

$$g_{\mathrm{fs}} = \frac{\Delta I_{\mathrm{D}}}{\Delta V_{\mathrm{G}}}\Big|_{V_{\mathrm{D}} = \mathrm{const}} \tag{9.14}$$

By differentiating Eq. (9.13) we obtain

$$g_{\mathrm{fs}} = \kappa\,(V_{\mathrm{G}} - V_{\mathrm{T}}) \tag{9.15}$$

According to Eq. (9.13), the current I_{Dsat} is independent of V_{D}. However, in reality the electric field penetrates into the p-zone, when V_{D} is strongly increased (see Fig. 9.8c) and the channel becomes shortened. This shortening of the channel length involves a slight ascent of the current at high voltages.

The current–voltage characteristics (9.10) can be found in numerous textbooks. However, comparing it with practically realized power devices, this equation is not very satisfactory. In the derivation it was not considered that a depletion zone is formed below the channel. This zone widens while the channel is narrowing, as it is indicated in Fig. 9.8c. A derivation of the current–voltage characteristics in consideration of the space charge can be found in [Gra89]. This leads to

$$I_{\mathrm{D}} = \kappa \cdot \left((V_{\mathrm{G}} - V_{\mathrm{T}}) \cdot V_{\mathrm{D}} - \frac{1}{2}\left(1 + \frac{C_{\mathrm{D}}}{C_{\mathrm{ox}}}\right) V_{\mathrm{D}}^2 \right) \tag{9.16}$$

with the area-specific capacity of the space charge region

$$C_{\mathrm{D}} = \sqrt{\frac{\varepsilon_0 \cdot \varepsilon_{\mathrm{r}} \cdot q \cdot N_{\mathrm{A}}}{2 \cdot \Delta V_{\mathrm{T}}}} \tag{9.17}$$

as it has been derived for the treatment of the pn-junction with Eq. (3.109). The voltage ΔV_{T} in Eq. (9.17) corresponds to the voltage necessary to develop a space charge region in the p-doped well with doping concentration N_{A}:

$$\Delta V_{\mathrm{T}} = 2 \cdot \frac{kT}{q} \cdot \ln\left(\frac{N_{\mathrm{A}}}{n_{\mathrm{i}}}\right) \tag{9.18}$$

ΔV_{T} is approximately 0.81 V (resulting from a typical doping of the p-well of 1×10^{17} cm^{-3}). Considering the space charge this way, little changes at a very small voltages V_{D}. The approximation for the ohmic region in Eq. (9.11) remains the same. However, I_{Dsat} and g_{fs} vary. Equation (9.16) is effective as long as the voltage drop across the channel is smaller than ΔV_{T}, that is, $V_{\mathrm{D}} < \Delta V_{\mathrm{T}}$.

Apart from that, the reduced mobility in the channel has to be considered. Even without a lateral electric field, the mobility is already reduced compared to the values indicated in Fig. 2.12. The reason is the influence of the semiconductor surface. If a voltage $V(y)$ is formed above the channel, then a significant electric field develops in lateral direction. Equation (2.38) has to be consulted for the velocity of the electrons. For the electron mobility it yields

$$\mu_e = \frac{\mu_{e0}}{1 + \theta \cdot (V_G - V_T)} \tag{9.19}$$

In [Gra89] a suitable agreement with experiments is achieved, when the values of 600 $cm^2V^{-1}s^{-1}$ and 0.02 V^{-1} are used for μ_{e0} and θ.

9.5 The Ohmic Region

For the MOSFETs ohmic resistance not only the channel resistance has to be considered. Indeed, the resistance of the low doped middle region already dominates in devices with a blocking voltage of 50 V upward. Because this layer is grown by epitaxy for vertical MOSFETs, the designation R_{epi} is commonly used. Figure 9.9 describes the structure of the MOSFET with a given path of the charge carriers (electrons) and with different parts of the resistance

$$R_{DS(on)} = R_{S*} + R_{n+} + R_{ch} + R_a + R_{epi} + R_s \tag{9.20}$$

For MOSFETs with blocking voltages <50 V, effort is made to reduce the channel resistance. The near-surface parts are reduced by increased cell density (larger W, see Eq. (9.1)). Most progress is achieved with the trench cell (Fig. 9.5), where additionally the resistance R_a is eliminated.

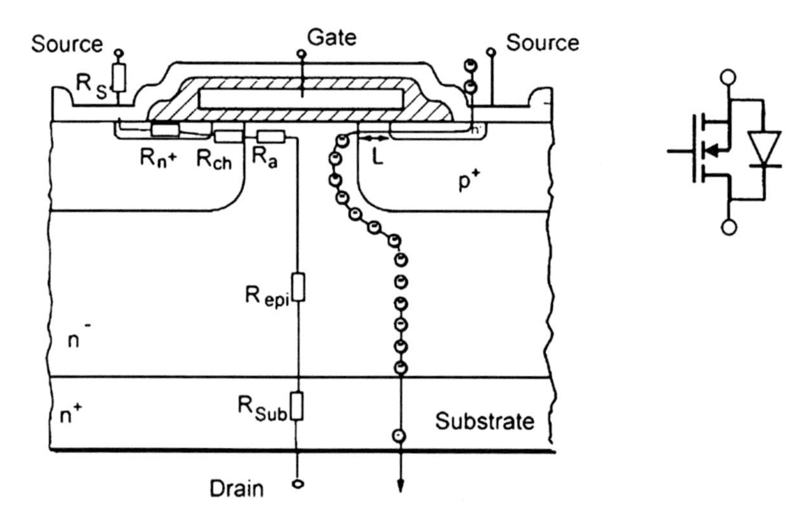

Fig. 9.9 Current path and resistances in a MOSFET

Table 9.1 Resistance $R_{DS(on)}$ for MOSFETs with different blocking voltages. Values from [Lor99]

		$V_{DS} = 30$ V (%)	$V_{DS} = 600$ V (%)
R^S*	Package	7	0.5
R_n+	Source layer	6	0.5
R_{CH}	Channel	28	1.5
R_a	Accumulation layer	23	0.5
R_{epi}	n^--layer	29	96.5
R_{Sub}	Substrate	7	0.5

In Table 9.1, the shares of the particular parts of the resistance are specified for a 30 V vertical MOSFET with planar cells together with a corresponding 600 V MOSFET.

The resistance R_{epi} of the low-doped region is identical with the voltage drop across the low-doped base region of a unipolar device, which has been given in Chap. 6 on Schottky diodes (Eq. 3.9):

$$R_{epi} = \frac{w_B}{q \cdot \mu_n \cdot N_D \cdot A} \tag{9.21}$$

If the device is designed for higher voltages, w_B has to be chosen larger as well as N_D smaller. Having a conventional MOSFET, the resistance can be calculated according to the shown approach in dependence of the voltage for which the device is designed. The lowest resistance can be obtained for a light PT design, as indicated in Eqs. (6.9), (6.10), (6.11), (6.12), (6.13), (6.14), and (6.15):

$$R_{epi,min} = 0.9 \frac{2 \cdot B^{\frac{1}{2}} \cdot V_{BD}^{\frac{5}{2}}}{\mu_n \cdot \varepsilon \cdot A} \tag{9.22}$$

Thus, the resistance increases more than with the square of the blocking voltage, namely with $V_{BD}^{2.5}$. Equation (9.22) or comparable equations (see Chap. 6) are handled as "unipolar limit" in the literature. Meanwhile, this limit has been broken by the principle of compensation structures.

9.6 Compensation Structures in Modern MOSFETs

The compensation principle for power MOSFETs has been introduced in commercially available products in 1998 with the 600-V CoolMOS^{TM} technology [Deb98]. The basic principle behind the drastic R_{ON}·A reduction compared to conventional power MOSFETs is the compensation of n-drift region donors by acceptors located in p-columns (also known as superjunction). Figure 9.10 shows the structure of a superjunction MOSFET compared to a conventional MOSFET. In the middle layer, p-columns are arranged. Their p-doping is adjusted to the value necessary for compensation of the n-regions. The compensating acceptors are located in lateral proximity to the drift region donors.

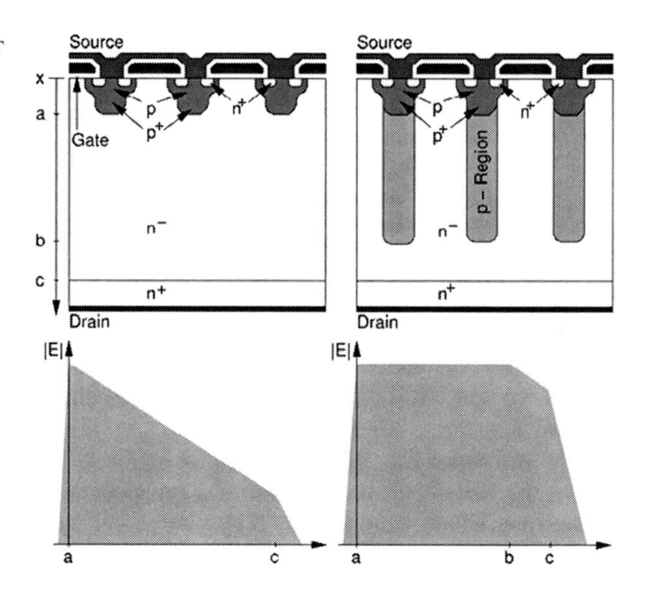

Fig. 9.10 Standard MOSFET and superjunction MOSFET

The result is a low effective doping in the entire voltage-sustaining region. An almost rectangular shape of the electric field is obtained, as can be seen in the lower part of Fig. 9.10. For this field shape, the highest voltage can be absorbed at a given thickness. The doping of the n-layer can be lifted in so far, as it is technologically possible to compensate it through an equally large p-doping. In this process it has to be considered that the area of the n-region, respectively n^--region, is decreased.

By means of the compensation principle, the coupling of the blocking voltage and the doping is neutralized and a degree of freedom for the adjustment of the n-doping is obtained. Since, according to Eq. (9.21), the n-doping determines the resistance in unipolar devices, the resistance can be lowered drastically.

In case of the rectangular shape of the electric field, the avalanche breakdown can be calculated by using the ionization integral (3.71), with the approach proposed by Shields and Fulop with $n = 7$ (see the sections about the PT diode, Eq. (5.12)), resulting in

$$w_B = B^{\frac{1}{6}} \cdot V_{BD}^{\frac{7}{6}} \text{ with } B = C/E_0^7 = 2.1 \times 10^{-35}\ \text{cm}^6/\text{V}^7 \tag{9.23}$$

Inserting Eq. (9.23) into Eq. (9.21) yields

$$R_{epi} = \frac{2B^{\frac{1}{6}} \cdot V_{BD}^{\frac{7}{6}}}{q \cdot \mu_n \cdot N_D \cdot A} \tag{9.24}$$

The factor 2 in the numerator of Eq. (9.24) is obtained due to the simplified account that the width of the p-columns is equal to the width of the n-zones. Only the n-areas contribute to the conduction; hence only half of the area is available.

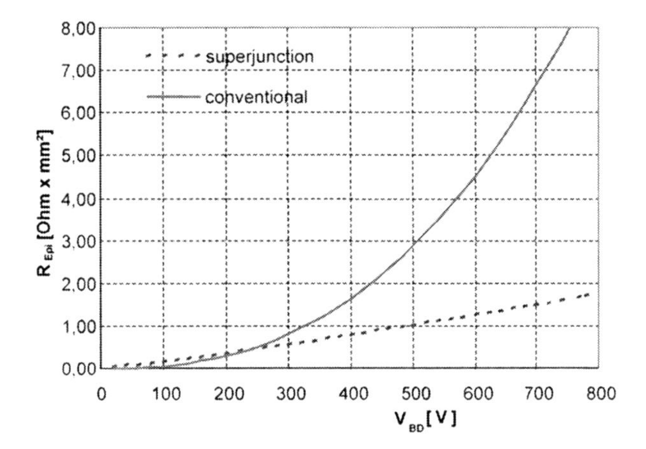

Fig. 9.11 Resistance R_{epi} as a function of the blocking voltage for the conventional MOSFET and for the superjunction device

Figure 9.11 compares the relation between R_{epi} and the blocking voltage for the conventional design (9.22) and for the superjunction device (9.24). Here, the doping $N_{\text{A}} = N_{\text{D}} = 2 \times 10^{15}\text{cm}^{-3}$ is chosen for the superjunction device. Furthermore, half of the total area is assumed to carry the electron current.

The following consequences can be derived for this very simplified case:

Under blocking conditions the space charge laterally penetrates into the n- and p-region. This is shown in Fig. 9.12. In Fig. 9.12 it is assumed that p- and n-regions are doped equally; the p-region has the doping N_{A}, the n-column the doping N_{D} with $N_A = N_D$. Moreover, both columns shall feature the same width, which constitute $2\,y_{\text{L}}$, respectively, in Fig. 9.12. When a voltage is applied in reverse direction, the

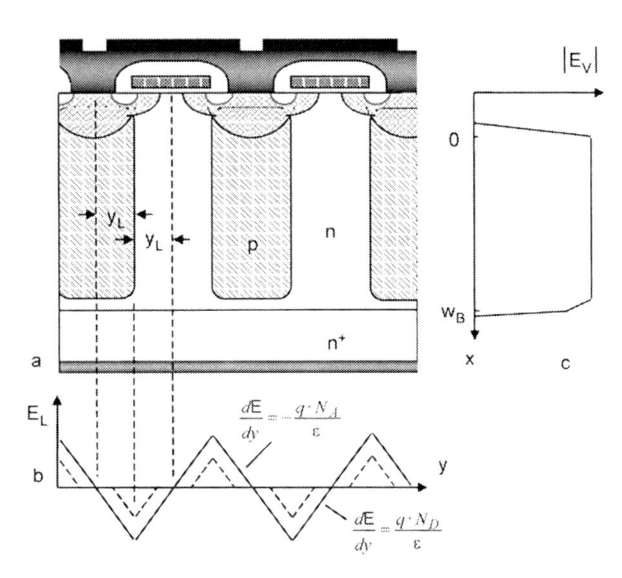

Fig. 9.12 Superjunction MOSFET. (**a**) Simplified structure, (**b**) electric field in lateral direction in the region of the columns, (**c**) electric field in vertical direction

space charge region penetrates only laterally into the columns. For low voltages, the dashed line in Fig. 9.12b indicates the magnitude of the electric field along a section in lateral direction in the region of the columns. With increasing voltage, the space charges will eventually meet in the center of the respective columns, as shown by the solid line in Fig. 9.12b. Now all acceptors and donors are ionized.

With further increase of the voltage the zigzag line in Fig. 9.12b is lifted. This results in a structure similar to a corrugated iron roof. In vertical direction the electric field as shown in 9.12c is obtained.

In lateral direction, the respective p- and n-regions have to be penetrated entirely by the field. The expansion of the electric field at the avalanche breakdown into an n-region with the doping N_D has been given in Eq. (3.85). The indicated width w is half the breadth of an n-region y_L. A doping $N_D = 2 \times 10^{15}\,\text{cm}^{-3}$ yields $y_L = 11\,\mu\text{m}$. The width of the p- as well as the n-regions has to be smaller than $2 \cdot y_L$, otherwise breakdown occurs in lateral direction. Therefore, the doping in Eq. (9.24) is connected to the width of the columns; a higher doping N_D demands a smaller y_L. Equation (3.85) inverted for N_D and inserted in Eq. (9.24) yields

$$R_{epi} = \frac{2 \cdot 2^{-\frac{3}{7}} B^{\frac{13}{42}} \cdot y_L^{\frac{8}{7}} \cdot V_{BD}^{\frac{7}{6}}}{\varepsilon \cdot \mu_n \cdot A} \tag{9.25}$$

Analogous considerations are leading to a similar result in [Zin01]. Equation (9.25) shows that the resistance can be reduced even more as shown in Fig. 9.11. This requires a yet smaller y_L. Note that the line for the superjunction in Fig. 9.11 can be shifted to a lower value for R_{epi} for an increased doping and a finer pattern. However, to realize this in the vertical structure with a depth $w_B >> y_L$ is a great technological challenge.

A more precise consideration, also including peaks of the electric field at the source and drain-sided border of the space charge region can be found in [Che01]. There, different arrangements of the columns are analyzed as well. Seen from the top of a device, the considerations of Figs. 9.10 and 9.12 would result in a stripe pattern of the p- and n-regions. However, a hexagonal arrangement would be better.

The requirement for precise lateral n- and p- dose compensation limits the n-drift region doping. The higher the deviation in the charge balance, the more loss in blocking capability occurs. This effect increases with higher doping N_D and smaller y_L. The process window for deviation from charge balance gets narrower [Kon06]. Process technology finally limits the possibility to reduce R_{on} in this type of compensation power MOSFETs.

For breakdown voltages below 200 V, field plate or oxide-bypassed MOSFETs are an excellent alternative [Lia01, Sie06c]. The device comprises a deep trench penetrating most of the n-drift region. An isolated field plate provides mobile charges required to compensate the drift region donors under blocking conditions as shown in Fig. 9.13. A voltage source dynamically provides electrons on the field plate and therefore precise lateral drift region compensation is ensured under all operating conditions.

The field plate isolation has to withstand the full source-drain blocking voltage of the device at the trench bottom; therefore oxide layers with thickness in the micron

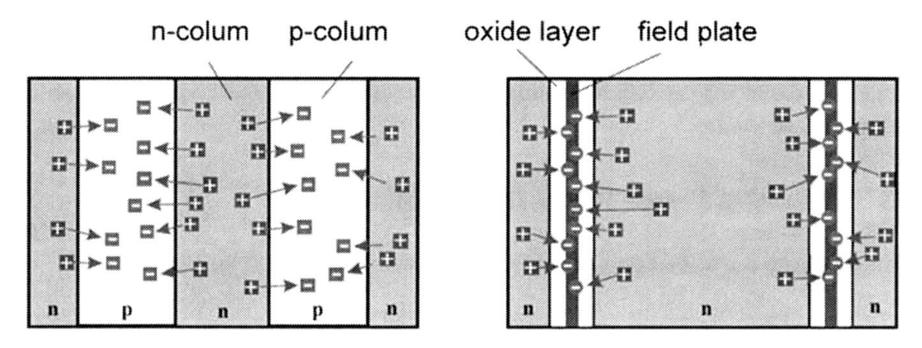

Fig. 9.13 Comparison of charge compensation by superjunction and by field plates

range have to be fabricated carefully with a special focus on avoiding thinning at the bottom trench corners and preventing generation of stress-induced defects. In contrast to standard trench MOS structures that exhibit a linearly decreasing electric field with a maximum at the body/drift region pn-junction, the field plate principle leads to a more constant field distribution. While in case of superjunction devices an almost homogeneous vertical field distribution is found, the shape of the electric field of a field plate device shows two peaks, one at the body/drift region pn-junction and the larger one at the bottom of the field plate trench [Che05]. The necessary drift region length for a given breakdown voltage is reduced and the drift region doping can be increased, leading to a significantly reduced on-state resistance.

As in case of superjunction devices, the on-state resistance is reduced below the "silicon limit." Depending on the feasible device geometry, either superjunction or field plate devices are advantageous in terms of device performance. In the voltage range between 30 and 100 V, the field plate compensation is superior to the superjunction compensation [Che05]. Figure 9.14 depicts the drift region resistance as a function of blocking voltage. A comparison is given between two-dimensional simulations [Paw08] for field plate compensation structures and the "silicon

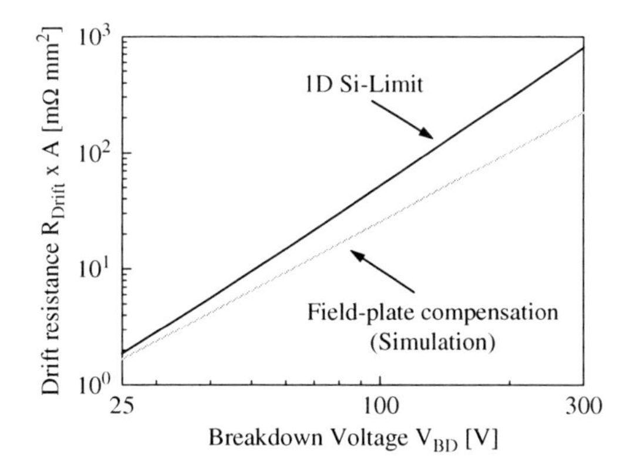

Fig. 9.14 Drift region resistance in dependency of breakdown voltage for field plate devices, compared to the "silicon limit"

limit " given by Eq. (9.22). Due to the small device dimensions of the field plate trench devices, the consequence is a larger doping density and thus a smaller drift region resistance [Paw08]. The simulation results in Fig. 9.14 agree well with experimental data.

9.7 Switching Properties of the MOSFET

Starting from the transition time of the charge carriers through the channel

$$\tau_t = \frac{L}{v_d} \tag{9.26}$$

with $v_d = \mu_n E$ and $E = V_{CH}/L$ yields

$$\tau_t = \frac{L^2}{\mu_n \cdot V_{CH}} \tag{9.27}$$

For instance, with $d = 2\ \mu m$, $V_{CH} = 1V$, and $\mu_n = 500\ cm^2/Vs$, we obtain the transition time $\tau_t \approx 80$ ps. It corresponds to a transition frequency of

$$f_t \approx 12.5 \text{ GHz}$$

In practice, this is accomplishable for a power MOSFET, since parasitic capacitances exist, leading to time constants that determine the limiting frequency:

$$f_{co} = \frac{1}{2\pi \cdot C_{iss} \cdot R_G} \tag{9.28}$$

with $C_{iss} = C_{GS} + C_{GD}$ and $R_G = R_{Gint} + R_{Gext}$

C_{iss} as well as the recommended gate resistance R_{ext} can be found in the data sheets. The internal gate resistance has to be asked for from the manufacturer.

Example: IXYS XFH 67 N10
$C_{iss} = 4500$ pF
$R_{ext} = 2\ \Omega$, $R_{int} \approx 1\ \Omega$ (assumed)
$\Rightarrow f_{co} = 12$ MHz

Figure 9.15 shows the structure of the MOSFET in which the parasitic capacitances are indicated. On the right, the equivalent circuit diagram of the MOSFET with its parasitic capacitances is shown. The inverse diode and some of the resistances are illustrated as well, only R_{CH} and R_{epi} are charted.

The turn-on and turn-off behavior shall be dealt with now under the condition of an inductive load, as an inductive load is usually existent in practice. The circuit corresponds to the one in Fig. 5.19. Figure 9.16 shows the turn-on waveform of the MOSFET with an inductive load. The characteristic quantities for the turn-on are

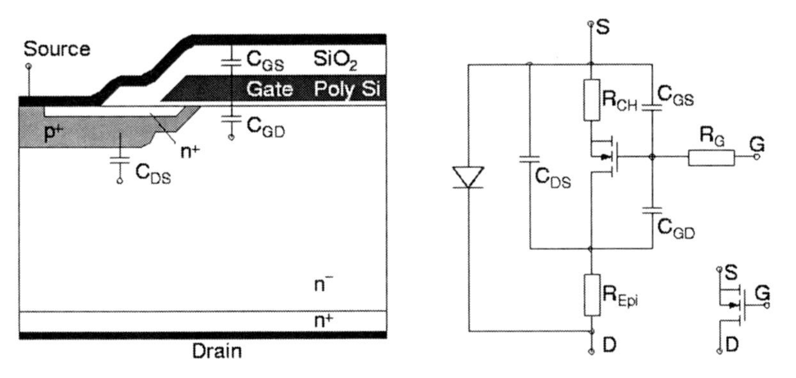

Fig. 9.15 MOSFET with parasitic capacitances, structure, and electrical equivalent circuit according to [Mic03] © 2003 Springer

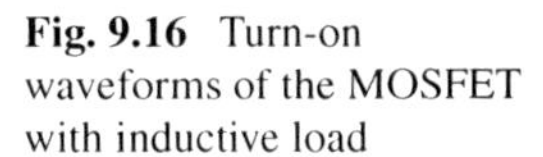

Fig. 9.16 Turn-on waveforms of the MOSFET with inductive load

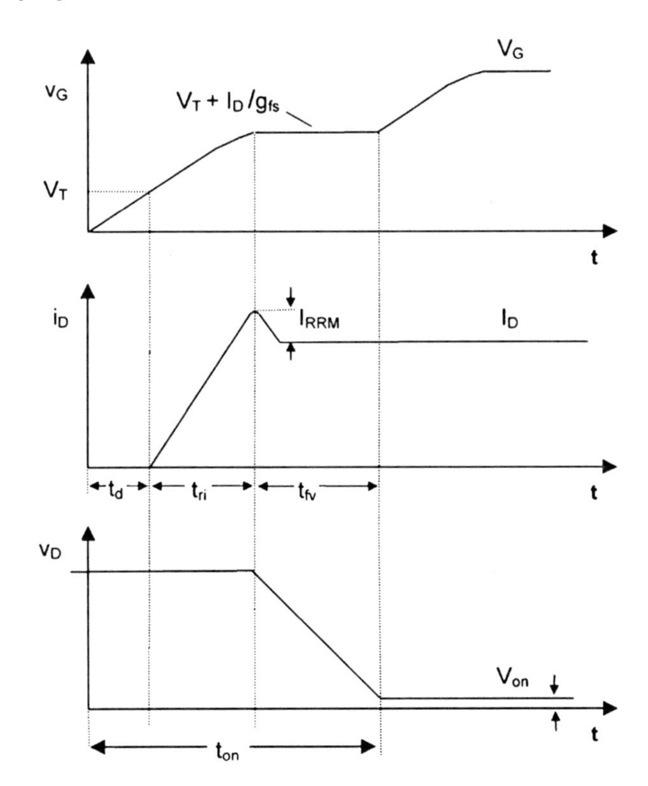

t_d: *Turn-on delay time*

Time until V_{GS} reaches the threshold voltage $V_{GS(th)}$

$$t_d \sim R_G(C_{GS} + C_{GD})$$

t_{ri}: *Rise time*

During this time the current increases

$$t_{\text{ri}} \sim R_{\text{G}}(C_{\text{GS}} + C_{\text{GD}})$$

Due to the freewheeling diode above the inductive load, the reverse current peak $I_{\text{RRM}}$is added, see Figs. 5.20 and 5.21. During this time the voltage remains virtually unaltered.

t_{fv}: *Voltage fall time*
Now the freewheeling diode takes over voltage and the voltage across the MOSFET drops. The capacitance C_{GD} (Miller capacitance) is being charged:

$$t_{\text{fv}} \sim R_{\text{G}} C_{\text{GD}}$$

In this phase, V_{G} remains at the value of the Miller plateau

$$V_{\text{G}} = V_{\text{T}} + I_{\text{D}}/g_{\text{fs}}$$

The voltage V_{D}falls to the value of the forward voltage

$$V_{\text{on}} = R_{\text{on}} I_{\text{D}}$$

The entire turn-on time t_{on} amounts to

$$t_{\text{on}} = t_{\text{d}} + t_{\text{ri}} + t_{\text{fv}}$$

Figure 9.17 shows the turn-off behavior for an inductive load. Characteristic quantities of the turn-off are

t_{s}: *Storage time*
In the driver the voltage signal is reset to zero or a negative value. However, the gate has to be discharged to the value at which the gate voltage corresponds to the value at which the on-state current I_{D} equals the saturation current, which means

$$V_{\text{G}} = V_{\text{T}} + I_{\text{D}}/g_{\text{fs}}$$

The capacitances C_{GS} and C_{GD}, which are in parallel to the channel, have to be discharged (see Fig. 9.15), For the storage time holds:

$$t_{\text{s}} \sim R_{\text{G}}(C_{\text{GS}} + C_{\text{GD}})$$

t_{rv}: *Voltage rise time*
The voltage increases to the value given by the circuit. The current remains constant at the initial value. The gate voltage persists at the Miller plateau. The Miller capacitance C_{GD} has to be discharged and, therefore

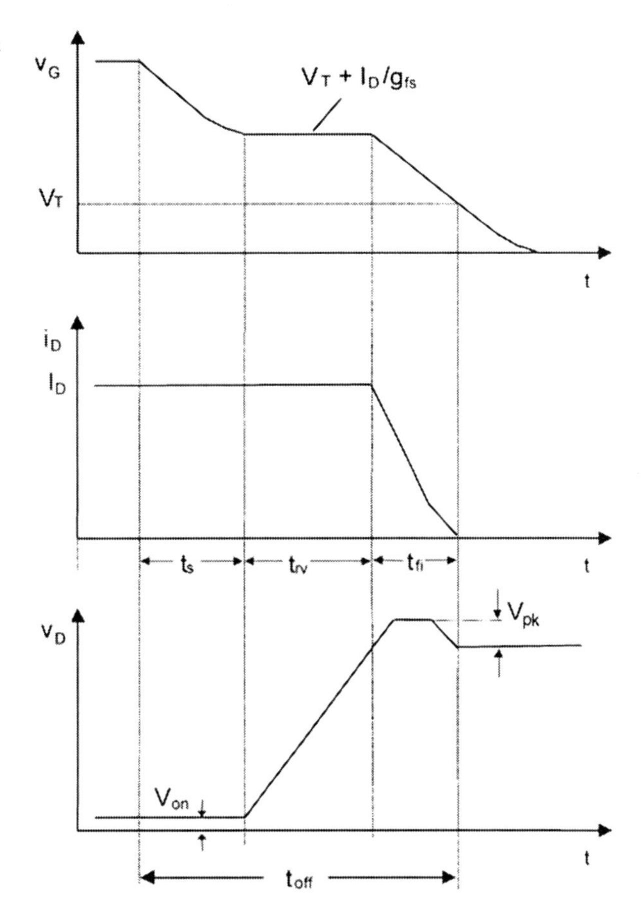

Fig. 9.17 Turn-off waveform of the MOSFET with inductive load

t_{fi}: *Current fall time*
The gate capacitance $C_{GS} + C_{GD}$ is discharged and the current decreases. The current becomes zero (or, more exact, it attains the value of the off-state leakage current), when V_{GS} has fallen to V_T:

$$t_{fi} \sim R_G(C_{GS} + C_{GD})$$

In this phase a spike V_{pk} is added to the applied voltage. This spike consists of

- the inductive voltage, which is generated by the current slope di/dt at the parasitic inductance L_{par}, L_{par} is indicated in Fig. 5.18.
- the turn-on voltage spike V_{FRM} of the diode

Consequently $V_{pk} = \left| L_{par} \cdot \frac{di}{dt} \right| + V_{FRM}$

The entire turn-off time is

$$t_{\mathrm{off}} = t_{\mathrm{s}} + t_{\mathrm{rv}} + t_{\mathrm{fi}}$$

The switching edges of the turn-on and turn-off can be controlled by means of the gate resistance under the conditions described. With a smaller R_{G} the switching time can be reduced and, thus, the switching losses in the device may be lowered as well (see below).

From the switching times a frequency limit can be derived:

$$f_{\max} = \frac{1}{t_{\mathrm{on}} + t_{\mathrm{off}}} \tag{9.29}$$

Taking the data sheet values of the above-mentioned MOSFET IXYS IXFH 67 N10 and adding all switching times typically 220 – 340 ns is obtained, which corresponds to a frequency of 3 – 4 MHz. This is considerably lower than f_{co}. But the example is not unproblematic, as the switching times in the data sheets are mostly specified for an ohmic load, which is rarely the case in practice.

9.8 Switching Losses of the MOSFET

The maximum attainable switching frequency of a power MOSFET depends on the switching losses. The energy loss per pulse can be calculated, like for other devices, by integrating the product $v(t)i(t)$ during turn-on and turn-off. During turn-on it can be calculated with

$$E_{\mathrm{on}} = \int_{t_{\mathrm{on}}} v_{\mathrm{D}}(t) \cdot i_{\mathrm{D}}(t)\,\mathrm{d}t \tag{9.30}$$

In practice, the energy loss per pulse is determined from oscillograms. Modern oscilloscopes are able to calculate the product of current and voltage and to integrate it over the selected time. An example for an IGBT is given in Fig. 5.21. For an estimation, Fig. 9.16 can be used from which emanates

$$E_{\mathrm{on}} = \frac{1}{2} \cdot V_{\mathrm{D}} \cdot (I_{\mathrm{D}} + I_{\mathrm{RRM}}) \cdot t_{\mathrm{ri}} + \frac{1}{2} \cdot V_{\mathrm{D}} \left(I_{\mathrm{D}} + \frac{2}{3} I_{\mathrm{RRM}} \right) \cdot t_{\mathrm{fv}} \tag{9.31}$$

assuming that the peak reverse current I_{RRM} caused by the diode decays linear during the time t_{fv}.

At turn-off, the energy loss is calculated with

$$E_{\mathrm{off}} = \int_{t_{\mathrm{off}}} v_{\mathrm{D}}(t) \cdot i_{\mathrm{D}}(t)\,\mathrm{d}t \tag{9.32}$$

which can be estimated according to Fig. 9.17 with

$$E_{\text{off}} = \frac{1}{2} \cdot V_{\text{D}} \cdot I_{\text{D}} \cdot t_{\text{rv}} + \frac{1}{2} \cdot \left(V_{\text{D}} + V_{\text{pk}}\right) \cdot I_{\text{D}} \cdot t_{\text{fi}} \tag{9.33}$$

The total switching losses follow from

$$P_{\text{on}} + P_{\text{off}} = f \cdot \left(E_{\text{on}} + E_{\text{off}}\right) \tag{9.34}$$

Conduction losses and blocking losses add to the switching losses. For power MOSFETs the off-state leakage current is in the order of few microamperes, so that the blocking losses may be neglected. The conduction losses cannot be ignored. Defining the duty cycle d as the ratio of the interval in which the MOSFET conducts versus the switching period, the conduction losses can be calculated according to

$$P_{\text{cond}} = d \cdot V_{\text{on}} \cdot I_{\text{D}} = d \cdot R_{\text{on}} \cdot I_{\text{D}}^2 \tag{9.35}$$

For the total losses one obtains

$$P_{\text{V}} = P_{\text{cond}} + P_{\text{on}} + P_{\text{off}} = d \cdot R_{\text{on}} \cdot I_{\text{D}}^2 + f \cdot \left(E_{\text{on}} + E_{\text{off}}\right) \tag{9.36}$$

These losses have to be led out as heat flux through the case of the device. The maximum allowable losses are determined by the cooling conditions, the acceptable temperature difference, and the thermal resistance. Details are provided in Chap. 11.

For the MOSFET IXYS IXFH 67 N10 used as an example, it can be estimated from the data sheet values of the thermal resistance that switching frequencies up to 300 kHz can be realized. Clearly, the MOSFET, as it is a unipolar device, is the fastest Si power semiconductor switch available.

The potential switching frequency depends, on the one hand, on the thermal parameters and considerably on the other devices in the circuit as well. The entire circuit has to be optimized accordingly. From Eqs. (9.31) and (9.33) arises that the switching losses depend on the switching times. By a reduction of the switching times due to smaller gate resistances R_{G}, switching losses can be lowered. On the other hand, the steepness of the slopes is limited in practice

- by motor windings which are not to be stressed with too high dv/dt
- even more by freewheeling diodes, which are required in inductive circuits. Inappropriate freewheeling diodes, in the presence of increased di/dt, lead to a snappy switching behavior, voltage spikes, and oscillations.

9.9 Safe Operating Area of the MOSFET

Between source and drain, the structure of the MOSFET contains a parasitic bipolar npn-transistor in parallel to the MOS channel, as shown in Fig. 9.18. This parasitic npn-transistor would lead to many problems:

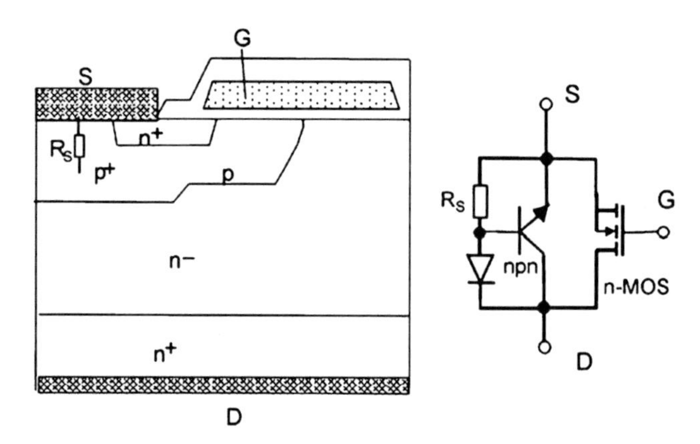

Fig. 9.18 MOSFET with its equivalent circuit, containing the parasitic npn-transistor and the parasitic diode

– the blocking voltage would be decreased by such an open-base transistor
– when applying a voltage with a high dv/dt, a displacement current could be generated due to the charging of the depletion layer of the base–collector junction. This current triggers the transistor

and, finally, the safe operating area of a transistor is limited by the second breakdown effect.

Therefore, the base–emitter junction of the npn-transistor has to be shorted by a low resistance R_S. This resistance is chosen preferably small by increasing the doping in this region with an additional p$^+$-ion implantation (p$^+$-doping) and by choosing the length of the n$^+$-source region as small as possible, as the photolithographic process allows it.

In today's MOSFETs the parasitic transistor is effectively made inoperative. Thus, the safe operating area is no longer limited by the second breakdown. The safe operating area of today's MOSFETs is rectangular, as it is shown in Fig. 9.19.

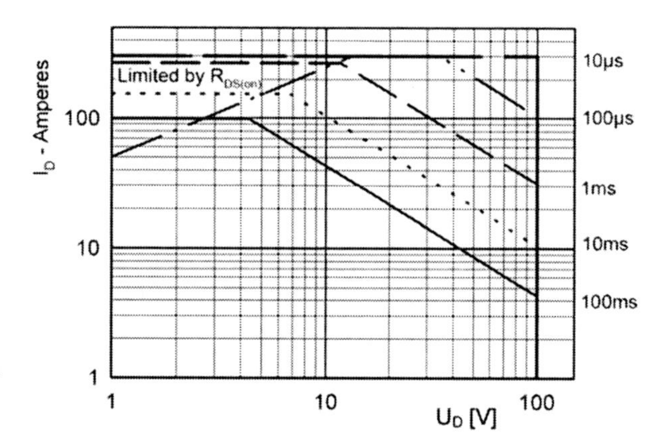

Fig. 9.19 Safe operating area (SOA) of a MOSFET, example IXYS IXFH 67 N10

It is only limited by the blocking voltage and the occurring losses. The SOA curves for pulse times above 10 μs in Fig. 9.19 are limited due to the maximum power losses, at which the junction temperature remains below 150°C.

9.10 The Inverse Diode of the MOSFET

Because of the contact between the p-well and the source metallization, a pin-type diode structure is built by the p-well, the n^--region, and the n^+-substrate, as shown in Fig. 9.18. Consequently, for the application in a bridge topology in a voltage source converter circuit, a freewheeling diode is intrinsically present. The characteristic of this diode corresponds to the current–voltage characteristic of the MOSFET in the third quadrant (see Fig. 9.7 for $V_G = 0$). However, the turn-off behavior of this intrinsic diode is relatively poor, compared to optimized pin-type or Schottky barrier diodes for the same blocking voltage. Figure 9.20 shows a snappy turn-off event of the inverse diode in a 200 V MOSFET.

The MOSFET manufacturing technology usually leads to a high carrier lifetime. Therefore, a high stored charge and a high peak reverse current of the diode occurs in conventional MOSFETs. This is an impediment for many applications.

A carrier lifetime adjustment is applicable to reduce the stored charge. It has to be carried out in a separate production step. As a first approximation, the insertion of recombination centers in the n^--region does not affect the properties of the MOSFET, because the MOSFET is a unipolar device. During the on-state of the MOSFET, carrier recombination cannot take place. Hence, the resistance R_{on} should remain unaffected. However, secondary effects have to be taken into account. Recombination centers, which are inserted for the reduction of the carrier lifetime,

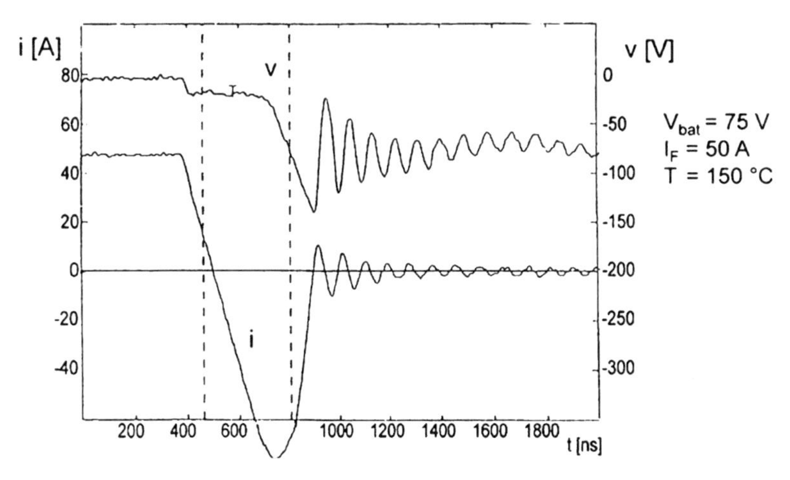

Fig. 9.20 Switching behavior of the inverse diode of a 200 V MOSFET with high-frequency LC oscillations

can retroact on the effective doping. Thus, the utilization of gold is ruled out, because gold, being an acceptor, will compensate the base doping and increase the resistance R_{on}. This effect does not occur when platinum or electron irradiation is used. With electron irradiation it has to be considered that it affects the charge in the gate oxide. Electron irradiation reduces the threshold voltage V_T. By means of adequate annealing, the threshold voltage can partially be restored.

MOSFETs with platinum diffusion or electron irradiation, which are used for the reduction of the stored charge of the inverse diode, are known as "FREDFET" (fast recovery diode field effect transistor). Here, the stored charge of the inverse diode is reduced. The reverse recovery behavior is improved slightly, so that these diodes can be used in circuits with low parasitic inductance.

Primarily, the reverse recovery behavior is problematic. Though the area of the p-region is considerably smaller – a measure, which is also used with an MPS diode to achieve soft recovery behavior – fundamental requirements on the MOSFET are contradictory to what is necessary to achieve soft recovery turn-off behavior:

- To keep R_{on} as low as possible, the base of the MOSFET has to be designed as thin as possible.
- To obtain an effective short-circuit R_S, the p^+-doping is chosen as high as possible.

Both measures result in a snappy switching behavior of the diode, and they limit the possibility for both the MOSFET and the diode to be optimized.

In many hard-switching applications the inverse diodes are unusable, and they are often called parasitic diodes. By inserting a Schottky diode in series to the MOSFET and in reverse direction to the inverse diode, they can be suspended, and an optimized soft recovery diode can be connected in parallel. However, further losses occur due to the threshold voltage of the additional junction.

An advantage for the body diode is if a part of the current in the diode mode flows via the n^+-source layer, as shown in [Zen00]. This is more pronounced in a

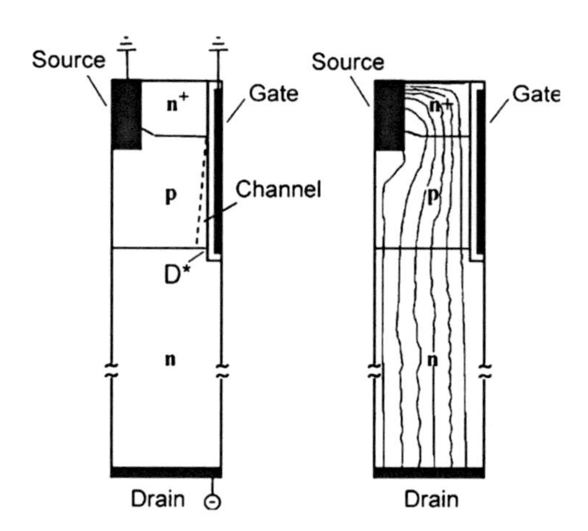

Fig. 9.21 Inverse diode in a Trench MOSFET: potential and channel formation (*left*), simulated forward conduction current distribution (*right*)

Trench MOSFET [Dol04]. The effect is shown in Fig. 9.21. As already mentioned in the discussion of the threshold voltage at the beginning of the MOSFET chapter, a potential difference between gate and semiconductor exists due to the different positions of the Fermi level in the heavily n^+-doped polysilicon gate and in the p-type semiconductor, which acts as a small positive gate voltage. Further, if the current flow is in reverse direction, a potential drop builds up and the point D^* in Fig. 9.21 becomes negatively biased against the source. If $V_G = 0$ (= source potential), then this leads to a positive potential of V_G compared to D^*. This is of same effect as a positive voltage between gate and p-body and an inversion channel can be formed, respectively, and the threshold voltage is dynamically reduced, as expressed by [Dol04]. Although the outside gate potential is set to zero, an electron-conducting channel becomes present. In devices with high channel density, as in Trench MOSFETs, it can become the dominant current component. Under these bias conditions, more than 90% of the total current is confined to the channel region [Dol04]. An example is given in Fig. 9.21.

The effect is somewhat similar to Fig. 9.8, where a positive potential at D^* to source led to the channel pinch-off, now the potential at D^* is of opposite sign.

The occurrence of an electron current is of positive effect to the reverse diode characteristic. In its on-state, significant conduction takes place at a voltage below the conventional diode turn-on voltage at ca. 0.7 V, reducing conduction losses. Additionally, since this current is primarily due to majority carriers, this effect will positively impact the diode recovery behavior, since less stored charge will need to be removed during the reverse recovery process. Only a fraction of the total current is conducted by the pure diode part in the structure, i.e. the body–epilayer junction, as shown in Fig. 9.21.

Therefore, only the fraction of the total current which is conducted in parallel by the pure diode part causes an excess carrier concentration in the base region of the pin-diode and contributes to the reverse recovery charge. The ratio between the channel-conducted electron current part and the hole current injected by the body of the inverse diode depends on the device structure (thickness of gate oxide, body doping). It also depends on the current density. The larger the current density, the larger becomes the share of injected holes and the larger will be the stored charge. With said measure, diodes of some modern Trench MOSFETs are significantly improved. Figure 9.22 shows a strong reduction of the stored charge of a new generation of trench MOSFETs (Device B) compared to an older design (Device A).

In applications with low blocking voltages and high currents (e.g., blocking voltage <100 V and output current >10 A), like switched power supplies, the technique of synchronous rectification is very common (see chapter MOS controlled diodes). Pin-diodes or Schottky barrier diodes used as freewheeling devices are replaced by MOSFETs, which are utilized for inverse conduction and allow a reduction of conducting losses due to the significantly lower voltage drop across the device.

When the channel of the synchronous rectifier is open, its intrinsic diode is bypassed. In the I–V characteristic in Fig. 9.7 this mode is given by the dotted line in the third quadrant for $V_G > V_T$. In this case, the voltage drop across the device is

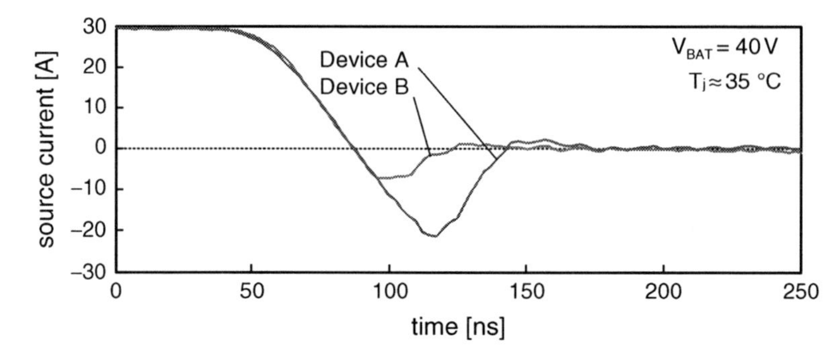

Fig. 9.22 Turn-on of MOSFETs in a half-bridge configuration with reverse recovery of the respective inverse diodes. Both devices rated for 75 V and for $I_D > 100$ A. Device A: IRF3808S planar technology, Device B: IRFS3207 trench technology. $di/dt = 800 A/\mu s$

lower compared to the case of a closed channel, particularly when low currents are applied.

Figure 9.22 shows the circuit of a buck converter, which is a typical application of low-voltage power MOSFET. To avoid shoot-through currents, which would occur if the low-side MOSFET gate still has an on-signal while the high-side MOSFET turns on, the gate control scheme for the switches must contain dead times during the current commutation, wherein both channels are closed. In the example shown in Fig. 9.23, the low-side MOSFET must be turned off before the high-side MOSFET is turned on. The same holds for the high-side switch. Thus, the intrinsic diode of the synchronous rectifier is conducting during these necessary dead times.

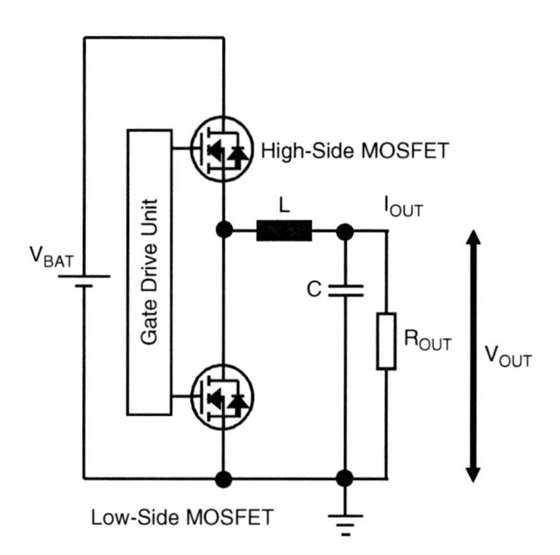

Fig. 9.23 Basic schematic of a buck converter with MOSFETs and their intrinsic diodes

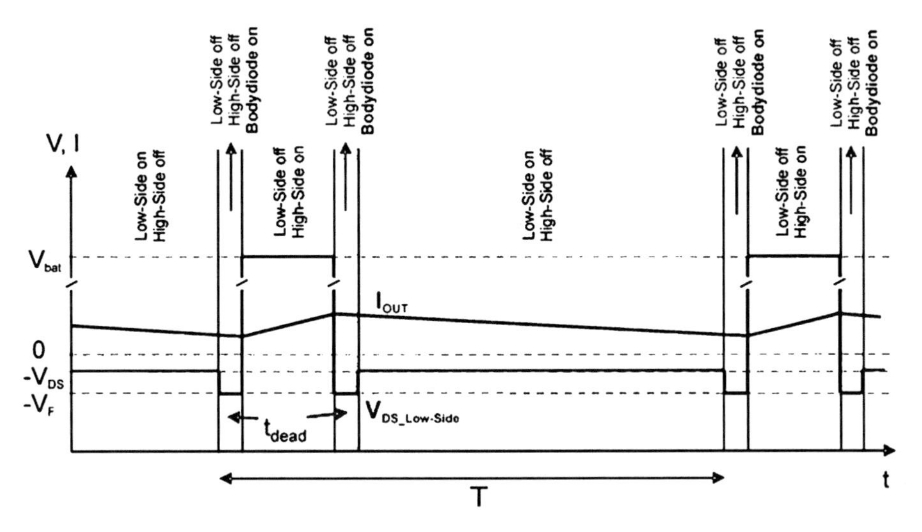

Fig. 9.24 Timing scheme for the buck converter as shown in Fig. 9.22

A control scheme for this type of operation is given in Fig. 9.24. During the diode conducting time, the gate of the low-side MOSFET has the signal "on," and the voltage drop across the device is $-V_{DS}$, a value below the junction voltage of the intrinsic diode. During the dead times, the voltage drop is $-V_F$, a value which is in its absolute value above the junction voltage of the intrinsic diode.

Besides the effort of precise gate control, the turn-off behavior of the inverse diode becomes again one of the main drawbacks in this kind of application. The dead times cannot be reduced below a certain value, thus the losses related to body diode conduction and body diode turn-off are responsible for an increasingly significant part of the total losses with increased switching frequencies.

The losses caused by the inverse diode of the MOSFET will share a larger part of the total losses in case of devices with low breakdown voltages. Inserting a Schottky barrier diode in parallel to the synchronous rectifier is a way to reduce or to prevent the conduction of its body diode during the dead times. While the effect of this measure is limited by parasitic inductances of the circuit when discrete devices are used [Pol07], it shows significant improvement of the switching behavior if the Schottky barrier diode is integrated monolithically into the MOSFET die [She90, Cal04, Bel05].

9.11 SiC Field Effect Devices

SiC unipolar devices allow a very thin drift zone and a higher doping of the base, they attain a much lower resistance R_{epi} compared to Si, see Fig. 6.9. Therefore the SIC MOSFET is a very attractive device, and research and development in SiC MOSFETs is being done since several years.

SiC MOSFETs are of n-channel type and have basically the similar structure as the vertical DMOS transistor (see Fig. 9.4). However, a problem of the SiC MOSFET is the channel conductivity, since the effective mobility of electrons in the channel is low. This is due to a high density of electron traps in the SiC–MOS interface, which results in the capture of channel electrons and an increased Coulomb scattering due to these captured electrons [Ima04]. The gate oxide is typically grown in an atmosphere containing NO or N_2O to reduce MOS interface state density. Channel mobilities in the range of 5 – 10 $cm^2V^{-1}s^{-1}$ are typical; a channel mobility of 13 cm^2/V s has also been reported [Ryu06]. It must be considered that also in Si MOSFETs the channel mobility is lower than the bulk mobility due to the influence of the surface (in the range of 500 $cm^2V^{-1}s^{-1}$ see Eq. (9.19)), but the effect is much worse in SiC. Channels can be designed very short and additional effort is done to improve the channel conductivity, e.g. by an additional thin n-doped layer. Prototypes of SiC MOSFETs reached remarkable low R_{DSon}, e.g. 5 mΩcm^2 for a 1200-V type [Miu06], but they are still far above the theoretically possible low values for SiC (see Fig. 6.9), because of the low electron mobility in the channel.

A further challenge for the SiC MOSFET is the control of the threshold voltage V_T [Aga06]. In particular, V_T is strongly temperature dependent and decreases with temperature. Finally, the requirements of long-term stability of the gate oxide at high electric fields must be accomplished.

A possible alternative could be the SiC junction field effect transistor (JFET). The structure of a JFET half cell [Mit99], which is in this configuration fabricated by Infineon, is shown in Fig. 9.25. With $V_G = 0$, there is a path from source to drain and the structure is conducting. For the blocking mode, a negative voltage must be applied at the gate, this builds a space charge between gate and source, and the channel is interrupted.

The I–V characteristic of a JFET is shown in Fig. 9.26. For $V_G = 0$ V the JFET is in the forward conducting mode. With $V_G < 0$, the channel is narrowing, and for $V_G = -20$ V this sample is in the forward blocking mode. The cut-off voltage V_{CO} of the specific sample was at $V_G = -17.5$ V. To ensure reverse blocking of the JFET, a gate voltage smaller than the cut-off voltage must be applied; however, the negative gate voltage is limited by the breakdown voltage V_{BD}(GS) of the gate–source diode, which occurs, e.g. at -25 V. Due to tolerances in the manufacturing process, there is a spread of the values of V_{CO} as well as V_{BD}(GS),

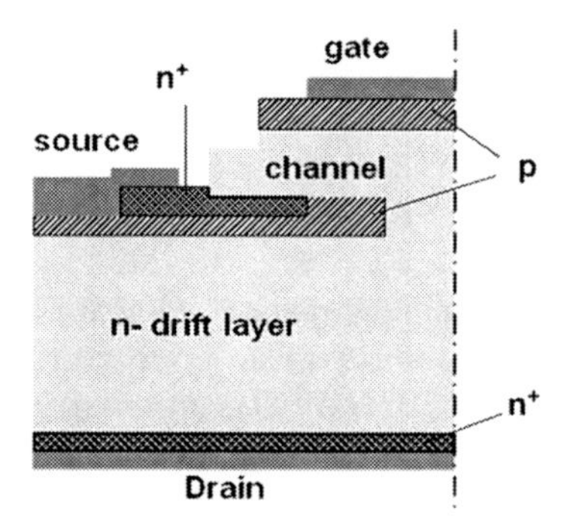

Fig. 9.25 SiC JFET half cell

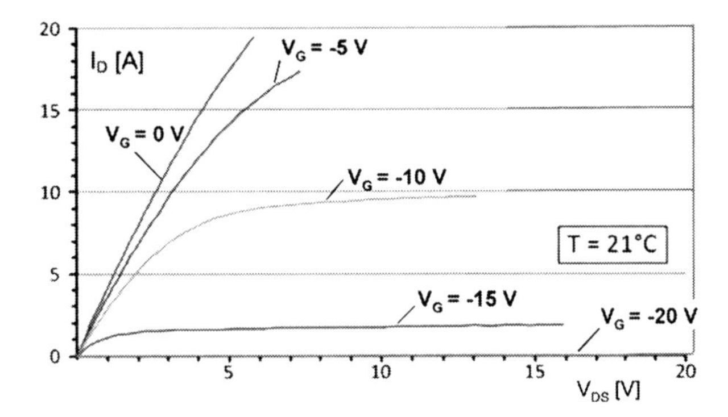

Fig. 9.26 Forward *I–V* characteristic of a JFET

this spread should be small. To ensure blocking capability, the JFET needs a negative voltage between V_{CO} and $V_{BD}(GS)$.

A "normally on" device is unwanted for voltage source converter applications. However, this can be solved by a MOSFET in series to the JFET in the cascode mode, as shown in Fig. 9.27.

If a voltage $V_G > V_T$ is applied at the MOSFET, it is in the conduction mode. The normally on JFET in series is conducting and the cascode configuration is in the conduction mode. If the MOSFET is turned off, a voltage up to its blocking capability is built up. This voltage, which has a negative polarity to the source of the JFET, is applied to the JFET gate. If it is lower than the pinch-off voltage of the JFET, the JFET is turned off. The JFET is in the blocking mode. The behavior of the cascode configuration is as that of a MOSFET and it can be operated widely as known for MOSFET.

In the conduction mode, the R_{on} of the MOSFET is in series to the R_{on} of the JFET. However, MOSFETs in the 30 V range can meanwhile be designed with extremely low on-resistances. They are available with R_{on} down to 0.1 mΩcm^2. What remains is the effort of packaging an additional device.

In some applications the normally off behavior of the JFET is of no disadvantage. These are, e.g. current source inverters. Also in the matrix converter the pure JFET may be an applicable device.

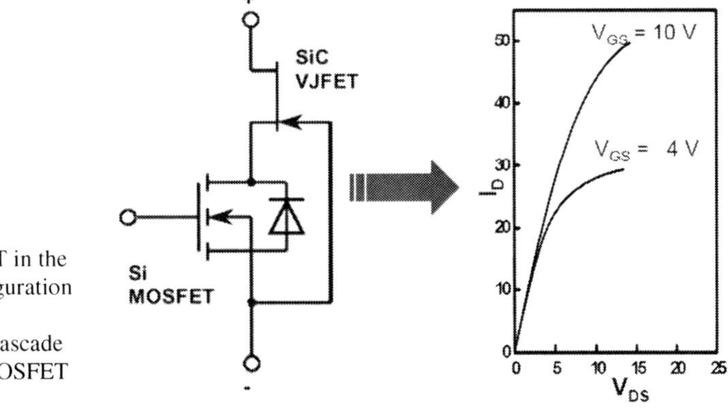

Fig. 9.27 JFET and low-voltage MOSFET in the cascode mode. Configuration (*left*), forward *I–V* characteristic of the cascade with V_{GS} at the Si-MOSFET gate (*right*)

9.12 Outlook

The MOSFET is a unipolar device. In forward direction, a threshold voltage does not occur. The MOSFET offers many advantages: Its control is easy and requires only low power. The switching slopes are adjustable via gate resistances. At turn-off no tail current exists, the MOSFET has minor switching losses, and high switching frequencies are possible. Moreover, it is short-circuit resistant and possesses a rectangular safe operating area.

Therefore, a MOSFET will always be used in application, if possible. MOSFETs can be paralleled without problems and the series connection is possible as well. Even high-power applications (50 kV, some kA), which require fast switching behavior and where costs are of minor importance, have been realized with parallel and series connection of MOSFETs.

A disadvantage of the MOSFET, however, is the inverse diode, which has insufficient properties.

When designed for higher blocking voltages, the resistance R_{on} increases strongly. The introduction of the superjunction principle, which does not show this correlation, represents an important development. Nowadays, superjunction MOSFETs are available for 600 and 900 V, even 1000 V devices are possible in principle. But the technological complexity is increasing. Consequently, applications requiring higher voltages are still dominated by bipolar devices.

With superjunction devices in the range of 600 V, it is expected that future generations will achieve a further reduced R_{on} per area.

In the low-voltage range (<100 V), the trench technology leads to a reduction of the resistance R_{on}. Further development in microelectronics will make finer structures and, thus, a higher cell density possible. This will further reduce the on-state losses. Contemporaneously, measures for the reduction of the capacitances are taken to reduce switching losses.

Field-controlled devices made of SiC have become available. Due to the possibly very thin drift zone and the feasible higher doping of the base, they attain a much lower resistance R_{epi}, see Fig. 6.9. They extend the range of unipolar field-controlled devices above the range of 1000 V; several kilovolts are possible. Also below 1000 V they will compete with silicon devices, because of the possibly low R_{on}. SiC MOSFETs and SiC JFETs are in development; which of the devices will be more successful in the application is an open question at the moment.

References

[Aga06] Agarwal A, Ryu SH: "Status of SiC Power Devices and Manufacturing Issues" CS MANTECH Conference, April 24–27, Vancouver, Canada, pp. 215–218 (2006)

[Bel05] Belverde G, Magrì A, Melito M, Musumeci S, Pagano R, Raciti A, "Efficiency Improvement of Synchronous Buck Converters by Integration of Schottky Diodes in Low-Voltage MOSFETs", Proc. of the IEEE ISIE 2005, pp. 429–434 (2005)

[Cal04] Calafut D, "Trench Power MOSFET Lowside Switch with Optimized Integrated Schottky Diode", Proc. of the International Symposium on Power Semiconductor Devices & ICs 2004, pp. 397–400 (2004)

[Che01] Chen Xing-Bi, Sin JKO.: "Optimisation of the Specific On-Resistance of the COOLMOS", IEEE Trans. Electron Devices, Vol. 48 No 2 (2001)

[Che05] Chen Y, Liang Y, Samudra G: "Theoretical Analyses of Oxide-Bypassed Superjunction Power Metal Oxide Semiconductor Field Effect Transistor Devices", Japanese Journal of Applied Physics, Vol. 44, No. 2, pp. 847–856 (2005)

[Deb98] Deboy G, März M, Stengl JP, Sack H, Tihanyi J, Weber H: "A new generation of high voltage MOSFETs breaks the limit line of silicon", Proc. IEDM, 683–685, (1998)

[Dol04] Dolny GM, Sapp S, Elbanhaway A, Wheatley CF: "The influence of body effect and threshold voltage reduction on trench MOSFET body diode characteristics" Proceedings of the International Symposium on Power Semiconductor Devices & ICs, Kitakyushu, 217–220 (2004)

[Gra89] Grant D.A., Gowar J., Power MOSFETS - Theory and Application, John Wiley and Sons New York 1989

[Hof63] Hofstein SR, Heiman FP: "The silicon insulated-gate field-effect transistor" Proceedings of the IEEE vol 51, Issue 9, pp1190–1202 (1963)

[Ima04] Imaizumi M, Tarui Y: "2 kV Breakdown Voltage SiC MOSFET Technology" Mitsubishi Electric R&D Progress Report March 2004, global.mitsubishielectric.com/pdf/advance/vol105/08_RD1.pdf (2004)

[Kon06] Kondekar PN, Oh H, Kim YB: "Study of the degradation of the breakdown voltage of a super-junction power MOSFET due to charge imbalance", Journal of Korean Physical Society, vol. 48, no. 4, pp. 624–630, 2006

[Lia01] Liang Y, Gan K, Samudra G: "Oxide-Bypassed VDMOS (OBVDMOS). An Alternative to Superjunction High Voltage MOS Power Devices", IEEE Elec. Dev. Let., 2001, 22, 407–409

[Lor99] Lorenz L, März M: "CoolMOS^{TM} - A new approach towards high efficiency power supplies", Proceedings oft the 39th PCIM, Nuremberg, S. 25–33 (1999)

[Mic03] Michel M, Leistungselektronik, 3rd edition, Springer-Verlag Berlin 2003

[Mit99] Mitlehner H, Bartsch W, Dohnke KO, Friedrichs P, Kaltschmidt R, Weinert U, Weis B, Stephani D: "Dynamic characteristics of high voltage 4H-SiC vertical JFETs" Proceedings of the 11th International Symposium on Power Semiconductor Devices & IC′s, Pages 339–342 (1999)

[Miu06] Miura N et al: "Successful Development of 1.2 kV 4H-SiC MOSFETs with the Very Low On-Resistance of 5 mΩcm^2", Proceedings of the 18th International Symposium on Power Semiconductor Devices & IC′s June 4–8, 2006 Naples, Italy (2006)

[Paw08] Pawel I, Siemieniec R, Born M: "Theoretical Evaluation of Maximum Doping Concentration, Breakdown Voltage and On-state Resistance of Field-Plate Compensated Devices", Proc. ISPS'08 (Prague 2008)

[Pol07] Polenov D, Lutz J, Pröbstle H, Brösse A, "Influence of Parasitic Inductances on Transient Current Sharing in Parallel Connected Synchronous Rectifiers and Schottky-Barrier Diodes", IET Circuits, Devices and Systems, Vol.1, No.5, pp. 387–394 (2007)

[Ryu06] Ryu SH et al: "10 kV, 5A 4H-SiC Power DMOSFET", Proceedings of the 18th International Symposium on Power Semiconductor Devices & IC′s, June 4–8, 2006 Naples, Italy (2006)

[She90] Shenai K, Baliga BJ, "Monolithically Integrated Power MOSFET and Schottky Diode with Improved Reverse Recovery Characteristics", IEEE Trans. Electron Devices, Vol.37, No.3, pp. 1167–1169 (1990)

[Sie06c] Siemieniec R, Hirler F, Schlögl A, Rösch M, Soufi-Amlashi N, Ropohl J, Hiller U: "A new fast and rugged 100 V power MOSFET", Proc. EPE-PEMC 2006, Portoroz, Slovenia (2006)

[Sod99] Sodhi R, Malik R, Asselanis D, Kinzer D: "High-density ultra-low R_{dson} 30 volt N-channel trench FETs for DC/DC converter applications", Proceedings ISPSD ′99 pp. 307–310 (1999)

[Ste92] Stengl JP, Tihanyi J, Leistungs-MOSFET-Praxis, Pflaum-Verlag München 1992

[Zen09] Zeng J, Wheatley CF, Stokes R, Kocon C, Benczkowski S, "Optimization of the body-diode of power MOSFETs for high efficiency synchronous rectification" Proceedings of the ISPSD, S. 145–148 (2000)

[Zin01] Zingg RP "New Benchmark for RESURF, SOI, and Super-Junction Power Devices", Proceedings of the ISPSD, Osaka, S. 343–346 (2001)

Chapter 10
IGBTs

10.1 Mode of Function

A lot of work was spent to combine bipolar devices with their superior current density with the possibility of voltage control as given in MOSFETs. Early works tried to combine thyristor-related structures with MOS gate control. However, a transistor-based device won the race. The insulated gate bipolar transistor (IGBT) was invented in the United States by Wheatley and Becke [Bec80]. The advantage compared to the bipolar transistor and MOSFET was described by [Bal82]. About 10 years later, IGBTs were introduced in the market by manufacturers from Japan and Europe. In a short time, IGBTs won an increasing share of applications and they replaced the formerly used bipolar power transistors, and nowadays even GTO-thyristors in the high-power range.

In a first rough approach, the IGBT is a MOSFET, in which the n^+-layer at the drain side is replaced by a p-layer. Figure 10.1 shows the structure of the IGBT.

The notations collector and emitter have been taken from the bipolar transistor, anode (for collector) and cathode (for emitter) also make sense.

If at the IGBT a positive voltage between collector C and emitter E is applied, the device is in the blocking mode. If now a voltage V_G higher than the threshold voltage V_T is applied between gate and emitter, an n-channel is created; electrons flow to the collector (Fig. 10.1). At the collector side pn-junction, a voltage in forward direction is generated and holes from the p-collector layer are injected into the lowly doped middle layer. The injected holes allow an increased charge carrier density; the increased carrier density lowers the resistance of the middle layer and the conductivity of the middle layer is modulated. The IGBT was first referred to as COMFET[1] (conductivity modulated FET) [Rus83, Rog88], also the term IGT (insulated gate transistor) [Bal83] was used. Like in a MOSFET, the turn-on and turn-off of the IGBT happen by the creation and removal of an n-channel by applying a gate voltage. Regarding threshold voltage and channel resistance, the same holds as described in Chap. 9 for the MOSFET.

[1] Meanwhile the name COMFET is usual for the composite field effect transistor, a device which is used in analogous circuits.

J. Lutz et al., *Semiconductor Power Devices*, DOI 10.1007/978-3-642-11125-9_10,

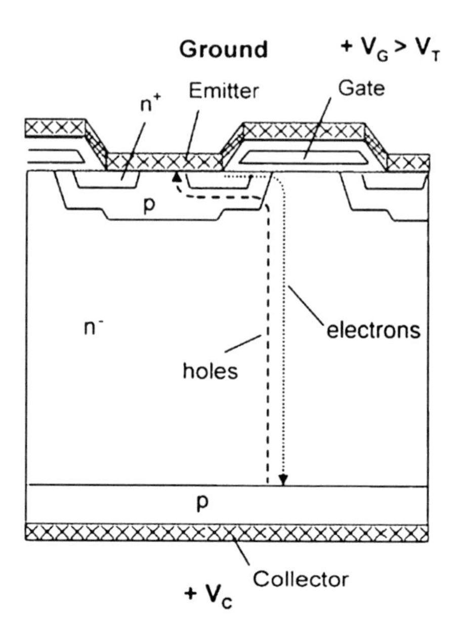

Fig. 10.1 IGBT in conduction mode at $V_G > V_T$, electron and hole current

Figures 10.1 and 10.2 show a pnpn-structure as in a thyristor; however the action of this thyristor is strictly avoided by a high-conductive emitter short R_S. Figure 10.2b shows the equivalent circuit of the four-layer structure. With the pnp- and npn-partial transistor, a parasitic thyristor structure is visible. The emitter of the npn partial transistor and its base are shortened by the resistor R_S. With this, the current gain of the npn partial transistor is eliminated at low current. But at very high current, the npn transistor can be activated and the parasitic thyristor can be triggered into the

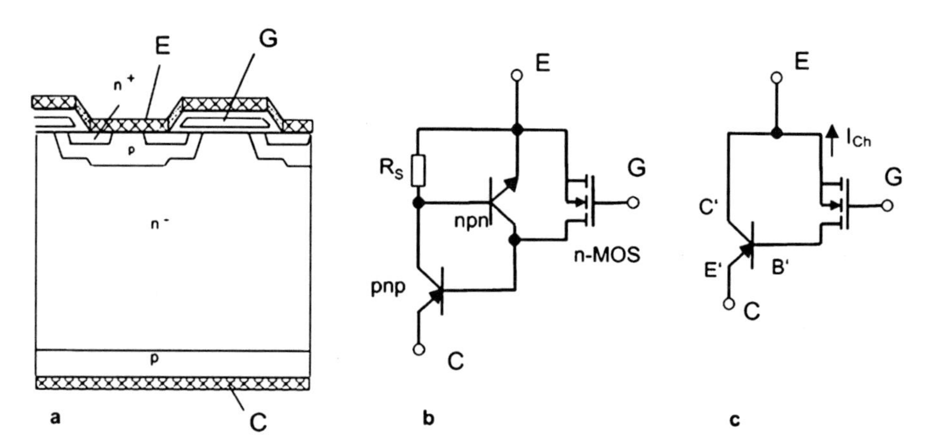

Fig. 10.2 IGBT (**a**) simplified structure, (**b**) equivalent circuit with parasitic npn-transistor and resistor R_S, and (**c**) simplified equivalent circuit

on-state mode with the internal feedback loop with the partial pnp-transistor. This effect is denoted as latch-up: the device can no longer be controlled by the MOS gate. Latching of the parasitic thyristor is a destructive effect for the IGBT.

For sufficiently low R_S, the npn partial transistor can be neglected and the simplified equivalent circuit, as shown in Fig. 10.2c, is obtained. This is the most important equivalent circuit for understanding the IGBT. The terminals of the pnp-transistor are denoted with C′, E′, and B′. The collector C of the IGBT is the emitter E′ of the pnp-transistor. Concerning the physics of the IGBT, it is an emitter.

The channel is created for a gate voltage V_G higher than the threshold voltage; in the base terminal of the pnp-transistor the channel current I_{CH} is flowing. For the current $I_{C'}$ at C' then holds $I_{C'} = \beta_{pnp} \cdot I_{CH}$, or, with the relation $\alpha = \beta/(\beta + 1)$ known the from the sections on bipolar transistors

$$I_{C'} = \frac{\alpha_{pnp}}{1 - \alpha_{pnp}} \cdot I_{CH} \tag{10.1}$$

For the collector current of the IGBT holds

$$I_C = I_{C'} + I_{CH} = \frac{\alpha_{pnp}}{1 - \alpha_{pnp}} \cdot I_{CH} + I_{CH} = \frac{1}{1 - \alpha_{pnp}} \cdot I_{CH} \tag{10.2}$$

Therefore the collector current of the IGBT is always higher than its channel current. The saturation current of the IGBT will also be much higher than that of the MOSFET. With the parameter of the channel conductivity, κ, defined for the MOSFET in Eq. (9.4), results for the IGBT

$$I_{Csat} = \frac{1}{1 - \alpha_{pnp}} \cdot \frac{\kappa}{2} (V_G - V_T)^2 \tag{10.3}$$

However, α_{pnp} must be adjusted not too high, and to meet all the different requirements, it should be adjusted very exactly. The current necessary for latch-up must be shifted to such a high value which will not occur in application. To achieve this, different measures are used in the design of the IGBT, which will be explained in the following.

10.2 The *I*–*V* Characteristic of the IGBT

The I–V characteristic of an IGBT in forward direction is shown in Fig. 10.3. The characteristic has some similarities to that of a MOSFET.

For a gate voltage V_G higher than the threshold voltage V_T, the channel is open. The IGBT differs from the MOSFET by the junction voltage of the additional pn-junction at the collector side. The IGBT is operated in the saturation region, as is usual for power devices and as it was already the case with the MOSFET and the bipolar transistor. The operation point is on the branch of the characteristic for

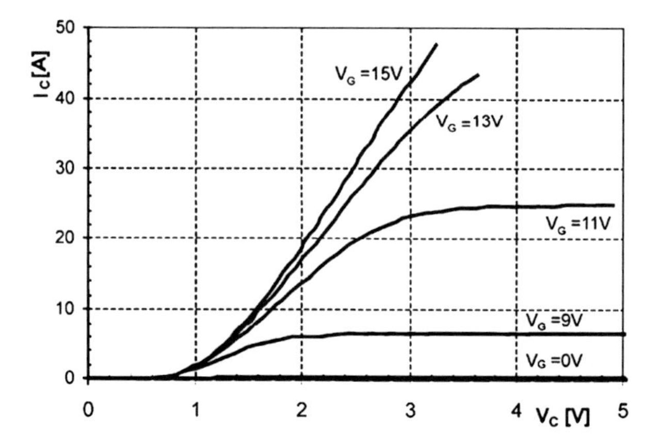

Fig. 10.3 *I–V* characteristic of a 20 A/600 V IGBT

$V_G = 15\,\text{V}$. At this branch for operation at a given current I_C, the generated voltage drop V_C is read off.

Figure 10.4 compares the characteristic of an IGBT for $V_G = 15\,\text{V}$ with that of a bipolar transistor, also at a high drive level of $I_B = 2.5\,\text{A}$. Both modules are specified for 600 V and have a comparable area. The threshold voltage caused by the reverse pn-junction is recognizable in the characteristics of the IGBT. At low current densities, the bipolar transistor has the inferior voltage drop because no threshold voltage occurs with it. However, at higher current densities, above 14 A, the forward voltage of the IGBT is much lower than that of the bipolar transistor. The shown IGBT is rated to a nominal current of 20 A; this current level is not reached with the used bipolar transistor even at a high base current.

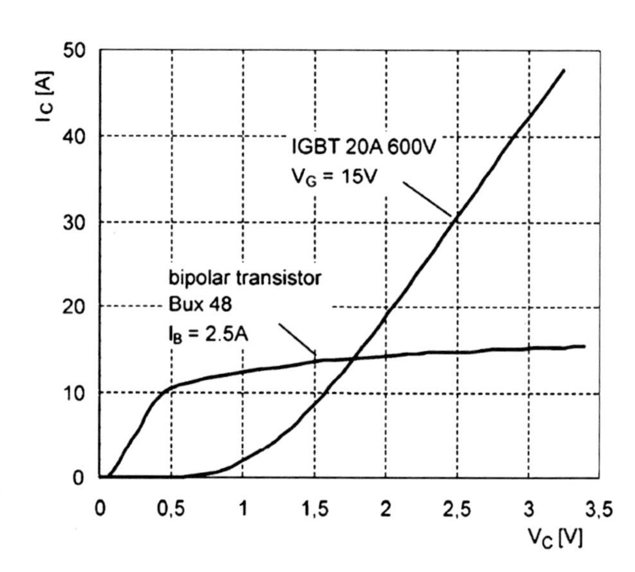

Fig. 10.4 Forward voltage–current characteristic of an IGBT in comparison to a bipolar transistor

A comparison of power devices of a higher voltage range would show the difference even more drastically. The IGBT especially can be designed also for higher voltages; it is not restricted like the MOSFET or bipolar transistor by physical mechanisms. This is described in the following in more detail. Meanwhile, IGBTs have been produced for voltages up to 8 kV [Rah02]; they have been commercially available in 2010 for voltages up to 6.5 kV.

10.3 The Switching Behavior of the IGBT

The determination of the switching behavior of the IGBT is done in a circuit according to Fig. 10.5 with inductive load. The time constant of the load $\tau = L/R$ is chosen so high that courses of voltage and current can be assumed as constant before the switching instant.

The turn-on process of a field-controlled device was discussed already with the MOSFET. The relation between the rise time of the current, the fall time of the voltage, the internal capacities, and the chosen gate resistors are the same as were discussed in the context with Fig. 9.15. The IGBT is used with a freewheeling pin-diode in almost all applications; at the turn-on process, it additionally has to take over the reverse current peak and the stored charge of the freewheeling diode. The processes are shown in Fig. 5.20 and 5.21 and are described in context with Eqs. (5.82), (5.83), and (5.84). For the turn-on energy per pulse in the IGBT, it can be given under the same simplifications:

$$E_{\mathrm{on}} = \frac{1}{2} \cdot V_{\mathrm{bat}} \cdot (I_{\mathrm{C}} + I_{\mathrm{RRM}}) \cdot t_{\mathrm{ri}} + \frac{1}{2} \cdot V_{\mathrm{bat}} \left(I_{\mathrm{C}} + \frac{2}{3} I_{\mathrm{RRM}} \right) \cdot t_{\mathrm{fv}} \tag{10.4}$$

A more exact determination is done with the oscilloscope, see Eq. (9.30).

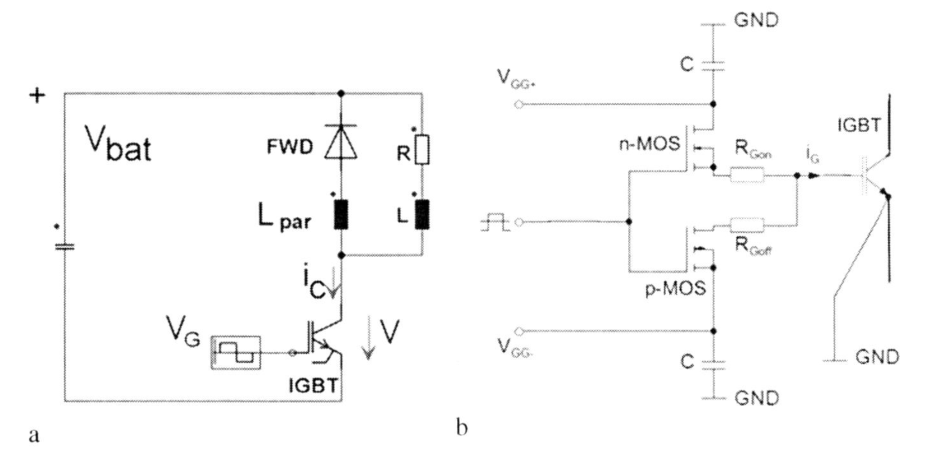

Fig. 10.5 Measurement of the IGBT switching behavior. (**a**) Power circuit (**b**) simplified output stage of a gate drive circuit containing gate resistors R_{Gon} and R_{Goff} adapted from [Nic00]

At turn-off of the IGBT, the positive gate voltages are put on zero or a negative value and in the first time interval, processes appear like those described with the MOSFET in the context of Fig. 9.17. As long as the stored charge in the IGBT is not too high, similar relations between the rise time of the voltage, the internal capacities, and the chosen gate resistance are valid. If the gate capacity is discharged abruptly, the channel current is interrupted. Usually a gate resistor is applied, and V_G is reduced to the value of the Miller plateau at $V_G = V_T + g_{fs} I_D$. While the voltage increases, the current in the IGBT in a circuit with inductive load flows on continuously until the voltage is higher than the applied battery voltage V_{bat}. During the voltage rise time, the channel current is reduced and the hole current flowing through the p-well is increased in the same amount, carriers from the n-base are extracted by the hole current. The hole current is increased compared to the steady-state conduction mode. The problem of possible latch-up is most serious at the turn-off event. The requirement that latch-up must be avoided limits the maximal possible current which an IGBT can turn off.

The hole current leads to the removal of the charge carriers from the n-base, a space charge is built-up, and the device takes over the applied voltage. After the voltage has increased to V_{bat}, the current falls steeply. With IGBTs, the slope of the decreasing current di_c/dt can be adjusted by the gate resistor only with limited effect. The slope di_c/dt causes an inductive voltage peak at the parasitic inductance, and the forward recovery voltage peak V_{FRM} of the freewheeling diode adds to this. The voltage peak amounts to

$$\Delta V = L_{par} \cdot \frac{di_c}{dt} + V_{FRM} \tag{10.5}$$

If devices in the voltage range > 1700 V are used, the part of V_{FRM} can be significant.

As the main difference to the MOSFET and to the bipolar transistor, the IGBT features a tail current at turn-off. A measurement of the turn-off of the IGBT including the tail current is shown in Fig. 10.6. The current falls down to the value I_{tail}, and then it goes down slowly during the time interval t_{tail}. The measurement of the end point of the tail current is difficult since it decreases very slowly. The time of the tail current t_{tail} is determined by the recombination of the remaining charge carriers in the device. With the typical high charge carrier lifetime in the NPT IGBT, t_{tail} can amount to several microseconds, while t_{rv} is in the order of some 100 ns and t_{fi} is in the range of 100 ns.

During the time of the tail current, the voltage is high and the losses created in this interval cannot be neglected. In practice, the determination of the turn-off energy per pulse is mostly done with an oscilloscope; the product of current waveform and voltage waveform is executed and integrated over the turn-off time interval. A simplified estimation can be made with

$$E_{off} = \frac{1}{2} \cdot V_{bat} \cdot I_C \cdot t_{rv} + \frac{1}{2} \cdot (V_{bat} + \Delta V)\, I_C \cdot t_{fi} + \frac{1}{2} I_{tail} \cdot V_{bat} \cdot t_{tail} \tag{10.6}$$

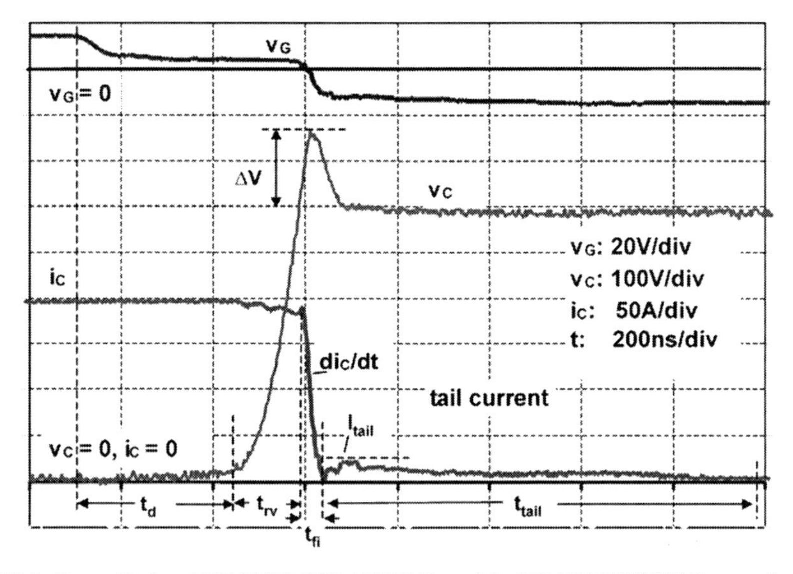

Fig. 10.6 Turn-off of an NPT-IGBT (200 A/1200 V module BSM200GB120DN2, manufacturer Infineon). $T = 125°\text{C}$, $R_{\text{Goff}} = 3.3\,\Omega$

IGBTs are usually used in bridge configurations. The IGBT is turned off mostly with a negative gate voltage, as in Fig. 10.6. In the blocking mode, a voltage of −15 V is applied, sometimes also a smaller voltage of −8 V. This does not have, besides the waveform of the gate voltage, much effect on the waveforms of current and voltage at turn-off.

10.4 The Basic Types: PT-IGBT and NPT-IGBT

In the very first IGBT structures the n^+-substrate of the MOSFET was replaced with a p^+-substrate. These structures were very sensitive to latch-up of the parasitic thyristor. The behavior could be improved by adding a medium-doped n-layer, a so-called n-buffer, between p^+-substrate and lowly doped n^--layer. This n-buffer must have a sufficient doping N_{buf} [Nak85]. The electric field can penetrate into the n-buffer, a trapezoidal shape of the electric field is given. From this characteristic, the notation punch-through IGBT or *PT-IGBT* was derived (In the proper meaning of the word, this notation is not correct, see Sect. 5.1). The structure is shown in Fig. 10.7.

As already mentioned, α_{pnp} must be adjusted and must not be too high. The doping of the buffer N_{buf} is of influence on it. In the section on the bipolar transistor, α was composed of two terms

$$\alpha = \gamma \cdot \alpha_{\text{T}} \tag{10.7}$$

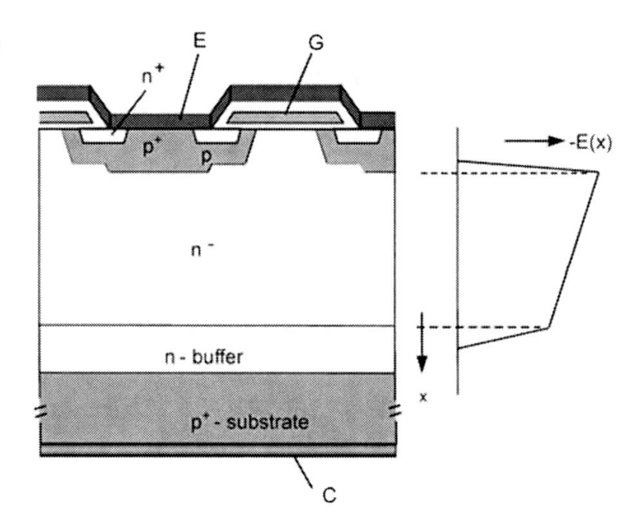

Fig. 10.7 PT-IGBT, structure and field shape

For the estimation of the emitter efficiency γ, Eq. (7.23) can be used; for the given case of a p-emitter it can be written as [Mil89]

$$\gamma = \frac{1}{1 + \frac{\mu_n}{\mu_p} \cdot \frac{N_{buf}}{N_{sub}} \cdot \frac{L_p}{L_n}} \tag{10.8}$$

Since L_p, the diffusion length of holes in the n-buffer, and L_n, the diffusion length of holes in the substrate, are in the same order of magnitude, and since μ_n/μ_p amounts to 2 – 3, it is difficult to make the denominator in Eq. (10.8) significantly larger than 1, as long as the doping of the buffer N_{buf} is not in the same range of magnitude as the doping of the p$^+$-substrate N_{sub}. Any control of γ will be difficult. For the PT-IGBT, we can therefore assume $\gamma \approx 1$.

Thus the adjustment of α_{pnp} in the PT-IGBT is done by the transport factor α_T. According to Eq. (7.29), the width of the base layer w_B and the diffusion length L_p in the substrate are important factors for this[2]:

$$\alpha_T = 1 - \frac{w_B^2}{2 \cdot L_p^2} \tag{10.9}$$

and L_p was given in Eq. (2.99):

$$L_p = \sqrt{D_p \cdot \tau_p} \tag{10.10}$$

With a low carrier lifetime, L_p, α_T, and finally α_{pnp} are reduced. For the reduction of the carrier lifetime, the technologies for the creation of recombination centers

[2]Equation (10.9) is usually derived for low injection. The considerations, however, will also be valid for high injection, qualitatively.

as described in Chap. 4 are used. Platinum diffusion, electron irradiation, irradiation with He^{2+}-ions or protons, or a combination of two of these processes are used; the specific process is different for different suppliers and their special device generations. The devices feature low turn-off losses.

The fabrication of the basic material for PT-IGBTs is done with an epitaxy process. n-buffer and n^--layer are deposited on a p^+-substrate. This technology is well applicable in the voltage range up to 600 V. For 1200 V devices, the epitaxial technology demands an increased effort because of the necessary thick epitaxial layer. PT-IGBTs dominated the applications at blocking voltages up to 600 V over several years.

As an alternative concept, the so-called *NPT-IGBT* (non-punch-through IGBT), was introduced. It is based on a suggestion of Jenö Tihanyi [Tih88] and was first realized by Siemens (today Infineon) [Mil89]. The structure is shown in Fig. 10.8. The space charge is triangular. The device for the same blocking voltage, which is given by the area under the line $E(x)$, must therefore be designed with a much thicker base width w_B. For the very first types of NPT-IGBTs, additionally a relatively high distance between the end of the space charge at $x = w$ and the p-collector layer at $x = w_B$ was chosen. Thus the effective base of the pnp-transistor at high voltage is $w_B - w$. A widened effective base, as expressed with Eq. (10.9), can somewhat reduce α_T and therewith α_{pnp}.

The main control of α_{pnp} is done via the emitter efficiency γ. The p-collector layer is lowly doped and its penetration depth is very shallow; with this the low emitter efficiency is adjusted. In Sect. 3.4, Eq. (3.96) was derived for the emitter efficiency; for p-emitter with $p_L \approx n_L$ it is written as

$$\gamma = 1 - q \cdot h_p \frac{p_L^2}{j} \tag{10.11}$$

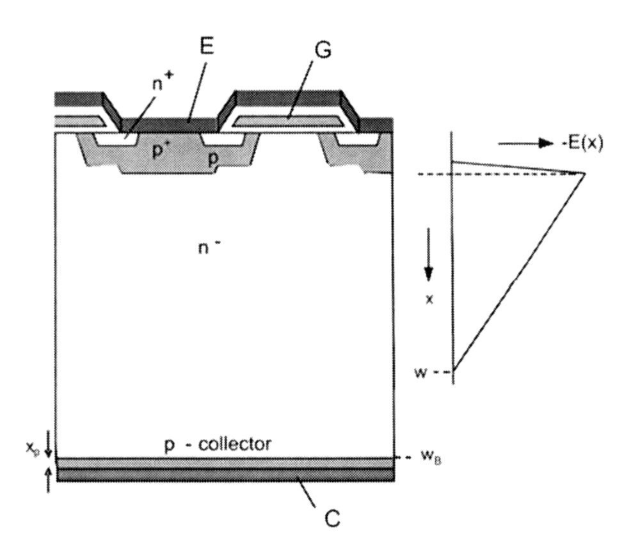

Fig. 10.8 NPT-IGBT. Structure and field shape

To achieve low γ, the emitter parameter h_p must be large. For the p-emitter, it can be expressed by Eq. (3.94)

$$h_p = \frac{D_n}{p^+ \cdot L_n} = \frac{D_n}{p^+ \cdot x_p} \tag{10.12}$$

For small x_p, the diffusion length L_n can be replaced by the very shallow penetration depth x_p of the p-emitter. At low p^+ and $x_p < 1$ for the given structure, the emitter efficiency γ will be low. High h_p means a large contribution of the emitter recombination at the total recombination. Special measures to reduce the carrier lifetime are no longer necessary.

The NPT-IGBT is therefore very robust against latch-up and it features high short-circuit robustness [Las92]. A further advantage results from this type of control of the plasma modulation. The temperature dependency on the forward voltage is very suitable for parallel connection. The voltage drop V_C at a constant collector current I_C and a constant gate voltage V_G is increasing with temperature at the typical operation current range of I_C.

The voltage drop across the lowly doped middle region was estimated for a pin-diode by means of Eq. (5.47) as

$$V_{drift} = \frac{w_B^2}{(\mu_n + \mu_p) \cdot \tau_{eff}} \tag{10.13}$$

with the effective carrier lifetime

$$\frac{1}{\tau_{eff}} = \frac{1}{\tau_{HL}} + \frac{h_p \cdot p_L^2}{w_B \cdot \bar{p}} + \frac{h_n \cdot p_R^2}{w_B \cdot \bar{p}} \tag{10.14}$$

In Eq. (10.14) the two last terms on the right-hand side stand for the contribution of the emitter regions. Compared to pin-diodes, a different profile of the plasma is given for the basic IGBT types (see Fig. 10.8). However, Eq. (10.13) may be used for estimation. For a PT-IGBT structure, the lifetime τ_{HL} dominates the effective lifetime τ_{eff}. τ_{eff} increases with increasing temperature. Depending on the used recombination centers, τ_{HL} increases by a factor of 2 to 4 for a temperature increase from 25 to 125°C. The influence of τ_{eff} dominates in Eq. (10.13) and V_{drift} and therewith V_C decrease with increasing temperature.

For an NPT-IGBT the carrier lifetime τ_{HL} in the lowly doped middle region is adjusted as high and both terms of emitter recombination dominate in the influence on τ_{eff}. With increasing temperature, τ_{HL} also increases, but this is of low effect to τ_{eff}, since the emitter terms dominate and their temperature dependency is weak. The temperature dependency on voltage across the drift region according to Eq. (10.13) is therefore dominated by the temperature dependency on the mobilities. The mobilities decrease strongly with increasing temperature, it holds $(\mu_n + \mu_p)_{(125°C)} \approx 0.5(\mu_n + \mu_p)_{(125°C)}$. V_{drift} and therewith V_C increase with increasing temperature.

This V_C temperature dependency, also assigned as "positive temperature coefficient of V_C", on the one hand leads to increased conduction losses at high operation temperature. But on the other hand, this is of advantage, if devices are connected in parallel: if one of the devices has a lower V_C because of production-induced scattering of V_C and therefore takes more current, its temperature will increase. Then its V_C increases and its current is reduced. With this a negative feedback is given, the system of parallel-connected devices is stabilized.

The manufacturing process of the NPT-IGBT can be controlled more exactly; the adjustment of the emitter efficiency at the collector side can be executed very accurately with ion implantation technology. The NPT-IGBT has become the dominating device in application, because of its robustness and because of the suitable behavior at parallel connection. Meanwhile, also NPT-IGBTs for 600 V have been developed and established in the market.

At turn-off, the NPT-IGBT has a long tail current, see Fig. 10.5. The PT-IGBT has a shorter, but higher tail current. This will be more understandable after the internal plasma distribution has been considered.

10.5 Plasma Distribution in the IGBT

Using the example of the NPT-IGBT, the internal distribution of the charge carriers, the plasma distribution, shall be investigated. Figure 10.9 shows a simulation of this shape of the plasma in the on-state mode for a 1200 V IGBT at different forward-current densities. The lowly doped n-base of the IGBT is flooded with free carriers. Because of neutrality, $n \approx p$ holds for the bipolar device. The shape of the hole distribution is therefore almost identical to the electron distribution shown in Fig. 10.9. The cell structures are on the left-hand side in Fig. 10.9; the vertical coordinate x is the same as in Fig. 10.8. The right-hand side p-collector layer has a penetration depth of less than 1 μm for the NPT-IGBT, it cannot be recognized.

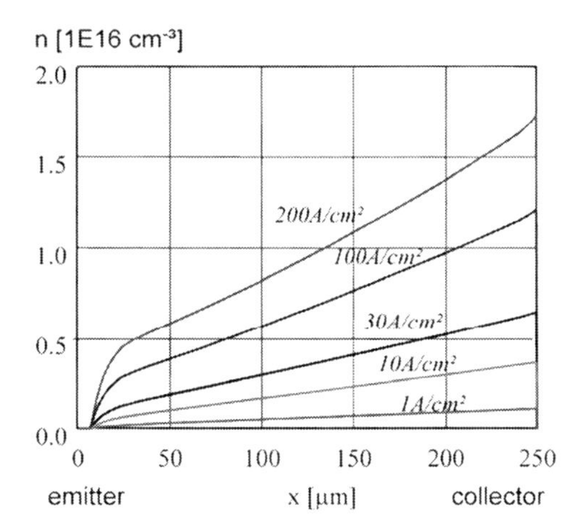

Fig. 10.9 Electron density at on-state, 1200 V NPT-IGBT. Figure from [Net99] © 2008 isle Steuerungstechnik und Leistungselektronik GmbH

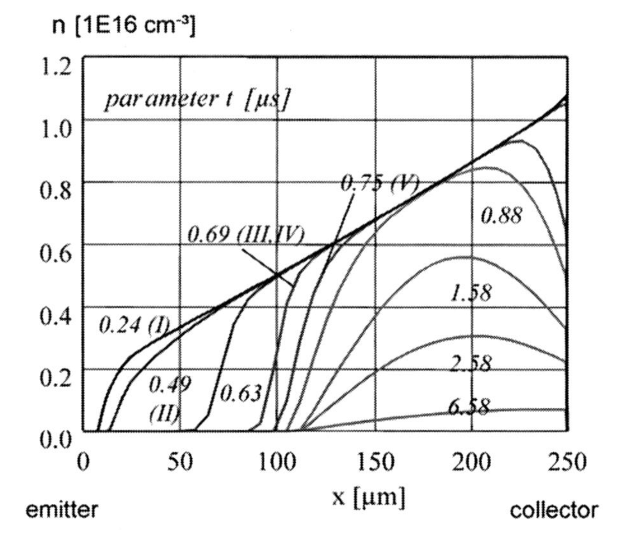

Fig. 10.10 Electron density in a 1200 V NPT-IGBT at turn-off of 80 A/cm^2. At the instant (III, IV), the device has taken the applied battery voltage and the current decreases. Figure from [Net99] © 2008 isle Steuerungstechnik und Leistungselektronik GmbH

Compared to the plasma distribution of a diode (see Fig. 5.6), the plasma distribution is strongly reduced at the side of the cell structures for the conventional IGBT. It corresponds to the plasma shape in a pnp-transistor; the collector of the IGBT in this case is the emitter of the pnp-transistor, compare Fig. 10.2. In a transistor, typically the plasma density decays from the emitter side to the collector side, see Fig. 7.6 for comparison.

Considering the rated voltage of 1200 V, the device is designed very thick with the base width w_B of 250 μm. According to Eq. (5.1) and the respective Fig. 5.5, a base width of 110 μm would be sufficient for the condition of a triangular field shape. However, a very wide n-base was typical for the first generations of NPT-IGBTs.

At turn-off with inductive load, the device must first take over the voltage, while the load current is still flowing. The development of the internal plasma at the turn-off process is shown in Fig. 10.10.

Up to the time $t = 0.69$ μs the voltage increases up to the value of the applied battery voltage. During the voltage increase, the charge in the left part of the base is extracted quickly. After the battery voltage is reached, the current decays steeply down to the value of the tail current, similar as in Fig. 10.6. The extracted part of the base extends to approximately $x = 90$ μm. After $t = 0.75$ μs, the remaining charge on the right-hand side of the base is removed. In Fig. 10.10, this process is no more driven by the electric field. The applied voltage amounts to 600 V, the space charge has extended up to 100 μm and has taken up the voltage. There is only a very low increase in the extension of the space charge in the further process. For the removal of the carriers in the part close to the collector, recombination is the determining mechanism. Because of the high carrier lifetime, there is still a considerable charge in the device even up to $t = 6.58$ μs during all this time t_{tail}, the tail current flows.

The voltage across the device has the value of the battery voltage during the time interval t_{tail}. Therefore, a main part of the turn-off losses is created by the tail current.

The PT-IGBT shows a shorter, faster decaying tail current because of its lower width of the base layer. However, PT-IGBTs have a more hanging-down plasma distribution in the conduction mode, caused by the reduced carrier lifetime. Compared to Fig. 10.9, the plasma density is lower at the emitter side, but higher at the collector side. As a result, the tail current in PT-IGBTs is shorter, but higher.

At the turn-off process described before, in which first the voltage increases before the current decays – switching with inductive load and also assigned as "hard switching" – the turn-off losses in the PT and NPT-IGBT are similar. However, the long tail current of the NPT-IGBT is a disadvantage in a turn-off process of the type "soft switching", at which at a circuit-induced zero crossing of the voltage or close-to-zero crossing the current is turned off, and the voltage increases slowly. Only a low share of the carriers is then removed in the first phase of the turn-off process. During the tail current time, the voltage increases and can lead to an additional increase in the current at the time t_{tail}. With this, additional losses are created.

10.6 Modern IGBTs with Increased Charge Carrier Density

The first IGBT generations had a plasma distribution as shown in Fig. 10.9. This distribution is expected for a bipolar pnp-transistor, compare Fig. 7.6. Therefore, it was first supposed that the IGBT will have similar limits as a bipolar transistor: high voltage drop V_C in the on-state and purely suitable for a blocking voltage above 1700 V. However, the IGBT could be improved.

Figure 10.10 shows that the plasma stored at the emitter side of the IGBT is quickly removed by the electric field while the voltage grows at the turn-off process. The plasma distribution, which is high at the collector side, low at the emitter side (see Fig. 10.9), leads to the effect that the main part of the charge carriers is removed in the tail phase. At the emitter side, the density of charge carriers could be significantly enlarged, without much increase in the turn-off losses. A higher density of carrier plasma will lead to a reduction of the voltage drop V_{drift} in the base layer and consequently to a lower on-state voltage drop V_C. To achieve this, it was supposed during a long time that a new device will be necessary. Research and development activities on MOS-controlled thyristors (MCTs) and similar devices were started. However, it was discovered that the IGBT is capable to achieve the desired internal profile of the carrier plasma and that the new devices might not be necessary.

10.6.1 Plasma Enhancement by High n-Emitter Efficiency

The effect of a possible increased plasma density at the emitter side was shown by Kitagawa et al. in 1993. They designated such a device as injection enhanced insulated gate bipolar transistor (IEGT) [Kit93]. Kitagawa et al. discovered the effect

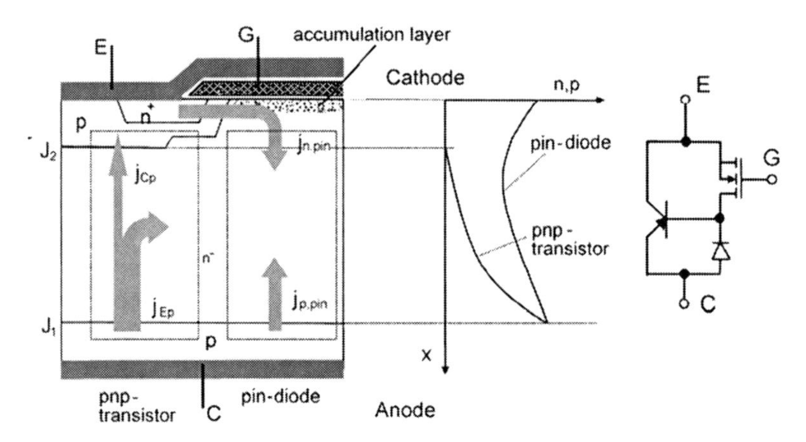

Fig. 10.11 Partition of an IGBT in a pnp-transistor area and a pin-diode area, carrier plasma distribution in both areas, equivalent circuit containing the pin-diode with MOS-switched n-emitter

at a Trench-IGBT, designed for 4.5 kV. The device had an increased charge carrier density at the emitter side and a surprisingly low voltage drop at forward conduction. The principal effect can also be explained with a planar IGBT, however [Lin06].

Figure 10.11 shows a partial section of an IGBT structure. In a first simplified investigation, the IGBT can be divided into two areas: an area of a bipolar pnp-transistor and an area of a pin diode.

In the *transistor area*, the IGBT behaves as a pnp-transistor in the saturation mode. The collector side of the IGBT corresponds to the emitter side of the pnp-transistor, see Fig. 10.2. The density of free carriers is high at the junction J_1 and decreases toward the junction J_2; at the junction J_2 it approximates to zero. For comparison, see Chap. 7 on the bipolar transistor (Fig. 7.6) and its description. From this, a shape of the carrier plasma as drawn in Fig. 10.9 results. In this figure, the junction J_2 is located at $x \approx 7\,\mu\text{m}$. This shape of the plasma holds if it is determined by the bipolar transistor.

In the diode area beneath the gate, similar conditions are given in the IGBT as in a pin diode. The electrons are supplied by the MOS channel. The MOS channel behaves as an ideal emitter; the current at this point is pure electron current. At the semiconductor surface between the p-wells, an accumulation layer of electrons will build up, created by the positive voltage at the gate above this area. Holes coming from the junction J_1 will not find a path here. The plasma distribution in this area will approximate the plasma distribution of a pin diode, as it was described in Chap. 5 with Fig. 5.6 and its discussion.

The voltage drop in the diode area, which is – in first approximation – the voltage drop V_C of the IGBT in forward conduction, is given by

$$V_C = V_{Ch} + V_{drift} + V_{J1} \tag{10.15}$$

whereby V_{Ch} is the voltage drop across the channel, V_{drift} is the voltage drop in the base layer, and V_{J1} is the built-in voltage of the pn-junction J_1. To make V_{drift} low, the relation of diode area to pnp-transistor area should be as high as possible. This can be achieved if the cells have a high distance. V_{drift} will decrease with increased cell distance (cell pitch). But then the cell density of the IGBT is decreased, and therefore also the voltage drop across the channel V_{Ch} will increase, as explained in Chap. 9 on the MOSFET. For the planar IGBT in Fig. 10.11, one finds a cell distance at which the result for V_{C} has a minimum.

A more detailed investigation is given by [Omu97]. The whole top-side cell structure is considered to be an n-emitter. In the area between the p-wells, beneath the gate oxide, an accumulation layer of free electrons is built up, caused by the positive voltage at the gate. This accumulation layer is summarized with the p-wells to an effective n-emitter. An n-emitter of high efficiency shall be created. For this, the emitter must inject a high electron current. For the efficiency of an n-emitter, results are in analogy to Eq. (3.95)

$$\gamma = \frac{j_{\mathrm{n}}}{j} \tag{10.16}$$

Here, j_{n} is the electron current delivered by the channel. Equation (10.16) can be written in a new form if $j = j_{\mathrm{n}} + j_{\mathrm{p}}$ is used

$$\gamma = \frac{j - j_{\mathrm{p}}}{j} = 1 - \frac{j_{\mathrm{p}}}{j} \tag{10.17}$$

To achieve a high γ, it is necessary

(a) to increase the distance between the cells. In [Omu97] it is shown that the share of j_{n} at the total current increases with increasing cell distance.
(b) to take care that j_{p} is only a small share of the total current density j. The hole current j_{p} flows across the p-well, see Fig. 10.11. If the area of the p-wells is reduced, also j_{p} is reduced. In the planar structure in Fig. 10.11, this happens by reduction of the p-well area.

So it is possible to increase the density of the plasma and therewith to decrease the voltage drop V_{drift} if the flow of the hole current j_{p} is reduced. The plasma distribution close to an emitter can be strongly influenced by its efficiency. In the example of the MOS-controlled diode (MCD), we have seen in Fig. 5.39 that the voltage drop in the drift layer can be controlled by the emitter injection. There, in the same way, we had an effective emitter efficiency of the p-well *and* the channel. In the MCD, the combination forms a p-emitter, and the current via the n-channel is the minority current. By injection of minority carriers from the channel, the emitter efficiency was reduced with the aim to reduce the plasma concentration at the side of the p-emitter. In the IGBT, the p-well, the channel, and the electron accumulation layer form an n-emitter. The holes flowing from the p-well are the minority carriers. Reducing the minority carrier current j_{p} will increase the efficiency of the n-emitter; with this, the plasma density close to the emitter is increased. More charge carriers

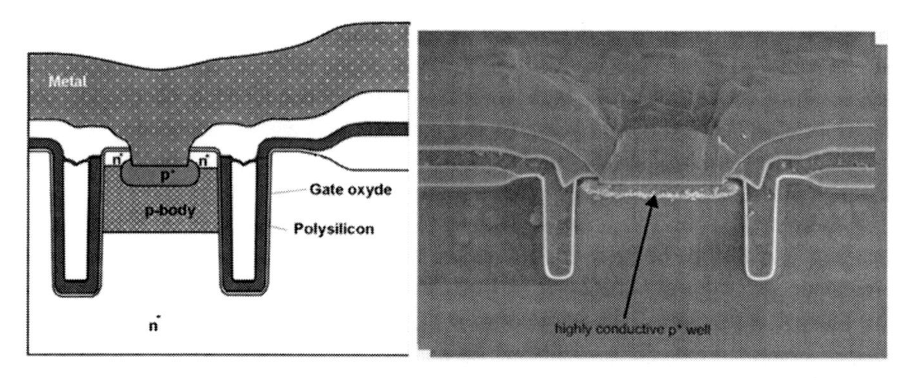

Fig. 10.12 IGBT trench cell. Structure (*left*), picture of a cross section of a cell made with a raster electron microscope (*right*). Figures from T. Laska, Infineon Technologies.

are available for current transport in the wide base layer, and the voltage drop V_C is decreasing.

This principle can be realized with planar structures and with trench structures. The trench structure offers special advantages for this. An example is the Infineon trench IGBT [Las00]. The trench cell is shown in Fig. 10.12. Emitter layers of the n^+-type and channels are arranged only in the middle of the cell between the two trenches. Outside the cell, a layer without contact to the emitter is arranged. Compare the IGBT trench cell with the MOSFET trench cell in Fig. 9.6. The mode of action of the trenches is different for the IGBT and for the MOSFET. At the MOSFET, a high part of the surface had to be equipped with channels, since the channel resistance R_{Ch} determines a main part of the total voltage drop; it is made low by arranging many channels in parallel. In the IGBT, the channel resistance is of minor effect to the voltage drop in the on-state. This voltage drop in the IGBT is determined by the plasma density in the middle layer, since the IGBT is a bipolar device. This plasma density must be high to reduce the voltage drop. Therefore, one has to reduce j_p to increase γ.

A comparison of the plasma density in the trench IGBT with the conventional NPT-IGBT according to Fig. 10.8 is given in Fig. 10.13 [Las00b]. The emitter is on the left-hand side of Fig. 10.13, the collector is on the right-hand side. The line for

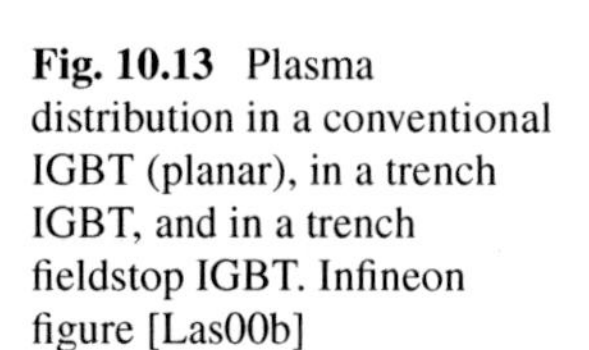
Fig. 10.13 Plasma distribution in a conventional IGBT (planar), in a trench IGBT, and in a trench fieldstop IGBT. Infineon figure [Las00b]

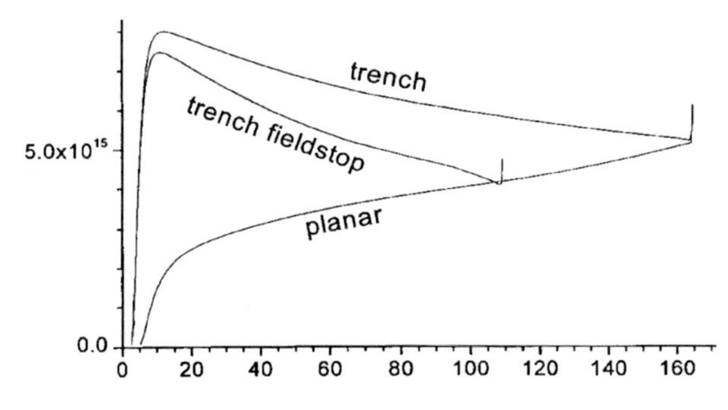

"planar" is the conventional IGBT, there the plasma distribution is as expected for a pnp-transistor with its emitter on the right-hand side. With the trench IGBT, we get an enhanced plasma concentration at the IGBT emitter side. It approximates the plasma distribution in a pin-diode.

In [Tak98], it was even shown that the trench IGBT voltage drop V_C decreases if some of the cells are not contacted. The plasma density increases in the same amount, as the part of contacted cells is reduced, as long as the number of not contacted cells is not too high. V_C is reduced, because the effect of plasma enhancement is much higher than an increased voltage drop across the channel.

10.6.2 The "Latch-Up Free Cell Geometry"

Essential for modern IGBTs is to avoid latch-up of the parasitic thyristor at turn-off. This requirement must now be met under the condition of increased plasma density. The cell must have a structure which is optimized for this requirement. Figure 10.14 shows a detail of a trench cell. The hole current will flow mainly close to the electron current because of the condition of neutrality. Therefore, it flows close to the channel and in its further way to the contact at the p^+-region it must flow underneath the n^+-source region. The n^+p-junction is biased in forward direction. If the voltage drop V_p that is generated by the hole current below the n^+-layer across the length L reaches the order of magnitude of the built-in voltage V_{bi} of this n^+p-junction, then the n^+-region will inject holes. The consequence will be latch-up of the parasitic thyristor, the turn-off capability is lost, the control of the device is lost, and destruction of the IGBT will follow.

Using the depictive representation in Fig. 10.14, this voltage drop V_p can be simplified according to [Ogu04] with

$$V_p = \int_0^L \rho \cdot j_p \cdot y \cdot dy = \frac{1}{2} \cdot \rho \cdot j_p \cdot L^2 \tag{10.18}$$

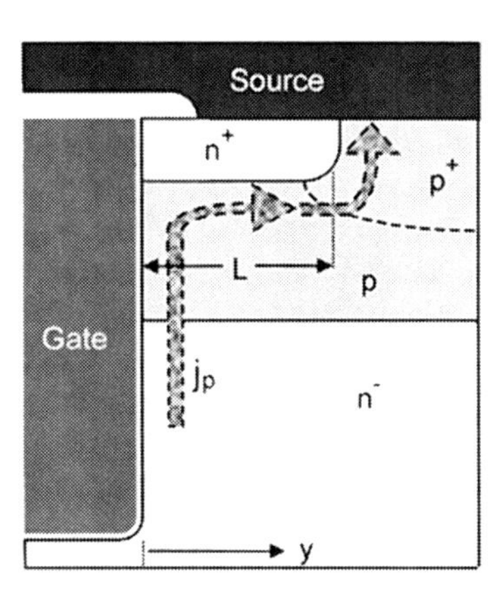

Fig. 10.14 Detail of an IGBT trench cell

where ρ is the sheet resistivity of the p-layer below the n^+-source in $\Omega/\square$ and L is the length of the source layer. To keep V_p below the built-in voltage (≈ 0.7 V at 25°C and decreasing with temperature) even at very high current density, not only L must be small but also ρ must be kept low. As one can recognize in Fig. 10.12, L is designed very short in modern trench IGBTs. A highly doped p^+-layer spreads as far as possible in direction of the trench below the source layer. With these design measures, destructive latch-up can be avoided even at high current density.

This design measure, termed to be a "latch-up free cell geometry" [Las03], is realized by a highly conductive p^+ well adjusted on a submicron scale (Fig. 10.14) concerning the distance of this p^+ well to the trench sidewall. This layer forms a resistance R_S (see equivalent circuit Fig. 10.2b). If R_S is low enough, the parasitic npn-bipolar transistor is effectively suppressed, and latch-up of the thyristor, containing both the npn- and the pnp-bipolar transistors, will not occur. The discussed measure is not restricted to the trench cell. The same measure is possible for a planar structure.

We will come back on this later in Chap. 12. In devices for high blocking capability, a mode of operation with dynamic avalanche occurs at turn-off, an additional hole current is generated and the current density may even be locally increased. IGBTs with well-designed cell structures overcome even these high-stress conditions.

10.6.3 The Effect of the "Hole Barrier"

The possibility to increase the plasma density below the emitter cells is also given by the implementation of an additional n-doped layer. This was first shown by [Tai96] on the example of a trench IGBT, the structure was denominated as "carrier stored trench gate bipolar transistor" (CSTBT). The effect is explained in [Tai96] in the following way: at the n^-n^+-junction a diffusion potential of approximately 0.17 V is built up, which hinders the outflow of holes. This additional layer was designated as a hole barrier.

This n-doped layer below the p-well acts in the same way in a planar IGBT [Mor07, Rah06], as shown in Fig. 10.15. In the n-doped layer, the hole current is the minority carrier current. It decreases strongly before it enters the p-well. In Eq. (10.17), this reduces j_p and increases the emitter efficiency γ of the total top-side cell structure, in which we again summarize p-well, n-channel, and n-accumulation layer to an effective n-emitter. The consequence is an increased plasma density, as shown in right-hand side of Fig. 10.15.

The hole barrier hinders the outflow of holes. To hold the condition of neutrality, additional electrons are delivered by the n-channel. The density of plasma increases, see right-hand side of Fig. 10.15.

A disadvantage of this measure is that the increased doping below the blocking junction J_2 decreases the blocking capability. This must be compensated with a slightly increased thickness of the n-base of the IGBT and this increases the forward voltage drop. A part of the achieved advantage is lost again, but this effect can be kept minimal [Lin06].

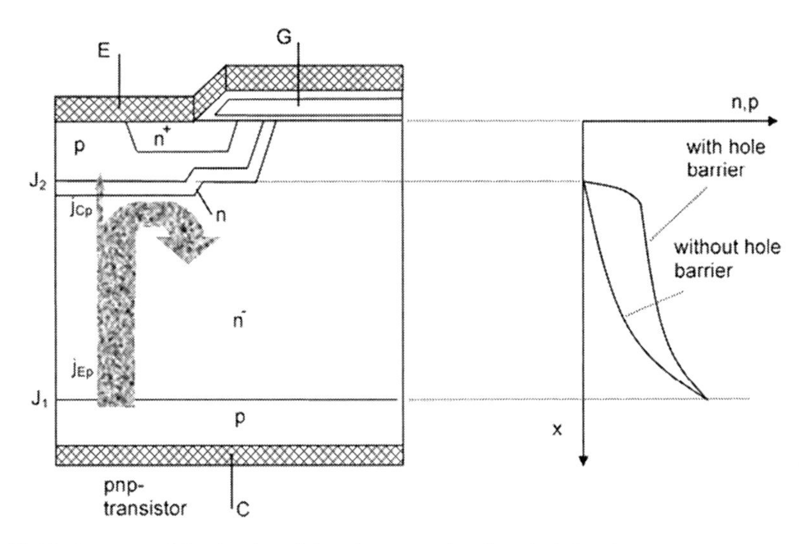

Fig. 10.15 Increase of the density of free charge carriers by a hole barrier

A combination of the hole barrier with the trench structure is done in the "carrier stored trench gate bipolar transistor" (CSTBT) of the manufacturer Mitsubishi (see Fig. 10.16). The n-layer as a hole barrier is arranged below the p-layer within the trench structure. Below the hole barrier, holes accumulate; to maintain neutrality, additional electrons are delivered effectively by the channel. The density of plasma is increased locally.

The hole barrier can be combined with the before-described measures of increasing the distance between the cells and decreasing the lateral extension of the p-layers. In a trench IGBT, this can be done very easily by debarring a part of the cells. These cells are denominated as "plugged cells" [Yam02]. The polysilicon in these cells, which forms the gate area, is shorted to the emitter metallization and

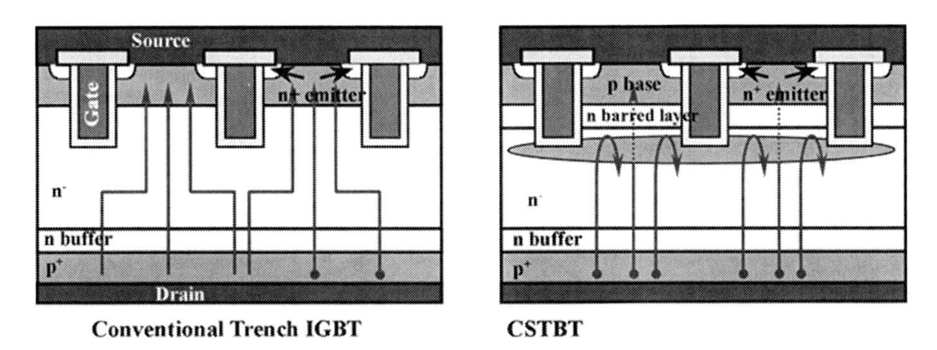

Fig. 10.16 Carrier stored trench gate bipolar transistor (*right-hand side*) in comparison to a conventional trench IGBT (*left-hand side*). Pictures from Mitsubishi Electronics

the cell is prevented from building an n-channel. This measure additionally has the advantage that the current I_{Dsat} in the active area is reduced and therewith the current at a short-circuit event is reduced.

10.6.4 Collector Side Buffer Layers

Besides the described increasing of the plasma density, every modern IGBT additionally uses the effect of a reduced width of the n-base layer by designing a trapezoidal electric field instead of a triangular one. This is achieved with an n-layer with increased doping in front of the p-collector layer. The denomination differs for different manufacturers: "fieldstop", "soft punch through", "light punch through", etc., but it is more or less the same.

The plasma density for the "Trench Fieldstop" IGBT has already been shown in Fig. 10.13. In a trench-fieldstop IGBT, the width of the n-base is shortened compared to a trench IGBT. Figure 10.17 shows the new structure. In front of the collector layer, the doping density is increased (n-fieldstop). The space charge is trapezoidal if a voltage close to the specified blocking voltage is applied. This is sketched in Fig. 10.17, left-hand side. In reality, it is only a moderate trapeze, it is closer to a triangle and not almost rectangle, as one could conclude from Fig. 10.17.

From the viewpoint of the shape of the electric field, the fieldstop IGBT is similar to a PT-IGBT. The base width w_{B} is significantly reduced at the same blocking voltage in an IGBT with collector side buffer layer compared to an NPT-IGBT. The voltage V_{drift} that drops across the base layer is proportional to $w_{\mathrm{B}}{}^2$ according to Eq. (10.13). Hence the voltage drop V_{C} can be reduced significantly with this design.

However, this is the only common feature with the PT-IGBT. The trench-fieldstop IGBT is not fabricated like the PT-IGBT on a p^+-substrate with an epitaxial n-layer; it is rather fabricated from a homogeneously n-doped wafer. The fabrication of a collector layer with exactly adjusted emitter efficiency is done similarly as with an

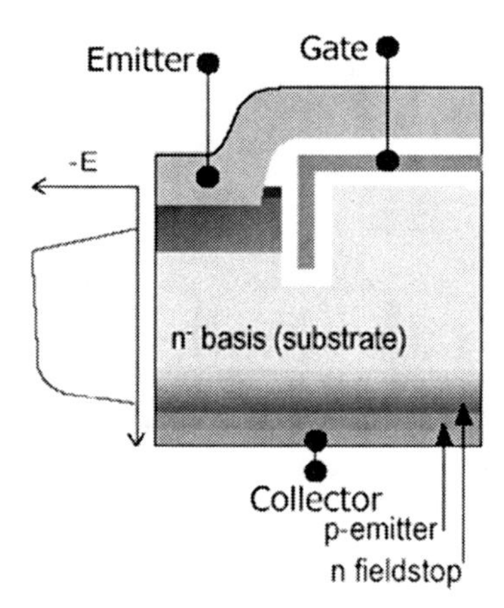

Fig. 10.17 Structure of the Infineon trench-fieldstop IGBT. Figure from Infineon

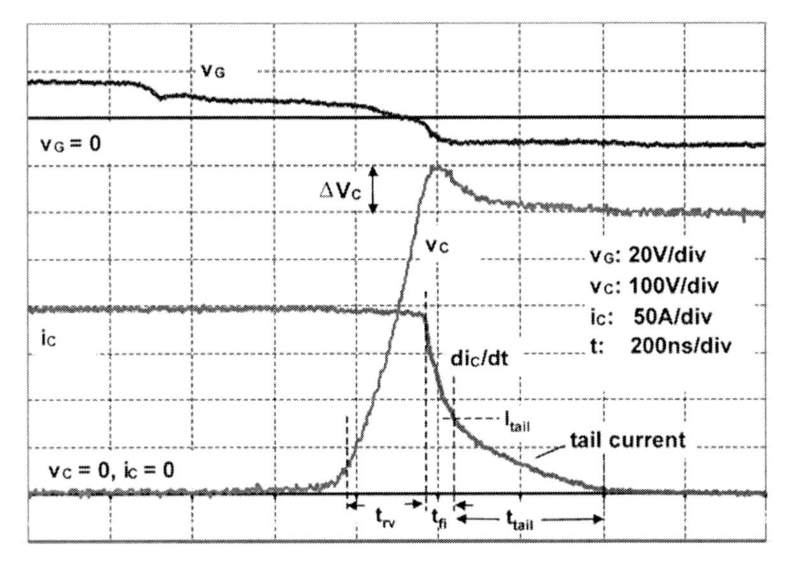

Fig. 10.18 Turn-off of the Trench Fieldstop IGBT (FF200R12KE3 from Infineon, 200 A 1200 V module). $T = 125°\text{C}$, $R_{\text{Goff}} = 5\,\Omega$

NPT-IGBT [Las00b]. α_{pnp} is adjusted by the emitter efficiency; an emitter of low penetration depth and low doping is used. No charge carrier lifetime reduction is done. The mode of function of the trench-fieldstop IGBT is closer to the NPT-IGBT than to the PT-IGBT. As a result, the temperature dependency on V_{C} is similar to that of the NPT-IGBT. The desired "positive temperature coefficient" remains.

Figure 10.18 shows the turn-off behavior of the trench-fieldstop IGBT. If one compares this figure with the turn-off behavior of the NPT-IGBT in Fig. 10.6, one can see a significantly shortened tail current. Additionally, the fall time of the current t_{fi} is increased and $\text{d}i_{\text{C}}/\text{d}t$ is decreased; this leads to a lower voltage peak ΔV_{C}.

The shortening of the tail current happens by the effect that a lower part of the stored plasma in the on-state remains close to the collector side at turn-off. The removal of the stored charge is done in a big amount during the voltage rise time t_{rv}. If the applied voltage V_{bat} is increased above the voltage of 600 V, as applied in Fig. 10.18, the tail current is shortened further and finally it can disappear completely. Then the space charge spreads across the whole middle layer. If V_{bat} is increased further, a current snap-off similar to the behavior at reverse recovery of a snappy diode will occur at the end of the fall time t_{fi}. This current snap-off can lead to high overvoltage peaks.

10.7 IGBTs with Bidirectional Blocking Capability

For some applications of power electronics, e.g. the matrix converter, a both-side blocking device is necessary. The IGBT structure contains a p-layer forming a pn^{-}-junction at the collector side. The basic structure, as drawn in Fig. 10.1, has the capability to take an electric field at the bottom-side pn$^-$-junction as well as at the top-side n$^-$p-junction. It has similarities to the structure of a thyristor which has a blocking capability in both directions. However, the back-side pn$^-$-junction has no defined junction termination. Such planar junction terminations for shallow pn-junctions need microstructures. The processes for the fabrication of junction terminations are only possible at the top side of the wafers, since semiconductor technology has been optimized for microstructures on one side of a wafer.

A possible solution is to lead the back-side pn-junction to the front side of the wafer. This can be done by the diffusion of a deep p-layer which reaches through the whole wafer; this technology is called "diffusion isolation" [Tai04]. A schematic drawing of an IGBT with this deep edge diffusion is shown in Fig. 10.19. With a deep diffusion in the area where the wafer is cut in the last production step, the collector-side pn$^-$-junction is connected to the front side of the wafer. Now it is possible to apply a junction termination for both directions. This is done in Fig. 10.19 for the forward direction with potential rings, as known from Fig. 4.19. The junction termination for the forward direction ends at the channel stopper. As junction termination for the reverse direction, a field plate structure is applied (see Fig. 10.19).

The device blocks the voltage in both directions like a thyristor. The bidirectional blocking IGBT must be dimensioned as NPT type. A buffer layer in front of the p-collector would reduce or even eliminate the blocking capability in reverse direction. Therefore, a minimal thickness w_B of the base layer for a triangular field shape is required. In [Nai04], this is given with 200 μm for a 1200 V reverse blocking IGBT; the electric field in the volume is similar to that in Fig. 10.8.

Since a buffer layer cannot be applied, the reverse blocking IGBT cannot have a forward voltage drop V_C as low as other modern IGBTs. But V_C is surely lower than the one of a serially connected IGBT and diode.

In the intended matrix converter applications, the reverse blocking IGBT works as freewheeling diode at some switching events; it is passively turned off like a diode. Because of the high carrier lifetime in IGBTs, a high reverse recovery peak I_{RRM} and a high reverse recovery charge occur. In one concept [Nai04], electron irradiation was used and a reduction of I_{RRM} of about 10% was achieved. The shorter carrier lifetime increases the on-state losses and a trade-off must be made.

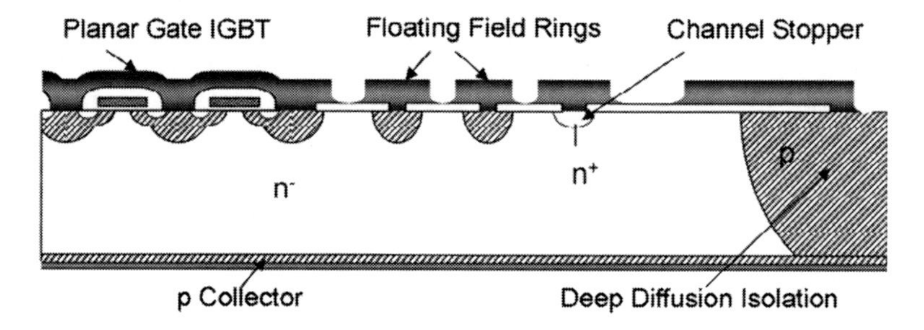

Fig. 10.19 Symmetrically blocking IGBT with edge diffusion. Figure from [Ara05] © 2005 EPE

Further work is done to improve the reverse blocking IGBT. The deep diffusion on the right-hand side of Fig. 10.19 has, however, also the consequence of lateral diffusion which will be approximately spread by 0.8× of the diffusion depth to the side. With a wafer thickness in the range > 100 μm, this deep p-layer will also be very wide. This leads to a loss of area for the region which takes the current, a loss of active area, and a high share of area for the junction termination which cannot be used for current transport. Research and development work is done for improved structures; for example, to replace the deep diffusion zones by deep trenches [Ara05].

10.8 Reverse Conducting IGBTs

If in a part of the collector area an n^+-layer is implemented, see Fig. 10.20, a diode is formed by this n^+-layer and the p-body. There are several advantages of the monolithic integration of the diode. Primarily, it is the reduction of the chip size. Thermal calculations of the diode and the IGBT in a three-phase inverter application [Rut07] have led to the result that a great saving of the area is possible, while the RC-IGBT stays almost at the same size like the IGBT and the diode area can be saved.

The concept of an RC-IGBT in a productive volume was first realized with an optimization for lamp ballast applications [Grie03]. This application needs 600 V blocking voltage, a current in the range of 1 A, and there is no hard commutation of the diode. These devices could be realized successfully without optimization of the diode.

To satisfy the requirements of hard switching applications, the reverse recovery behavior of the integrated diode has to be optimized. For this diode, the n-doped regions on the backside act as a cathode emitter (see Fig. 10.20), while the p-body of the IGBT and the highly p-doped anti-latch-up region near the front side act as an anode emitter of a freewheeling diode integrated into the chip in this way.

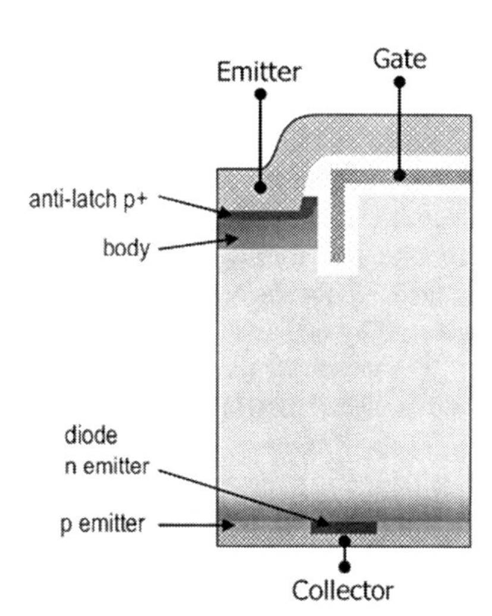

Fig. 10.20 Reverse conducting IGBT with trench gate. Figure from [Rut07] © 2007 IEEE

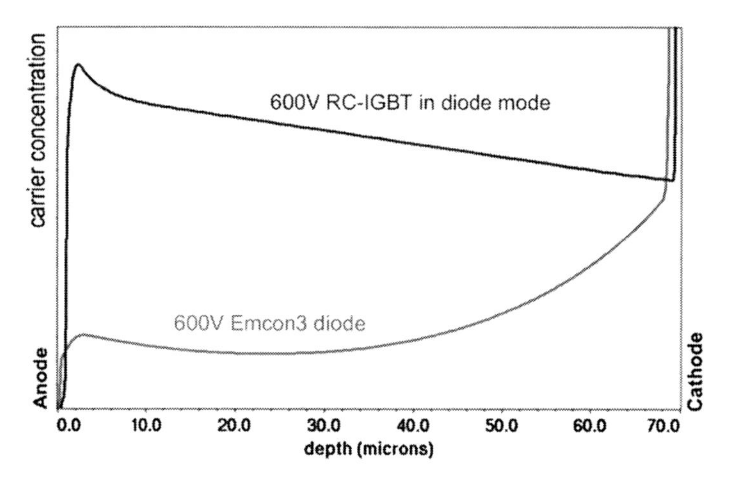

Fig. 10.21 Carrier concentration profile in a 600 V EMCON3 compared to a 600 V RC-IGBT in diode mode. Figure from [Rut07] © 2007 IEEE

Unfortunately, such a diode with a highly doped p-emitter has a plasma distribution as shown in Fig. 10.21. It is very high at the anode side and lower at the cathode side. This will lead to a high reverse recovery peak and snappy reverse recovery behavior, as discussed in detail in Chap. 5. The internal plasma distribution as it must be for a soft recovery diode is shown for comparison in Fig. 10.21, too.

To reduce the plasma density at the cathode side, different methods have been tested. Irradiation of He^{2+}-ions is applied for the reverse conducting IGBT in [Tai04b]. The reverse recovery current peak I_{RRM} was reduced close to the value of a commercial freewheeling diode. But irradiation technologies can have a disadvantage. In [Rut07], it is reported that particle irradiation causes also undesirable trapped charges in the gate oxide and the silicon–gate oxide interface. These charges give rise to a drop of the threshold voltage and a broader parameter distribution. Therefore, the way used in [Rut07] is to reduce the efficiency of the anti-latch p^+-emitter by reducing the implantation dose. The advantage of this method is that there is no effect on the V_C. However, a reduction of this doping dose is limited by the overcurrent turn-off robustness (latch-up robustness). It was found that the potential for a reduction of Q_{RR} by decreasing the p^+-emitter dose at the same robustness as today's devices is between 15 and 25% [Rut07]. With platinum diffusion, Q_{RR} was reduced by 60 %, and the increase in V_C of the IGBT was only 0.1 V [Rut07].

The measures to improve the IGBT and to improve the freewheeling diode contradict each other in some aspects. However it is expected that reverse conducting IGBTs will become available also for switching with inductive load, since a reduction of costs in the power electronic system is possible by saving half of the number of power dies. These solutions might be successful in applications in which the requirements to the diode reverse recovery are medium, e.g. for low current applications up to 1 kW, since the energy stored in the parasitic inductance is small

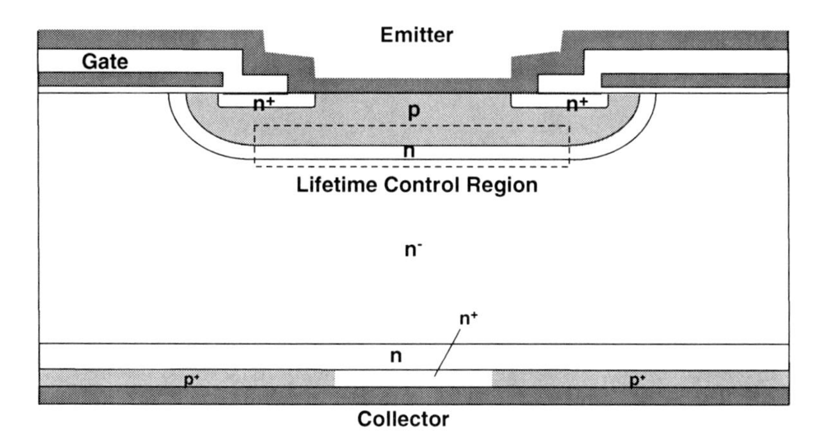

Fig. 10.22 3.3 kV RC-IGBT with optimized integrated diode. Figure following [Rah08]

at low current. There is a wide field of such applications, such as inverters in air conditioners, refrigerators, and others.

A very interesting concept for a reverse conducting IGBT rated 3300 V was presented in [Rah08]. The requirements to the freewheeling diode are highest in the intended high power motor drive application. The new structure is shown in Fig. 10.22.

This structure uses the fact that there are measures which are not conflicting in their effect on the plasma distribution as well in the IGBT as in the diode. An n-layer below the p-well is used, as it was named as "hole barrier" in Fig. 10.15. It decreases the injected hole current and for the effective emitter of the IGBT it acts in an enhanced emitter efficiency. For the diode, it acts as reduced p-emitter efficiency.

On the collector side, the n^+-layer acts as a kind of anode short, known from GTO thyristors, and it will reduce the IGBT collector side emitter efficiency. Doping of the collector side p-layer, as well as width and doping of the n^+-layer, must be carefully adjusted to fulfill the requirements for the diode and for the IGBT. For the diode, additional local lifetime control region below the p-well is applied, as known from the CAL diode in Chap. 5. The generation of recombination centers is executed with masked irradiation of light ions. The reverse recovery of the diode can further be improved by opening the channel. Injected electrons by the channel in the diode conducting mode will reduce the p-emitter efficiency, as it is described in hand of the MOS controlled diode (see Sect. 5.7).

Investigations in [Rah08] show that the IGBT as well as the diode could even profit in some aspects due to the integration. For the application it is a high advantage. The effective IGBT area is increased by 50% and the effective diode area even by 200%. A potential for rating 50% higher current for IGBT in modules with the same area is given, however only a part of it can be used in practice because of the thermal resistance of the housing sets limits. In the typical motor drive application conditions, in one sinus half-wave the IGBT, in the other the diode is heated up. The absence of inactive periods can reduce the temperature ripple and will increase the lifetime of the module. For details, see Chap. 11. In the beginning of 2010, the reverse conducting IGBT is still an object of research and development.

10.9 Outlook

It has been shown that with modern IGBTs the optimal, in terms of conduction losses, thyristor-like distribution of the internal plasma can be achieved. Therefore the IGBT applications could be extended to voltage ranges which were formerly only possible with thyristors. IGBTs are commercially available in 2010 up to 6.5 kV blocking voltage. Development work is done on IGBTs for the voltage range 8 or 10 kV. The development of devices with new structures as IGBT followers has been placed back or suspended, since the new structures were not necessary. The desired advantage has been possible with the IGBT to a great extent.

However, further intensive work in research and development has to be done to decrease the conduction and switching losses of IGBTs. In 2007, IGBTs for the most frequently used 1200 V voltage range were introduced whose voltage drop V_C at rated current is typically as low as 1.7 V. It is expected that a further reduction down to less than 1.5 V will be possible in the next years. The losses in inverters for motor control will be reduced strongly.

IGBTs with additional integrated functions, as the reverse blocking IGBT and the reverse conducting IGBT, are expected to become commercially available.

With the progress in IGBTs, it is possible to control more and more power in a small device area. But with this, the dissipated losses per area are also increasing. These losses have to be extracted by the package and the challenges and requirements for the packaging technology are increasing. This is the subject of the next chapter.

References

[Ara05] Araki T: "Integration of Power Devices - Next Tasks", Proceedings of the EPE, Dresden, (2005)

[Bal82] Baliga BJ, Adler MS, Grey PV, Love RP, The insulated gate rectifier (IGR): A new power switching device", Proceedings of the IEDM, pp. 264–267 (1982)

[Bal83] Baliga BJ: "Fast-switching insulated gate transistors", IEEE Electron Device Letters, Volume 4, Issue 12, pp. 452–454 (1983)

[Bec80] Becke HW, Wheatley Jr CF: "Power MOSFET with an anode region", United States Patent Nr. 4,364,073, Dec. 14, 1982 (filed March 25, 1980)

[Gri03] Griebl E, Hellmund O, Herfurth M, Hüsken H, Pürschel M, "LightMOS - IGBT with Integrated Diode for Lamp Ballast Applications", PCIM 2003, p. 79ff (2003)

[Kit93] Kitagawa M, Omura I, Hasegawa S, Inoue T, Nakagawa A: "A 4500 V Injection Enhanced Insulated Gate Bipolar Transistor (IEGT) in a Mode Similar to a Thyristor" IEEE IEDM tech.digest pp. 697–682 (1993)

[Las92] Laska T, Miller G, Niedermeyr J: "A 2000 V Non-Punchtrough IGBT with High Ruggedness", Solid-St. Electron., Vol 35, No 5, pp. 681–685, 1992

[Las00] Laska T, Lorenz L, Mauder A: "The Field Stop IGBT Concept with an Optimized Diode", Proceedings of the 41th PCIM, Nürnberg (2000)

[Las00b] Laska T, Münzer M, Pfirsch F, Schaeffer C, Schmidt T: "The Field Stop IGBT (FS IGBT) - A New Power Device Concept with a Great Improvement Potential", Proceedings of the ISPSD, Toulouse, pp. 355–358 (2000)

[Las03] Laska T. et al: "Short Circuit Properties of Trench/Field Stop IGBTs Design Aspects for a Superior Robustness", Proc. 15th ISPSD, pp. 152–155, Cambridge (2003)

[Lin06] Linder S, Power Semiconductors, EPFL Press, Lausanne, Switzerland, 2006

[Mil89] Miller G, Sack J: "A new concept for a non punch through IGBT with MOSFET like switching characteristics", Proceedings of the PESC ′89 Vol 1, pp. 21–25 (1989)

[Mor07] Mori M et al: "A Planar-Gate High-Conductivity IGBT (HiGT) With Hole-Barrier Layer" IEEE Trans. Electron Devices. vol 54, no 6 pp. 2011–2016 (2007)

[Nai04] Naito T, Takei M, Nemoto, M, Hayashi T, Ueno K: "1200 V reverse blocking IGBT with low loss for matrix converter"; Proceedings of the ISPSD ′04. pp. 125–128 (2004)

[Nak85] Nakawaga A, Ohashi H: "600-1200 V Bipolar Mode MOSFETS with High-Current Capability", IEEE-EDL 6, No. 7 pp. 378–380. (1985)

[Net99] Netzel M, Analyse, Entwurf und Optimierung von diskreten vertikalen IGBT-Strukturen, Dissertation, Isle-Verlag Ilmenau 1999

[Nic00] Nicolai U, Reimann T, Petzoldt J, Lutz J: Application Manual Power modules, ISLE Verlag 2000 http://:www.semikron.com/skcompub/en/application_manual-193.htm

[Ogu04] Ogura T, Ninomiya H, Sugiyama K, Inoue T: "4.5 kV Injection Enhanced Gate Transistors (IEGTs) with High Turn-Off Ruggedness," IEEE Trans. Electron Devices, vol. 51, pp. 636–641 (2004)

[Omu97] Omura I, Ogura T, Sugiyama K, Ohashi H: "Carrier injection enhancement effect of high voltage MOS-devices - Device physics and design concept", Proceedings of the ISPSD, Weimar (1997)

[Reh02] Rahimo MT, Kopta A, Eicher S, Kaminski N, Bauer F, Schlapbach U, Linder S: "Extending the Boundary Limits of High Voltage IGBTs and Diodes to above 8 kV" Proc. ISPSD 2002, Santa Fe, USA, pp. 41–44

[Rah06] Rahimo MT, Kopta A, Linder S: Novel Enhanced–Planar IGBT Technology Rated up to 6.5 kV for lower Losses and Higher SOA Capability" Proc. ISPSD 2006, Naples, Italy, 33–36 (2006)

[Rah08] Rahimo M, Schlapbach U, Kopta A, Vobecky J, Schneider D, Baschnagel A: "A High Current 3300 V Module Employing Reverse Conducting IGBTs Setting a New Benchmark in Output Power Capability" Proc. ISPSD, Orlando, Florida (2008)

[Rog88] Rogne T, Ringheim NA, Odegard B, Eskedal J; Undeland TM.: "Short-circuit capability of IGBT (COMFET) transistors" 1988 IEEE Industry Applications Society Annual Meeting, vol.1 pp. 615–619 (1988)

[Rus83] Russell JP, Goodman AM, Goodman LA, Neilson JM, "The COMFET - A new high conductance MOS-gated device", IEEE Electron Device Letters, Volume 4, Issue 3, pp. 63–65 (1983)

[Rut07] Rüthing H, Hille F, Niedernostheide FJ, Schulze HJ, Brunner B: "600 V Reverse Conducting (RC-) IGBT for Drives Applications in Ultra-Thin Wafer Technology" 19th International Symposium on Power Semiconductor Devices and IC′s, ISPSD ′07, pp. 89–92 (2007)

[Tai96] Takahashi H, Haruguchi H, Hagino H, Yamada T: "Carrier stored trench-gate bipolar transistor (CSTBT) - a novel power device for high voltage application" ISPSD ′96 Proceedings., 8th International Symposium on Power Semiconductor Devices and ICs 20–23 May 1996 pp. 349–352, 1133 (1996).

[Tai04] Takahashi H, Kaneda M, Minato T: "1200 V class reverse blocking IGBT (RB-IGBT) for AC matrix converter" Proceedings of the 16th ISPSD, pp. 121–124 (2004)

[Tai046] Takahashi H, Yamamoto A, Aono S, Minato T, "1200 V Reverse Conducting IGBT", Proceedings of the 16th ISPSD, pp.133–136, (2004)

[Tak98] Takeda T, Kuwahara M, Kamata S, Tsunoda T, Imamura K, Nakao S: "1200 V trench gate NPT-IGBT (IEGT) with excellent low on-state voltage", Proceedings of the ISPSD, Kyoto (1998)

[Tih88] Tihanyi J: "MOS-Leistungsschalter", ETG-Fachtagung Bad Nauheim, 4.-5.Mai 1988, Fachbericht Nr. 23, VDE-Verlag, pp. 71–78 (1988)

[Yam02] Yamada J, Yu Y, Donlon JF, Motto ER: "New MEGA POWER DUAL^{TM} IGBT Module with Advanced 1200 V CSTBT Chip", Record of the 37th IAS Annual Meeting Conference, Vol. 3, pp. 2159–2164 (2002)

Chapter 11
Packaging and Reliability of Power Devices

11.1 The Challenge of Packaging Technology

The operation of a power semiconductor device produces dissipation losses. The order of magnitude of these losses shall be estimated in the following example:

IGBT module BSM50GB120DLC (Infineon) mounted on an air-cooled heat sink
Operation conditions: $I_C = 50$ A, $V_{bat} = 600$ V, $R_G = 15\ \Omega$,
$f = 5$ kHz, duty cycle $d = 0.5$
The following parameters can be extracted from the data sheet:
Forward voltage drop: $V_C = 2.4$ V
Turn-on energy loss per pulse: $E_{on} = 6.4$ mWs
Turn-off energy loss per pulse: $E_{off} = 6.2$ mWs

For details on E_{on} see Fig. 5.20 and Fig. 5.22. A simplified calculation can be done with Eq. (10.4), a more exact determination is done with the oscilloscope, see Eq. (9.30). For details on E_{off} see Fig. 10.5; for simplified calculation Eq. (10.6) is useful.

Losses created by the leakage current can usually be neglected in modern IGBT and MOSFET applications. Therefore total power dissipated in the device is the sum of on-state and switching losses:

$$P_V = P_{cond} + P_{on} + P_{off} = I_F \cdot V_C \cdot d + f \cdot E_{on} + f \cdot E_{off} \tag{11.1}$$

This amounts to 123 W for the given example. These losses are marginal compared to the controlled power of approximately 30 kW. To calculate the efficiency an additional freewheeling diode has to be taken into account; for most applications a half-bridge configuration of two switches in series must be considered. Nevertheless, the efficiency of the power control circuit is in the range of 98%.

However, 123 W of power losses have to be extracted from an IGBT switch with an area of about 1 cm^2 which requires a heat flux density of 123 W/cm^2 or 1.23 MW/m^2. The heat flux density can even amount to the 2–3-fold value for an

J. Lutz et al., *Semiconductor Power Devices*, DOI 10.1007/978-3-642-11125-9_11,

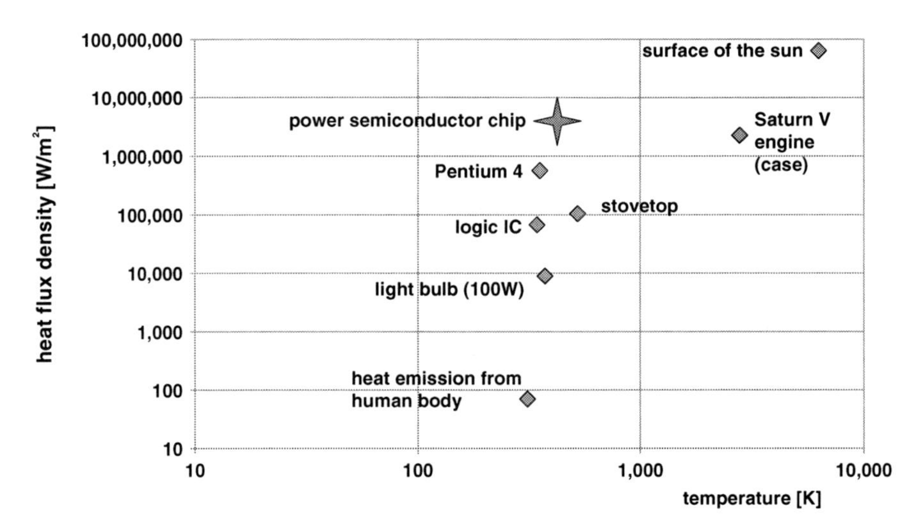

Fig. 11.1 Heat flux density of different heat sources, inspired by Dr. W. Tursky, Semikron

assembly on a water-cooled heat sink and with maximum utilization of the module capability. Figure 11.1 relates this power loss to that of other heat sources.

The heat flux density of a power semiconductor chip exceeds that of a stovetop of a conventional kitchen stove by more than one order of magnitude and outranges a Pentium 4 microprocessor. Therefore, a power module has to provide a high thermal conductivity. Additionally, a power device package has to meet a number of requirements:

- High reliability, i.e. a long lifetime in application and therefore a high durability under alternating load conditions (power cycling stability)
- High electrical conductivity of the components to achieve low undesirable (parasitic) electrical properties (parasitic resistance, capacity and inductivity)
- For power modules additional electrical insulation between switches and between circuit and heat sink

The solution to this problem is by no means trivial and it is today one of the most exciting challenges for engineers. Power modules are the prevalent types of packages in power electronic applications and they will be discussed in detail in the following chapters.

11.2 Package Types

A substantial criterion for the selection of an appropriate package type is the power range of the semiconductor device. A survey of power ranges is given in Fig. 11.2.

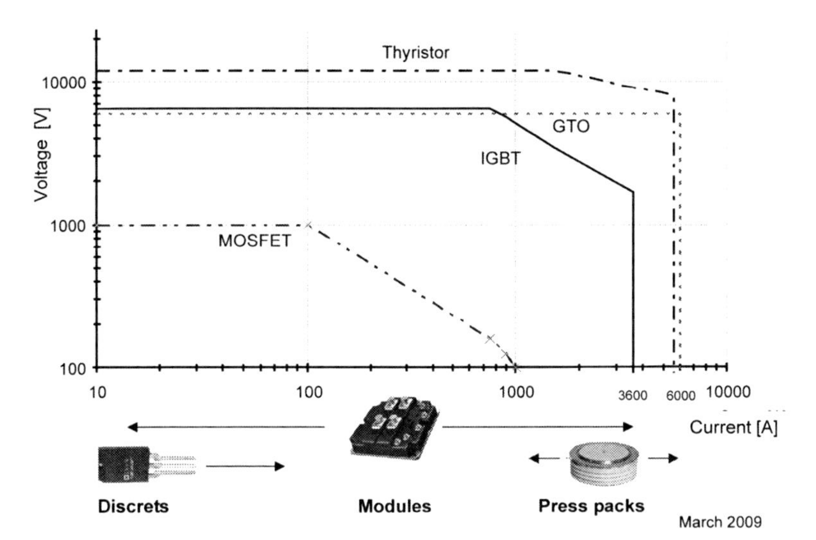

Fig. 11.2 Power range of modern semiconductor devices (2009) together with the predominant package type

Discrete packages are prevailing in the range of small power. These packages are soldered to a laminated "power circuit board" (PCB) for application. Since the generated power losses are relatively small, the requirements for heat dissipation are unincisive. These packages are mostly designed without internal insulation with the consequence that only a single switch can be integrated in one package. The most common package of this type is the "transistor outline" (TO) package.

The discrete design has to fulfill the following functions:

– Conduction of load current and control signals
– Dissipation of heat
– Protection against environmental influences

Capsules also belong to the discrete packages. They are applied for the high-power end of the range that is not jet reached by power modules. Capsules are not equipped with an internal insulation. They can be cooled from two sides. A power chip can have the size of a whole wafer in the peak performance range. Therefore, the circular footprint of the capsule is the ideal package for circular chips. Because of its shape, this package is also known as "hockey puk".

A thyristor from Mitsubishi for 1.5 kA with 12 kV blocking voltage is packaged in a capsule. Thyristors in capsules from Infineon are specified for 3 kA with 8.2 kV and recently developed thyristors for HVDC applications are rated for 5.6 kA with 8 kV blocking voltage. The "chip" in these packages consists of a complete 6 in. wafer with a diameter of approximately 150 mm.

Mitsubishi offers a gate turn-off thyristor (GTO) in a capsule with a chip fabricated from a single 6 in. wafer (150 mm) which is specified for 6 kA with 6 kV blocking voltage.

In contrast to discrete packages power semiconductor modules are characterized by

- an insulated architecture in which the components of the electrical circuit are dielectrically insulated from the heat dissipating mounting surface
- several single functions (phase leg circuit), often with paralleling of chips

Power semiconductor modules are dominating in the range of more than 10 A for blocking voltages of 1200 V and above. They are characterized by the integration of multiple functions (for example, converter–inverter–brake topologies) in the lower power range. In the high-power area, Infineon offers a module with 6.5 kV IGBTs and the associated freewheeling diodes with a continuous maximum current of 900 A. For 1200 V blocking voltage, Infineon produces a module specified for 3.6 kA continuous current which contains 24 IGBT chips in parallel and 12 freewheeling diodes in parallel. These examples show that modules have penetrated deep into the high-power range which was formerly dominated by capsules. This trend will continue.

11.2.1 Capsules

Figure 11.3 displays the internal construction of a capsule in a simplified schematic view. The silicon device (e.g., a thyristor) is mounted between two metal discs in order to homogenize the pressure and to avoid pressure peaks. Molybdenum is the ideal material for this purpose because of its great hardness and its well-adapted coefficient of thermal expansion. In the architecture shown in Fig. 11.3 the silicon device is rigidly coupled to one of the molybdenum discs on the anode side and pressed to the second molybdenum disc on the cathode side. Alignment features to

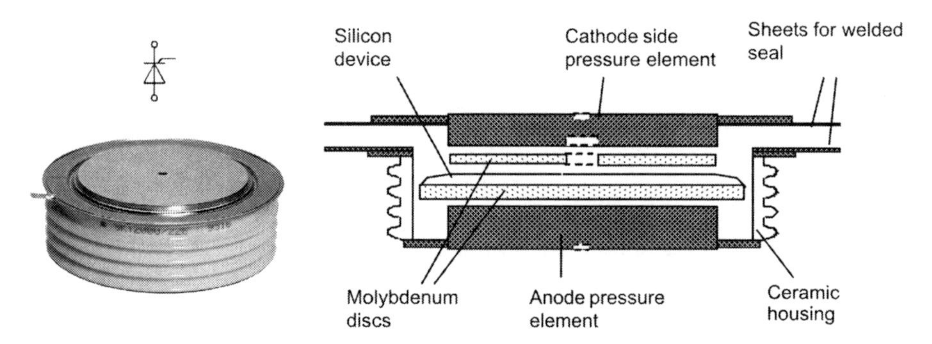

Fig. 11.3 Internal construction of a capsule (simplified)

center the chip inside the package are not shown in Fig. 11.3 for clarity, neither is the gate contact spring displayed, which is guided by a slot in the cathode compression piece to the center of the silicon device. The package is hermetically sealed by welding together the two metal latch rings.

A complete electrical and the thermal contact will only be established by the application of a defined pressure to the package, which is typically in the range of 10–20 N/mm^2.

The interconnection between the silicon device and the molybdenum discs can vary for different package sizes and manufacturers. For small chip diameters up to 5 cm, solder interfaces are feasible. But care must be taken to select a solder material, which will show only little plastic creep under high pressure. For larger diameters of the device alloyed interfaces are generally preferred. Also designs without any rigid connection between Mo and Si are available, which allow a floating of the power device. A progressive technology for the interconnection of silicon and molybdenum is a diffusion sinter technique: The partners to be connected are equipped with a noble metal plating, a silver powder is applied to the connecting surfaces, and a very reliable connection is established by sintering the interface layer at a high pressure and temperatures of approximately 250°C.

Mostly conventional devices are packaged in capsules: Diodes, thyristors, GTOs, and the GCTs derived from the GTO. The advantages of capsules are

- Compact design with good relation between device surface area and package surface area
- Cooling of both device surfaces
- No wire bonds – wire bonds generally represent a reliability constrictive feature
- Few or no rigid interconnection between materials with different coefficients of thermal expansion

A high reliability can be expected from the last two factors. The disadvantages of capsules are

- No dielectric insulation – the user has to provide for insulation in the application
- Higher effort in the mounting assembly – a defined uniaxial high pressure must be established and maintained

Due to their advantages the capsule package was also adapted as a package for IGBTs. But IGBTs are today produced only in a small chip size compared to thyristors. The reason is the high cell density of modern IGBT chips which would result in a yield problem caused by single-cell defects with increasing chip size. The largest commercially available IGBT has an area of 300 mm^2. Furthermore, the simplicity of paralleling fast switching IGBTs compared to the difficulty of paralleling slow-switching thyristors, together with the thermal disadvantage of large-area chips, does not produce a market pressure to develop larger area IGBTs. However, for the adaptation of the capsule package for IGBTs, the parallel arrangement of quadratic chips in a so-called presspack IGBT is a technological challenge.

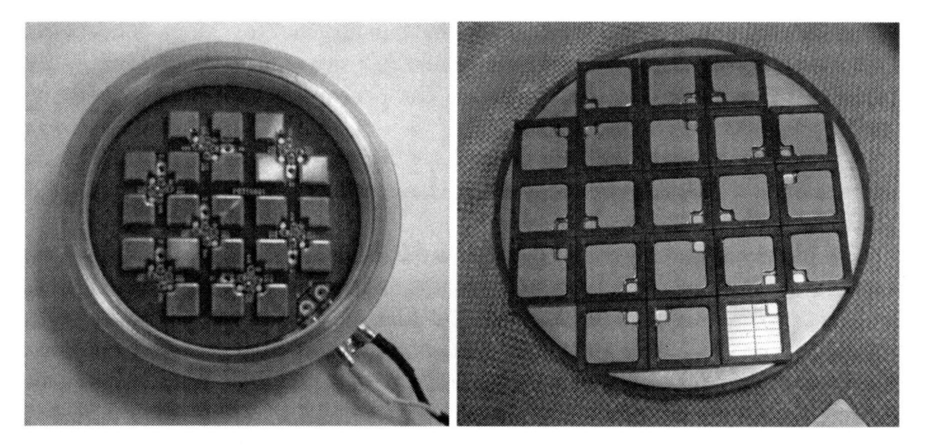

Fig. 11.4 Presspack IGBT: emitter pressure element (*left*), arrangement of the chips (*right*)

An example for a presspack IGBT is illustrated in Fig. 11.4. The chips are assembled on a large molybdenum disc, each chip equipped with a collector side small Mo square. Alignment frames are positioning the chips relative to each other. Small Mo squares with cutouts for the gate contact area are placed on the emitter contacts. The gate connection is implemented by springs, which are guided by another alignment structure. The upper pressure element has to transmit a uniform pressure to each of the chips below. To press each of the 21 paralleled IGBTs with an identical pressure requires maintaining very tight tolerances for every part of the package.

Integrated in the upper pressure element, a printed circuit board – carrying the gate resistors in "surface-mounted device" (SMD) technology – is installed. The complex construction of a presspack IGBT results in a considerable increased demand on the precise alignment of a multitude of parts and on the allowable part tolerances compared to a semiconductor module. Whether the expectable higher reliability in active power cycling will outperform the drawbacks in package assembly and application and thus will replace other package types is yet unresolved.

11.2.2 The TO Family and Its Relatives

Discrete packages are also very common in the lower power range. Today, this field is dominated by the "transistor outline" (TO) family. The principle design is shown in Fig. 11.5.

The TO package family comprises an extensive set of standardized package outlines, the most popular representatives are the TO-220 and the TO-247 package. In these standard packages, the power silicon chip is soldered directly to a solid copper base, which serves as mounting surface. Therefore, the package has no inherent electrical insulation. The contact leads or contacts legs are fixed by a "transfer mold"

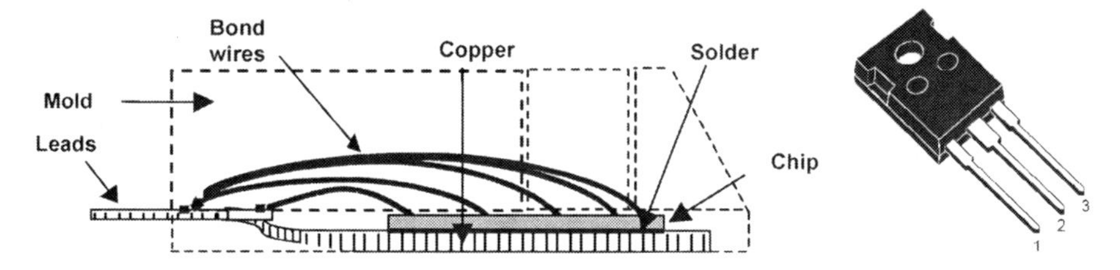

Fig. 11.5 TO package, principal design

housing. One of the leads is directly connected to the copper base, the others are connected to the load and control contact areas on the silicon chip by aluminum wire bonds (Fig. 11.5).

The difference in thermal expansion between the silicon chip and the copper base limits the reliability of this package. An improvement in this respect is the ISOPLUS package introduced by IXYS. As illustrated in Fig. 11.6, a ceramic substrate replaces the solid copper base, thus adapting a technology which is successfully applied in power modules. This design exhibits a number of advantages compared to the standard TO package:

- better adaptation of thermal expansion resulting in higher reliability
- internal insulation
- smaller parasitic capacity compared to a standard TO package mounted on a heat sink with an external insulation polyimide foil (for details refer to Sect. 11.5).

At the first glance, the smaller thermal conductivity of the ceramic layer with respect to copper appears as a serious drawback. However in a system, where several discrete packages – which typically show different potentials at their copper base – are mounted on a single heat sink, the ceramic insulation system generally is superior to the insulation of standard TOs by externally applied electrically insulating foils.

The prevalent power device packaged in a TO housing is the MOSFET. For this device, a drastic reduction of the electrical on-state resistance R_{on} has been accomplished in the last years. As a consequence, a fundamental weakness of this package design is emerging: The TO package has a parasitic electrical resistance in the same order of magnitude as the on-state resistance of a modern MOSFET device.

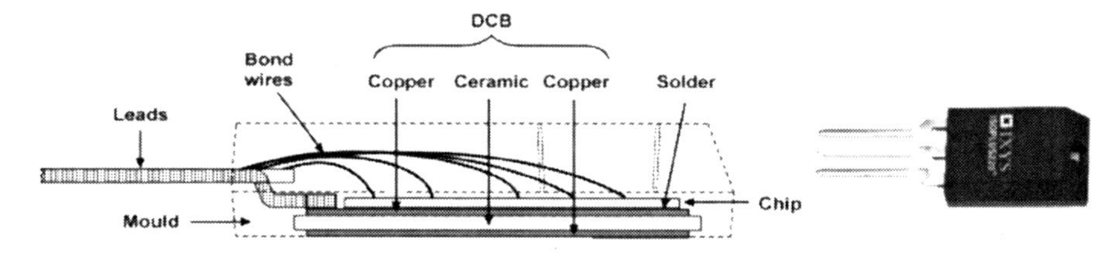

Fig. 11.6 ISOPLUS package with TO outline, but with insulated base

The contact leads are a major limiting factor. Their electrical resistance can be calculated by

$$R_Z = \rho \cdot \frac{l}{A} \tag{11.2}$$

Considering a copper input lead and a copper output lead with a cross section of 0.5 mm^2 and a length of 5 mm each, the specific electrical resistance of copper $\rho_{Cu} = 1.69\,\mu\Omega$ cm yields a total resistance of 0.34 mΩ. For a mean current of 50 A, the power loss

$$P_Z = R_Z \cdot I^2 \tag{11.3}$$

dissipated in these leads amounts to approximately 0.85 W. Since the contact leads are cooled only marginally, they are heated up by the ohmic losses to temperatures, which can get close to the melting temperature of the solder alloy applied for the PCB solder contact [Swa00]. This effect damages the solder contacts and reduces the reliability.

Since the through holes for PCB mounting are standardized and since insulation requirements demand to maintain minimal clearance distances between the leads, the lead cross section cannot be increased by simply implementing wider leads (Fig. 11.7). But it was possible to increase the cross section by improving the shape of the leads as shown in the right schematics in Fig. 11.7 and thus enhance the current capability of the TO package by 16%. This upgraded version of a TO-247 package is labeled as "super-247" package by the manufacturer.

Another weakness of the TO package is the use of aluminum wire bonds. Improvements of this weakness are attempted by the implementation of thicker wires and/or by an increase of the number of wire bonds. In this context not only the ohmic resistance, but also the inductive effects of the bond wires have to be taken into account.

Figure 11.8 shows a package development, designed for "surface mounted device" (SMD) technology. This architecture is also suitable for multilayer PCBs. For the "super" version of this package, not only the contact leads are designed as short as possible – allowing more area to package larger silicone devices – but also the wire bond connections were optimized. This improvement results in a reduction of the parasitic inductance of the package of 33% according to [Swa00].

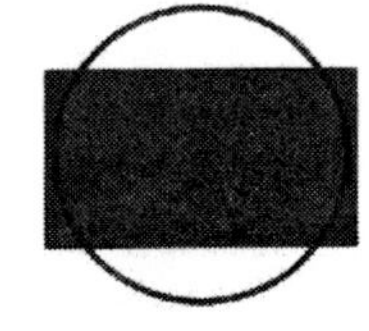

Fig. 11.7 Reduction of the electrical resistance of contact leads in a TO package [Swa00]

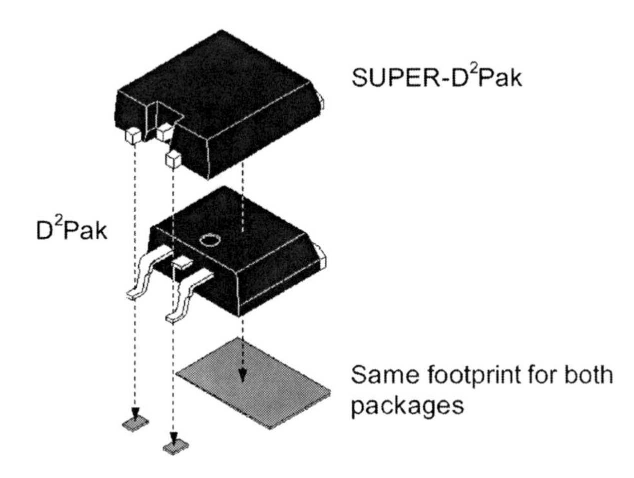

Fig. 11.8 For SMD technology optimized package design [Swa00]

A revolutionary solution was introduced by the US American company International Rectifier, which completely eliminates the problematic contact leads, as well the bond wires. This "DirectFET" package is displayed in Fig. 11.9. The emitter and gate contacts of the silicon device are equipped with a solderable surface metallization. A so-called drain clip is attached to the drain contact of the device by a solder connection. This package is mounted on the PCB surface in a "flip chip" fashion, where the SMD compatible solder connections for the gate, emitter, and drain are established in a single reflow solder step.

Besides the low-effort mounting procedure, the advantages of this package concept stem from the facts that virtually no limitation of the current capability originates from contact leads and that parasitic inductance generated by bond wires is completely eliminated. Furthermore, a double-sided cooling of the package is possible, whereas the drain clip can dissipate significantly more heat than can be extracted through the PCB.

Nevertheless, no complete encapsulation is provided by this package, which leaves the sensitive silicon device unprotected against humidity and corrosive atmosphere influence. Furthermore, the visual access to the solder interconnection

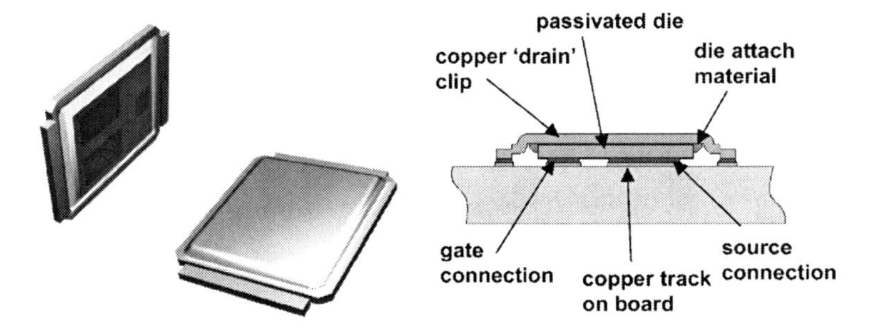

Fig. 11.9 DirectFET package [Swa01]

Fig. 11.10 Transfer mold DIP-IPM package from Mitsubishi

beneath the package is nearly impossible, which impedes the quality control of the assembled PCB. The application and field experience with this new package design will show, if this concept will prevail.

However, the combination of the "lead frame" construction with the "transfer mold" technology, as was developed and optimized with the discrete TO packages, has led to a powerful group of descendants: the transfer mold "intelligent power module" (IPM) packages. In these packages, the advantages of both technologies were merged with the integration of various functions in a single package. Figure 11.10 shows an example of an IPM transfer mold package, containing a three-phase inverter together with their driver ICs.

The internal structure of such a transfer mold IPM device as shown in Fig. 11.11 allows to comprehend the high potential of this packaging concept, which today is

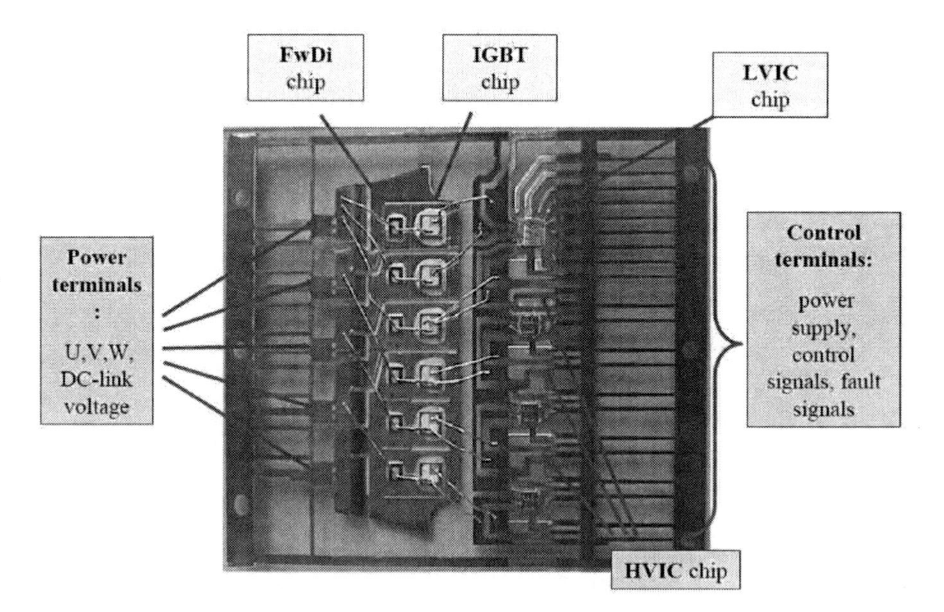

Fig. 11.11 Internal structure of the DIP-IPM package from Mitsubishi

dominating the field of low-power IPMs worldwide. Despite all the weaknesses of this package design discussed before, the manufacturing process of these lead frame packages is highly optimized and very competitive. In production, the lead frames are connected to each other by the frame elements, forming a continuous band of such lead frames, ideal for automated assembly. After the chip soldering and wire bonding, the band is separated into single lead frames and subjected to a transfer mold process, which completely encapsulates the internal structure. Now the leads are fixed by the plastic encapsulation and the remaining supporting lead connections, which connect the ductile contact leads during the assembly, are stamped out.

Nowadays, more than 10 million of these transfer mold-type IPM packages are produced every month, dominating the field of low-power applications in the power semiconductor market.

11.2.3 Modules

As a result of the isolated construction, power modules provided substantial advantages in application. Soon after the first insulated power module was introduced by Semikron in 1975, this new architecture penetrated the market, even though the first design was rather complicated with a multitude of interfaces. Figure 11.12 shows the successor of this first power module, which today is still manufactured in large quantities. Shown here is the fifth generation design with an identical package outline as the first power module, but with an upgraded inner construction.

The thyristor chip, which is equipped with solderable metallization on the anode, cathode, and gate contact areas, is connected to the contact leads by solder interfaces. The cathode connector consists of a composite material with a coefficient of

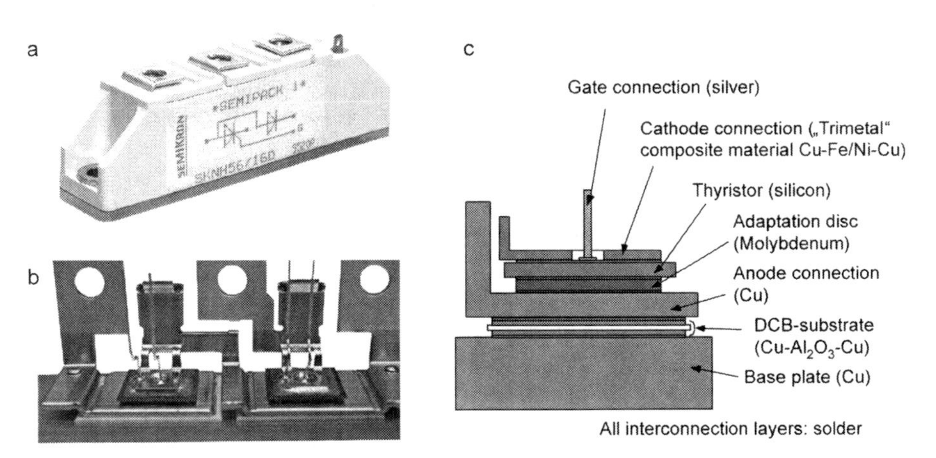

Fig. 11.12 Architecture of a classical thyristor power module: exterior view (**a**), inner construction (**b**), and cross section schematics showing layer sequence (**c**)

thermal expansion adapted to that of silicon. The anode contact of the silicon chip is joined to a molybdenum disc. This intermediate layer is required to accommodate the difference in thermal expansion between silicon and copper. The molybdenum disc is then soldered to a compact copper terminal, which conducts the current to the anode. The copper terminal is again soldered to the copper surface of a ceramic "direct bonded copper" (DBC) substrate that provides the electrical insulation. The substrate is attached to the base plate by another solder layer. In summary, the construction contains five solder layers. Despite the complexity of this construction, this power module is manufactured these days in a high production quantity on an automated assembly line.

The cross section image in Fig. 11.12 visualizes the numerous interfaces that the heat flow has to overcome on his passage from the silicon device to the base plate and further into the heat sink, which is not shown in the schematics. Since every solder layer has a small risk of the formation of solder voids, the multitude of solder interfaces represent an increased risk factor for potential sources of error.

The introduction of advanced power devices like IGBTs or MOSFETs has induced the development of a package concept capable of housing multiple chips per electrical function in parallel. This architecture has emerged to the "standard" or "classical" module design in power electronics. The example displayed in Fig. 11.13

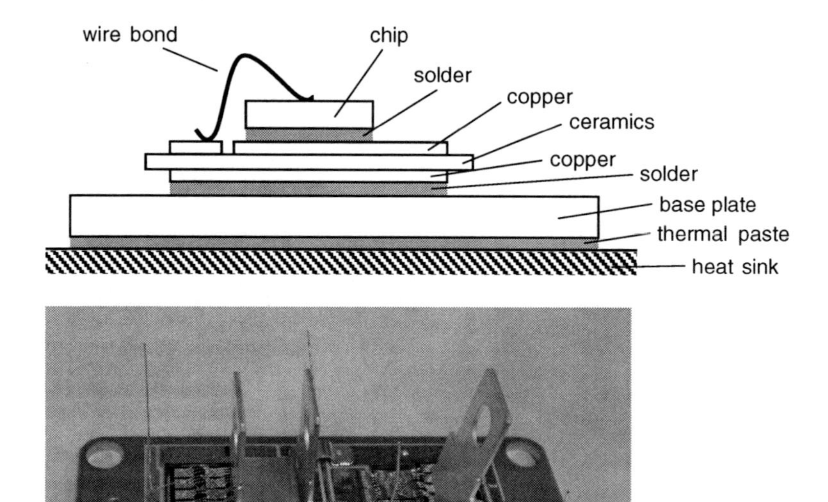

Fig. 11.13 Classical base plate module in schematic cross section (*top*) and as half-bridge IGBT module with two chips per switch (*bottom*)

Table 11.1 Layer thickness in module designs with base plate

	Standard module Al_2O_3 ceramic Cu base plate, *d* (mm)		High-power module AlN ceramic Cu base plate, *d* (mm)		High-power module AlN ceramic AlSiC base plate, *d* (mm)	
Solder	0.05		0.05		0.05	
Copper	0.3		0.3		0.3	
Ceramics	Al_2O_3	0.381	AlN	0.635	AlN	1.0
		0.635		1.0		
Copper	0.3		0.3		0.3	
Solder	0.1		0.1		0.1	
	0.07		0.2			
Base plate	Cu	3	Cu	5	AlSiC	5
Thermal paste	0.05		0.04		0.04	

illustrates the general features of this concept: The top side contact of the silicon device is connected via aluminum wire bonds. The molybdenum adaptation disc and the bottom side copper terminal are completely eliminated. Trenches to form current tracks comparable to the familiar PCB in low-power electronics structure the upper copper layer of the DBC substrate. Several power chips are directly soldered to these copper tracks and connected to other tracks by aluminum wire bonds. Powerful load current terminals are soldered to the load current tracks of the substrate.

Table 11.1 lists the layer thicknesses typically implemented in standard modules displayed in Fig. 11.13. This generic construction is found in 70–80% of all power modules produced by European manufacturers (Infineon, Semikron, IXYS, Danfoss, Dynex) and is also common in modules produced by Asian manufacturers.

The thickness of the ceramic layer is 0.63 mm in older generation modules; newer generation modules with base plate have a ceramic thickness of only 0.38 mm for improvement of thermal resistance. The substrate is attached to the base plate by a solder interface. Differences in solder thickness between 0.07 and 0.1 mm have a marginal impact on the thermal resistance.

For modules requiring a higher thermal conductivity or higher insulation strength, Al_2O_3 ceramics are replaced by AlN ceramics. The standard thickness of AlN is 0.63 mm, but for assemblies with extreme requirements with respect to insulation strength, ceramics with a thickness of 1 mm are applied. The fabrication of AlN substrates requires an additional process step compared to Al_2O_3 substrates, since no oxides are available at the surface to form an oxide–oxide interface as in the DBC production. Therefore, an oxide layer has to be generated first or other bonding techniques have to be applied. This increases the cost for AlN substrates. Furthermore, the coefficient of thermal expansion of AlN (and thus also the CTE of an AlN-DBC) is smaller than that of Al_2O_3. This amplifies the difference in CTE between the substrate and a copper base plate and reduces the lifetime of this interface under thermal stress. For this reason, the copper base plate was replaced in some

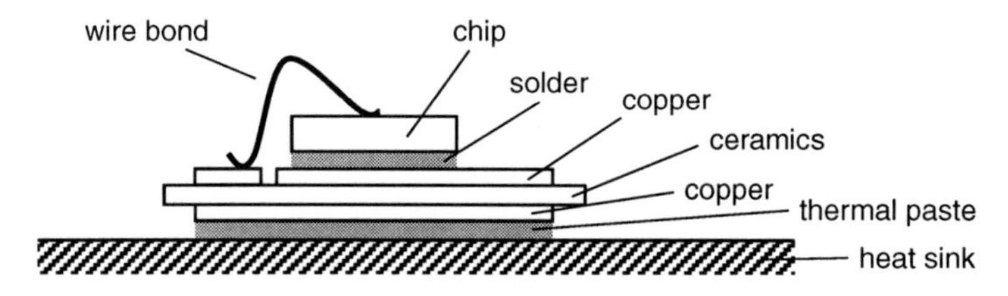

Fig. 11.14 Schematic cross section of a module without base plate

high-performance power modules by AlSiC, a metal matrix composite material. An AlSiC plate is manufactured by first forming a matrix of SiC with a controlled porosity and second by filling the pores with aluminum. The material parameters are determined by the ratio of both components and can therefore be tailored to the application.

While the implementation of AlSiC as material for the base plate has the advantage of an adaptable thermal expansion, it has the disadvantage of a reduced thermal conductivity compared to copper. This was the major reason that a module design concept without a base plate emerged. Even for copper as base plate material the base plate adds to the total thermal resistance in the vertical direction between the chip and the heat sink, so that systems without base plate should be advantageous. Figure 11.14 shows a schematic cross section of this architecture.

The applied substrates and solder materials are essentially the same as in modules with base plate. A value of 0.38 mm is the standard ceramic thickness for Al_2O_3 substrates, but thicknesses of 0.5 or 0.63 mm are also encountered. Especially for substrates with increased copper thickness for high-current applications, thicker ceramics are preferred to enhance the mechanical robustness of the substrate (e.g., 0.4 mm Cu on both sides of a 0.5 mm Al_2O_3 ceramics). For AlN substrates 0.63 mm is the standard ceramics thickness. Other ceramic materials can easily be substituted in a module without base plate.

Modules without base plate are provided, for example, in the SKiiP-, Mini-SKiiP-, and Semitop module families from Semikron and in the EasyPIM series from Infineon. The packaging concept was available in several modules provided by IXYS for a long time. It is also applied in the insulated version of the TO-247 package. Since the solder interface between a base plate and the substrate is eliminated, only one single solder interface between the chip and the heat sink remains in this package type.

In contrast to base plate modules, no limitation of the substrate size must be obeyed in non base plate constructions. Therefore, complex circuits can be realized on a single substrate. The example of a substrate from a highly integrated module without base plate is shown in Fig. 11.15. It contains a single-phase input rectifier and a three-phase output inverter for a medium power frequency converter. Shunt resistors and a temperature sensor are also integrated.

The load, control, and sensor terminals in this package concept are accomplished by identical springs. With this spring contact technology, a multitude of load and

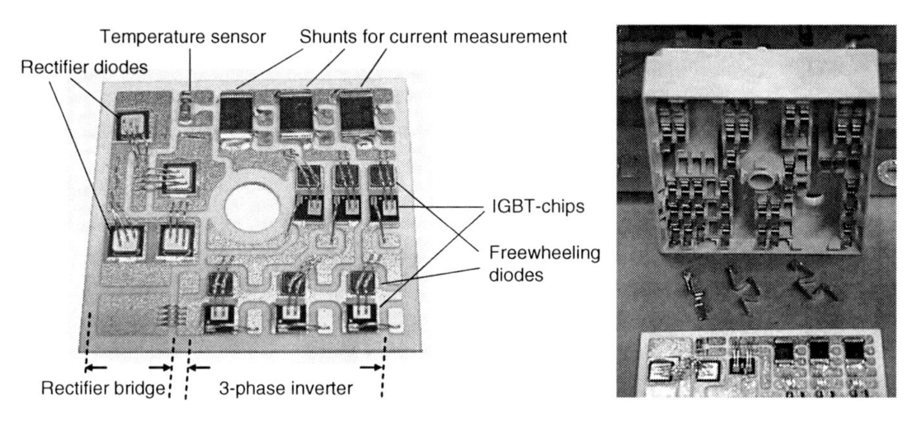

Fig. 11.15 Internal structure of a modern power module with input rectifier, output inverter, and sensors (*left*) and housing with spring contacts (*right*) (MiniSKiiP by Semikron)

control contacts can be positioned at almost any location on the substrate, thus allowing a very flexible contact technology to realize a multitude of different circuits on the same package platform. Each spring can continuously conduct 20 A; higher currents can be attained by paralleling of springs.

Modules without base plate are not limited in footprint size. They require minimum effort in production and also minimize the number interconnections in complex circuits, thus reducing the number of potential sources of failure. On the other hand, sophisticated pressure systems have to be integrated into modules with large footprints to ensure an optimum thermal contact between the substrate and the heat sink. The impact of the capability of this pressure system on the thermal interface between the module case and the heat sink results in different thicknesses of the thermal interface layer shown in Table 11.2. Finally, modules without base plate not only have advantages like the improved thermal gradients inside the module, which reduce the thermal stress and therefore increase the reliability under active thermal cycling [Scn99]. The absence of a base plate also has some drawbacks. First, the

Table 11.2 Layer thickness in module designs without base plate

	Al_2O_3 substrate, *d* (mm)		AlN substrate, *d* (mm)	
Solder	0.05		0.05	
Copper	0.3		0.3	
	0.4			
Ceramics	Al_2O_3	0.381	AlN	0.635
		0.5		
		0.635		
Copper	0.3		0.3	
	0.4			
Thermal paste	0.02–0.08		0.02–0.04	

thermal spreading of the base plate no longer helps to reduce the temperature distribution across the chip. Therefore, smaller chip sizes are preferred in non base plate designs. The second drawback is the missing heat capacity of the base plate, which increases the thermal impedance of the module in a range between 50 and 500 ms for isolated overload events in this time range.

A common problem for all module designs is the interface between the module case and the heat sink surface. Due to the geometric tolerances of the contact surfaces, no perfect metal-to-metal contact can be achieved. The gaps have to be filled with a "thermal interface material" (TIM), which typically has a specific thermal conductivity in the range of 1 $Wm^{-1}K^{-1}$. Even though this conductivity is a factor 30 better than air, the conductivity is more than a factor of 100 worse than that of most metal layers. Therefore, the thickness of the thermal grease has to be kept as small as possible without the risk of air gaps.

The application of the optimum thermal grease thickness during the mounting process on a heat sink is a serious quality issue for many users of power modules. Semikron therefore delivers the SKiiP module family already mounted on the customer heat sink with a controlled thickness of thermal grease. For modules with base plate, the thermal resistance of the interface between module case and heat sink is specified by a typical value in the data sheet $R_{th(c-h)}$, which amounts to roughly 50% of the internal thermal resistance from chip to case. While this interface is difficult to establish in a controlled process, it is of greatest importance for the thermal characteristic of the power module in application.

11.3 Physical Properties of Materials

The properties of the materials used in a package design are fundamental for the characteristics of the module. The most important parameters are the thermal conductivity and the coefficient of thermal expansion of a material, but the electrical conductivity and the heat capacity are also of great interest. It is therefore inevitable to know and consider the properties of the materials prior to their implementation into a power module package.

A survey of the thermal conductivity of the most important materials in power electronic packaging is shown in Fig. 11.16. The best of the ceramic materials used for insulation features thermal conductivities in the range of metals. Beryllium oxide, which exhibits the highest thermal conductivity, had been used in power module designs in the early days of module history. Nowadays, this material is implemented no more due to the toxicity of BeO dust and the resulting threats and limitations in handling and disposal of this material. Second in line of the ceramic insulators in this survey is AlN. But substrates with AlN are a couple of times more expensive than standard Al_2O_3 substrates, so that this material is implemented only when high power density requirements make it inevitable. Organic insulators like epoxy or polyimide (Kapton®) only provide a comparable low specific thermal conductivity.

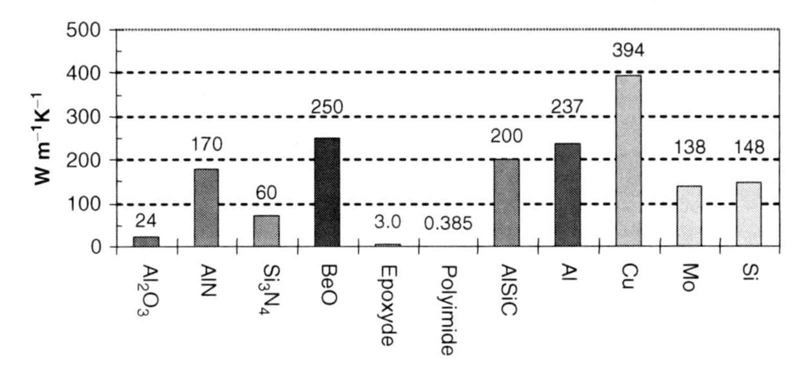

Fig. 11.16 Thermal conductivity of different materials frequently used in packaging technologies

Inherent to the performance of a power module in application are varying load conditions, which generate temperature swings. Differences in the thermal expansion of different materials stress the package. To minimize the stress induced by the thermal expansion between different adjacent layers, their coefficients of thermal expansion (CTE) should be comparable (or more precisely in the presence of thermal gradients inside a stack of layers: the difference of the product of layer temperature and the CTE in adjacent layers should be as small as possible).

Figure 11.17 illustrates the fact that the CTE of Si and Cu are quite different. It is therefore very unfavorable to connect both materials directly, as is the case in standard TO packages (Fig. 11.5). The CTE of ceramic materials is better adapted for an interconnection with Si, especially that of AlN. However, the combination of Cu/AlN/Cu as used in AlN DBC substrates shows an even greater mismatch in combination with a solid Cu-base plate, shifting the stress to this interface. Therefore, AlSiC is implemented in high-performance power modules as base plate material. The ratio of the two components of this metal matrix compound allows to adjust the

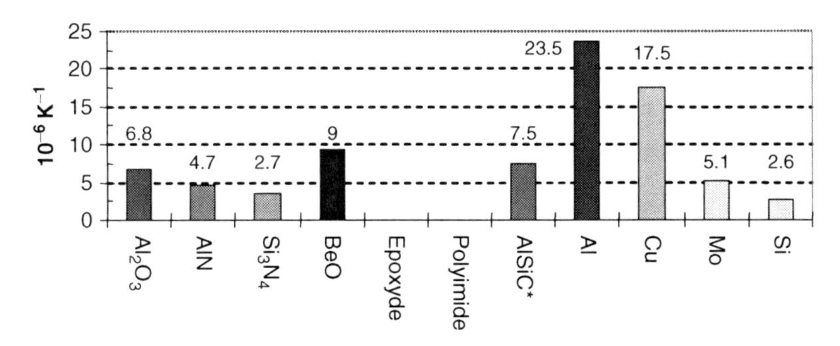

Fig. 11.17 Coefficient of thermal expansion (CTE) of materials frequently used in package technologies. (*) Depending on the composition of compound

Table 11.3 Standard layer thickness of insulators and emanating properties

Material	Standard Thickness (μm)	Heat transfer coefficient (WK^{-1} cm^{-2})	Capacity per unit area (pF cm^{-2})	Breakdown voltage (kV)
Al_2O_3	381	6.3	22.8	5.7
AlN	635	28.3	12.5	12.7
Si_3N_4	635	11.0	12.8	8.9
BeO	635	39.4	11.8	6.4
Epoxy	120	2.5	52.4	7.2
Polyimide	25	1.5	138.1	7.3

CTE of the material to an optimal value for AlN substrates. On the other hand, a considerably reduced thermal conductivity is the consequence as shown in Fig. 11.16. Al_2O_3 as the prevailing ceramic material for power (DBC) substrates is from the thermal expansion point of view the best compromise to attach to silicon on one side and to copper on the other.

The organic insulation materials epoxy and polyimide (Kapton®) have a wide elastic deformation range, so that the coefficient of thermal expansion is not of interest and therefore omitted in Fig. 11.17. Otherwise, these organic insulators are characterized by a much higher breakdown voltage (refer to Appendices C and D) and thus can be implemented in very thin layers. Table 11.3 gives a compendium of standard material parameters and standard thicknesses, which are established in the packaging technologies. The comparison shows that a polyimide layer has more than 10 times smaller thickness compared to ceramic insulators for an equivalent breakdown voltage.

Despite the small layer thickness, substrates on the bases of organic insulators exhibit a smaller thermal conductivity than ceramic substrates [Jor09]. Additionally, the small layer thickness provokes a high electrical capacity, which as parasitic capacity interacts detrimental with the power circuit.

The comparison of all properties of the insulation materials delivers that AlN is technically the best choice as insulating material for power semiconductor packages, if BeO is abandoned due to its toxic characteristics. AlN possesses the highest thermal conductivity and it is indispensable by virtue of its high breakdown voltage for modules with a blocking capability > 3 kV. However, AlN exhibits due to its brittle structure an increased risk of fracture and thus inflicts a greater challenge for the industrial production of modules.

11.4 Thermal Simulation and Thermal Equivalent Circuits

11.4.1 Transformation Between Thermo-dynamical and Electrical Parameters

The differential equations describing the physical process of one-dimensional heat conduction have the same form as the set of equations characterizing the one-dimensional electrical conduction. By exchanging the corresponding parameters,

Table 11.4 Equivalent electrical and thermal parameters

Electrical parameter	Thermal parameter
Voltage V (V)	Temperature difference ΔT (K)
Current I (A)	Heat flux P (W)
Charge Q (C)	Thermal energy Q_{th} (J)
Resistance R (Ω)	Thermal resistance R_{th} (K/W)
Capacity C (F)	Thermal capacity C_{th} (J/K)

a thermal problem can therefore be transformed into an electrical problem and vice versa. Due to the equivalence of the differential equations, all operations performed for electrical networks can be transferred to thermal networks, especially the approximation of a continuous conduction line by a set of discrete elements in a lumped network. Since a variety of tools is available today for the simulation of electrical networks, thermal problems can be calculated by solving the equivalent electrical circuit.

The standard procedure is to first transform the thermal parameters into the corresponding or analog electrical parameters. Then the corresponding equivalent network can be solved by applying advanced electrical network simulation tools. Finally, the results are transformed back into the thermal parameters. Table 11.4 gives a list of the fundamental corresponding parameters [Lap91].

From these fundamental parameters, other corresponding parameters can be derived. The electrical time constant as the product of resistance and capacity, for example, has its correspondence in the thermal time constant, defined as the product of thermal resistance and thermal capacity.

While, at the first glance, the correspondence between electrical and thermal parameters seems to be perfectly symmetrical, there is a difference which destroys that perfect symmetry. This difference is the explicit appearance of the temperature in the thermal equations. To examine this difference closer, let us consider the definition of the thermal resistance R_{th} between the geometrical locations a and b:

$$R_{th(a-b)} = \frac{T_a - T_b}{P_v} = \frac{\Delta T}{P_v} \tag{11.4}$$

In the electrical theory, Ohm's law postulates that the ohmic resistance is constant and therefore independent of the voltage, if the boundary condition of a constant temperature is fulfilled. This boundary condition reflects the fact that material properties are generally dependent on temperature. But since the temperature is an explicit parameter in the definition of the thermal resistance, a correspondence to Ohm's law in the thermal theory with the boundary condition $T = \text{const.}$ is not reasonable. This means that the thermal resistance is always temperature dependent [Scn06].

The temperature dependence of the specific thermal conductivity of silicon, aluminum, and copper is shown in Fig. 11.18 according to [EFU99]. Following [Poe04], the temperature characteristic of silicon can be approximated between −75 and +325°C by the expression

$$\lambda = 24 + 1.87 \times 10^6 \cdot T^{-1.69}\,\text{Wm}^{-1}\text{K}^{-1} \tag{11.5}$$

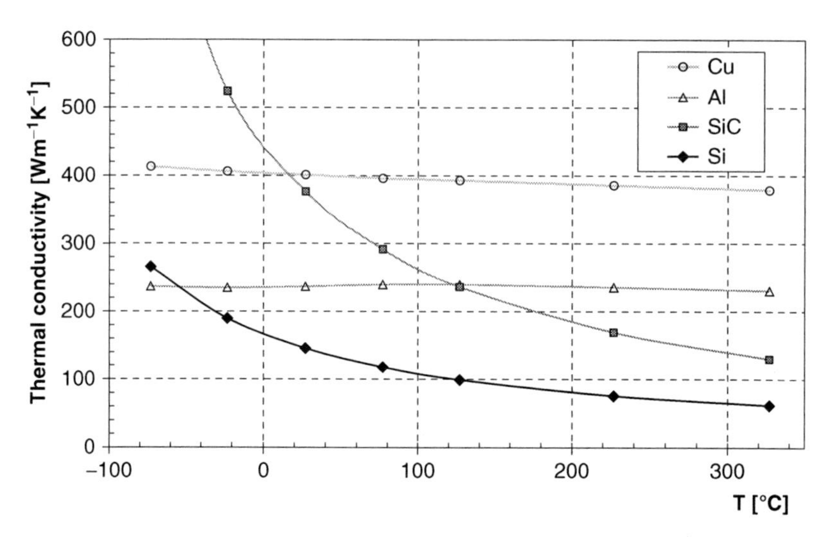

Fig. 11.18 Temperature dependence of the thermal conductivity of Si, SiC, Al and Cu. Data according to [EFU99] and [Fel09], data for Si calculated by Eq. (11.5)

The thermal resistance (11.4) is constant only if λ is temperature independent. For Al and Cu and for most other materials a good approximation between -50 and $+150$°C is obtained. In power electronic systems, the thermal resistance of silicon amounts to only 2–5% of the total resistance, so that a negligence of its temperature dependence results in most cases only in a small error. To eliminate this fundamental problem, a temperature-dependent resistance can be simulated by using a voltage-dependent resistor in the equivalent network, which is possible in most electrical network simulation tools.

Another general problem in thermal simulation is the interpretation of temperatures. Conventional reference points are the ambient temperature T_a, the heat sink temperature T_s, sometimes the case temperature T_c, and the so-called virtual junction temperature T_{vj}. Three-dimensional systems, that are not in a state of thermal equilibrium, exhibit pronounced gradients of temperature in every layer of the system. So a single temperature T_c or T_s is only a rough estimation for a real system, in which the base plate (as the module case) and the heat sink surface are characterized by temperature distributions.

Especially, this holds true for the junction temperature T_j. In the power device where the power is dissipated, the greatest gradients of temperature are present. Thus, it is expedient to postulate a virtual junction temperature T_{vj} as a characteristic temperature of the silicon device. This parameter is defined by the measured voltage drop over a pn-junction for a small sense current, as was already discussed in Sect. 3.2.

The forward voltage drop of a pn-junction at very small current depends strongly on temperature. It is always decreasing with temperature. To use this effect to determine the temperature, the sensing current must be small enough that a

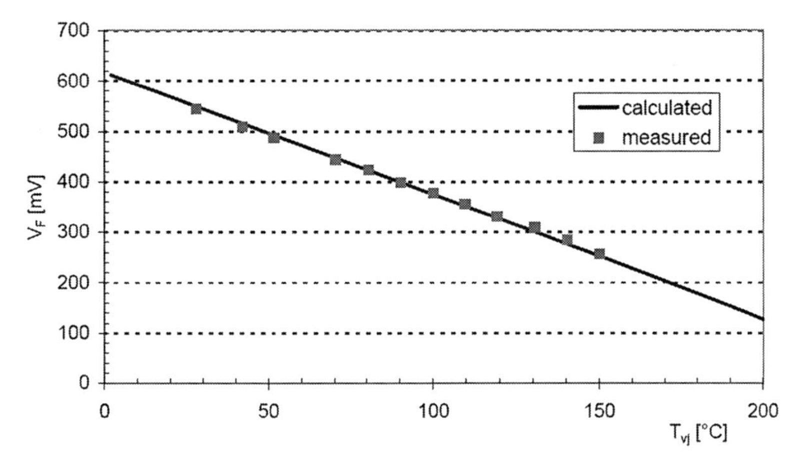

Fig. 11.19 Calibration of the pn-junction of a 50 A 1200 V Si-diode for use of the pn-junction as temperature sensor. Forward voltage drop at the pn-junction of the 50 A diode measured at 50 mA. Calculation according to Eqs. (3.53) and (3.55)

temperature influence of the sense current can be neglected. Typically a current density of 100 mA/cm^2 or lower is selected. Figure 11.19 shows a measurement of the forward voltage drop at the pn-junction of a 50 A diode measured at 50 mA as a function of temperature. After determination of this calibration function, the junction temperature of the device can be measured by applying a sense current of 50 mA and measuring the voltage drop at an instant where the device is exposed neither to a forward current nor to a blocking voltage. Figure 11.19 compares the measured calibration function with the calculation according to Eqs. (3.53) and (3.55), resolved for V. As ideality factor, $n = 1.05$ was used in Eq. (3.55) for this fast recovery diode.

Together with the calibration of this voltage drop at different ambient temperatures, this method delivers a convenient technique to determine the virtual junction temperature of a device without intrusion into the package. This technique works well with diodes and IGBTs. The pn-junction between the gate and the cathode can be used for this method for thyristors. For the MOSFET, the inverse diode can be utilized for the measurement of the virtual junction temperature; in this case the sense current is applied in reverse direction.

As pointed out before, there is no constant temperature on the surface of a real power device in non-equilibrium condition. The edges of the silicon chip have a lower temperature than the center of the chip, because the heat flux can propagate not only vertically toward the heat sink, but it can also spread out away from the chip center, which can be envisaged by the cross-sectional illustrations in Figs. 11.13 and 11.14. This phenomenon is called heat spreading. The exemplary simulation in Fig. 11.20 illustrates the impact of a power dissipation of 200 W, homogeneously generated in the volume of a 12.5 $\times$ 12.5 mm^2 IGBT chip. The calculated temperature distribution reveals a center temperature, which is approximately 20°C higher than the cooler edges of the chip.

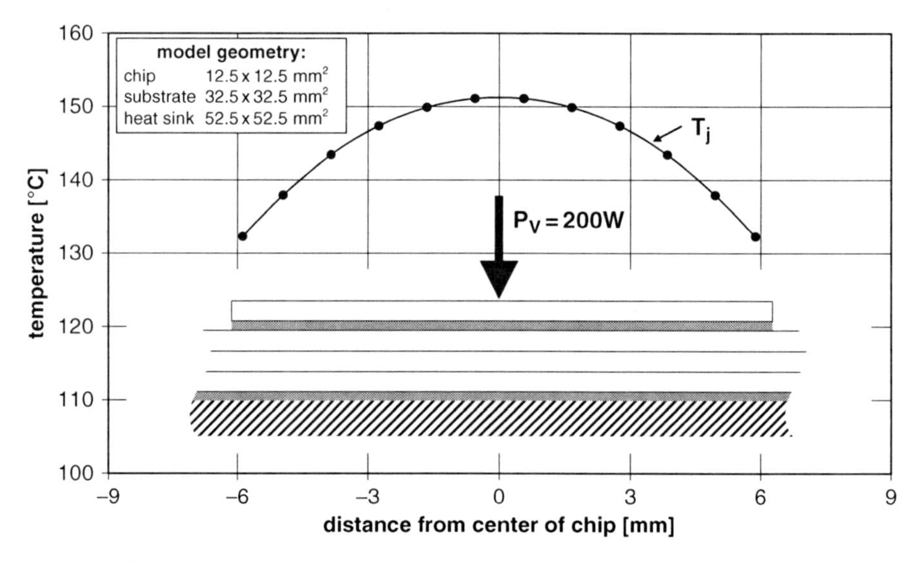

Fig. 11.20 Simulated temperature distribution in a silicon chip with layers according to Fig. 11.14. Illustration from [Scn06]

An experimental validation of the simulated temperature distribution was presented in [Ham98]. The temperature was measured using a potential separated sensor, consisting of a phosphorescent powder at the end of a silica glass rod, which was excited by a laser. The temperature dependence of the phosphorescent radiation was used to measure the temperature at the tip of the silica glass rod. With this potential separated sensor, the surface temperature of an IGBT chip could be measured at different locations. The measured temperatures at the center and at the edge of the chip were related to the virtual junction temperature determined by the voltage drop for a sense current of 100 mA (Fig. 11.21). For a high load current, the temperature difference between center and edge was also found to be in the range of 20°C. The results also show that the temperature T_{vj} is an average value for the real temperature distribution, which is shifted toward the hotter chip center temperature.

The reason for this shift toward the hot chip center temperature is found in the temperature characteristic of the voltage drop across the backside pn-junction of the IGBT (refer Fig. 10.10, junction J_1) for small currents. Although the temperature coefficient of the forward voltage drop of an IGBT is positive in the range of the nominal current, the voltage drop for small currents is only determined by the physical properties of the pn-junction and thus exhibits a negative temperature coefficient. This results in a smaller resistance of the hot chip areas and therefore implies a greater weight of the hot areas in the averaging process by the sense current flow. This is a desirable feature of the temperature measurement because the emphasis on the areas of higher temperature reduces the difference to the maximum temperature in the real chip. But it should be kept in mind that the maximum temperature can

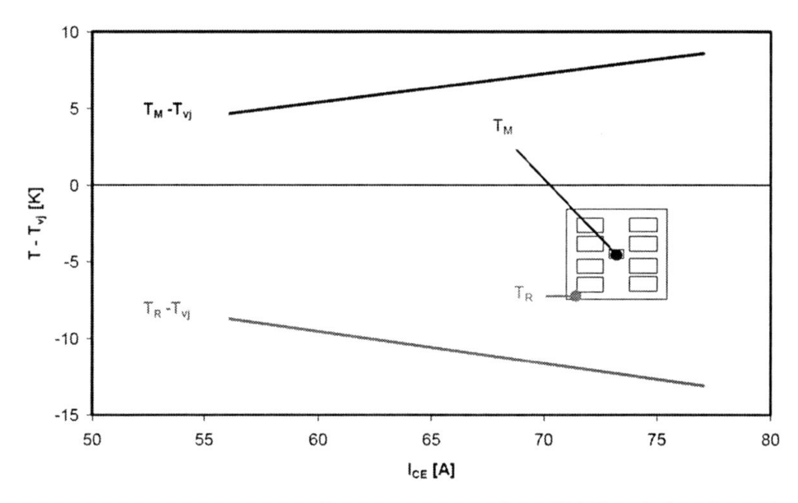

Fig. 11.21 Relation between the surface temperature of an IGBT and the virtual junction temperature T_{vj}, according to [Ham98]

still be considerably higher than the virtual junction temperature for large chips and high load currents.

If the restrictions discussed above are obeyed, this method of measuring the virtual junction temperature seems to be an ideal method, since the power chip itself is used as a sensor and the temperature of the device can be measured down to less than 100 μs after the turn-off of a load current pulse for fast switching devices like IGBTs. An extended investigation of the averaging effect of the sense current in [Scn09] shows that the measured temperature value corresponds to the area-related average value of the temperature.

However, there are two major drawbacks of this measurement technique. The first drawback is the fact that this method cannot directly be used to determine the chip temperature in real switching applications like a PWM operation in a frequency inverter. There is no simple possibility of applying a defined sense current in a real inverter operation.

The second drawback is attributed to the averaging process of this technique. While this effect can be accounted for in the measurement of a single chip, the ambiguity of the measured value is growing for paralleled chips. If one of the parallel chips has a higher thermal resistance – for example, due to a deficient solder quality – its higher temperature will only have a small impact on the average value. The virtual junction temperature is an average parameter, which delivers no information on the distribution of temperatures in case of parallel chips.

The measurement technique for determining the other reference point temperatures T_c and T_s is also not trivial. For the measurement of the case temperature T_c, which is a common reference point for classical modules with base plate, a drilling has to be incorporated into the heat sink exactly in the center of the silicon device generating the power losses as displayed in Fig. 11.22. This measurement therefore requires the knowledge of the exact position of the chips inside the module.

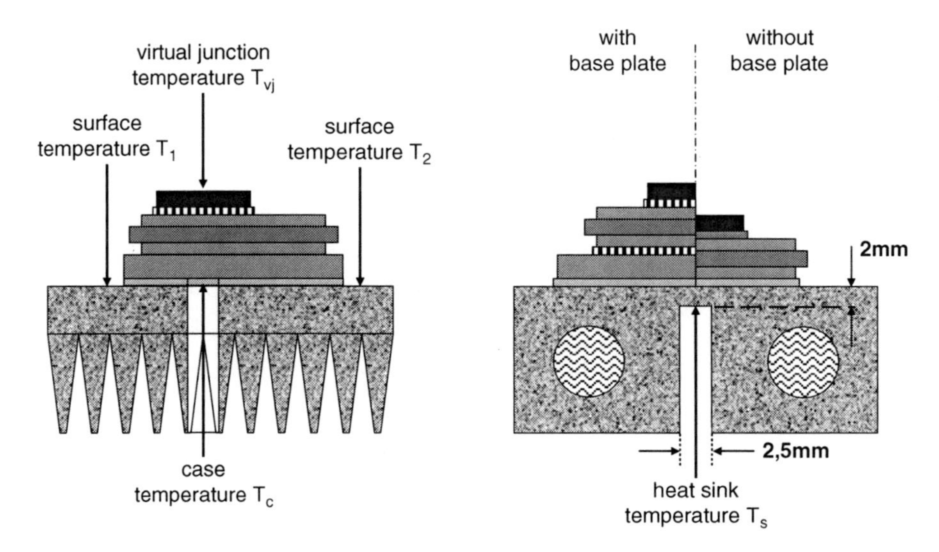

Fig. 11.22 Definition of the case temperature T_c and the heat sink temperature T_s

This drilled hole interferes with the heat flux into the heat sink. However, due to the thermal spreading in the base plate of classical modules, this disturbance results only in a deviation of $\leq 5\%$ from the undisturbed value, as was verified by thermal simulation.

The impact of such a drilled hole for temperature sensing on a module without base plate is much more severe because of the non-existing spreading of a base plate. It was proposed by [Hec01] to replace the through hole in the heat sink by a blind hole, which only reaches up to 2 mm underneath the heat sink surface. This measurement configuration has the advantage that the thermal interface between module and heat sink is integrated in the heat path. This method can be applied for any type of module. The reference temperature defined by this geometry is called heat sink temperature T_s (in older publications often indicated by T_h).

In contrast to the measurement of the virtual junction temperature, the measurement of the case or heat sink temperature is mostly restricted to equilibrium state conditions. The transient response of thermocouples is in the range of 100 ms and more, so they cannot be utilized for the measurement of fast temperature evolutions in power modules, which have a typical time constant for the internal thermal resistance of approximately 1 s.

These considerations shall illustrate that the thermal characterization of a power electronic system is never easy – neither by simulation nor by measurement. A lot of experience and a critical mindset are necessary to select the right model and to interpret simulation results. Temperature measurements have to be carefully reviewed as well. The thermal characterization is one of the most difficult tasks in power electronic systems and only succeeds by combining experimental skill with correctly applied thermal simulation.

11.4.2 One-Dimensional Equivalent Networks

In a one-dimensional equivalent network, the power dissipated in a thermal heat source is represented by a current source. A network of resistors and capacitors represent the thermal resistances $R_{th,i}$ and the thermal capacities $C_{th,i}$ of the analog thermal system. The ground potential is equivalent to the ambient temperature. In the physically correct Cauer model, the thermal capacities are connected from each node of the model to the ground potential. If power losses are generated in the system, the temperature will rise in the nodes and thermal energy is stored in the capacitors. The stored energy is proportional to the temperature difference to the situation before the power losses were applied; therefore, this network correctly describes the physical reality (Fig. 11.23).

In contrast to the Cauer model, the capacitors are connected in parallel to the resistors in the Foster model. It should be noted that the values of the resistors and capacitors are different in both networks. The equivalent Foster model has in total the same transient behavior as the Cauer model with respect to the temperature of the first node next to the power source. While the internal nodes in a Cauer model can be interpreted as geometrical locations in the system, this is not possible for internal nodes of the Foster model. The feature that pairs of *R*s and *C*s in a Foster model can be exchanged without altering the transient response of the whole system might help to remember this important fact. The exchange of pairs of *R*s and *C*s in the Cauer model will on the other hand alter the transient response of the system, as

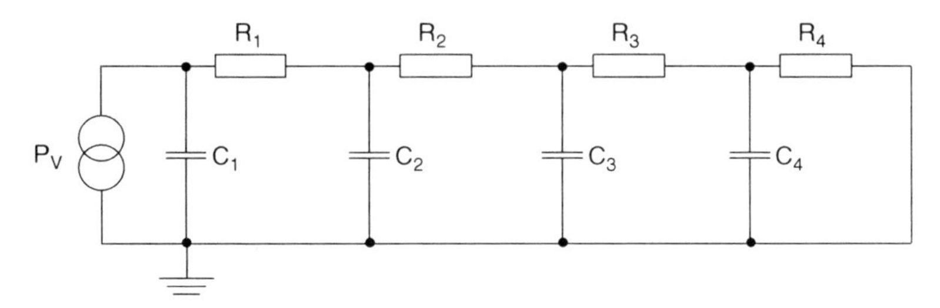

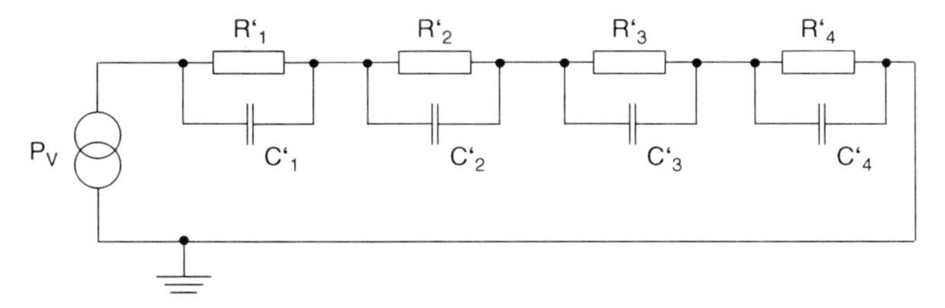

Fig. 11.23 One-dimensional thermal equivalent networks

the exchange of layers in a real system would do. This missing link to the system geometry also implies that the values of the resistors and capacitors in the Foster model cannot be calculated from material constants, as is the case for Cauer models. Finally, a Foster model can neither be divided nor can two Foster models be connected together, while both operations are possible for Cauer models.

Those severe restrictions in the application of Foster models lead to the question, why we use this type of model at all. The answer is that the time-dependent thermal resistance – often referred to as the "thermal impedance" Z_{th} of a system – can be expressed by a simple analytical expression. The model parameters R'_i and C'_i in this expression can be determined from the system response to a step function in power losses. By applying a least square fit algorithm, the parameters in the analytical expression can be optimized until the time response matches the transient system response, for example, measured by a heating or cooling curve.

$$Z_{th} = R_{th}(t) = \sum_{i=1}^{n} R'_i \cdot \left[1 - \exp\left(-\frac{t}{\tau_i}\right)\right] \quad \text{with } \tau_i = R'_i \cdot C'_i \tag{11.6}$$

The values of the R'_i and τ_i are often explicitly listed in data sheets of power packages. They allow a fast calculation of the transient response of a package to complex power profiles for application engineers.

The resistances and capacitors in the Cauer model can be calculated straightforward from the material parameters:

$$R_{th} = \frac{1}{\lambda} \cdot \frac{d}{A} \tag{11.7}$$

$$C_{th} = c \cdot \rho \cdot d \cdot A \tag{11.8}$$

with layer thickness d, cross section area A, specific thermal conductivity λ, specific heat capacity c, and specific density ρ, (Fig. 11.24).

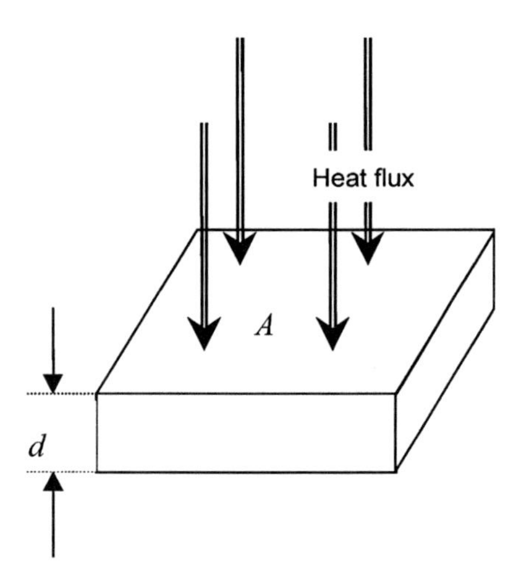

Fig. 11.24 Extraction of R_{th} and C_{th} values from geometry and material constants of a layer

With these extraction rules, a Cauer model of a layer system can be derived, which allows a complete geometrical interpretation and all geometrical operations such as combining or dividing systems.

11.4.3 The Three-Dimensional Thermal Network

The one-dimensional Cauer model is only a rough approximation for the complex geometry of layers in a real power module. The layers typically exhibit different cross sections and the virtual resulting lateral heat spreading cannot be described by a one-dimensional model (refer Sect. 11.2, especially Figs. 11.13 and 11.14). The thermally high-conductive copper layers extend the effective heat conduction area above the layers with a high thermal resistance (ceramic, thermal paste). These three-dimensional features can be accounted for by extending the Cauer model to three dimensions as shown in Fig. 11.25. The nodes are arranged in a three-dimensional lattice; each node is connected to its neighbors via resistors.

The nodes in Fig. 11.25 are located in the center of each cuboid element. The resistors between the nodes are determined by the material parameters along the path, so that across the interface of two adjacent layers the resistors are determined by the material parameters of both layer materials.

Figure 11.25 also exemplifies that the number of elements is increasing fast in the three-dimensional model. Even though this simple system contains only 2 layers with 9 nodes for layer 1 and 25 nodes for layer 2, a total number of 86 resistors are necessary to connect the nodes. Additionally 34 capacitors have to be connected from every node to the ground potential if the transient response of the system is of interest. For a realistic model of a power semiconductor package, several hundred of nodes with more than a thousand elements are adequate. Such complex networks require fast network simulation tools and a suitable pre-processor to generate the input data from a given geometry. However, the simulation results reveal

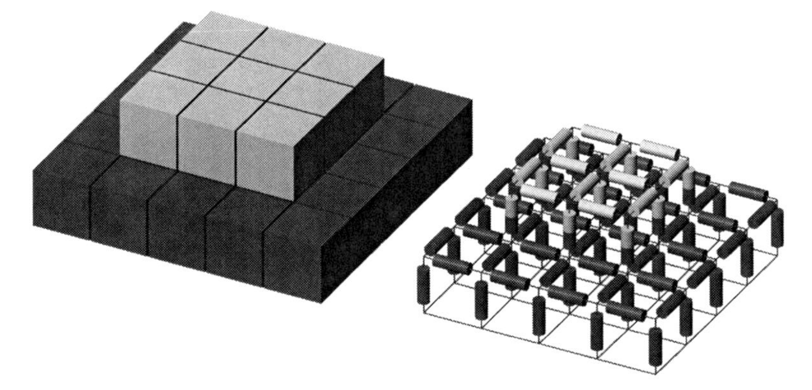

Fig. 11.25 Three-dimensional model for a simple two-layer system showing the lattice of nodes with the interconnecting resistors; diagram from [Scn06]

the temperature evolution in layers, which are not accessible to measurement without considerable interference with the system. Furthermore, the three-dimensional simulation is the only way to calculate the impact of interference by a measurement setup and allows quantifying the offset with respect to an undisturbed system. Finally, fast transient temperature evolutions in all layers other than the silicon device are virtually not accessible by measurement [Hec01] and only simulation models deliver the accurate data necessary for the thermal characterization of a power module package.

11.4.4 The Transient Thermal Resistance

A simulation of the transient thermal resistance or thermal impedance Z_{th} based on a three-dimensional model is shown in Fig. 11.26. Three different power module packages using AlN-substrates are calculated for comparison: two modules with base plate according to Fig. 11.13 with layer dimensions as listed in Table 11.1 and one module design without base plate according to Fig. 11.14 with layer thicknesses as given in Table 11.2.

The thermal impedance is small for a short single pulse below 50 ms and it is independent of the module design, because the heat is almost completely stored in silicon chip and the substrate. For large pulse lengths, the thermal impedance approaches the equilibrium value of the thermal impedance, which is the thermal resistance R_{th}. The comparison of the different designs reveals that the thermal resistance of the module without base plate is moderately smaller than the thermal resistance of the module with a Cu base plate. This is a consequence of the thermal resistance of the base plate in vertical direction. The thermal resistance of the module with AlSiC base plate is considerably higher due to the inferior thermal conductivity of AlSiC. This inferior thermal conductivity also increases the thermal impedance in the intermediate range between 50 and 500 ms compared to the copper

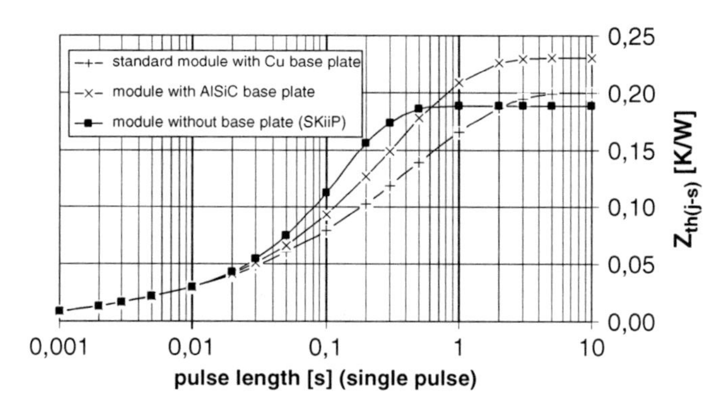

Fig. 11.26 Transient thermal resistance simulated for different module designs based on AlN substrates. Figure from [Scn99]

base plate modules. However, the Z_{th} of the module without base plate features the highest values in this interval. This is a consequence of the missing thermal capacity of the base plate. Therefore, the module design without base plate possesses less buffer capacity for single pulse overload conditions between 50 and 500 ms.

However, the single-pulse response of a system is relevant for a limited number of applications only. A single-pulse event draws maximum advantage out of additional thermal capacity in the system, because there is an infinite time to dissipate the heat after the end of the pulse. The situation is different in case of repetitive pulses. Then there is only limited advantage of the additional thermal capacity, in our case the base plate.

The temperature evolution in a system (described by Eq. (11.6)) for an infinite series of constant power pulses can be calculated analytically. For such a series of constant pulses of the constant power P_{on} during the time t_{on} followed by no power for the time t_{off}, we can calculate the stationary maximum temperature swing:

$$\Delta T_{\text{max, stationary}} = P_{on} \sum_{i=1}^{n} R'_i \frac{1 - \exp\left(-\frac{t_{on}}{\tau_i}\right)}{1 - \exp\left(-\frac{(t_{on} + t_{off})}{\tau_i}\right)} \tag{11.9}$$

If we define the duty cycle as the ratio $D = t_{on} / (t_{on} + t_{off})$, we can discuss the impact of the thermal capacity on the system response in more detail. For very low duty cycles, the single-pulse characteristic as shown in Fig. 11.26 can be used as good approximation. In real applications, this applies to welding applications or some induction heating applications with long pauses between high-current pulses. For motor drive applications, which are still the majority of all power electronic applications, the IGBTs as well as the diodes are under load during one-half wave of the output current and under no load during the next half wave. It shall be simplified in a first approximation as a load duty cycle of 50%. Then the thermal impedance of the three systems of Fig. 11.26 are compared as shown in Fig. 11.27. Here we see that the system without base plate is still inferior to the system with a copper base plate, but it is superior to the AlSiC base plate design in the whole frequency range. Due to the much smaller thermal conductivity of AlSiC, the heat stored in the base plate cannot be dissipated fast enough into the heat sink to give an advantage for the 50% duty cycle.

Equation (11.9) is also very useful to determine the ripple amplitude in equilibrium state. The difference between the maximum temperature swing defined by this equation and the temperature value obtained by multiplying the stationary thermal resistance with the average power dissipated $P_{av} = P_{on} \cdot D$ delivers half of the amplitude of the temperature ripple in steady-state condition. Figure 11.28 illustrates these dependencies.

An application example shall be given. We consider a dissipated power of $P_{av} = 300\,\text{W}$ in an application with an output power frequency of 5 Hz and a duty cycle of 50%. This is, for example, a slowly rotating electric motor fed by a variable

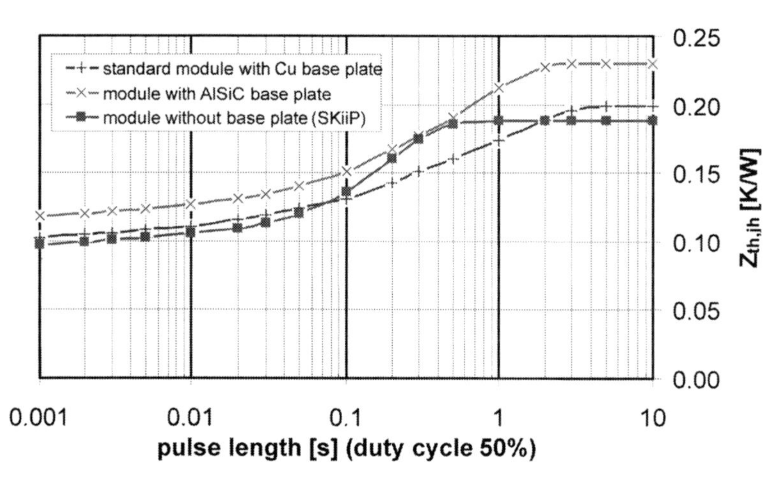

Fig. 11.27 Transient thermal resistance simulated for different module designs based on AlN substrates for a duty cycle of 50%

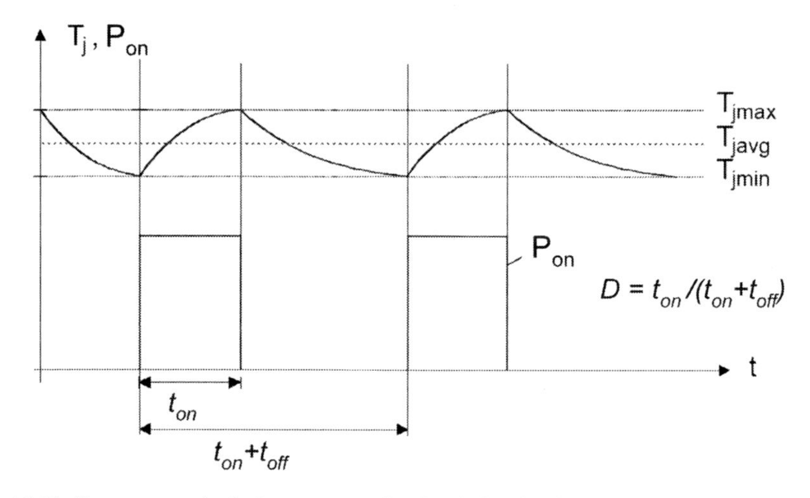

Fig. 11.28 Temperature ripple for a power pulse P_{on} during the time t_{on}

speed drive with IGBTs. For the module with Cu base plate we get from Eq. (11.4) a temperature increase $\Delta T_{\text{jav}} = 60\,\text{K}$ using the thermal resistance of 0.2 K/W as follows from Figs. 11.26 and 11.27 for the stationary case. During the pulse length of 100 ms we have to calculate

$$\Delta T_{\text{jmax}} = Z_{\text{thjc}} \cdot P_{\text{on}} \tag{11.10}$$

with $Z_{\text{thjc}} = 0.13\,\text{K/W}$ from Fig. 11.27 and $P_{\text{on}} = 600\,\text{W}$ we get $\Delta T_{\text{jmax}} = 78\,\text{K}$. The difference of ΔT_{jmax} and ΔT_{av} is half of the temperature ripple, the temperature ripple amounts to 36 K. For the module without base plate we get, using

0.19 K/W, as result $\Delta T_{jav} = 57\,K$; and with $Z_{thjc} = 0.135\,K/W$ from Fig. 11.27 we get $\Delta T_{jmax} = 81$ K, the temperature ripple amounts to 48 K. For the AlSiC base plate results $\Delta T_{jav} = 69\,K$ and $\Delta T_{jmax} = 90\,K$, the temperature ripple amounts to 42 K.

The Cu-base plate module has a high-thermal mismatch between Cu and AlN; therefore, in the viewpoint of high reliability only the other systems shall be considered. In the module without base plate we have a higher temperature ripple; in the AlSiC base plate module we have a lower temperature ripple, however, at a significant higher temperature. Maximal temperature affects reliability as well as temperature ripple. For details on lifetime estimation refer to Sect. 11.6.

The used simplification of constant losses during t_{on} is not realistic in motor drive applications, in fact the on-state losses have a term proportional to $\sin^2(\omega t)$, the switching losses are proportional to $\sin(\omega t)$. The result of the comparison, however, will be similar.

11.5 Parasitic Electrical Elements in Power Modules

Every power module contains parasitic resistances and parasitic inductances caused by internal conduction tracks, as well as parasitic capacities provoked by parallel conductors separated by dielectric layers. Their influence is not negligible, especially during fast switching operation.

11.5.1 Parasitic Resistances

The significant contribution of external and internal connectors to the total voltage drop in discrete packages was already addressed in Sect. 11.2. Figure 11.29 illustrates the evolution of package designs produced by International Rectifier (IR). Table 11.5 lists the characteristic parameters of these package types.

The substitution of wire bonds by a copper strap in the transition from the SO-8 package to the Copperstrap design reduces the parasitic resistance and the

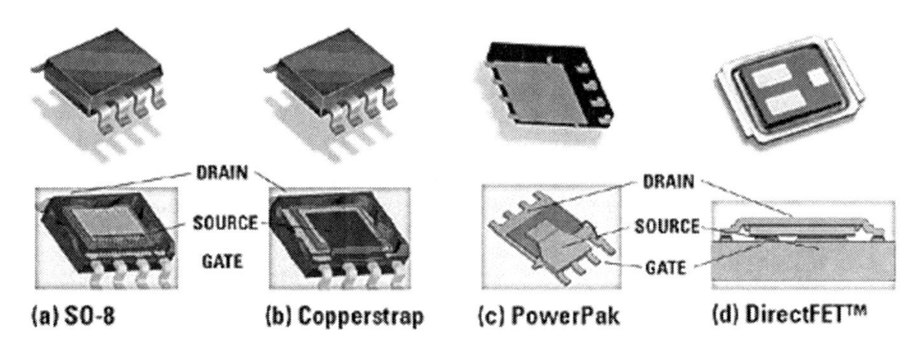

Fig. 11.29 Optimization of discrete package architecture concerning the reduction of parasitic resistance and induction, as well as the improvement of the thermal resistance, according to [Zhg04]

Table 11.5 Characteristic parameters of the package designs shown in Fig. 11.29 according to [Zhg04]

Package type	Electrical parasitic resistance (mΩ)	Parasitic inductance (nH)	R_{th} junction-to-PCB (K/W)	R_{th} junction-to-case (top surface) (K/W)
SO-8	1.6	1.5	11	18
Copperstrap	1	0.8	10	15
PowerPak	0.8	0.8	3	10
DirectFET	0.15	<0.1	1	1.4

parasitic inductance. Since the progress in chip technology succeeded in reducing the on-state resistance to approximately 1 mΩ for a 40 V MOSFET, this package improvement was mandatory. The progress toward the Power-Pak housing is marked by a substantial improvement of the thermal resistance by implementing a solid copper base, which supplies an effective thermal path and at the same time represents the electrical drain contact. The ultimate progress was the development of the DirectFet package, which reduces the parasitic effects and the thermal resistance to a minimum.

The parasitic resistance in power modules is also considerable. For advanced power modules, the manufacturers often explicitly specify the parasitic resistance induced by the package in data sheets. Infineon indicates a value of 0.12 mΩ for the high-performance power module FZ3600R12KE3, which is a 1200 V IGBT module with a rated current of 3600 A. Thus, the parasitic resistance implies an additional voltage drop of 0.43 V at the nominal current of 3600 A. The on-state voltage drop of the IGBT has a typical value of 1.7 V. Therefore, the package provokes roughly 20% of the total voltage drop. Other high-current modules feature comparable values.

If the 36 IGBT chips rated 100 A each in the Infineon power module would be replaced by 75 V 100 A MOSFET chips with an on-state resistance $R_{DS, on}$ of 4.9 mΩ, the voltage drop of the package would be in the same range as the voltage drop of the MOSFETs. Additionally, it is very difficult to design the current leads for all chips completely symmetrical, i.e. to arrange the chips in a pattern that results in an equal resistance in the lead to each chip. The smaller the on-state voltage of the power chip, the more pronounced is the impact of unbalance in parasitic electrical resistance.

This example shows that it is an important goal to reduce the parasitic resistance in power modules, especially for high-current and low-voltage modules. However, the internal parasitic inductance has an even greater impact on the performance of the package.

11.5.2 Parasitic Inductance

Every current lead is associated with a parasitic inductance. For the estimation of the magnitude of inductance, a rule of thumb applies for the inductance of current tracks:

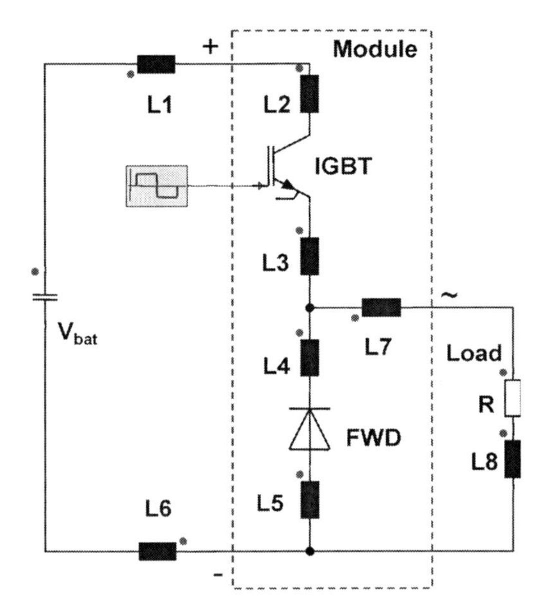

Fig. 11.30 Parasitic inductances in the commutation circuit

$$L_{\text{par}} \approx 10\,\text{nH/cm} \tag{11.10}$$

The inductance can be reduced by parallel arrangement of the plus and minus tracks; this technique is applied frequently in modern power modules. The module inductance is typically in the range of 50 nH for classical module designs. For more advanced packages, this value reduced to 10–20 nH. These parasitic inductances affect the commutation circuit as shown in Fig. 11.30.

L_1 and L_6 represent the inductances of the DC-link capacitors and of the current tracks to the DC link.

L_2 is the inductance formed by the plus terminal and the current track on the substrate to the collector of the IGBT soldered to the substrate.

L_3 is composed by the bond wires on the emitter of the IGBT and by the current tracks on the substrate to the AC terminal.

L_4 is synthesized by the current tracks from the AC terminal to the cathode contact of the freewheeling diode soldered to the substrate.

L_5 consists of the bond wires on the anode contact of the freewheeling diode and the track to the minus terminal, including the terminal itself.

L_8 represents the inductance of the load, which acts as the major current source during commutation. This inductance, as well as the series inductance of the AC terminal (L_7), is not affecting in the commutation circuit. The effective internal parasitic inductances are all connected in series, so that they can be merged into a single-module inductance L_{pm}:

$$L_{\text{pm}} = L_2 + L_3 + L_4 + L_5 \tag{11.11}$$

The total parasitic inductance L_{par} is then given by the series connection of the module inductance with the DC-link inductances:

$$L_{\mathrm{par}} = L_{\mathrm{pm}} + L_1 + L_6 \tag{11.12}$$

The impact of this parasitic inductance on the dynamic properties of a power module will be illustrated by two examples. In the first example, we will consider a frequency converter for a three-phase motor drive. This converter is formed by three IGBT half-bridge modules of the voltage class 1200 V with a nominal current of 800 A each.

The maximum DC-link voltage V_{DC} is 800 V and the parasitic inductance of each phase leg L_{par} is assumed to be 20 nH. The rate of current rise $\mathrm{d}i_{\mathrm{r}}/\mathrm{d}t$ shall be 5000 A/μs. The voltage characteristic during commutation under these assumptions can be calculated by Eq. (5.72):

$$V(t) = -V_{\mathrm{DC}} - L_{\mathrm{par}} \cdot \frac{\mathrm{d}i_{\mathrm{r}}}{\mathrm{d}t} + V_{\mathrm{tr}}(t)$$

Evaluation of this equation delivers an over-voltage peak of 100 V. Therefore, the maximum occurring voltage would amount to 900 V, which lies safely within the specification limits of the power modules. Furthermore, it is a typical feature of IGBTs that the voltage $V_{\mathrm{tr}}(t)$ is not exhibiting an abrupt cutoff, but rather decreases slowly after turn-on. Since $V_{\mathrm{tr}}(t)$ has the opposite polarity compared to the voltage spike generated by parasitic inductance, no voltage spike above 800 V can be detected in an actual measurement. An example is shown in Fig. 5.21.

The second example considers half-bridge modules in an integrated starter–generator application for a 42 V vehicle power system of an automobile. The MOSFET power switches have rated current of 700 A each and blocking voltage of 75 V. As before, the parasitic inductance is assumed 20 nH and the $\mathrm{d}i_{\mathrm{r}}/\mathrm{d}t$ shall be 5000 A/μs. Again, the over-voltage peak generated by the parasitic inductance would be 100 V, resulting in a total maximum voltage of 142 V. In contrast to an IGBT, the voltage decay in a MOSFET is rather abrupt after turn-on, so that it does not assist to reduce the total over-voltage spike. The resulting 142 V spike is clearly exceeding the maximum blocking voltage of the MOSFETs.

These examples illustrate a general feature in power electronic applications: Systems with high currents at a low voltage are most sensitive to parasitic inductance. Additionally, the problem of symmetric current paths is more severe in these applications.

To investigate this problem further, a parallel configuration with five IGBTs and the associated freewheeling diode with 1200 V blocking capability on a single DBC substrate is considered (Fig. 11.31a). The positions of the load terminals are indicated. A schematic circuit diagram for this design is depicted in Fig. 11.31b. The current tracks on the substrate are represented by the inductances L_1–L_9, whereas L_{10}–L_{15} are symbolizing the wire bond connections.

While the current path relevant for commutation from the terminals via IGBT3 contains only four parasitic series inductances, the relevant current path via IGBT1

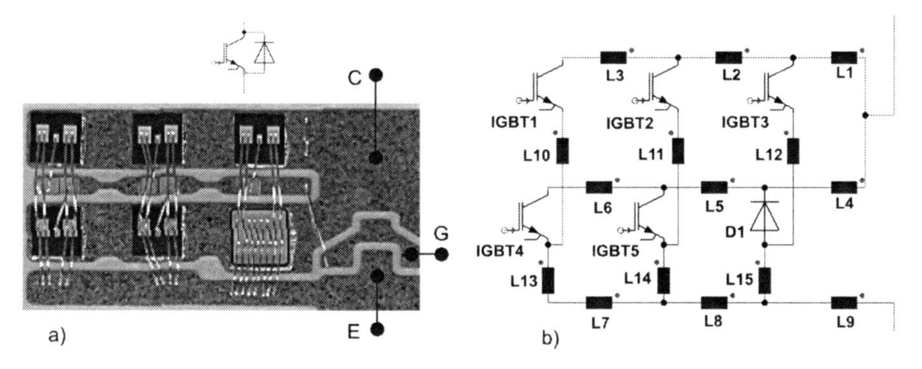

Fig. 11.31 (**a**) Realistic power circuit consisting of five parallel IGBT chips and one anti-parallel freewheeling diode chip (**b**) schematic circuit diagram showing the power devices plus the parasitic inductances formed by the current tracks

contains eight parasitic series inductances. Since the values of the parasitic inductances are in the same order of magnitude for the given geometry, a factor of 2 can be assumed between the parasitic inductance for the two IGBTs. This will result in a pronounced dynamical unbalance in the current distribution of this circuit during commutation. Moreover, the parasitic inductances can lead to oscillations between the chips, which will be investigated in Chap. 13.

It is difficult to find a symmetrical arrangement for a multitude of parallel chips in a high-current power module; the example in Fig. 11.31 is all in all one of the better solutions. In this context, designs combining a chip connected via a geometrically short current track with chips connected in parallel via geometrically long tracks are especially problematic. In induction measurements, the low parasitic inductance along the short path will dominate the result, while internally extensive differences generate a severe dynamical imbalance.

Solutions to this problem have been proposed [Mou02], which denote a substantial progress especially for applications with high currents at low voltages. Figure 11.32 shows an example.

The elementary cell is a half-bridge configuration of two MOSFET chips. The function of the freewheeling diodes is adopted by the internal diode of the MOSFET switches. The design of the elementary cell was optimized by numerical simulation using a Fast-Henry algorithm [Kam93], which allows to calculate the dynamical current distribution during high-frequency commutation in the three-dimensional model, with skin effect and eddy currents taken into account. The optimized cell design in Fig. 11.32a exhibits a parasitic inductance of 1.9 nH for a single elementary cell. By symmetrically paralleling seven of these elementary cells and by connecting the DC-link bus bar as laminated metal sheets directly on top of the plus and minus terminals, a parasitic inductance in the sub-nanohenry range was achieved. This module architecture is suitable for the application in an integrated starter–generator system as described in example 2 above.

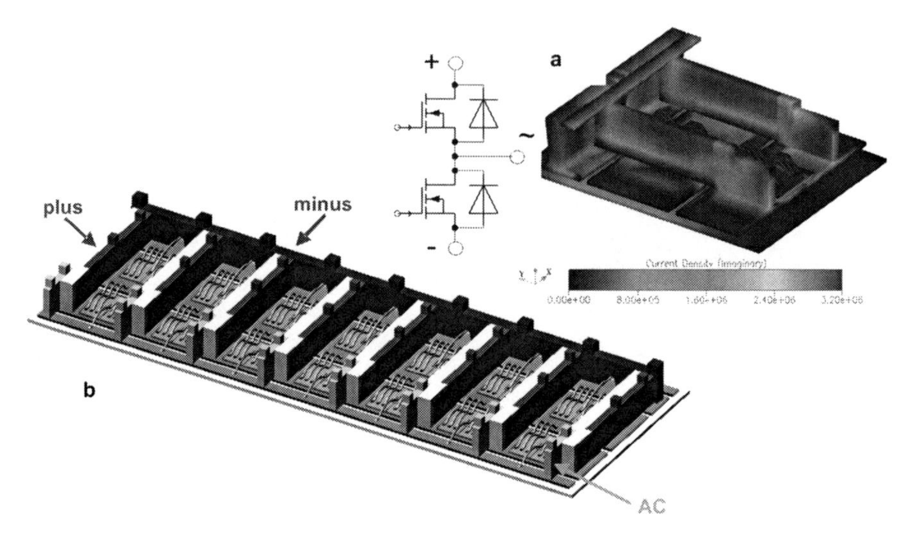

Fig. 11.32 Advanced MOSFET half-bridge configuration: (a) elementary cell with a simulated parasitic induction of 1.9 nH and (b) highly symmetrical circuit design with seven elementary cells in parallel per half-bridge [Mou02]

11.5.3 Parasitic Capacities

Insulated substrates in a power module create a capacitor, which will also influence the dynamical characteristics of the power circuit, as illustrated in Fig. 11.33 for a simple construction.

The copper tracks on the substrate generate two capacities C_{PA} and C_{PK} connected to the ground contact, which is represented by the module case. The series connection of C_{PA} and C_{PK} is connected in parallel to the internal junction capacity of the diode. In more complex circuits, these capacities will also form capacitive coupling links to other parts of the circuit. The dimension of these capacities depends on the insulator material and the thickness of the insulating layer. Characteristic parameters are shown in Table 11.3, Sect. 11.3.

Since insulation layers of epoxy and polyimide have only a marginal thermal conductivity, but at the same time exhibit a very high breakdown voltage, layers of these materials are rendered very thin. This results in a high capacity per unit area and limits the applications for components with these insulation materials (i.e. IMS substrates) for fast switching devices.

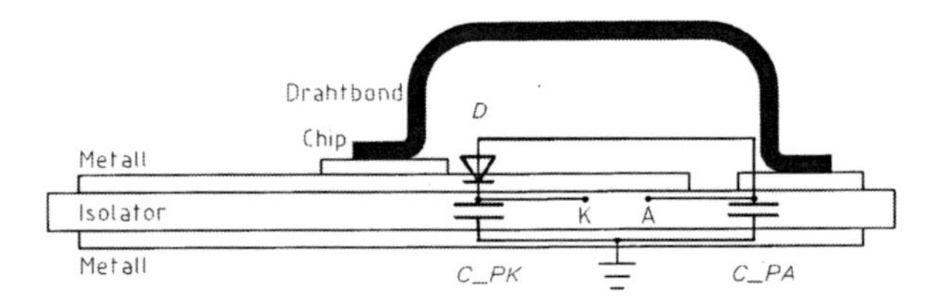

Fig. 11.33 Parasitic capacities in a package with a diode chip mounted on an insulated ceramic substrate [Lin02]

In an insulated TO-220 package, the cathode current contact features an area of 8 mm × 12.5 mm. Assuming a ceramic insulation layer of 0.63 mm thickness with a relative dielectric constant $\varepsilon_r = 9.8$ delivers a parasitic capacity on the cathode side C_{PK} according to

$$C_{PK} = \varepsilon_0 \cdot \varepsilon_r \cdot \frac{A}{d} \quad (11.13)$$

The evaluation of this equation yields a parasitic capacity of 14 pF for the given geometry, for a thinner ceramic layer of 0.38 mm the value rises to 23 pF. If the diode used in this example is a GaAs Schottky diode DGS10-18 A, then the (voltage-dependent) junction capacity is specified at 100 V with $C_J(100\,\text{V}) = 22\,\text{pF}$ [Lin02]. This value lies in the same range as the cathode parasitic capacity of the package. Therefore, the dynamic characteristic will not be determined by the junction capacitance alone, it will rather be modified by the external parasitic capacity.

However, as stated before, the parasitic capacity parallel to the junction capacitance is determined by the series connection of C_{PK} and C_{PA}. This diminishes the problem, since the area of the anode side current track is generally much smaller than the area of the cathode track. The total parasitic capacity C_{PG} parallel to the junction capacity is given by

$$C_{PG} = \frac{C_{PK} \cdot C_{PA}}{C_{PK} + C_{PA}} \quad (11.14)$$

If C_{PA} is only 1/5 of C_{PK}, the total parasitic capacity parallel to the junction capacitance C_{PG} is only 1/6 of C_{PK}. This favorable condition is generally fulfilled in packages of the TO family.

Now the example circuit is extended. In order to increase the blocking capability, two diodes in TO packages are connected in series, while the geometry for each single diode is still in accordance with Fig. 11.31. The equivalent schematic circuit diagram for the extended example is depicted in Fig. 11.34. The parasitic capacity parallel to the junction capacity C_{J1} of diode D_1 is formed by C_{PK1} in series with the parallel connection of C_{PA1} and C_{PK2}. With two identical packages and the assumptions of the relation of areas discussed above, the parasitic capacity parallel to C_{J1} amounts to 6/11 C_{PK1} or 0.54 C_{PK1}.

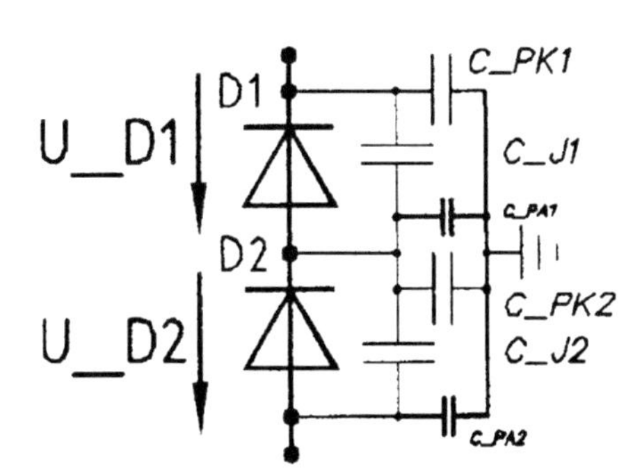

Fig. 11.34 Parasitic capacities in a series connection of two TO-220 diodes [Lin01]

The parasitic capacity parallel to C_{J2} of diode D_2 is composed by the small capacity C_{PA2} in series with the parallel connection of C_{PK2} and C_{PA1}. This capacity calculates to 0.17 C_{PK1}.

Therefore, the parasitic capacities form an asymmetrical dynamical voltage divider, which generates different voltage drops over the two diodes during high-frequency switching processes. This example illustrates that parasitic capacities can lead to unfavorable effects, which do not become obvious at the first glance.

If TO packages without internal insulations are mounted on a common heat sink, external insulation layers like polyimide foils have to be applied. These external foils also establish parasitic capacities, which exhibit even higher values, according to Table 11.3.

While the presented simple examples can be evaluated by analytical inspection, real multi-chip packages with a variety of chips and current tracks exhibit a much higher complexity, which makes it almost impossible to investigate by an analytical approach.

The situation is actually even more complex, since parasitic resistances, inductances, and capacitances, as well as the junction capacities of the power devices have to be considered at the same time. Their interaction during dynamic switching processes can form resonant circuits, which can cause oscillations [Gut01]. This will be discussed in Chap. 13. Today, software tools like the Fast-Henry algorithm allow simulating the electrical characteristics of complex three-dimensional systems in detail. The analysis and optimization of power modules with respect to parasitic effects are possible and necessary to increase the reliability of power electronic systems.

11.6 Reliability

The reliability of power electronic devices and components has been mentioned several times in the previous chapters. It is so important, because it is a prerequisite for the performance in applications: Reliability is the ability of a system or component to perform its required functions under stated conditions for a specified period of time [SAE08]. The requested lifetime of power electronics systems is seldom below 10 years and can reach up to 30 years.

11.6.1 The Demand for Increasing Reliability

Applications of power electronic devices face an increasing requirement for high reliability for several reasons:

- Power electronics faces a continuous demand for an increase in power density – often expressed in terms of controlled power per unit volume. This demand results in an increasing current density in power chips and in an increasing package density in power modules, with the consequence of higher temperatures and temperature gradients in the package.

- New fields of applications define more severe ambient conditions for the power packages, i.e. automotive hybrid traction systems, in which the combustion engine cooling system, featuring cooling liquid temperatures up to 120°C, extracts the losses of power electronic components. This requirement has led to an extension of the operation temperature range from a maximum $T_j = 150°C$ to a specification limit of 175°C.
- The number of interdependent frequency inverters is continuously growing in some areas of industrial automation. In automobile assembly, for example, several hundred process steps are linked together in a single production line, each has to remain functional to keep the line running. For the same operational availability of the assembly line as for a single inverter, this results in mean time between failure values for each inverter divided by the number of interlinked inverters, which reduces the acceptable failure rate easily by orders of magnitude.

These general trends have been the boundary conditions for the progress of power electronics packages in the past and will continue to do so in the future.

It is obviously impossible to test the reliability of power modules under field conformal stress conditions, because these tests would last as long as the expected service lifetime in the field – 10 to 30 years. Thus, manufacturers of power modules have developed a canon of accelerated test procedures during the last 30 years, which are derived from experience and which are considered as a base line for the product qualification to verify the expected functionality over the total field lifetime.

This historic dimension might make it more comprehensible that the general test categories seem identical for all power module manufacturers at the first glance, but exhibit considerable differences at a closer look. International standards are defining the general test setup, but the procedural details remain ambiguous. Therefore, every manufacturer of power modules has established its internal test philosophy, which is capable of defining, maintaining, and improving the internal quality level, but which makes it difficult to compare qualification test results between different manufacturers.

Table 11.6 collects a common set of (accelerated) qualification tests for IGBT and MOSFET power modules and highlights some differences for two selected power module manufacturers.

This compilation also indicates the progress of modern IGBT and MOSFET modules compared to conventional diode and thyristor modules: By way of example is the "high temperature reverse bias" test performed at 100% of the nominal blocking voltage and the power cycling requirement is increased from 10,000 to 20,000 cycles.

Before we will discuss each test in more detail, we have to add an important definition for every test: the failure criteria. It must be emphasized that the exact definition of failure criteria is essential for the evaluation of any test. Table 11.7 states the common failure criteria for qualification and endurance tests, as they are specified by international standards.

The failure criteria in Table 11.7 allow a certain increase relative to specification limits or initial measured values. Since the accelerated test conditions aim to

Table 11.6 Reliability tests for qualification of IGBT/MOSFET modules for industrial applications with reference to conventional modules

	Name	Conditions	Standards
HTRB	High temperature reverse bias test[a]	MOS/IGBT: 1000 h, T_{vjmax}, $V_{CEmax}(\leq 2.0\,kV)$, $0.8 \times V_{CEmax}(> 2.0\,kV)$	IEC60747-9:1998
		Conv: 1000 h, $T_j = 125°C$, $V_{RM} = 0.9 \times V_{RRM}$, $V_{RM}/V_{DM} = 0.8 \times V_{RRM}/V_{DRM}$ [c]	IEC 60747-2/6 Kap. V
HTGS (HTGB)	High temperature gate stress test[a]	1000 h, $\pm V_{GEmax}$, $T_j = 125°C$	IEC60747-9:1998
H3TRB (THB)	High humidity high temperature reverse bias test[a]	1000 h, 85°C, 85% RH, $V_{CE} = 0.8 \times V_{CEmax}$; however, max . 80 V, $V_{GE} = 0\,V$	IEC60749:1996
LTS	Low temperature storage test[b]	$T = T_{stgmin}$, 1000 h	IEC60068-2-1
HTS	High temperature storage test[b]	$T = T_{stgmax}$, 1000 h	IEC60068-2-2
TST	Thermal shock[a]	$T_{stgmin} - T_{stgmax}$; −40°C to +125°C, but $\Delta T_{max} \leq 165\,K$ $T_{storage} \geq 1\,h$ $T_{change} \leq 30\,s$ High power, standard: 20 cycles High power, traction: 100 cycles Medium power: 50 cycles Conv.: 25 cycles[c]	IEC60749:1996
TC	Temperature cycling[a]	External heating and cooling 2 min . < $t_{cycl.}$ < 6 min; $\Delta T_C = 80\,K$, $T_{cmin} = 5°C$ High power, standard: 2000 cycles	IEC60747-9:1998
		Medium power: 5000 cycles Conv.: 5000 cycles[c]	IEC60747-2/6 Kap. IV
TC	Temperature cycling[b]	Two-chamber air system, $T_{change} \leq 30\,s$ $T_{stgmin} - T_{stgmax}$: 100 cycles Conv. modules: 25 cycles[c]	IEC60068-2-14 Test Na
PC	Power cycling[a]	Internal heating and external cooling $0.5 < T_{cycl} < 10\,s$; $\Delta T_j = 60\,K$ $T_{jmax} = 125°C$, 130,000 cycles	IEC60747-9:1998
PC	Power cycling[b]	Internal heating and external cooling $\Delta T_j = 100\,K$, 20,000 cycles Conv. modules: 10,000 cycles[c]	IEC60749-34
V	Vibration	Sinusoidal sweep: 5 g, 2 h per axis (x, y, z)	IEC60068-2-6 Test Fc
MS	Mechanical shock	Half-sine pulse, 30 g, three times each direction ($\pm x$, $\pm y$, $\pm z$)	IEC60068-2-27 Test Ea

[a]Manufacturer Infineon
[b]Manufacturer Semikron
[c]Conventional devices – thyristors, diodes

Table 11.7 Failure criteria for acceptance after endurance tests

Failure criteria IEC60747-9(2001)	
I_{GSS}/I_{GES}	+ 100% USL
I_{DSS}/I_{CES}	+ 100% USL
$R_{DS(on)}/V_{CE(sat)}/V_F$	+ 20% IMV
$V_{GS(th)}/V_{GE(th)}$	+ 20% USL − 20% LSL
$R_{th(j\text{-}c)}/R_{th(j\text{-}s)}$	+ 20% IMV
V_{ISOL}	Not below specification limit

USL, upper specification limit; LSL, lower specification limit; IMV, initial measured value

simulate the stress applied in the total service lifetime, the specification limits are permitted to be exceeded within certain limits. For parameters less critical for the performance of the module, a greater increase can be allowed as shown for the leakage current. Other more critical parameters for the performance such as the forward voltage drop or the thermal resistance have to remain within closer limits, because they have a direct impact on the chip temperature.

The qualification tests in Table 11.6 can be classified into three groups. The first three tests are chip-related qualification tests, which are also part of every chip qualification. But since the chips are exposed to different substances during the module assembly process (i.e. solder flux, cleaning solvents, and silicone soft mold), a confirmation of the chip reliability in the assembled module is inevitable. This set of chip-related tests is followed by a group of seven tests related to stability of the package in the specified operation and storage temperature range and under external and internal temperature swings. Especially the power cycling test is important for the lifetime of power modules in application. The last two tests are confirming the mechanical integrity of the package.

11.6.2 High Temperature Reverse Bias Test

The high temperature reverse bias test (HTRB) – sometimes also referred to as hot reverse test – verifies the long-term stability of the chip leakage currents.

During the HTRB test, the semiconductor chips are stressed with a reverse voltage at or slightly below the blocking capability of the device at an ambient temperature close to the operational limit. No degradation can be expected in the bulk silicon of the devices at these temperatures, but the test is able to reveal weaknesses or degradation effects in the field depletion structures at the device edges and in the passivation.

The electrical field has to be expanded at the edges of a power device to reduce the tangential field at the chip surface by a field depletion structure. This can be achieved by a field ring structure, by a variation of lateral doping or by a suitable geometric contour. Nevertheless, the electrical fields at the surface are in a range of 100–150 kV/cm. Movable ions can accumulate in these high field areas and can generate a surface charge. The source of these ions can be contaminations during the assembly process or residues of process agents, for example, solder flux. The high temperature accelerates the process. The surface charge can alter the electrical field in the device and generate additional leakage currents. It can even produce inversion channels in device regions with low doping profiles and produce short-circuit paths across the pn-junction.

The failure criteria limit the allowed leakage current increase after the test – when the device is disconnected from the voltage supply and cooled down – to prevent such degradation effects. Additionally, most manufacturers of semiconductor devices also continuously monitor the leakage current during the 1000 h test and require a stable leakage current throughout the test.

Figure 11.35 shows an example of the recorded leakage current in a high-temperature reverse bias test. Eight devices were monitored for the test duration. The devices are initially stable but after approximately 200 h the leakage current starts to increase. The test was aborted after 920 h due to the massive increase of leakage current of some devices. The test failed for these devices, because the applied junction passivation was not capable of fulfilling the requirement.

The test conditions apply a considerably higher stress than the typical application. The nominal DC-link voltage will be in the range of 50–67% of the specified device blocking voltage in real systems; it will be exceeded only by temporary voltage peaks. Furthermore, the device will reach the maximum operational temperature

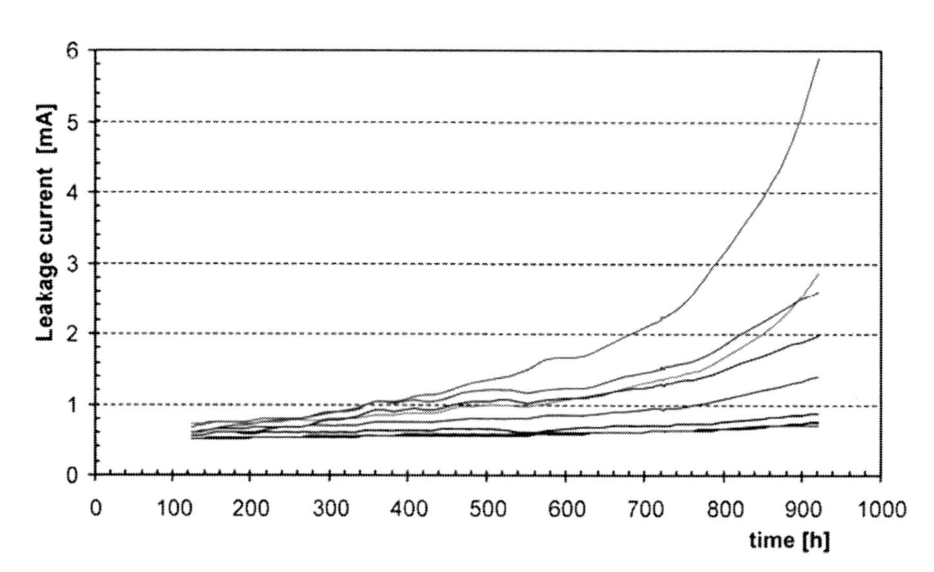

Fig. 11.35 Recorded leakage current during a high temperature reverse bias test – an example for a failed test

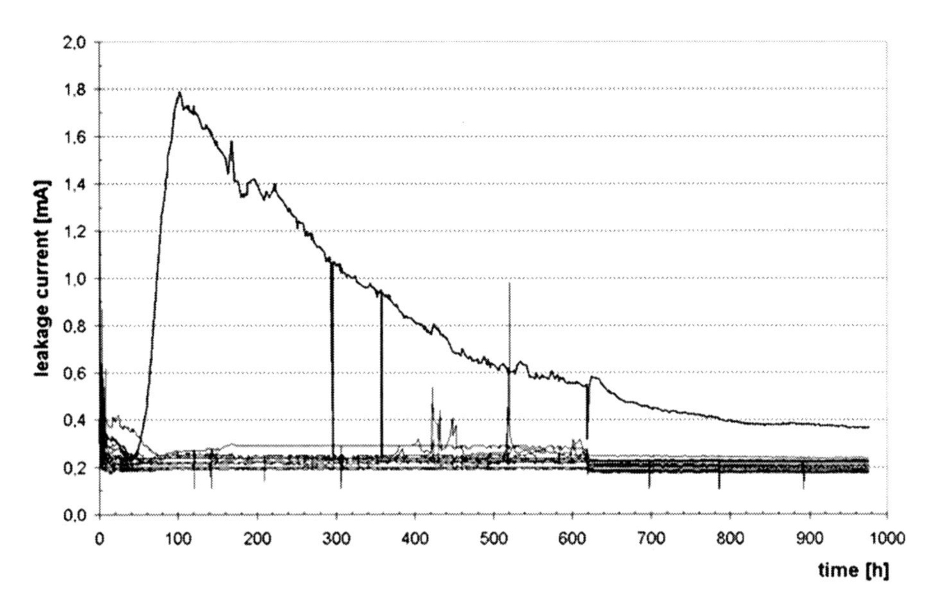

Fig. 11.36 Recorded leakage current during a high temperature reverse bias test – the test was passed but revealed a design flaw of the wire bond geometry

only occasionally in normal applications. Thus, the test is a highly accelerated procedure to generate stress within a test duration of 6 weeks for applications designed for a lifetime of 20 years and more.

But even if the junction passivation is compatible with the requirements and the assembly process, the test can reveal flaws in the package design. Figure 11.36 shows an example of a temporarily increased leakage current for a single device after nearly 100 h of test time. After reaching its maximum, the leakage current decreased and showed only a small increase at the end of the test. The following investigation showed that a single wire bond of this device had a wrong geometry: it had no loop and was directly laying on the guard ring passivation of the IGBT. The test was repeated after the correction of the bond wire layout and the formally observed leakage current increase was eliminated. This result confirms the relevance of the HTRB test for the package development.

11.6.3 High Temperature Gate Stress Test

The high temperature gate stress test or high temperature gate bias test confirms the stability of the gate leakage current. Even though the maximum allowed gate voltage is limited to ± 20 V, this voltage is applied to not more than 100 nm thick gate oxide layer in state-of-the-art IGBTs and MOSFETs. This results in an electrical field of 2 MV/cm across the gate oxide. For a stable leakage current, the gate oxide must be free of defects and only a low density of surface charges is tolerable. The boundary condition of maximum operational temperature for the test again accelerates the test.

Since the leakage currents are very small (< 10 nA), this test is also extremely sensitive to surface contaminations on the chip. In test modules, where thermocouples were glued to the emitter contact of an IGBT for measurement purposes, the gate leakage was found to be considerably increased. This increase was caused by the residues of the solvent of the glue, which remained on the chip surface between the gate and the emitter and caused a measurable increase of the leakage current. Therefore, the gate leakage test also ensures the cleanness during the assembly process of a module.

11.6.4 Temperature Humidity Bias Test

The temperature humidity bias test, also known as high humidity high temperature reverse bias test, is focusing on the impact of humidity on the long-term performance of a power component.

Capsules are – when defect-free assembled – hermetically sealed against the environment. This is not the case for the majority of power module packages. Although bond wires and chips are completely embedded in silicone soft mold, this material is highly permeable for humidity. Therefore, humidity can intrude the package and can reach the chip surface and junction passivation. This test aims to detect weaknesses in the chip passivation and to initiate humidity-related degradation processes in the packaging materials.

Sometimes, proposals are made to suppress the humidity access by an additional protective layer. But two strong arguments have to be taken into account: First, the silicone soft mold exhibits a high linear coefficient of thermal expansion (~300 ppm/K). This yields a considerable increase in volume during temperature swings and makes it difficult to apply a hermetically tight layer on top of the soft mold. Moreover, if a hermetically sealing is not achievable, then an additional layer can only reduce the diffusion rate of the humidity. However, a reduced diffusion rate works in both ways: It increases the time for the humidity to soak in, but it will also increase the time for the intruded humidity to be driven out. Therefore, a non-hermitical protection layer would increase the duration for the humidity to influence the device in operative condition. This is not desirable. If the humidity cannot be kept from intruding the package, then it should be driven out fast when the module is going in operational mode. For this reason, a highly penetrable embedding compound should be preferred.

The applied electrical field during the test acts as a driving force to accumulate ions or polar molecules at the semiconductor surface. On the other hand, the power losses generated by the leakage current must not heat up the chip and its environment and thus reduce the relative humidity. Therefore, standards require a limitation of the self-heating of the chip to not more than 2°C. Consequently, the reverse voltage had been limited to 80% of the blocking voltage for low-blocking voltage MOSFETs and was restricted to a maximum of 80 V for higher blocking capabilities in the past.

A number of field experiences in the past years have shown that this test condition is not sufficient for all application conditions. Field failures, which could clearly be attributed to the influence of humidity, have raised a discussion about this 80 V maximum applied voltage. Since the leakage currents of modern semiconductor chips are low enough to maintain within the allowed 2°C temperature increase even at 80% of the nominal blocking voltage for blocking voltages of 1200 V and more, the restriction to 80 V seems to be outdated. It can be anticipated that restriction of the test voltage will be abandoned in the near future, resulting in a higher reliability of non-hermetical power modules in high-humidity application environments.

11.6.5 High Temperature and Low Temperature Storage Tests

The storage test at the maximum and minimum storage temperatures have been implemented to verify the integrity of the plastic materials, rubber materials, organic chip passivation materials, glues, and silicone soft molds utilized in most state-of-the-art power module packages. These materials must maintain their characteristics in the complete specified storage temperature range.

At this point, a remark is necessary to prevent a misunderstanding of the term "storage temperature." The denomination "storage temperature" has been established in the early days of semiconductor power modules. It refers to the non-operational temperature limits for power modules assembled in a power electronic system. It does not refer to storage conditions of the unassembled power module, as the denomination might suggest. The description "non-operational temperature limits" would be more appropriate, but the heritage of the early days of power modules impedes this transition.

Long-term storage at high temperatures is critical for the mechanical strength of all thermoplastic housing materials. It is also minatory for the flame retardant additives to the thermoplastic materials required for the resistiveness against fire hazards. Silicone soft mold starts to degenerate at temperatures above 180°C, so that this will be a limit for the classical module construction.

Long-term storage at low temperatures is critical for the softening agents in plastic material and rubber materials; they can lose their function and can destroy the elastic capabilities of plastic materials. The silicone soft mold compound is also limited in the minimum storage temperature; most standard soft molds are limited to −55°C. Below this temperature, cracks will appear in the soft mold, which will not be cured by increasing temperatures and thus will not be able to maintain an insulating environment for high blocking voltage devices.

The standard storage temperature limits are −40°C/+125°C and a variety of materials are available for reliable performance in this temperature range. Extensions of the temperature range, which are demanded by many applications today, will make these qualification tests more challenging in the future.

11.6.6 Temperature Cycling and Temperature Shock Test

Temperature swings are an essential stress condition for every power electronic component in application. The temperature cycling test and the temperature shock test are two test methods to simulate ambient temperature swings during the field lifetime.

The test conditions are discriminated by the change rate of the externally imprinted temperature. If the rate of temperature change is slow in the range of 10 – 40°C/min, the test is called temperature cycling test. In a temperature shock test, the ambient temperature is changed typically in less than 1 min. For power modules, this is typically achieved by a two-chamber equipment, in which the air is permanently heated or cooled to the maximum or minimum test temperature, while an elevator carrying the devices under test moves between the two chambers in a time interval below 1 min. Since the heat exchange rate is rather slow for a gas environment, the duration for reaching an equilibrium temperature distribution inside the module can vary from 30 min to 2 h, depending on the total thermal capacity of the devices under test.

A more extreme version of the temperature shock test is the liquid-to-liquid thermal shock test. In this test, the ambient is formed by appropriate liquids, heated or cooled to the desired temperature limits, for example, oil at 150°C or more and liquid nitrogen at −196°C. Such test conditions are not common for modules, but are often performed for package elements as DBC substrates. In a liquid ambient, the heat transfer is much faster than in a gaseous ambient, so that an equilibrium temperature distribution can be achieved in minutes rather than hours.

Due to the wide range of heat transfer rates, the denomination of temperature swing test is somewhat ambiguous. While Infineon refers to the two-chamber air ambient test as temperature shock test to distinguish it from the rather slow single chamber test with ambient temperature change rates in the range of 20°C/min, Semikron labels the two-chamber air environment test as temperature cycling test to discriminate it from the much faster liquid-to-liquid test. This ambiguity of denomination must be kept in mind when comparing qualification requirements and test results from different manufacturers.

A common boundary condition for all types of temperature cycling tests is the requirement that the cycle time must be long enough, so that all parts of the assembly reach the maximum or minimum temperature – which are typically the storage temperature limits – so that the assembly is in a thermal equilibrium condition. Since the test simulates the impact of temperature changes by external sources, for example, the change of ambient temperatures or an increase of heat sink temperature by other heat sources, the power modules are not actively stressed by current or voltage. The changes in parameters are checked by an initial and a final measurement and have to comply with the failure criteria.

The combination of different materials with different coefficients of thermal expansion results in high mechanical stress in the system. More so, the bimetal effect causes a cyclic deformation of the module. Simulations of the thermo-mechanical behavior of a power module have shown [Mik01] that if this bimetal bending is reduced – for example, by mounting the module on a heat sink – the stress is reduced and the lifetime is extended. Therefore, modules should be mounted on assembly plates during the test to simulate as close as possible the application conditions.

The cyclic mechanical deformation generated by temperature cycles due to the difference in coefficients of expansion of the material layers causes stress in the functional layers themselves and in the interconnection layers. This will lead over time to the initiation of cracks and cause growing delaminations in these layers. Scanning acoustic microscopy (SAM) is the appropriate detection method of identifying delaminations in power semiconductor modules.

An example of the damage caused during temperature cycling with classical 34 mm base plate modules show the SAM images in Fig. 11.37. Two different solder materials for the substrate-to-base plate interface were compared in this test: a RoHS compatible SnAg(3.5) solder and the classical SnPb(37) solder. Both solder interfaces – the interface between the substrate and the base plate and the interface between chip and substrate – were investigated by choosing the appropriate time-of-flight windows in the SAM signal for each solder type. The comparison between the initial measurement and the SAM image after 200 temperature cycles (−40/+125°C) in a two-chamber test equipment reveals growing delaminations in the substrate-to-base plate solder layer for both solder versions, indicated by the white areas (=regions of high reflection) which move inward from the corners and short edges of the substrates. The classical eutectic SnPb solder shows more damage along the outside short sides of the substrates, where the terminals are connected at the topside of the substrate.

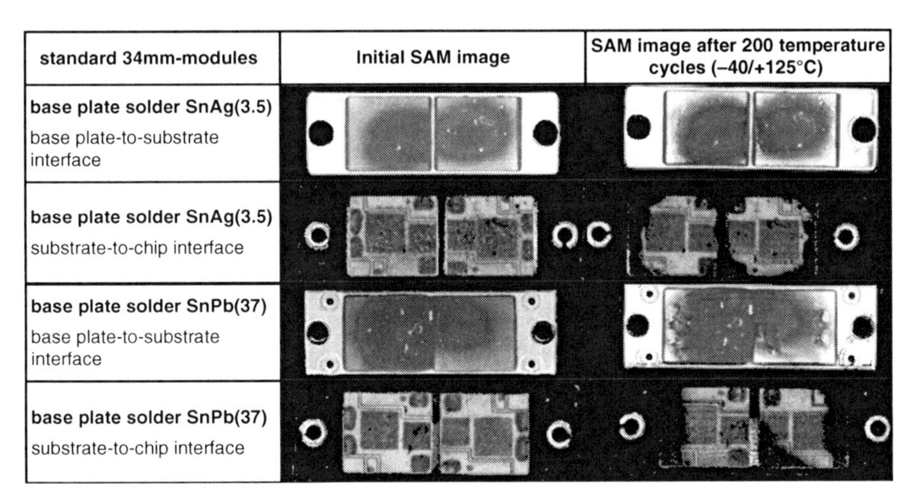

Fig. 11.37 Scanning acoustic microscope (SAM) images of standard 34 mm modules before and after 200 temperature cycles (−40/+125°C) – different delamination patterns are found for different base plate solder materials

The SAM images of the chip-to-substrate interface show no indications of any fatigue in the chip solder interfaces, but it presents black areas in the regions, where solder delaminations are found in the substrate-to-base plate solder layer. This artifact is produced by a lack of acoustic energy in these regions, because most of the signal was already reflected in the delaminations found in the substrate-to-base plate solder layer. Since the SAM signal is injected from the base plate surface, reflections nearer to the base plate reduce the signal propagating into deeper layers. However, this common artifact allows conveniently evaluating how close the delamination has come to the chip position. As clearly shown in Fig. 11.37 the delaminations have propagated much deeper under the chip areas for the SnPb solder system as for the SnAg solder system with a stronger unfavorable impact on the thermal resistance of the chips.

The lifetime under temperature cycles is determined by the combinations of different materials (with different coefficients of thermal expansion) and the stability of the interconnect layers. Due to the mechanical deformation, smaller packages as the TO family are more stable than larger modules that are more complex. Since the source for the passive temperature cycles is located outside of the module, only the used materials and interconnection layer decide about the reliability of a package.

11.6.7 Power Cycling Test

In contrast to the temperature cycling test, the power chips are actively heated by the losses generated in the power devices themselves in a power cycling test. This accounts for a fundamental difference between the two tests: In a power cycling test, the amount of losses can be affected by the chip technology and by the silicon area implemented in the module. Therefore, any power cycling lifetime requirement can be met just by implementing sufficient silicon area to reduce the temperature swing generated by the chip losses. However, commercial aspects limit this option in practical applications.

During power cycling test, the device under test is mounted on a heat sink as in a real application. A load current is conducted by the power chips and the power losses are heating up the chip. When the maximum target temperature in the chip is reached, the load current is switched off and the system cools down to a minimum temperature. The reaching of the minimum temperature completes the cycle and the next cycle begins by starting the load current again. During each cycle, considerable temperature gradients are generated inside the module. An exemplary temperature evolution is shown in Fig. 11.38.

The control is executed using the heat sink temperature. When the upper limit of T_{h} is reached, the load current is turned off and the cooling is turned on. The temperature decreases. When the lower limit of T_{h}, identical to the lower limit of the junction temperature T_{low}, is reached, the load current is turned on again and the cycle is repeated. The characteristic parameter for power cycling tests, the temperature swing ΔT_{j}, is given by the temperature difference between maximal junction temperature T_{high} at the end of the heating phase and the minimal junction temperature at the end of the cooling interval:

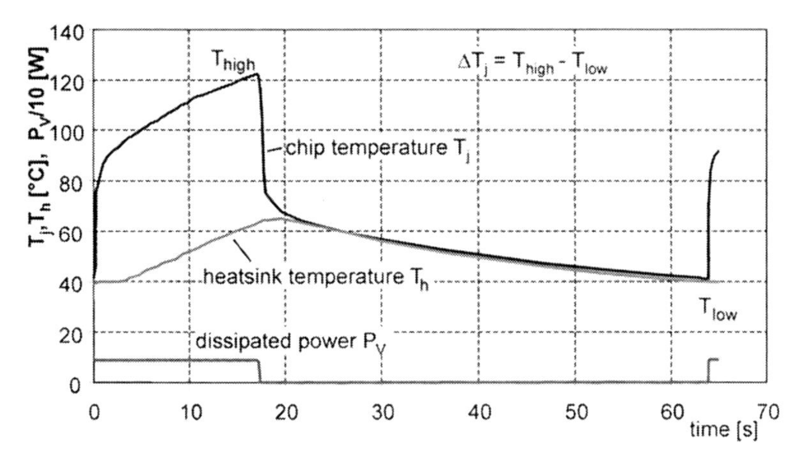

Fig. 11.38 Temperature profile of a power cycling pulse resulting from a constant load current

$$\Delta T_j = T_{\text{high}} - T_{\text{low}} \tag{11.15}$$

In Fig. 11.38 ΔT_{j} can be read as 82°C.

A further important parameter for the power cycling test is the medium temperature T_{m}:

$$T_{\text{m}} = T_{\text{low}} + \frac{T_{\text{high}} - T_{\text{low}}}{2} \tag{11.16}$$

Instead of T_{m} also T_{high} or T_{low} can be used as a characteristic parameter, because they are related by ΔT_{j}. Further parameters, e.g., the duration of the cycle, are of importance as is shown below. A long power cycle time (64 s in Fig. 11.38) usually represents a higher stress for the devices.

The different coefficients of thermal expansion of materials during the temperature swing create mechanical stress at the interfaces. This thermal stress leads in the long run to fatigue of materials and interconnections. Figure 11.39 shows the result of a power cycling test with a standard module. During the test, the forward voltage V_{C} of an IGBT is monitored. Additionally, it is possible to feed a defined sense current of some milliamperes through the device after the turn-off of the load current, which allows to determine the upper temperature T_{high} with the use of a calibration function (compare Fig. 11.19). The power losses P_{v} are also measured online. From junction temperature T_{high}, heat sink temperature T_{h}, and P_{v} the thermal resistance is calculated using Eq. (11.4). Since the measurement of the fast changing heat sink temperature is difficult due to the response time of the applied sensor, the thus measured thermal resistance can deviate from the true stationery value, especially for short cycles below 10 s. However, even then this relative value can be used to monitor relative changes in R_{thjh}.

In Fig. 11.39 one can recognize that the on-state voltage drop at the IGBT remains almost constant up to a large number of cycles, and the thermal resistance increases slowly after approximately 6000 cycles. This is an indication for

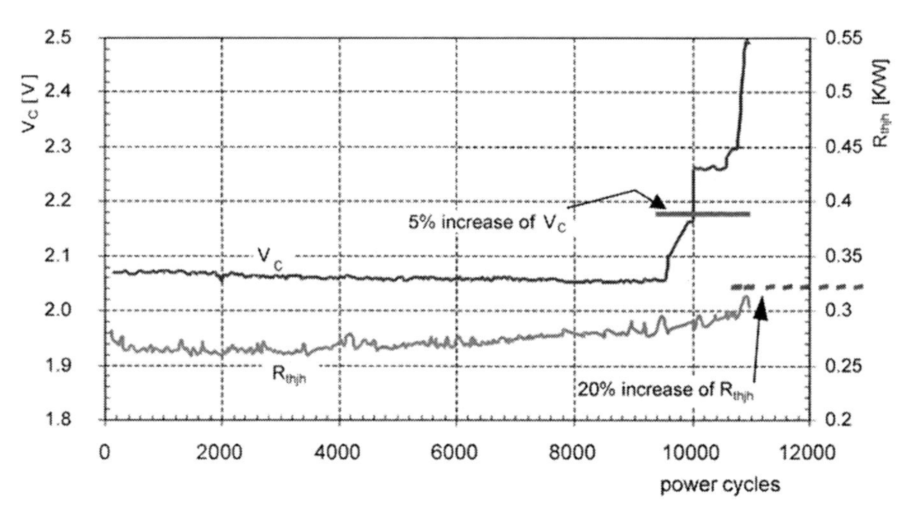

Fig. 11.39 Behavior of on-state voltage drop V_C and thermal resistance R_{thjh} at a power cycling test with $\Delta T_j = 123$ K

an increase of thermal resistance in the thermal path, mostly attributed to solder fatigue. After more than 9000 cycles one finds a first step in the V_C characteristic, which is due to lift-off of bond wires. Shortly after that, the next step is observed and finally all bond wires are lifted off, the power circuit is open and the test can no longer be continued.

Bond wire lift-off and solder fatigue are the main failure mechanisms in standard power modules. But from the shape shown in Fig. 11.39 it is difficult to determine the primary failure mechanism. The failure limit of R_{thjh} would be reached at after approximately 11,000 cycles. But increase of R_{thjh} results in increasing temperature T_{high} and this will escalate the thermal stress for the bond wires. Therefore, solder fatigue is a significant failure mechanism in this test; it could be even the main failure mechanism. On the other hand, bond wire lift-off leads to increased V_C, which together with the constant current causes increasing losses and raises the upper junction temperature T_{high}, resulting in more thermal stress in solder layers. Due to the interdependency of the failure modes, power cycling tests require a careful failure analysis.

Failure limits are defined as:

- an increase of V_C by 5% or by 20%, varying for different suppliers, depending on the measurement accuracy for V_C. Different failure limits, however, have only a negligible impact on the lifetime. Usually, after the first significant increase of V_C the bond wires will soon fail completely, as can also be seen in Fig. 11.39
- an increase of R_{th} by 20%
- failure of one of the functions of the device, e.g., failure of blocking capability or of the gate to emitter (gate to source) insulation capability for IGBTs and MOSFETs

11.6.7.1 Weibull Statistics for Power Cycling Analysis

An estimation of the lifetime of a power device as a series product is possible by evaluating the power cycling results using a Weibull statistics. The Weibull statistics is specially suited for the description of end-of-life phenomena. It is applicable for failure mechanisms determined by aging mechanisms of materials. An example is shown in Fig. 11.40. The test was executed until five of six devices under test failed. The numbers of cycles to failure are marked. The Weibull distribution is described by the probability density $f(x, \alpha, \beta)$ and the accumulated probability $F(x, \alpha, \beta)$ as follows:

$$F(x, \alpha, \beta) = 1 - exp\left(-\left(\frac{x}{\beta}\right)^{\alpha}\right) \tag{11.17}$$

The accumulated probability corresponds to the rate of parts, which have already failed. For a power cycling test analysis, x is the number of cycles to failure. The scale parameter β defines the range of the distribution; for $x = \beta$ the rate $1/e$ of the parts have survived. For $F = 1$, all parts have failed. The shape parameter α characterizes the spread of the distribution. The higher the α, the more concentrated are the numbers of cycles to failure of the different parts around the median of the distribution. The derivative $dF/dx = f$ corresponds to the probability density; it determines the probability that the failure occurs in an interval $x + dx$:

$$f(x, \alpha, \beta) = \frac{\alpha}{\beta^{\alpha}} x^{\alpha - 1}\, exp\left(-\left(\frac{x}{\beta}\right)^{\alpha}\right) \tag{11.18}$$

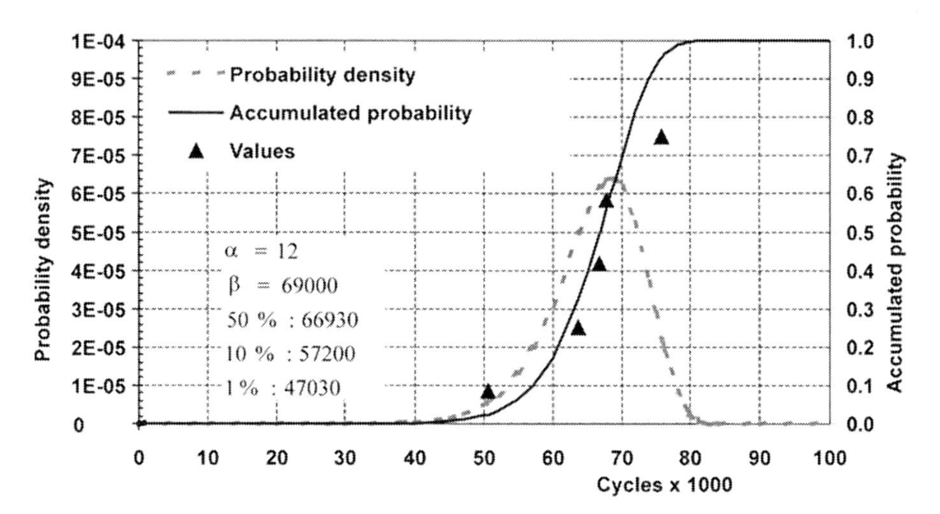

Fig. 11.40 Weibull analysis of a set of power cycling tests on a series product

Even though Weibull statistic on power cycling test results are based on a comparatively low number of devices, this statistical analysis allows to predict the survival rate under defined application conditions for a series product. The restriction of a limited number of failures is caused by the fact that these tests are very time consuming and that it is not easy to perform the test simultaneously on groups of devices since the test conditions will change when some of the devices fail during the test. Therefore, the first failure of a set of devices under test is often used to define N_f.

11.6.7.2 Models for Lifetime Prediction

Since a standard technology is established for the construction of power modules with base plate and technologies and materials are very similar even for different suppliers, a research program for determination of lifetime for standard power modules was implemented in the early 1990s. In this project named LESIT, modules from different suppliers from Europe and Japan have been tested; a common feature was the standard package according to Fig. 11.13 with the use of an Al_2O_3 ceramics according to Table 9.1, left row. Tests were executed at different ΔT_j and different medium temperatures T_m; the results have been summarized in [Hel97] and are shown in Fig. 11.41 in the form of characteristics of cycles to failure N_f depending on ΔT_j and for different medium temperatures T_m.

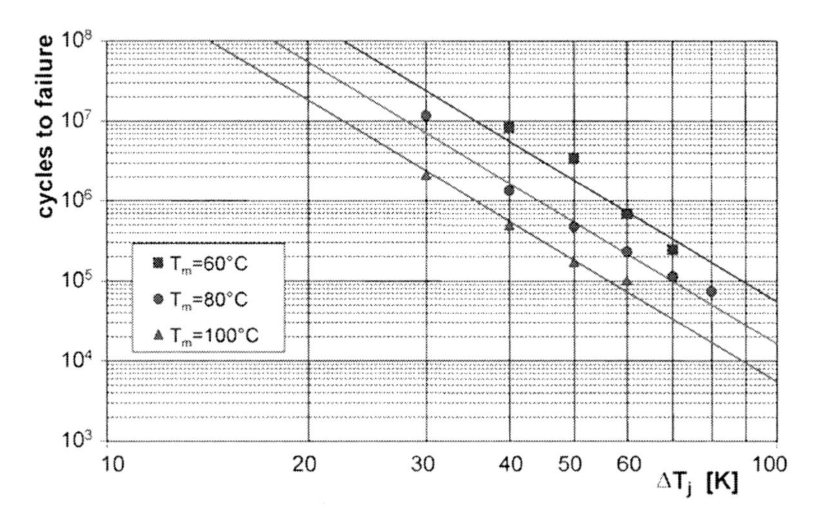

Fig. 11.41 LESIT results

The lines in Fig. 11.41 represent a fit from [Scn02b] to the experimental data: The expected number of cycles to failure N_f at a given temperature swing ΔT_j

and an average temperature T_m, the absolute medium temperature in K, can be approximated with the equation

$$N_f = A \cdot \Delta T_j^{\alpha} \cdot \exp\left(\frac{E_a}{k_B \cdot T_m}\right) \tag{11.19}$$

with k_B (Boltzmann constant) $= 1.380 \times 10^{-23}$ J/K, activation energy $E_a = 9.89 \times 10^{-20}$J, and the parameters $A = 302,500\,\mathrm{K}^{-\alpha}$ and $\alpha = -5.039$.

Equation (11.19) consists of a Coffin–Manson law, i.e. the number of cycles to failure (N_f) is assumed to be proportional to $\Delta T_j^{-\alpha}$ [Hel97]. It appears as a straight line when plotting $\log(N_f)$ over $\log(\Delta T_j)$. In addition, an Arrhenius factor containing an exponential dependency on an activation energy is added to the Coffin–Manson law [Hel97]. From models for solder fatigue the impact of dwell time, ramp time, or cycle time are also known as factors influencing the lifetime. However, in the past such models have been suggested rather for passive thermal cycling tests and not for power cycling tests.

Using Eq. (11.19) it is possible to calculate the number of cycles to failure for given ΔT_j and T_m according to the LESIT results. If the typical cycles in the application are known, it is possible to calculate the expectation for the lifetime of a module under these conditions.

Technologies for standard modules have been improved since 1997. Power cycling results for more recent power modules of two different suppliers are shown in Fig. 11.42. They are compared with Eq. (11.19) for the condition $T_{low} = 40°C$,

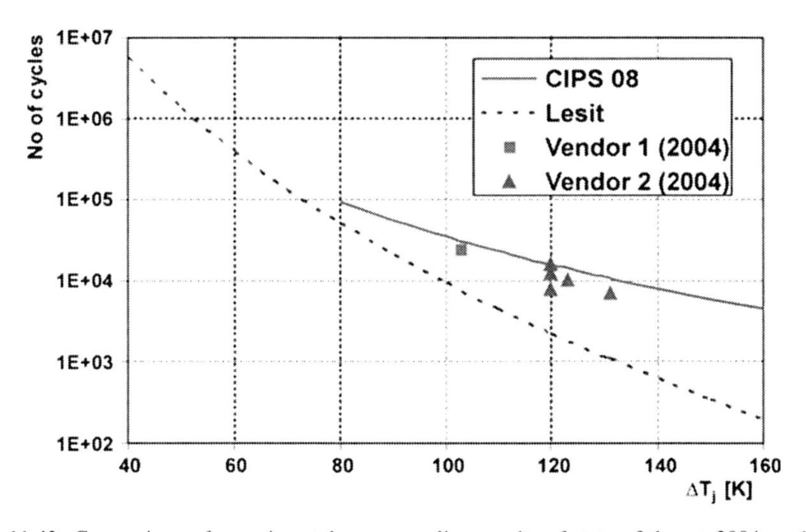

Fig. 11.42 Comparison of experimental power cycling results of state-of-the-art 2004 modules with predictions by the extrapolated LESIT model (11.19) and new CIPS 08 model (11.20); $T_{low} = 40°C$

extrapolated to higher temperature swings outside of the data in Fig. 11.41. It is visible that the number of cycles to failure is increased by a factor of 3 to 5 in the range of $\Delta T_j > 100\,\text{K}$ compared to the prediction using Eq. (11.19).

A fundamental problem in performing power cycling tests is the difficulty in the selection of test conditions. The target ΔT_j is a function of the dissipated energy – which for a given chip technology is determined by the forward current and the t_{on} duration in the test – and of the thermal resistance of the test setup. It is therefore very difficult to repeat a test with exactly the same test conditions, but it is even more difficult to select test parameters for different ΔT_j values with the same current and heating times. If these parameters have an impact on the test results, they have to be taken into account by a lifetime model.

This was the motivation to present an extended model for the lifetime of standard power modules [Bay08]. Based on a large number of power cycling results of modules, the following equation was derived:

$$N_f = K \cdot \Delta T_j^{\beta_1} \cdot \exp\left(\frac{\beta_2}{T_{low}}\right) \cdot t_{on}^{\beta_3} \cdot I^{\beta_4} \cdot V^{\beta_5} \cdot D^{\beta_6} \qquad (11.20)$$

As parameter K we use the value 9.30×10^{14}, the other parameters $\beta_2 - \beta_6$ are given in Table 11.8 [Bay08]. Equation (11.20), which we denote as CIPS 08 model, contains additionally the dependence on the heat-up time t_{on} in seconds, the current per bond stitch on the chip I in A, the voltage range of the device V in V/100 (reflecting the impact of the semiconductor die thickness), and the bond wire diameter D in micrometers. The prediction by the new CIPS 08 model is also shown in Fig. 11.42 for $t_{on} = 15$ s. The CIPS 08 model holds for standard modules with Al_2O_3 substrates; it is not valid for high-power traction modules which are built with the materials AlN and AlSiC, see Table 11.1.

Equation (11.20) was a result of purely statistical analysis and is not a result of physics-based models [Bay08]. The dependency on the cycling time t_{on} in Eq. (11.20) – higher number of cycles to failure for short cycling times – may be explained by the fact that in a short cycle time mainly the semiconductor itself is heated up. Thermal-mechanical stress occurs mainly at the interface between semiconductor and bond wire, while in layers closer to the heat sink the temperature increase is marginal and less thermal stress is applied. The dependency on current per bond stitch on the chip I can be attributed to improved current distribution on the chip with more bond stitches and presumably to a positive impact of the thermal capacity of the bond stitches. The dependency on the bond wire diameter D is

Table 11.8 Parameters for calculation of power cycling capability according to Eq. (11.20)

β_1	−4.416
β_2	1285
β_3	−0.463
β_4	−0.716
β_5	−0.761
β_6	−0.5

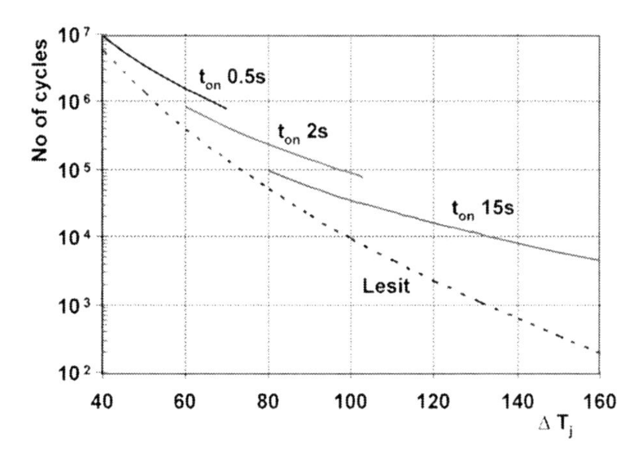

Fig. 11.43 CIPS 08 model according to Eq. (11.20) for different heating times t_{on}, compared to the LESIT model; $T_{low} = 40°C$

related to the greater mechanical stress applied to the bond stitch by thicker bond wires. The dependency on the voltage range V is in fact a dependency on device thickness, which increases from 600 to 1700 V. With thinner devices, the mechanical stress implied on the solder interface by the Si-material will be reduced. Note that the used devices are produced in a thin-wafer technology for 1200 V ($V = 12$) and 600 V ($V = 6$). For devices fabricated from epitaxial wafers, e.g., PT-IGBTs or Epi-diodes, Eq. (11.20) is not applicable.

As a consequence of the statistical approach of the CIPS 08 model, the parameters are not physically independent, which was pointed out and discussed by the authors themselves [Bay08]. For low ΔT_j, for example, a short heating time will be typical. The influence of different heating times t_{on} is shown in Fig. 11.43. Figure 11.42 gives the impression that for a $\Delta T_j < 60$ K the new model predicts less cycles to failure N_f than the former LESIT model. But the dependency on the heating time shows that the lifetime N_f is higher for state-of-the-art 2008 modules, if a short heating time t_{on} for low ΔT_j is assumed.

Despite the fact that data for Eq. (11.20) were only generated with modules of one manufacturer, the equation seems also useful for lifetime calculation of modules of other manufacturers. If lifetime calculations are of vital importance in an application with high reliability requirements, the manufacturing company should always be consulted.

11.6.7.3 Superimposition of Power Cycles

Additionally to the already discussed restrictions, lifetime models are derived from the repetition of identical power cycles, but in real applications various different cycles are superimposed.

To calculate the lifetime under realistic application conformal conditions with superimposed power cycles, a linear accumulation of damage is often assumed as discussed, for example, in [Cia08]

$$Q(\Delta T) = \frac{N(\Delta T)}{N_\mathrm{f}(\Delta T)} \tag{11.21}$$

If the number of cycles N reaches N_f – the number of cycles to failure calculated by a lifetime model – Q is equal to 1. For superimposition of n conditions Q_n, the failure is expected when the sum of $Q_n = 1$. Since this so-called Miner's rule [Min45] is independent of the considered lifetime model, the simple Eq. (11.21) which is only a function of ΔT_j can be extended using the model in Eq. (11.20):

$$Q(\Delta T_j, T_\mathrm{low}, t_\mathrm{on}, I, V, D) = \frac{N(\Delta T_j, T_\mathrm{low}, t_\mathrm{on}, I, V, D)}{N_\mathrm{f}(\Delta T_j, T_\mathrm{low}, t_\mathrm{on}, I, V, D)} \tag{11.19}$$

An investigation with two superimposed power cycles is reported in [Fer08]. A cycle with high ΔT_j of 140°C is superimposed with short cycles with low ΔT_j. However, the cycle with high ΔT_j was dominant for the failure under the chosen superposition. It is generally difficult to define two superimposed power pulses in such a way that each of them contributes with exactly 50% to the final failure.

A different approach was reported in [Scn02b], where first conventional power cycling tests with uniform cycles were performed at two different temperature swings and then a test with both cycling conditions interleaved was conducted. The test results showed that in contradiction to the assumption of linear fatigue accumulation, the number of cycles to failure was equal to the sum of cycles for each single test condition. The explanation for this phenomenon was that the test conditions are initiating different failure mechanisms, which do not interact with each other.

Before we discuss the failure mechanisms in power cycling test known today in more detail, a deficiency common to all of the above-presented lifetime models must be added. The models (11.19) and (11.20) are based on a purely statistical analysis of data under different conditions. No physical relationship is attributed to the impact of different test parameters. Currently, many research groups are working on physics-of-failure lifetime models, which promise to result in better lifetime prediction models and give more insight into the physical mechanisms that limit the lifetime of power modules for active power cycles.

A final remark addresses the problem of cycle counting in application conformal temperature evolutions known as mission profiles. A simple collection of maximum–minimum temperature swings is very sensitive to the resolution of the analysis. A more robust approach, which is also consistent with the strain–stress characteristic in physics-of-failure-oriented models, is the widely accepted rainflow counting method [Dow82]. It is less sensitive to resolution changes and evaluates the fundamental frequencies with a higher weight.

11.6.7.4 Bond Wire Lift-Off

A typical fault image after power cycling of a standard module with base plate is shown in Fig. 11.44. All bond wires are lifted off from the IGBT chip. The bond wire in the background was the last one to fail and the current was flowing in a short time via an arc flash-over which caused a crater below the bond wire stitch. The bond wire failure in the foreground shows a characteristic feature of the lift-off failure mode. The dissection did not occur at the interface between bond wire and chip metallization, but it emerges partially in the volume of the bond wire. Residues of bond wire material can still be detected on the surface of the chip metallization.

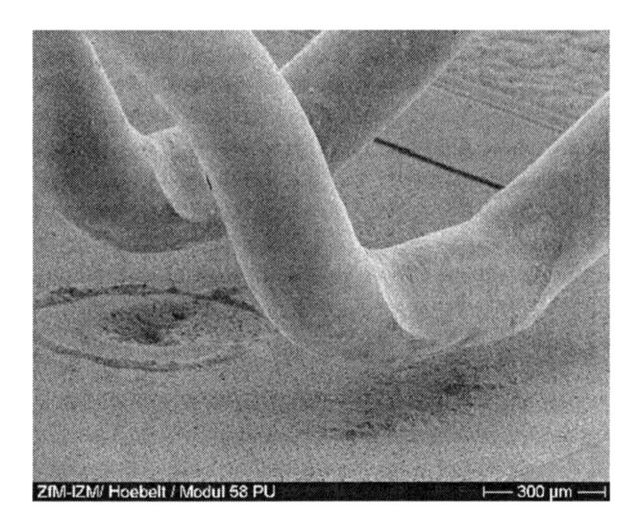

Fig. 11.44 Lifted bond wires after power cycling test with $\Delta T_j = 100\,K$ of a standard IGBT module. The total failure occurred between 10,791 and 13,000 cycles

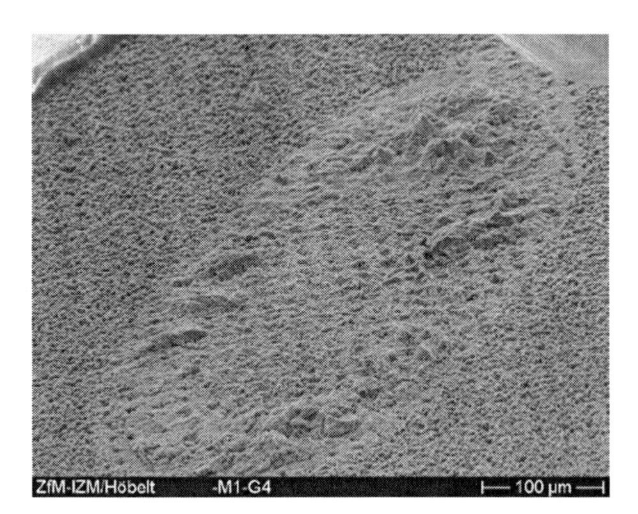

Fig. 11.45 Lift-off pattern as a result of power cycling

Figure 11.45 shows a magnified image of the lift-off area on the chip metallization. This example shows no adherence in the center of bond area. Improvements of the wire bond process can enhance the quality of the adherence significantly. In [Amr06] it was shown that an improved wire bond process in combination with a replacement of the chip solder by a silver sinter technology can achieve a very high-power cycling capability for power cycles up to $T_{high} = 200°C$. In the viewpoint of high-temperature applications, bond wires seem not to be the main limiting factor.

It was found that bond wires have a higher lifetime if they are coated with a polyimide cover layer [Cia01]. Additionally, optimization of the wire bond geometry and the wire bond material can increase the lifetime of bond wires during active power cycles.

Special attention has to be paid to gate wire bonds on chips with a center gate contact. For power devices with a field effect gate structure, the leakage current of the gate is so small that is takes days to discharge the gate via the leakage current. Therefore, a wire bond lift-off of the gate wire bond will not be noticed if the gate is continuously switched on and the load current is controlled by an external switch. In this case, a special functional test must be performed during the cooling phase of each cycle to verify the gate functionality. This can be done by switching of the gate voltage during the cooling phase. The constant current source, that supplies the sense current during this phase, must then go into voltage limitation mode. This technique ensures that a gate bond wire lift-off will not remain undetected.

11.6.7.5 Reconstruction of Metallization

A phenomenon observed during active power cycles with a high temperature swing is the reconstruction of the chip metallization. This contact metallization is conventionally made of a vacuum-metalized aluminum layer, which is formed in a grain structure. Due to the difference in thermal expansion between silicon and aluminum, this layer suffers from a considerable stress during repeated temperature swings. While the silicon chip is only marginally expanding with increased temperature (2–4 ppm/K), the grains of the Al metallization expand considerably (23.5 ppm/K). Thus, the metallization layer is subjected to a compressive stress during the heating phase of temperature cycles.

The surface reconstruction of aluminum films on silicon was first reported in the late 1960s [Pad68] followed by detailed investigations of this degradation effect. Comparison between temperature-cycled samples with annealed (uncycled) samples for equivalent time-at-temperature revealed that the surface reconstruction is increased by thermal cycling by a factor 2 to 5 depending on temperature and grain size of the aluminum film [San69]. Analysis of reconstruction phenomena at high temperatures (above 175°C) and low temperatures (below 175°C) suggests different fatigue mechanisms for these temperature ranges. Diffusional creep and plastic deformation involving conservative motion of dislocations are assumed as dominant contributions for high temperatures, while for low temperatures the only possible mechanism of mass transport is plastic deformation caused by compressional fatigue [Phi71].

While the former investigations were triggered by surface reconstruction observed in integrated circuits, the same effect was found in power electronic devices [Cia96]. The periodical compressive stress during the heating phase in temperature cycles results in plastic deformation of grains when the elastic limits are exceeded. According to [Cia01] this is the case for junction temperature above 110°C. The plastic deformation can lead to the extrusion of single grains. This process leads to an increasing surface roughness of the metallization with the macroscopic observable effect of a dull non-reflective surface appearance. In the cooling phase of the temperature cycle, tensile stress can lead to cavitation effects at the grain boundaries if the elastic regime is exceeded. This can explain the observed increase in electrical resistance of the surface metallization [Lut08].

Reconstruction of the aluminum contact layer was also found in failed devices after repetitive short-circuit operation of IGBTs [Ara08]. Recently, a similar effect of increasing surface roughness was observed also in thick aluminum layers of "direct bonded aluminum" (DBA) substrates after temperature cycling between −55 and 250°C. After 300 cycles the surface roughness of 300 μm thick aluminum layers on AlN substrates increased by more than a factor of 10, while a multitude of voids were observed in a cross section. The authors attribute this effect to grain boundary sliding [Lei09].

Figure 11.46 shows the optical image of such a metallization reconstruction after the power cycling test. The reconstruction appears as a milky white discoloration of the diode metallization. It is concentrated at the center of the chip. Especially interesting is the area around the solder void, which can be seen in the X-ray image. The reconstruction is also very pronounced in this region, corroborating that the reconstruction is generated by the temperature swing. The maximum temperature has its peak in the center of the die, while the thermal resistance is locally increased in the area of the solder void. The dissymmetry of the reconstruction clearly follows this temperature profile [Scn99].

Fig. 11.46 Optical image (*left*) and X-ray image (*right*) of a diode after active power cycling – the reconstruction of the metallization appears as a milky non-reflecting discoloration, which is most pronounced in the chip center and in the area of the solder void [Scn99]

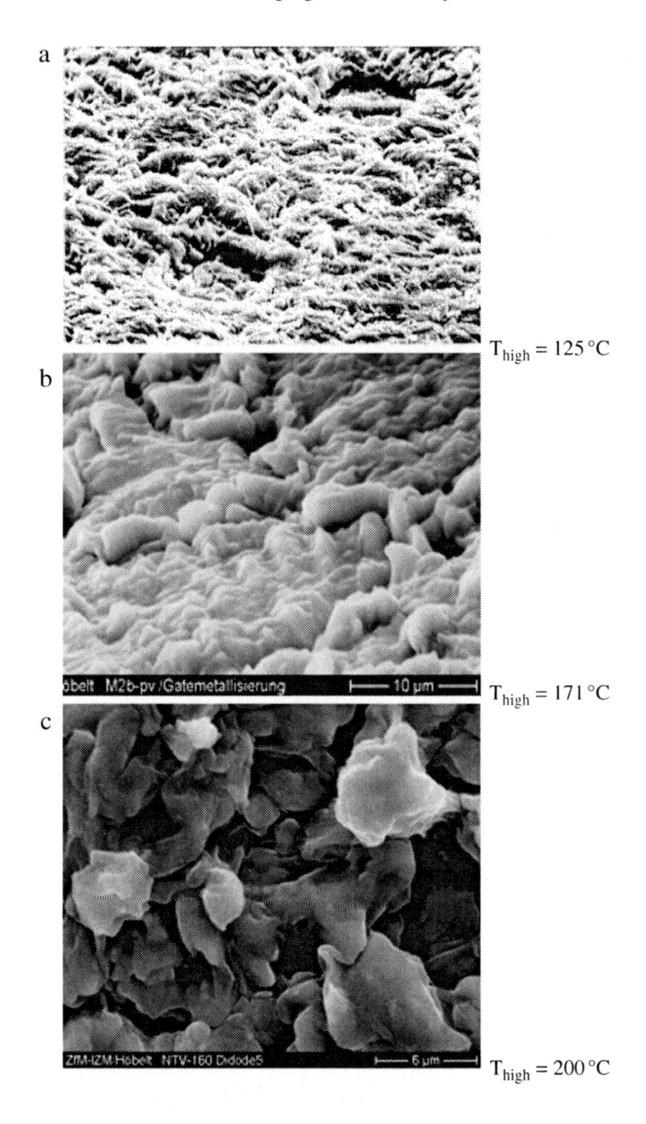

Fig. 11.47 REM images show the augmentation of contact reconstruction with increasing maximum cycle temperature T_{high} during power cycling

The impact of the maximum temperature during power cycling is illustrated in Fig. 11.47. Figure 11.47a shows an IGBT metallization after 3.2 millions of power cycles between 85 and 125°C; the figure is taken from [Cia02]. Figure 11.47b shows an IGBT metallization after 7250 power cycles with $\Delta T = 131\,\text{K}$ and $T_{\text{high}} = 171°\text{C}$. Figure 11.47c finally shows the metallization of a diode after 16,800 cycles at $\Delta T = 160\,\text{K}$ with $T_{\text{high}} = 200°\text{C}$; significant grains of approximately 5 μm diameter appear on the surface leaving voids in the metallization layer.

The reconstruction effect is suppressed beneath wire bond stitches. Figure 11.48 shows a detail of the metallization at the edge of a bond stitch after wire bond lift-off. This diode survived 44,500 cycles with $\Delta T_{\text{j}} = 130\,\text{K}$, $T_{\text{high}} = 170°\text{C}$. The high number of cycles was achieved because of single-side silver sinter technology

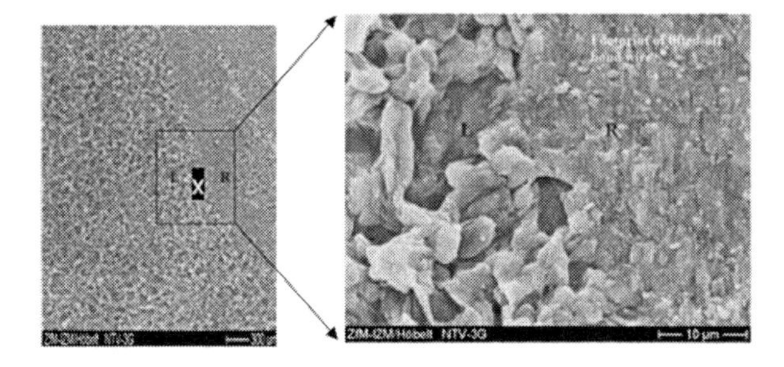

Fig. 11.48 REM image of the edge of bond stitch area after a wire bond lifted-off – the reconstruction is more pronounced outside of the stitch area. The diode failed after approximately 44,500 cycles with $\Delta T_{\mathrm{j}} = 130\,\mathrm{K}$ and $T_{\mathrm{high}} = 170°\mathrm{C}$

[Amr06]. Other investigations have shown that a polyimide cover layer suppresses the reconstruction effect as well [Ham01]. This is an expected phenomenon, because any cover layer will restrict the movement of the grains out of the contact layer. Nonetheless remains the high stress in the layer and it can be expected that the initiation and the growth of fractures in the interface of bond stitches is driven by the same CTE mismatch that generates the metallization reconstruction.

The reconstruction of the chip metallization reduces the density of the contact layer and therefore increases the specific resistivity of the contact. Since the layer thickness is typically in the range of 3–4 μm, the movement of grains sized in the range of the layer thickness as in Fig. 11.47c can be expected to change the conductivity of the layer considerably.

The layer resistivity can be measured according to the method of van der Pauw [Pfu76]. For the example of the power cycled diode in Fig. 11.47c, a specific resistance of 0.0456 mΩ m was measured. The resistivity of an unstressed diode of the same type was found to be 0.0321 mΩ m. This is close to the literature value for pure bulk Al of 0.0266 mΩ m; a slightly higher value can be expected because the chip metallization has a grain structure and contains some small admixture of Si. The observed resistivity increase of 42% during power cycling [Lut08] is a severe change of the device characteristics. Even though only a very small increase of V_{F} was detected during the power cycling tests of the devices in Figs. 11.47c and 11.48, an impact on the current distribution in the device must be expected which will reduce the lifetime of the power device under high stress conditions.

Latest investigations on device failure under repetitive short-circuit exposure below the critical thermal destruction level have shown that the increase in contact resistance caused by reconstruction of the contact layer is the root cause of the failure [Ara08]. It must be expected that this reconstruction also has an impact on the surge current capability of freewheeling diodes.

11.6.7.6 Solder Fatigue

The degradation of the solder interface is a fundamental failure mode during active power cycling. The so-called solder fatigue is caused by the formation of fractures in the solder interface, which lead to an increase in thermal resistance and are thus accelerating the total failure of the device. In devices with a positive temperature coefficient of the forward voltage drop (i.e. IGBT and MOSFET), the increasing

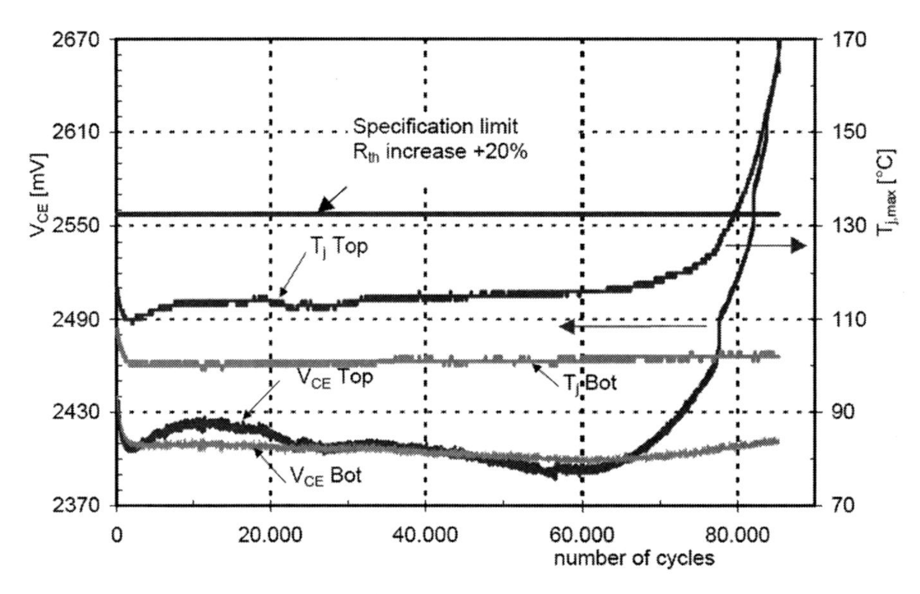

Fig. 11.49 Power cycling results in an IGBT phase leg in a module without base plate with $\Delta T_\mathrm{j} = 80$ K [Scn02b]. The temperature evolution is a strong indication of solder fatigue

temperature leads to increasing power losses and therefore expedites the degradation in a positive feedback loop. However, the final end-of-life failure of the conventional module design will typically be the breakdown of the wire bond contacts. If no investigation of the solder interface is performed by scanning acoustic microscopy (SAM) after a power cycling test, the impact of solder fatigue cannot be evaluated and the wire bond breakdown is often mistakenly assumed to be the root cause of failure.

Figure 11.49 shows the evolution of the forward voltage drop and the maximum temperature reached at the end of each heating phase of the IGBTs in the center phase leg of a six pack module without base plate. The maximum temperature starts to increase after 65,000 cycles and generates increasing power losses in the device due to the positive temperature coefficient. After 85,000 cycles, the temperature swing has already increased to $\Delta T_\mathrm{j} = 125$ K and it would continue to grow fast until the liquidus temperature of the solder is reached, if the wire bond contacts would not have failed first. This example illustrates that only marginal lifetime can be gained by improving the top side chip contact (i.e. the wire bonds) without improvement of the bottom side contact (i.e. the solder layer), because the increasing temperature amplitude would destroy any top side chip contact.

The geometry of the solder fatigue depicted in the SAM image in Fig. 11.50 follows the expected deterioration pattern. The discontinuity at the chip edges is responsible for a stress peak at the edges and especially at the chip corners. Therefore, the fractures start at the outside corners and edges and propagate toward the chip center. In this case, the arrangement of the four chips in parallel generates

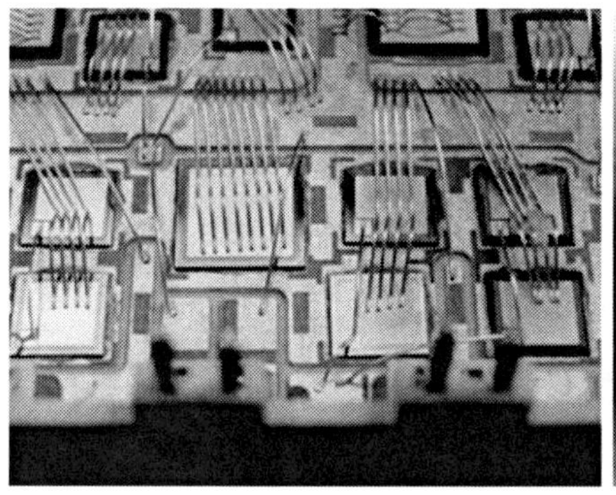
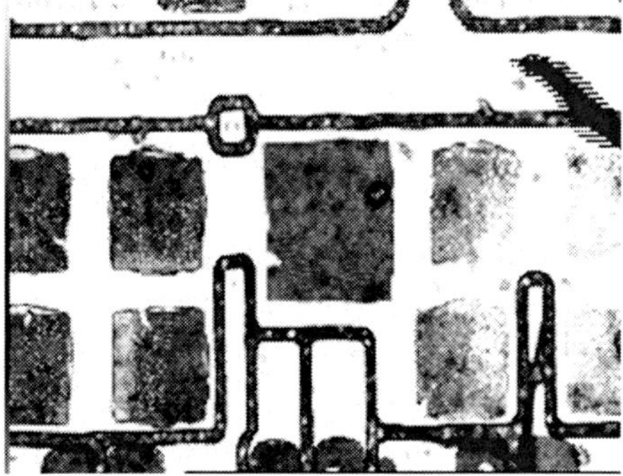

Fig. 11.50 Optical image and SAM image of the IGBT module in power cycling test shown in Fig. 11.46. The four IGBTs in parallel of the failed TOP switch (right side of the images) show the effect of solder fatigue

a temperature distribution with the maximum temperature being located at the chip corners close to the center of this quadruple chip assembly. Thus the degradation of the solder layers starts at the corners pointing toward the center of the group and moves outward.

The continuous improvement of the package thermal resistance and the enhanced heat extraction capability of advanced cooling systems have considerably increased the power density in modern IGBT chips. Assisted by the trend to minimized chip thicknesses (down to 70 μm for latest generation 600 V trench chips), which reduces the lateral thermal conductivity and thus diminishes possible heat spreading effects, the lateral temperature gradient in the chip has become more and more pronounced. Figure 11.51 illustrates this effect, where the diagonal temperature gradient exceeds 40°C for a 12.5×12.5 mm^2 IGBT on a water-cooled copper heat sink with 9°C cooling liquid temperature [Scn09].

For such pronounced lateral temperature gradients, the stress in the chip center generated by the high temperature exceeds the stress induced by the edge discontinuity and the degradation begins in the center instead of the edges and corners. This is confirmed by the SAM image of a state-of-the-art power module after the power cycling test in Fig. 11.52. Here the damaged region is located in the chip center, indicated by the light areas of high reflection, while the chip edges seem unaffected.

This observation was also reported from other authors. While some publications attribute this phenomenon to special solder types used [Mor01], other authors have observed this effect with lead-free and lead-rich solder systems [Her07]. The characteristics of the solder interface layer might have a small impact on the solder fatigue progress; however, the driving force for the deterioration of the solder interface results from the high stress induced by the differences in thermal expansion.

This change in the failure mechanism could have a considerable impact on the evolution of the degradation process. Fractures in solder interface increase the local thermal resistance of the affected chip region and thus raise locally the chip temperature. If the fractures start at the edges, the temperature of relatively cool chip

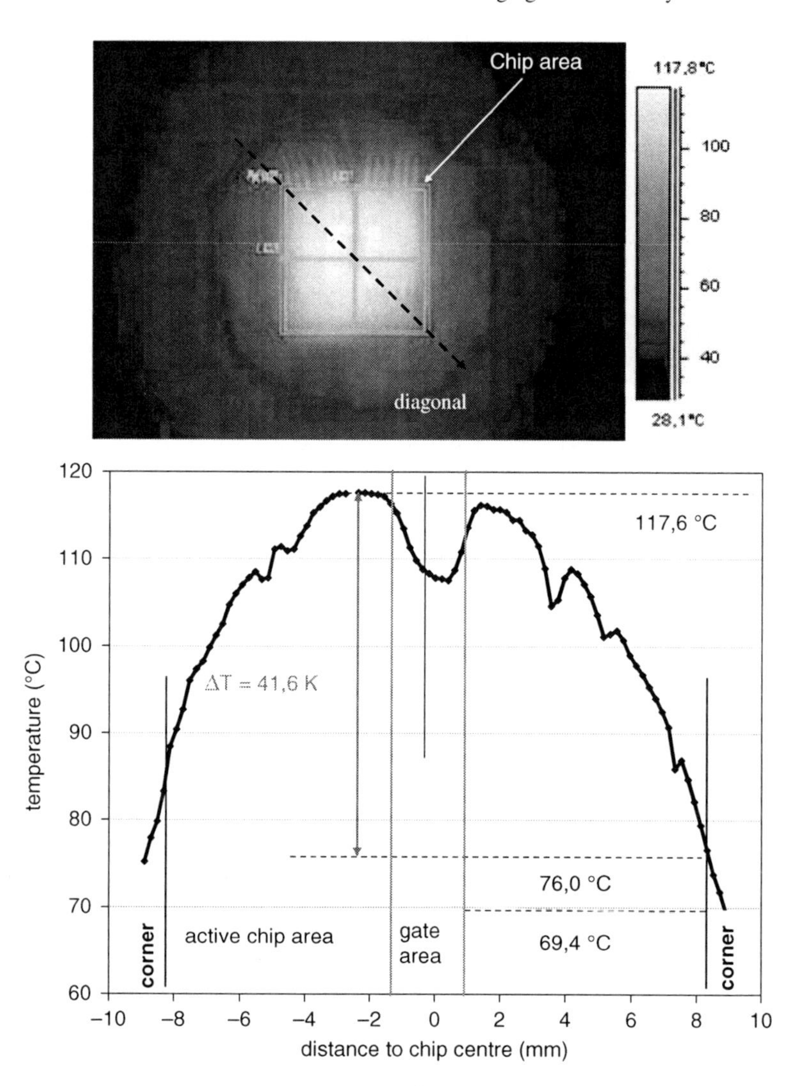

Fig. 11.51 Infrared image of a 12.5 × 12.5 mm^2 1200 V IGBT chip heated by a continuous DC current of 150 A. The module was mounted on a Cu water cooler with 9°C cooling liquid temperature. The equilibrium temperature distribution along a diagonal line through the chip center (*bottom* diagram) shows a temperature gradient between center and corner of more than 40°C

regions is increasing, while the maximum temperature remains unchanged in the chip center. The situation is different, when the fractures start at the center, which has the highest temperature to begin with. Therefore, with fractures in the center of the chip, the maximum chip temperature is immediately increased and it can be expected that this positive feedback loop will accelerate the fatigue progress and thus reduce the power module lifetime.

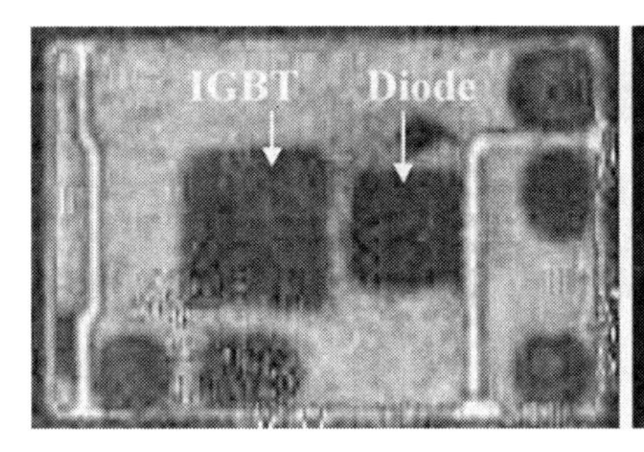

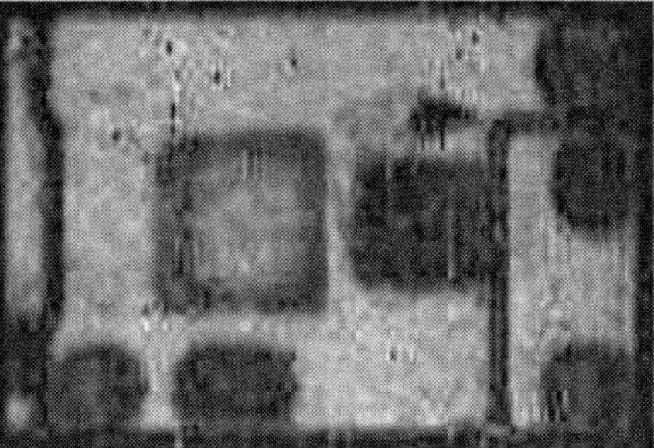

Fig. 11.52 SAM images of the initial solder layer (*left*) in comparison to the end-of-life image (*right*) after 113,400 power cycles of the IGBT with $T_{high} = 150°C$, $\Delta T_j = 70\,K$, heating phase ~2 s – the unstressed diode solder remains unchanged

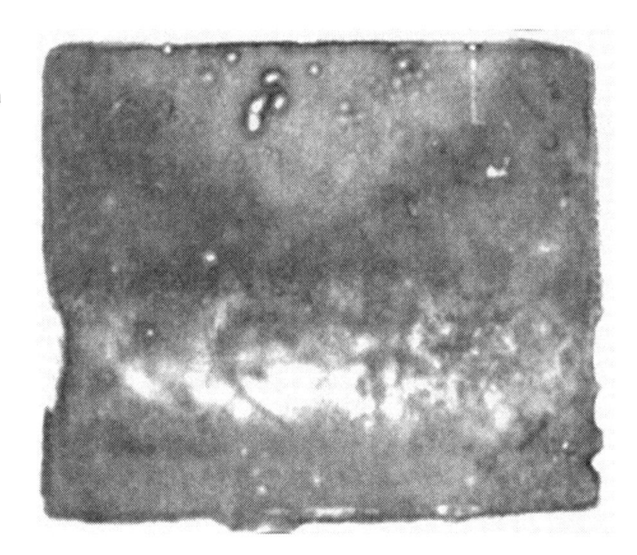

Fig. 11.53 SAM image of solder degradation below active power cycled IGBTs in a lead-free substrate solder joint of an AlSiC base plate module

The considerable temperature gradients on the chip level also generate temperature gradients in the solder layer between the substrate and the base plate. Test results from a power cycling test at $\Delta T_j = 67\,K$ and $T_{high} = 150°C$ on a module with AlSiC base plate show degradation effects in the substrate-to-base plate interface that start below the chip position as shown in Fig. 11.53

These results illustrate the complexity of solder fatigue effects due to the interaction of thermal and mechanical characteristics of the interconnection interface. The problem of solder fatigue must be solved to exceed the limits in reliability of the classical module design.

11.6.7.7 Power Cycling Capability of Molded TO Packages

A high power cycling capability was found for DBC-based transfer molded TO housings, which were described in the context of Fig. 11.6 [Amr04]. The power

Fig. 11.54 Failure analysis of a DBC-based soft-molded TO package after 75,000 power cycles with $\Delta T = 105$ K, showing a footprint of a detached bond wire in the foreground (*left*) and a heel crack in a still attached bond wire

cycling results in Fig. 11.40 were gained in power cycling test with this package type for $\Delta T = 105°C$ and $T_m = 92.5°C$. The number of cycles to failure (Weibull 50% accumulated probability) is about a factor of 10 higher than predicted by the LESIT results (Eq. (11.19)) and still significantly above the CIPS 08 model (Eq. (11.20)). Figure 11.54 shows a picture of bond wires in a device which survived 75,000 power cycles under said conditions. On the dark area in the foreground, a bond foot was initially attached. In the first of the still attached bond wires, a heel crack is visible.

The stiff mold material has a similar effect as a bond wire coating; it additionally hinders mechanically the detachment of bond wires from the chip surface. Even though the bond wire shows signs of heavy deterioration, the electrical contact is still maintained. In contrast to standard TO packages with copper lead frames, the implementation of Al_2O_3 substrates accounts for a lower thermal mismatch between the semiconductor material Si and the assembly layer.

The classical TO package with Cu-lead frames (Fig. 11.55) exhibit a great mismatch in thermal expansion between copper and silicon. For large chip sizes, the

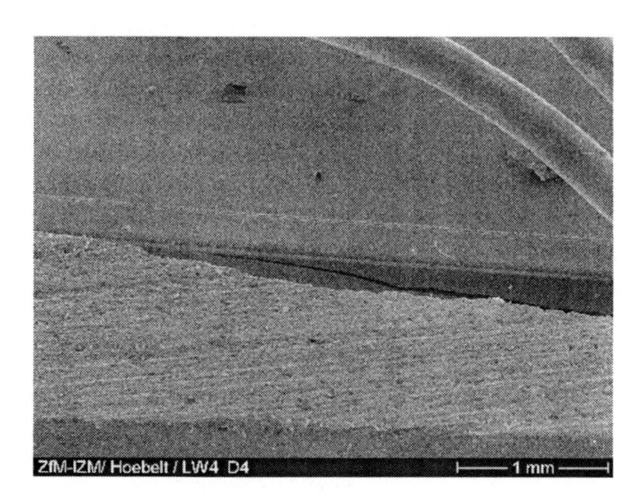

Fig. 11.55 Crack in the silicon diode chip in a transfer mold TO package on a Cu-lead frame after 3800 cycles with $\Delta T_j = 110$ K and $T_m = 95°$ C

power cycling capability was found to be clearly inferior to packages with ceramic substrates. In a power cycling test of standard TOs with chips of 63 mm^2 area, two of six chips lost their blocking capability after only 3800 cycles with $\Delta T_j = 110°C$ and $T_m = 95°C$. The failure analysis of these TOs revealed cracks in the silicon device as the root cause. Figure 11.55 shows such a fracture of a silicon chip.

Even though no more failures occurred up to more than 38,000 cycles when the test was continued with the four remaining samples, the early failures caused by fractures in the silicon device are alarming. Presumably, the relatively large area of the silicon devices is responsible for these early failures, because this effect was not observed for power cycling tests of smaller chips (< 30 mm^2) in the same package type.

11.6.7.8 Comparability of Power Cycling Lifetime Curves

The general statement on the comparability of qualification test results given at the beginning of this section holds even more for power cycling results of different manufacturers of power modules. The lifetime characteristics published by different manufacturers are of limited value for the evaluation of the lifetime in real application, because the test conditions and control strategies are mostly not disclosed.

In addition to the selection of the load current, the cooling system, the cycle duration, and the medium temperature – which all have an impact on the power cycling lifetime as discussed before – as well as the reaction to degradation effects during the test is of fundamental importance for the number of cycles to failure during a power cycling test. Four different strategies are possible for the control of a power cycling test, which account for degradation effects completely different:

- Cycle time control with fixed turn-on and turn-off times only. In this strategy, the desired ΔT_j is adjusted by selecting appropriate turn-on and turn-off times. These times are then kept constant during the test and no other control parameters are used to react to degradation effects. This is the most severe test condition, because degradation effects can increase the ΔT_j during the test and thus shorten the lifetime to failure considerably.
- Control of the turn-off and turn-off times by a reference temperature. This is the prevalent strategy used by European manufacturers, where either the case temperature or the heat sink temperature is used as control parameter. This strategy eliminates the impact of any changes of the cooling conditions (i.e. changes in coolant temperature or coolant flow) from the test progress. If the case temperature is used for the control, even changes in the case to heat sink thermal resistance have no impact on the test result.
- Control strategy that maintains constant power losses. This strategy, which is typically combined with fixed turn-on and turn-off times, keeps the power losses generated in each heating phase constant by controlling either the current or the gate voltage. Single bond wire failures increase the on-state voltage of a power device during the test and thus increase the power losses. For devices with a positive temperature coefficient (i.e. IGBT or MOSFET), solder fatigue may increase

the junction temperature and therefore will also increase the on-state voltage of the device (see Fig. 11.49, for example). The control for constant power losses will reduce the losses in these devices artificially and will thus lengthen the lifetime of the device. Some Japanese manufacturers use the control of the gate voltage during power cycling.

- Control strategy to maintain a constant temperature swing ΔT_j. In this strategy, the control parameter is the junction temperature itself, which is measured at the end of each heating phase. Typically, the turn-on time and the turn-off time during the cycling test are controlled to maintain a constant temperature swing, but a control of the current or the voltage drop by means of a gate voltage adaptation could alternatively be applied. In this case, no degradation effect in the module would alter the temperature swing. This is the least challenging test strategy and it will deliver the highest lifetime.

Lifetime curves for power modules always are based on accelerated test results, but they are of great importance for the estimation of lifetime in real applications. Therefore, the test results are extrapolated to real conditions in the field. However, no control strategy is implemented in real converter systems to account for degradation effects in the power module. Only a temperature sensor to monitor the heat sink temperature is incorporated into many power electronic applications. Thus, only the first two strategies should be applied for power cycling tests that are used as a basis for lifetime estimation.

11.6.8 Additional Reliability Tests

A test sequence similar to the example given in Table 11.6 is mandatory for all series products of a power module manufacturer. However, additional test sequences can be negotiated for specific applications.

For applications in extreme environmental conditions, special tests under corrosive atmospheres are recommended. Corrosive gases can interfere seriously with the reliable function of power modules. The silicone soft mold represents almost no protection against corrosive gases. SO_2 interacts with all metal surfaces except noble metals, H_2S is highly corrosive for silver and silver alloys, and Cl_2 together with high humidity will produce HCl, which corrodes non-noble metals, especially Al. The contribution of NO_x is not fully understood today, but its corrosion effect in connection with humidity is comparable to the impact of SO_2 and H_2S. The reliability of non-hermetically sealed modules must be verified in accelerated corrosive atmosphere tests in single or mixed corrosive gas environments for these applications.

Similar corrosion effects are connected with the impact of salt spray. This specific stress condition is found particularly in seaside or off-shore applications, which are typical environments for wind generator systems. The NaCl dissociates in aqueous solution and produces HCl, which over time can penetrate the soft mold cover layer and dissolve the Al wire bonds. If non-hermetically sealed modules are used in these environments, an additional protection must be implemented in the system to prevent corrosive degradation of power modules and driver PCBs.

Additional tests can be conducted for special application conditions to verify the suitability of power electronics modules in extreme environments.

11.6.9 Strategies for Enhanced Reliability

The standard test procedures discussed in the previous section always use new power modules fresh from production for every reliability test. Some experts in the field of reliability testing propose the combination of tests to increase the reliability of power electronic components. Even though the consecutive or simultaneous application of different stress conditions could accelerate the time to failure of a power module, the number of possible sequences and combinations is quite high and there is no basis of experience for such test conditions.

However, the idea was adapted in the concept of HALT/HASS testing. The "highly accelerated lifetime test" (HALT) is actually not a test, but much more a test philosophy that accompanies the development process. During the HALT procedure, a single device parameter is selected and it is stepwise increased even far beyond the specification limit until the sample finally fails. Typical examples are the stepwise increase of the insulation test voltage until failure or the stepwise increase of the temperature swing during passive temperature cycling test. The goal of this procedure is to increase the stress to the destruction limit. From the analysis of the failure, small modifications can sometimes improve the robustness of a design considerably.

In a second step, a "highly accelerated stress screening" (HASS) can be applied. Using the failure limits resulting from the HALT procedure, which typically are far beyond the specification limit, a HASS stress level can be defined, which is selected higher than the specification limit but a sufficient safety margin away from the destruction limit. For a fixed period of time 100% of a production is then subjected to an accelerated stress screening test, which allows to identify weak parts in a series production and improve the production process to enhance the product reliability. In this test sequence, combined stress levels are common. A typical example is the operation of a power module with high load currents under varying ambient temperature swings, sometimes even combined with the stress of mechanical vibration.

The general philosophy is developed further in the recently proposed concept of robustness validation [SAE08]. This concept combines experimentally determined failure limits and simulation results and compares these limits with the reliability requirements of applications, defined in mission profiles and specifications to increase the reliability of automotive electrical or electronic modules.

A general problem is the limited knowledge of the stress levels and the distribution of stress requirements in real applications. The implementation of the digital driver technology in power electronic systems gives the opportunity to monitor and record characteristic parameters during field operation and therefore has the potential to deliver more realistic data on the stress requirements of realistic field applications.

11.7 Future Challenges

Power semiconductor packaging has become a key technology for the progress of power electronic devices. There are four basic challenges to be met:

1. The current density in power devices is continuously increasing. Even today, the package-related voltage drop in a power module accumulates to a considerable percentage of the total voltage drop at nominal current. Thus, improved architectures with a reduced electrical resistance of the load current leads are required.
2. The increasing power density requirement of advanced power electronic applications enhances the power density per unit area. This development demands progressive technologies to extract the heat generated in modern power module designs.
3. The physics of silicon semiconductor devices allows maximum junction temperatures up to 200°C for selected applications. It can be expected that MOSFETs, IGBTs, and freewheeling diodes with a voltage rating of 600 V can be operated up to a maximum junction temperature $T_{\mathrm{j}} = 200°\mathrm{C}$ after the necessary improvements of leakage current levels and of the reliability of the passivation. Wide bandgap devices on the basis of SiC and GaN are capable of even higher operation temperatures. In consequence, the reliability under extended temperature swings and extended maximum temperature must be ensured, especially with respect to active power cycles. The established standard module architectures are not capable to meet these requirements today; new materials and interconnection technologies must be developed.
4. The parasitic inductances and capacities must be minimized or else controlled, so that they are transformed from undesirable obstacles into functional elements of power electronic circuits.

Finding a solution to these challenges is a task, which is intensively addressed by research and development groups all over the world. The "center for power electronic systems" (CPES), a consortium of five universities and several industry partners in the USA, has proposed to replace the aluminum wire bonds by a copper foil, which provides a larger effective cross section at the top side chip contact [Wen01]. The interconnection between this foil and the contact metallization of the chip is achieved by a "dimple array technique," where only localized indentations in the foil are soldered to the chip. However, this technology has so far not been implemented in a series production and the expected increased lifetime during active power cycling has not yet been demonstrated.

The integration of the cooling system into the base plate has been proposed in order to improve the heat transfer of power modules. This concept eliminates the need for thermal interface materials between the base plate and the heat sink, which accounts for a considerable contribution to the total thermal resistance. The technique of integrating the cooling system into the DBC substrate goes even further [Scz00], because it eliminates the interface between the substrate and the base plate as well. The substrate serves as an assembly layer for the chips and as a heat sink while providing an electrical insulation, thus combining tree functions in a single element. A drawback of this proposal is the comparatively small cross section of the cooling liquid channels, which is responsible for a high pressure drop in the cooling system. This makes the system vulnerable to pollution particles in the cooling system. Moreover, the reliability of such a high-tuned cooling system is of crucial importance for the device operation: A transgression of the maximum heat extraction capability would lead to the formation of a vapor layer between the cooled surface and the liquid flow, which would result in an instantaneous dramatic increase in the thermal resistance of the system. The short time constant of such a highly efficient system will result in an abrupt junction temperature increase, which will damage or possibly destroy the semiconductor device. This reliability issue is common to all highly effective cooling systems, which are currently investigated on the basis of heat pipes or based on impingement cooling methods.

A task of fundamental importance is the accomplishment of a sufficient power cycling lifetime at high maximum junction temperatures T_j. A very promising approach to reach this goal is the "low temperature joining technology" (LTJ). In this process, which is a diffusion sintering technique, a powder of silver particles is placed between the two surfaces to be joined. These surfaces require noble metal surface platings. An organic protection layer inhibits the silver particles to avoid diffusion of particles prior to the joining process. A heating process to approximately 250°C during the application of a high pressure to the sinter interface dissolves this protective coating and activates the diffusion of the silver particles. This results in a densification of the powder layer to a rigid interconnection layer of high reliability [Mer02].

The properties of this interconnection layer are superior to solder interfaces in all parameters. The specific thermal conductivity of the sinter layer can be as high as 220 $Wm^{-1} K^{-1}$ and is therefore almost a factor of four times higher than the thermal conductivity of a conventional SnAg(3.5) solder layer. Together with a characteristic layer thickness of < 20 μm, the sinter technology exhibits a reduced thermal resistance between the chip and the substrate compared to solder layers of typically 50 μm thickness. The electrical conductivity is also improved due to the low specific electrical resistance of silver.

However, the major advantage of the silver diffusion sinter interface is the high melting temperature of the interconnection. This advantage can be illustrated by the concept of homologous temperature. Mechanical engineers use this concept to evaluate the reliability of an interconnection under mechanical stress. The homologous temperature is the ratio of the operation temperature divided by the melting temperature of the material in absolute temperature. Figure 11.56 displays the homologous temperature for a conventional SnAg(3.5) solder interface,

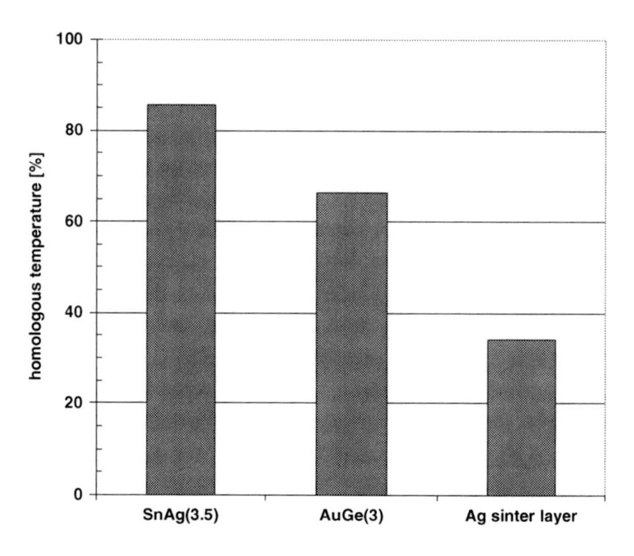

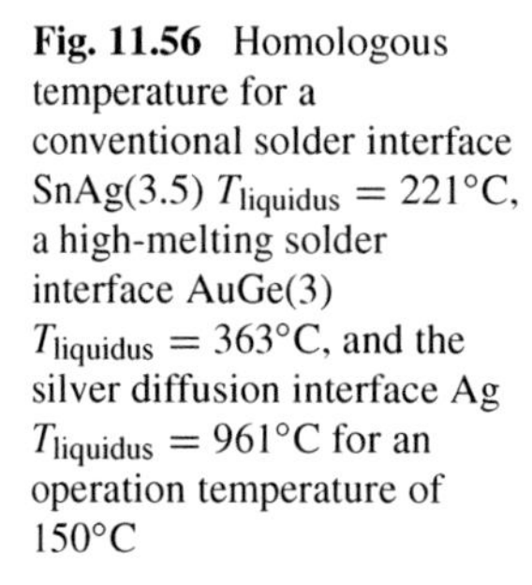
Fig. 11.56 Homologous temperature for a conventional solder interface SnAg(3.5) $T_{liquidus} = 221°C$, a high-melting solder interface AuGe(3) $T_{liquidus} = 363°C$, and the silver diffusion interface Ag $T_{liquidus} = 961°C$ for an operation temperature of 150°C

a high-temperature AuGe(3) solder interface, and the silver diffusion interface, assuming an operation temperature of 150°C.

Mechanical engineers consider interconnections operated below 40% of the homologous temperature as mechanically stable, between 40 and 60% as operated in the creep range, sensitive to mechanical strain, and above 60% as unable to bear engineering loads. Figure 11.56 illustrates clearly that even solder interfaces with a liquidus temperature of 363°C have a limited reliability for an operation temperature of 150°C, while the silver diffusion technology can be expected to be reliable under mechanical stress.

Another advantageous feature of the silver diffusion technology, which might be overlooked at the first glance, is the absence of a liquid phase during the connection process. In a solder process, the solder interface passes through a liquid phase, while the temperature exceeds the liquidus temperature. During this phase of the solder process, the chip swims on a liquid film with the consequence of a series of fundamental problems:

- The chip might shift or turn out of its desired position. Solder jigs or solder stop layers are necessary to minimize this effect. Both countermeasures require a considerable margin, so that the position accuracy in a solder process is limited.
- The solder layer can exhibit a wedge-shaped thickness distribution due to a variation of the surface wettability with considerable impact on the thermal resistance and thus on the reliability of the solder interface.
- Solder voids cannot be eliminated completely in an industrial series production.

Since the silver diffusion technology does not comprise a transition through a liquid phase, these problems well known from solder technologies are eliminated.

In a well-controlled silver diffusion process, the chips are perfectly aligned and the interface has a homogeneous thickness without any large-scale voids.

The diffusion sinter technology was adapted to the assembly of modern power devices like IGBTs, MOSFETs, and freewheeling diodes in the middle of the 1990s [Kla96]. Further process improvements verified that the simultaneous connection of multiple (different) power chips can be achieved in a single process step, which makes this technology compatible with modern series production [Scn97]. Recently, the first commercially available series power module was introduced, which contains not a single solder interface [Scn08]. This module design combines the silver diffusion technology with the pressure system technology and spring contacts.

Experimental results confirm the expected high reliability under extreme power cycles. A power cycling test with $\Delta T_j = 130$ K survived 30,000 cycles, which exceeds the estimated lifetime for classical base plate modules derived from an extrapolation of the LESIT curve (Eq. (11.19)) by more than a factor of 20 [Amr05]. This technology seems to be very promising even for maximum operation temperatures up to 200°C – as was investigated in active power cycling test with $\Delta T_j = 160$ K [Amr06] – which allows to extend the application of power modules to challenging environments, e.g., in the motor compartment of hybrid automobiles.

The silver diffusion sinter technology has the potential of replacing even the wire bonds by connecting a silver foil to the top side chip contact as shown in Fig. 11.57. This eliminates another weakness in the classical module architecture: the aluminum wire bond. This improvement reduces the parasitic resistance and inductance of the top side chip contact and further enhances the power cycling reliability [Amr05].

Evaluating potential improvements in progressive packaging technologies, it is inevitable to consider the complex interdependency of single optimizations with respect to the performance of the complete system. Especially the impact of parasitic influences cannot be omitted. This aspect will be addressed in more detail in Chap. 14.

Fig. 11.57 Silver diffusion sinter technology applied to the bottom and top side chip contact, eliminating the traditional solder interface and replacing the wire bonds with a silver foil. Source: TU Braunschweig

References

[Amr04] Amro R, Lutz J, Lindemann A: "Power Cycling with High Temperature Swing of Discrete Components based on Different Technologies", Proceedings of the PESC, Aachen (2004)

[Amr05] Amro R, Lutz J, Rudzki J, Thoben M, Lindemann A: "Double-Sided Low-Temperature Joining Technique for Power Cycling Capability at High Temperature" Proceedings of the EPE, Dresden (2005)

[Amr06] Amro R, Lutz J, Rudzki J, Sittig R, Thoben M: "Power Cycling at High Temperature Swings of Modules with Low Temperature Joining Technique", Proceedings of the ISPSD, Neapel, (2006)

[Ara08] Arab M, Lefebvre S, Khatir Z, Bontemps S: "Investigations on ageing of IGBT transistors under repetitive short-circuits operations", Proceedings PCIM, Nuremberg (2008)

[Bay08] Bayerer R, Licht T, Herrmann T, Lutz J, Feller M: "Model for Power Cycling lifetime of IGBT Modules – various factors influencing lifetime", Proceedings of the 5th International Conference on Integrated Power Electronic Systems, pp. 37–42 (2008)

[Cia96] Ciappa M, Malberti P: "Plastic-Strain of Aluminium Interconnections During Pulsed Operation of IGBT Multichip Modules", Quality and Reliability Engineering International 12, pp. 297–303 (1996)

[Cia01] Ciappa M: Some Reliability Aspects of IGBT Modules for High-Power Applications, Dissertation, ETH Zürich (2001)

[Cia02] Ciappa M: "Selected failure mechanisms of modern power modules", Microelectronics Reliability Vol. 42, pp. 653–667 (2002)

[Cia08] Ciappa M.: Lifetime Modeling and Prediction of Power Devices" Proceeding CIPS 2008

[Dow82] Downing S D, Socie D F: "Simple rainflow counting algorithms", International Journal of Fatigue 4, pp. 31–40 (1982)

[EFU99] eFunda engineering fundamentals, http://www.efunda.com/materials/

[Fer08] Feller M, Lutz J, Bayerer R: "Power Cycling of IGBT- Modules with superimposed thermal cycles". Proceedings of the PCIM, Nuremberg (2008)

[Fel09] Felsl HP: "Silizium-und SiC-Leistungsdioden unter besonderer Berücksichtigung von elektrisch-thermischen Kopplungseffekten und nichtlinearer Dynamik", PhD-Thesis, Chemnitz 2009

[Gut01] Gutsmann B, Silber D, Mourick P, Kolloquium Halbleiter-Leistungsbauelemente und ihre systemtechnische Integration, Freiburg (2001)

[Ham98] Hamidi A, Contribution à l'étude des phénomènes de fatigue thermique des modules IGBT de forte puissance destines aux application de traction, Dissertation, Grenoble (1998)

[Ham01] Hamidi A, Kaufmann S, Herr E: "Increased Lifetime of Wire Bond Connections for IGBT Power Modules", IEEE Applied Power Electronic Conference and Exhibition (APEC), Anaheim (2001)

[Hec01] Hecht U, Scheuermann U: "Static and Transient Thermal Resistance of Advanced Power Modules", Proceedings of the 43th PCIM, Nürnberg (2001)

[Hel97] Held M, Jacob P, Nicoletti G, Sacco P, Poech MH: "Fast power cycling test for IGBT modules in traction application", Proc. Power Electronics and Drive Systems (1997)

[Her07] Herrmann T, Feller M, Lutz J, Bayerer R, Licht T: "Power Cycling Induced Failure Mechanisms in Solder Layers" Proceedings EPE 2007, Aalborg

[Jor09] Jordà X, Perpiñà X, Vellvehi M, Millán J, Ferriz A: "Thermal Characterization of Insulated Metal Substrates with a Power Test Chip", Proc. ISPSD Barcelona, pp. 172–175 (2009)

[Kam93] Kamon M et al.: "Fast Henry - A Multipole Accelerated 3-D Inductance Extraction Programm", Proc. 30th ACM/IEEE Design Automation Conference, pp. 678–683 (1993)

[Kla96] Klaka S: Eine Niedertemperatur-Verbindungstechnik zum Aufbau von Leistungshalbleitermodulen, Dissertation, Braunschweig (1996)

[Lap91] Lappe, Conrad, Kronberg, Leistungselektronik, 2. Auflage, Verlag Technik Berlin 1991

[Lei09] Lei T G, Calata J N, Lu G-Q: "Effects of Large Temperature Cycling Range on Direct Bond Aluminum Substrate", CPES Power Electronics Conference, Virginia Tech (2009)

[Lin01] Lindemann A, Kolloquium Halbleiter-Leistungsbauelemente und ihre systemtechnische Integration, Freiburg (2001)

[Lin02] Lindemann A, Friedrichs P, Rupp R: "New Semiconductor Material Power Components for High End Power Supplies", Proceedings of the 45th PCIM, pp. 149–154, Nuremberg (2002)

[Lut08] Lutz J, Herrmann T, Feller M, Bayerer R, Licht T, Amro R: "Power cycling induced failure mechanisms in the viewpoint of rough temperature environment", Proceedings of the 5th International Conference on Integrated Power Electronic Systems (2008)

[Mer02] Mertens C, Sittig R: "Low Temperature Joining Technique for Improved Reliability", Proc. 2nd International Conference on Integrated Power Systems CIPS, pp. 95–100 (2002)

[Mik01] Mikkelsen JJ: "Failure Analysis on Direct Bonded Copper Substrates after Thermal Cycle in Different Mounting Conditions", Proc. Power Conversion PCIM Nuremberg 2001, pp. 467–471

[Min45] Miner M A: "Cumulative Damage in Fatigue", Journal of Applied Mechanics, v01.12 (1945)

[Mou02] Mourick P, Steger J, Tursky W: "750A 75 V MOSFET Power Module with Sub-nH Inductance", Proceedings of the ISPSD, pp. 205–208 (2002)

[Mor01] Morozumi A, Yamada K, Miyasaka T: "Reliability Design Technology for Power Semiconductor Modules", FUJI ELECTRIC JOURNAL 2001 Vol.74-No.2

[Pad68] Paddock A, Black J R: "Hillock Formation on Aluminum Thin Films", presented at the Electrochemical Society Meeting, Boston, May 5-9 (1968)

[Pfu76] Pfüller S: Halbleiter Messtechnik, VEB Verlag Technik, Berlin 1976, S. 89

[Phi71] Philofsky E, Ravi K, Hall E, Black J: "Surface Reconstruction of Aluminum Metallization – A New Potential Wearout Mechanism", IEEE International Reliability Physics Symposium 9, pp. 120–128 (1971)

[Poe04] Poech MH, Fraunhofer-Institut Siliziumtechnologie, Itzehoe, private communication (2004)

[SAE08] SAE/ZVEI: Handbook for Robustness Validation of Automotive Electrical/Electronic Modules, www.zvei.org/ecs, (2008)

[San69] Santoro C J: "Thermal Cycling and Surface Reconstruction in Aluminum Thin Films", Journal of the Electrochemical Society 116, pp. 361–364 (1969)

[Scn97] Scheuermann, U, Wiedl P: "Low Temperature Joining Technology - a High Reliability Alternative to Solder Contacts", Workshop on Metal Ceramic Composites for Functional Applications, Wien, pp. 181–192 (1997)

[Scn99] Scheuermann U: "Power Module Design for HV-IGBTs with Extended Reliability", Proceedings of the PCIM, Nuremberg, pp. 49–54 (1999)

[Scn02b] Scheuermann U, Hecht U: "Power Cycling Lifetime of Advanced Power Modules for Different Temperature Swings", Proceedings PCIM, pp. 59–64, Nuremberg (2002)

[Scn06] Scheuermann U: "Aufbau- und Verbindungstechnik in der Leistungselektronik", in Schröder D, Elektrische Antriebe Bd. 3 – Leistungselektronische Bauelemente, 2.Auflage, Springer Berlin 2006

[Scn08] Scheuermann U, Beckedahl P: "The Road to the Next Generation Power Module – 100% Solder Free Design", Proc. CIPS 2008, ETG-Fachbericht 111, pp. 111–120, Nuremberg, 2008

[Scn09] Scheuermann U, Schmidt R: "Investigations on the $V_{CE}(T)$ Method to Determine the Junction Temperature by Using the Chip Itself as Sensor", Proceedings PCIM, Nuremberg (2009)

[Scz00] Schulz-Harder T, Exel J, Meyer A, Licht K, Loddenkötter M: "Micro Channel Water Cooled Power Modules", Proceedings PCIM, pp. 9–14, Nuremberg (2000)

[Swa00] Swawle A, Woodworth A: "Innovative Developments in Power Packaging Technology Improve Overall Device Performance", Proceedings of the 41st PCIM, pp. 333–339 (2000)

[Swa01] Swawle A, Standing M, Sammon T, Woodworth A: "Directfet™ - a Proprietary New Source Mounted Power Package for Board Mounted Power", Proceedings of the 43rd PCIM, pp. 473–477 (2001)

[Wen01] Wen S, Huff D, Lu GQ, Cash M, Lorenz RD: "Dimple-Array Interconnect Technique for Interconnecting Power Devices and Power Modules", Proc. CPES Seminar, pp. 75–80, Blacksburg (2001)

[Zhg04] Zhang J, "Choosing the Right MOSFET Package", IR application note Feb 2004, http://www.eepn.com/Locator/Products/ArticleID/29270/29270.html

Chapter 12
Destructive Mechanisms in Power Devices

This chapter will deal with some destructive mechanisms in power devices, and typical failure pictures for them will be shown. Failure analysis requires a lot of experience, especially regarding the conditions in the power circuit at failure, which must be carefully considered. Although some of the failure pictures appear to be similar, it is difficult to draw conclusions only from pictures. However in practice the engineer often has the problem to find the reason for failures, and the following sections might be helpful.

12.1 Thermal Breakdown – Failures by Excess Temperature

In Chap. 2 the intrinsic carrier density n_i was explained, it strongly depends on the temperature as described in Eq. (2.6). In silicon, n_i amounts approx. to 10^{10} cm^{-3} at room temperature, and it is negligible compared to the background doping. However, n_i increases rapidly with increasing temperature. Therefore, at very high temperatures, the thermal generation becomes the dominant mechanism for the creation of carriers.

With the introduction of an intrinsic temperature T_{int} similar to [Gha77], one can estimate, when the rise of some critical mechanisms in a device with increasing temperature can be expected. T_{int} is that temperature, at which the density of carriers n_i generated by thermal generation is equal to the background doping N_D. It is drawn in Fig. 12.1 as a function of N_D. Below T_{int} the carrier density is only weakly dependent on temperature. Above T_{int} the carrier density increases exponentially with temperature according to Eq. (2.6). From Fig. 12.1 we can see that for a high-voltage device, which requires an N_D in the range of 10^{13} cm^{-3}, T_{int} will be reached at a much lower temperature compared to a device with a lower voltage rating, at which, e.g. a range of 10^{14} cm^{-3} is used for N_D.

However, this point of view is much too simplified. The intrinsic carrier density is defined for thermal equilibrium, while a device in application is usually never operating in thermal equilibrium. Therefore, one must also consider in which operation mode and by which effect an increased or locally increased temperature is reached.

For low-voltage MOSFETs operated in a short time interval in avalanche breakdown, it was found that destruction occurs if the temperature increases just below

J. Lutz et al., *Semiconductor Power Devices*, DOI 10.1007/978-3-642-11125-9_12,

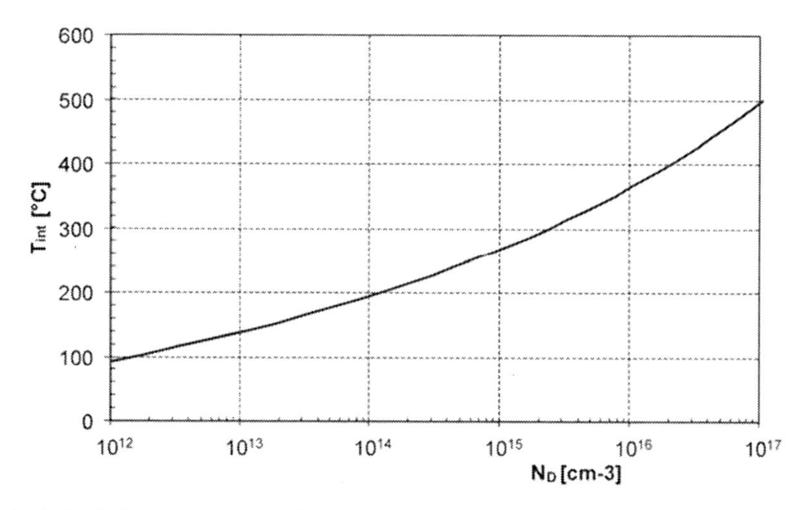

Fig. 12.1 Intrinsic temperature in silicon as a function of background doping

the temperature T_{int}, which was in the range 320°C for the used 60 V MOSFET [Ron97]. This agrees with Fig. 12.1, if the doping N_{D} of 4×10^{15} cm^{-3} is assumed which is reasonable for this voltage range.

If a bipolar device is in the forward conduction mode, e.g., in a surge current event, it is flooded with free carriers in a density in the range above 10^{16}, even slightly above 10^{17} cm^{-3} can occur. When the thermal generation amounts to a carrier density in this range, it becomes the dominating mechanism. Therefore, in such short time events, very high temperatures may occur without a failure in the device. In the surge current mode, a T_{int} up to the range of 500°C is to be expected.

In the blocking mode, a space charge region builds up and carriers are removed from the depleted zone. Their density is given by the leakage current, which is low for most modern power devices (except for gold-diffused devices). Thermal stability is now determined by the height of the leakage current.

Therefore the associated mechanism which leads to a high temperature must be taken into account. For the investigation of stability, one has to consider the temperature dependency of the electrical mechanism which leads to high losses and therewith high temperature.

If the heating is caused by a high leakage current, the leakage current will further increase in regions of high temperature. These regions will get hotter, which again leads to more increased leakage current. Such behavior shall be termed here as a *positive feedback*. If the high power loss density cannot be extracted by the cooling of the device, the device will be destroyed inevitably.

If the high losses are generated by a high voltage above the breakdown voltage V_{BD}, which drives the device in the avalanche breakdown, losses are created and the temperature increases. However, with increasing temperature, V_{BD} increases. The region, where avalanche breakdown occurs, will move to the regions of the device where the temperature is lower. Even if such electrical mechanisms lead to local filaments, the increase of temperature releases the local stress, leading to a *negative feedback effect*.

However, if the temperature reaches T_{int}, the thermal generation will become the dominating effect, and then the temperature increase acts as a positive feedback. Inhomogenities in current density, even if they are small, will be amplified rapidly. If the temperature reaches T_{int}, one has to expect current tubes or filaments, considering a device with an area above some square millimeter.

If P_{gen} is the generated power density and P_{out} is the power density that can be maximally be drawn out via the package and heat sink, one can express a condition for thermal runaway [Lin08]:

$$\frac{\partial P_{gen}}{\partial T} > \frac{\partial P_{out}}{\partial T} \tag{12.1}$$

If this condition is fulfilled for a stationary operation point, a fast exponential temperature increase will occur. Equation (12.1) is a general form of an equation that was used for thermal stability of bipolar transistors [How74]:

$$S = R_{th} \cdot V_C \frac{\partial I_C}{\partial T} \tag{12.1a}$$

If the stability factor S increases > 1, any temperature disturbance will grow and thermal runaway will occur.

Finally, the destruction of the device is always due to high temperature. Failed devices show local regions of molten semiconductor material. If the temperature increase occurs locally, in a very small point-like area of the device, one can find even cracks in the crystal lattice. However, it must be distinguished by which effect the temperature increase was generated. Some of these effects will be described below.

As a simple example, Fig. 12.2 shows an IGBT device, which has failed due to very high power losses. The IGBT was only stressed with forward current and the failure was caused by a too low gate voltage V_G. One can recognize a comparatively large molten area (>1 mm^2) at the emitter side. It is located typically close to the center of the device and close to the bond wires.

If overtemperature occurs in an application in which the device is switched between the blocking mode and the conducting mode at a high frequency, the destruction picture may be different. With increased temperature, the blocking capability is lost first. In almost all devices with planar junction termination the breakdown will occur at the edge. Therefore the point of destruction will be at the edge of the device or at least a small part of the edge should be included.

12.2 Surge Current

In the application of a diode or a thyristor in a rectifier, momentary high over-current pulses can occur. Therefore the possible surge current is determined and given in the datasheets for rectifier diodes, fast diodes, and thyristors. During the qualification

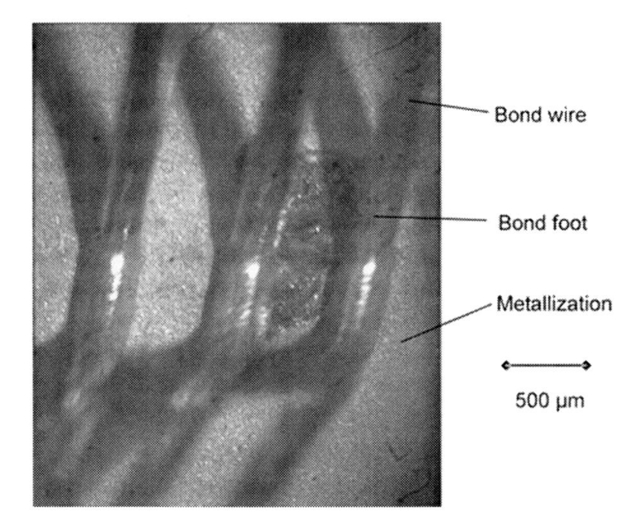

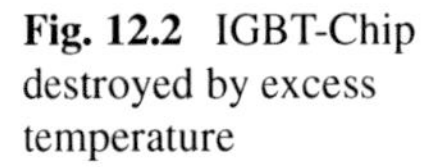
Fig. 12.2 IGBT-Chip destroyed by excess temperature

of a diode, a single sinus half-wave of the grid current is imposed in the forward direction of the diode. Figure 12.3 shows the surge current measurement of a fast 1200 V diode with an area of 7×7 mm^2, the waveforms of current, voltage, and additionally the power $p = v \cdot i$ are pictured as a function of time.

Due to the junction voltage of the measured diode and the junction voltage of a thyristor in the measurement equipment, the current pulse duration is not 10 ms as it should be for a grid frequency of 50 Hz, but it is 7.5 ms. In Fig. 12.4 the *I–V*

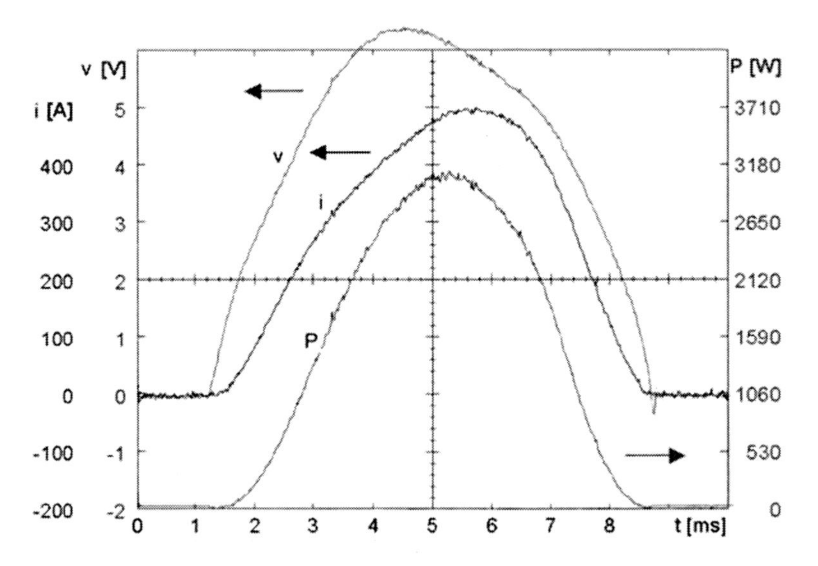

Fig. 12.3 Surge current load of a fast diode. Voltage, current, and power depending on time

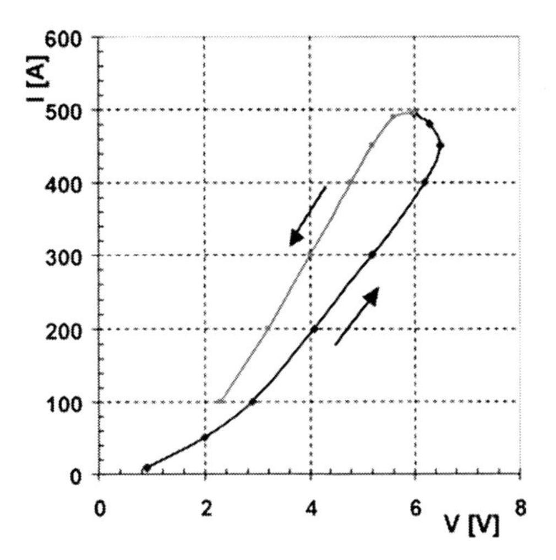

Fig. 12.4 *I–V* characteristic of the surge current load in Fig. 12.3

characteristic from the measurement in Fig. 12.3 is shown. Since a high temperature is reached in the device, the characteristic splits into an ascending and descending branch. In the descending branch the voltage drop is significantly lower.

For the dissipated power in Fig. 12.3 the temperature in the semiconductor is estimated in Fig. 12.5 with a thermal simulation using the simulator SIMPLORER. The diode is packaged as shown in Fig. 11.13, the thickness of the Al_2O_3 ceramic is 0.63 mm. The heat flux is fed in the volume of the low-doped n^--layer of the device in the form of a sinus-square function within a time of 7.5 ms and amplitude of 3060 W, according to the measurement in Fig. 12.3. The temperature dependency

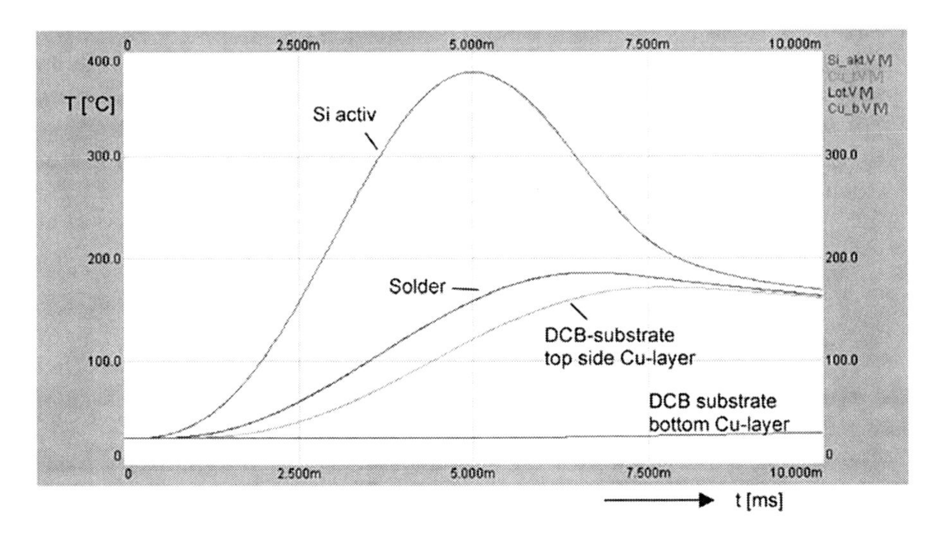

Fig. 12.5 Simulation of temperatures for the surge current event in Fig. 12.3

of thermal conductivity in silicon is considered according to Eq. (11.5), since this will be significant for the result at the expected high temperature increase. In the estimation in Fig. 12.5, the temperature in the n^--layer (Si-active) of the device increases up to 382°C.

This high temperature may explain the strong difference of the forward voltage in Fig. 12.4 in the descending branch compared to the ascending branch. The voltage drop V_F consists of $V_F = V_j + V_{drift} \cdot I_F$ decreases at high temperature because n_i is strongly increasing with temperature. For V_{drift}, the temperature dependency can be discussed using Eq. (5.47)

$$V_{drift} = \frac{w_B^2}{(\mu_n + \mu_p) \cdot \tau_{eff}} \tag{12.2}$$

where τ_{eff} contains the carrier lifetime τ_p as well as the emitter influence, see Eq. (5.52).

To this temperature dependency contribute the following effects:

- The carrier lifetime. It increases with temperature. This leads to a decrease of the forward voltage drop with temperature.
- The emitter recombination: For this, Auger recombination is important and at a high current density τ_{eff} will become very small. Additionally, in many modern devices the emitter depth is smaller than the diffusion length and then the emitter depth must be used in the emitter parameter. However, no strong temperature dependency of emitter recombination is to be expected.
- The mobilities. They decrease strongly with temperature. This effect leads to an increase of the forward voltage.
- The temperature dependency of the resistance of metallization and bond wires. This resistance increases with temperature.
- Finally the density of thermal-generated carriers n_i depends strongly on temperature. If n_i significantly contributes to the density of free carriers – this is in the range of 10^{17} cm^{-3} for surge current events – then a significant decrease of the forward voltage is to be expected. Now Eq. (12.2) is no longer valid. However, Eq. (5.34) which was used to derive the basics of forward conduction can be written as

$$V_{drift} = \frac{j}{q} \int_0^{w_B} \frac{1}{\left(\mu_n(x) + \mu_p(x)\right) p(x)} dx \tag{12.3}$$

The density of free carriers $p(x) \approx n(x)$ is strongly increasing and V_{drift} is decreasing.

The form of the I–V characteristic will therefore be very different for different diode fabrication technologies. The behavior like Fig. 12.4 is normally observed in some special fast diodes. Finally, the failure can be caused by the following mechanisms:

(a) Melting of the top side metallization. This occurs especially in bonded diodes in power modules.
(b) Mechanical destruction, cracks in the device, caused by very high temperature and mechanical stress resulting from thermal expansion.
(c) If finally n_i is dominating, the characteristic behavior of a resistor with a negative temperature coefficient occurs [Sil73]. The forward voltage decreases strongly. A positive feedback occurs. A filamentation of the current into tubes with very high current density is to be expected.

In [Sil73] it is estimated that the negative temperature coefficient of the resistance according to mechanism (c) occurs after the temperature has increased to an amount, at which n_i approximates to $0.3\,\overline{n}$. There are some hints that there is a dependency on the device area. Small diodes can bear a higher current density since there are fewer possibilities for the formation of filaments. If the failure occurs according to (c), then it will be typically close to the edge of the active area, since these are the locations with the highest current densities.

For wire-bonded diodes in IGBT modules, as shown in Fig. 11.13, the anode layer with a typical low junction depth is on the top. The heat dissipating volume is close to the metallization layer and bond wires. For such diodes, failure according to mechanism (a) is expected. In a more detailed simulation with the Sentaurus$_{TCAD}$ device simulator [Syn07], it was found that the metallization layer and the bond foot arrangement are of high influence to the occurring temperatures and the surge current capability [Hei08b]. The thicker the metallization the better is the capability to absorb the heat especially in short times, since it acts as additional thermal capacitance. Another important factor on the surge current capability of diodes is the location and size of the contact area of the bond wires. A high ratio of bond foot area to the diodes anode will increase the surge current capability.

A surge current stress below the destruction limit will not lead to irreversible modifications in the semiconductor itself. If the surge current is increased above the value in Fig. 12.4, the split of the characteristics will increase and higher temperature will occur in the device. Finally destruction of the device occurs.

However, already in the simplified temperature estimation in Fig. 12.5 the temperature in the chip solder layer grows up to 186°C. Such a temperature is already close to thermal softening of solder layers. Therefore irreversible modifications in solder layers, and also in metallization and bond wires, may occur. The surge current capability therefore is intended for singular overload events and it is not intended for regular operation of a power semiconductor.

The temperature in the bottom Cu-layer of the DCB substrate has only grown up to 27°C, this is negligible. Therefore the influence of further components of the package can be neglected at surge current conditions; hence all effects happen in the semiconductor and in the immediately adjacent layers.

The surge current capability of a fast diode is typically at 10 – 12 times of the rated current. The surge current capability of a diode for grid frequency operation or of a thyristor is typically in the range of 20 times of the rated current, since these

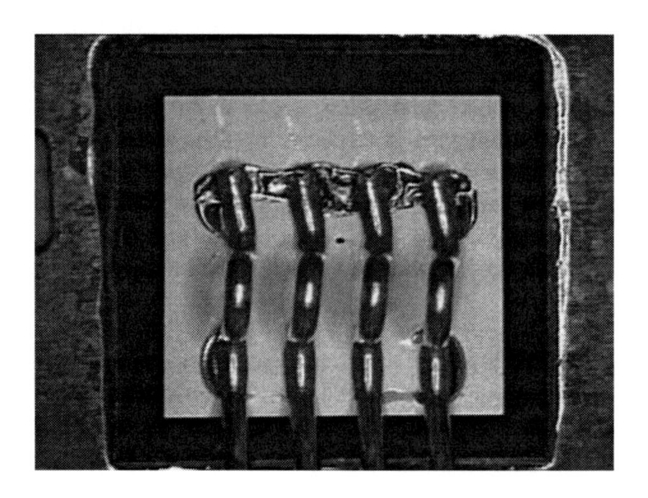

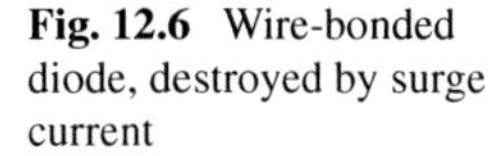

Fig. 12.6 Wire-bonded diode, destroyed by surge current

devices are manufactured with high carrier lifetime, and the forward voltage drop is lower.

A diode which failed due to surge current is shown in Fig. 12.6. The molten regions close to the bond feet are typical. The hottest surface spot at surge current is beside the bond feet [Hei08b]. The failure mechanism here is according to (a). The molten area for this case is always in the active area of the device. Such pictures allow a clear identification as a surge current fault during analysis in the quality department of the device manufacturer.

The failure in an application must not be during a sinus-shaped pulse. The current pulse might be of a different shape. A general description is done by the i^2t value in data sheets, which holds for arbitrary current pulses. The failures are due to exorbitant heating of the device by too high currents and the occurrence of one of the described mechanisms.

12.3 Overvoltage – Voltage Above Blocking Capability

The blocking capability of power devices is limited by avalanche breakdown. Avalanche breakdown occurs above the rated voltage of the device. Most power devices can sustain some current in the avalanche breakdown mode. However, the data sheet of the manufacturer excludes operation of the device in the avalanche mode, if the device is not avalanche rated.

Several MOSFETs and diodes in the range up to 1000 V are avalanche rated; this allows short-time operation in the avalanche mode. The maximum dissipated energy in the avalanche E_{av} is specified in the data sheets of the manufacturers, in general form it is given by :

$$E_{av} = \int_{t_{av}} V_{BD} \cdot i(t)\mathrm{d}t \tag{12.4}$$

where V_{BD} is the device breakdown voltage, $i(t)$ the current pulse in avalanche, and t_{av} is the time of the current pulse. Usually avalanche occurs at "unclamped inductive switching," where the MOSFET is turned off in a circuit with an inductor L in series. The voltage now rises up to the breakdown voltage and the current decreases with

$$\frac{\mathrm{d}i}{\mathrm{d}t} = \frac{V_{\mathrm{BD}} - V_{\mathrm{bat}}}{L} \tag{12.5}$$

If $i(t)$ decays linearly from $I_{\mathrm{av(peak)}}$ to zero during t_{av}, as it is the case when the energy of an inductor is dissipated, Eq. (12.4) can be expressed as

$$E_{\mathrm{av}} = \frac{1}{2} \cdot V_{\mathrm{BD}} \cdot I_{\mathrm{av(peak)}} \cdot t_{\mathrm{av}} \tag{12.6}$$

and the dissipated energy in this case is the stored energy in the inductor $0.5 \cdot L \cdot I^2_{\mathrm{av(peak)}}$ and additionally the energy $0.5 \cdot V_{\mathrm{bat}} \cdot I_{\mathrm{av(peak)}} \cdot t_{\mathrm{av}}$ which is delivered by the voltage source V_{bat} during t_{av}. Avalanche rating of low-voltage MOSFETs can be up to $E_{\mathrm{av}} = 1$ J, such energy can only be dissipated in single events and never in continuous mode at a high operation frequency.

The design of the MOSFETs for avalanche capability is in a way, that the breakdown occurs in the volume of the device, e.g., in a planar MOSFET at the p^+-layer in the center of the cell (Fig. 9.4) where the n^--base is narrowest. Avalanche capability is also possible for trench MOSFETs (Fig. 9.6), for these structures an additional design effort is necessary to avoid the location of avalanche at the trench corners [Kin05].

The occurrence of breakdown in the volume of the device is also possible for diodes with a beveled junction termination of positive angle (Fig. 4.17). The breakdown will occur at the edge for diodes with planar junction terminations with floating potential rings (Fig. 4.19). Well-designed potential rings can also bear current in the avalanche mode, even if it flows mainly at the edge. However, above 1200 V, one can only rarely find avalanche rating and if it is with more restrictive conditions.

For avalanche capability, as second condition is that branches with negative differential resistance (NDR) must be avoided. This will be explained more in detail in the following.

For a higher rated blocking voltage the device must have a lower doping, see Fig. 3.17. If avalanche occurs, the shape of the electric field is triangular or trapezoidal and electron–hole pairs are generated in the region with high electric field. The generation depends exponentially on the electric field and the main part is generated close to the pn-junction.

Holes are flowing to the anode and electrons to the cathode. Within the space charge, we can include the generated carriers p_{av} and n_{av} in the basic equation (2.90) according to their polarity, in one-dimensional expression as

$$\frac{\mathrm{dE}}{\mathrm{d}x} = \frac{q}{\varepsilon_0 \cdot \varepsilon_{\mathrm{r}}}(N_{\mathrm{D}^+} + p_{\mathrm{av}} - n_{\mathrm{av}}) \tag{12.7}$$

At the anode side border of the space charge, no electrons arrive and the reverse current is carried only by holes. We neglect the small contributions of diffusion current and recombination center-induced leakage current. Then at this position holds

$$p_{av} = \frac{j_R}{q \cdot v_{sat}(p)} \tag{12.8}$$

where j_R denotes the reverse current and $v_{sat}(p)$ denotes the saturation velocity of holes since at the given high electric field. At the cathode-side border the arriving reverse current is pure electron current. The density of generated electrons is

$$n_{av} = \frac{j_R}{q \cdot v_{sat}(n)} \tag{12.9}$$

with $v_{sat}(n)$ of 1.05×10^7 cm/s at $T = 300$ K.

The generated free carriers influence the effective doping N_{eff} and thereby the shape of the electric field. The feedback is shown with device simulation in Fig. 12.7. At the left-hand side, and at the pn-junction, it holds $N_{eff} = N_D + p_{av}$. With increasing j_R the gradient of the electric field becomes more steep according to Eqs. (12.7) and (12.8) and at the pn-junction occurs an increased field peak.

At the nn$^+$-side, N_{eff} is lowered since $N_{eff} = N_D - n_{av}$. If n_{av} is in the range of N_D, a part of the positively charged donors will be compensated. If we assume a high-voltage diode with N_D of 1.1×10^{13} cm^{-3} as in Fig. 12.7a, only a current density $j_R = 19$ A/cm^2 is necessary to compensate the background doping for this design (a) completely. The gradient dE/dx becomes flat. In Fig. 12.7b the same conditions are taken, but the doping N_D is increased to 1.7×10^{13} cm^{-3} for this design (b).

The voltage corresponds to the area below $E(x)$. For design (a) the negative voltage area is always larger than the positive voltage area that occurs at the nn$^+$-junction. For design (b), up to a current density of 40 A/cm^2 there is a larger positive voltage area at the nn$^+$-junction, resulting in a positive branch of the post-avalanche characteristics.

In the simulated I–V characteristic, branches with positive and with negative differential resistance (NDR) occur at increased carrier density in avalanche, depending on the respective design [Hei05]. Figure 12.8 shows the post-avalanche behavior. Diode design (a) with very low background doping $N_D = 1.1 \times 10^{13}$ cm^{-3} shows NDR already at 0.1 A/cm^2. Diode design (b) with increased doping shows as consequence a lower breakdown voltage, but then follows a branch with positive differential resistance.

For a further increased current density, a second voltage peak builds up at the nn$^+$-junction. Finally, at a very high avalanche current density, design (b) runs at the same line as (a), since at this condition the blocking capability is no longer determined by the doping, but only by the free carriers and the device thickness. Design (c) in Fig. 12.8 with thicker base has an extended branch of positive differential resistance.

The occurrence of branches with the NDR was first explained by Egawa [Ega66]. They are in conjunction with a hammock-like field shape, as to be seen in Fig. 12.7a

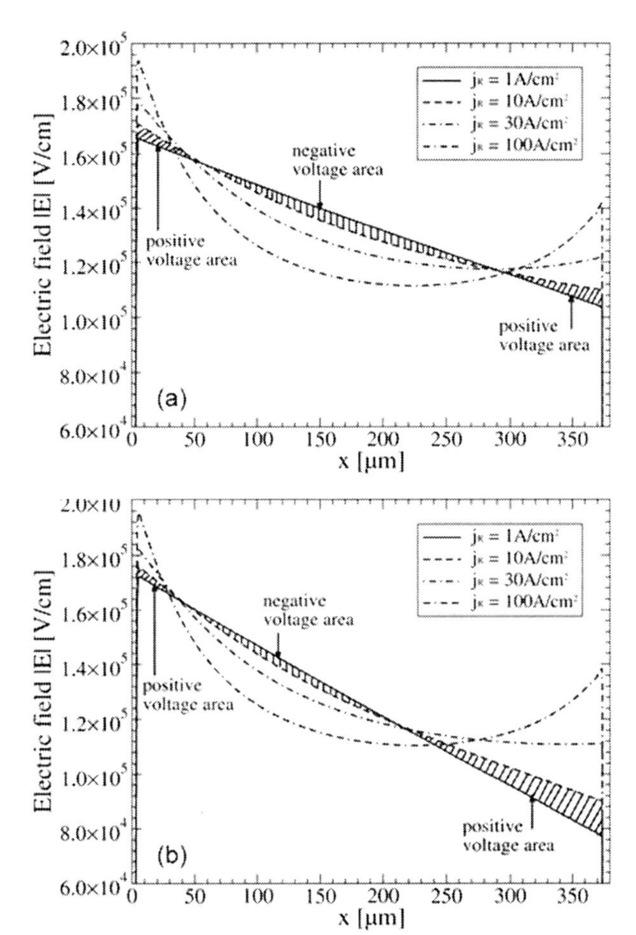

Fig. 12.7 Electric field $|E|$ at low and at increased avalanche current. (a) Design with $N_D = 1.1 \times 10^{13}\,\text{cm}^{-3}$, negative differential resistance, (b) design with $N_D = 1.7 \times 10^{13}\,\text{cm}^{-3}$, positive differential resistance up to some 10 A/cm². Fig from [Lut09] © 2009 IEEE

and b for high current density. This shall be called Egawa-type field. Such fields and branches with NDR have the precondition that the electron current n_{av}, calculated with Eq. (12.9), is higher than N_D. This condition is fulfilled for low-voltage devices only at very high current densities, but it may be reached at a moderate current density in high-voltage devices with low background dopings. This is one of the reasons why high-voltage devices are usually not avalanche rated.

At the nn$^+$-junction, avalanche is triggered by electrons and we have to use the multiplication factors for electrons, which are much higher than that for holes, see Fig. 3.15. Therefore impact ionization will occur at the nn$^+$-junction already at a lower electric field. Impact ionization at the nn$^+$-junction will create electron–hole pairs, the holes are flowing to the pn-junction and will increase the avalanche at the pn-junction. A positive feedback between impact ionization at the left and the right side will occur. Avalanche at an nn$^+$-junction has been described as a failure mechanism of devices in [Ega66, How70].

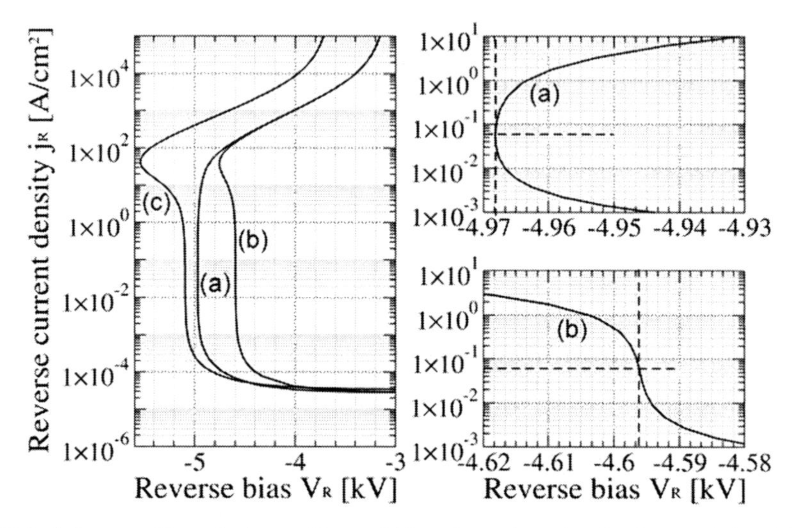

Fig. 12.8 Influence of base doping and base width to the static avalanche characteristics. (**a**) and (**b**) with $w_B = 375\,\mu$m and different doping of 1.1×10^{13} cm^{-3} (**a**) and 1.7×10^{13} cm^{-3} (**b**), (**c**) with 1.7×10^{13} cm^{-3} and wider w_B of 450 μm. Fig from [Lut09] © 2009 IEEE

In Fig. 12.7, the difference between (a) and (b) is only the increased doping in (b). The range of positive differential resistance can be extended by a thicker base w_B . Most effective are buffer layers, in which a region of increased doping is arranged in front of the nn$^+$-layer. With special buffers, fields at the nn$^+$-junction can be limited and branches of negative differential resistance can be widely avoided. Details are given in [Fel06].

In devices with planar junction termination, the edge usually limits the maximal possible blocking capability[1]. Avalanche breakdown first occurs at the edge. Therefore current densities high enough to lead to Egawa-type fields can occur locally at the edge. For a failure by overvoltage it is typical that the edge of the device is included in the destroyed area. Figure 12.9 shows the failure picture of a 1700 V diode. The device has a planar junction termination with potential rings similar to Fig. 4.19. Three potential rings can be seen in the upper part of the figure. The point of destruction is located between the p-anode layer and the first potential ring. This is one of the positions of the highest electric field, as it is marked in Fig. 4.19.

The occurrence of such a failure position indicates that the failure was caused due to voltage. However, Fig. 12.9 does not allow a clear decision, whether an overvoltage above the rated voltage of the device was applied or the particular device had a weak point induced in the manufacturing process. A failure picture as in Fig. 12.9 is only then occurring, if no high current was flowing across the point of destruction.

[1] Exception MOSFETS. In MOSFETs the cell geometry is adjusted that the avalanche breakdown occurs first below the cells and not at the edge, see above.

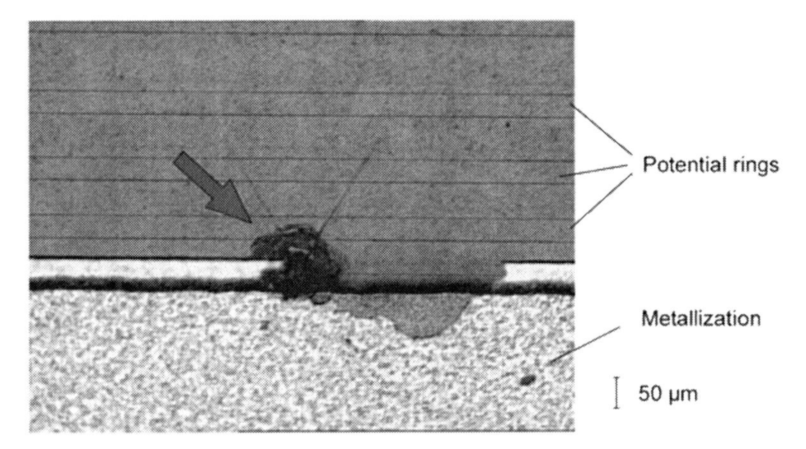

Fig. 12.9 1700-V diode destroyed by voltage

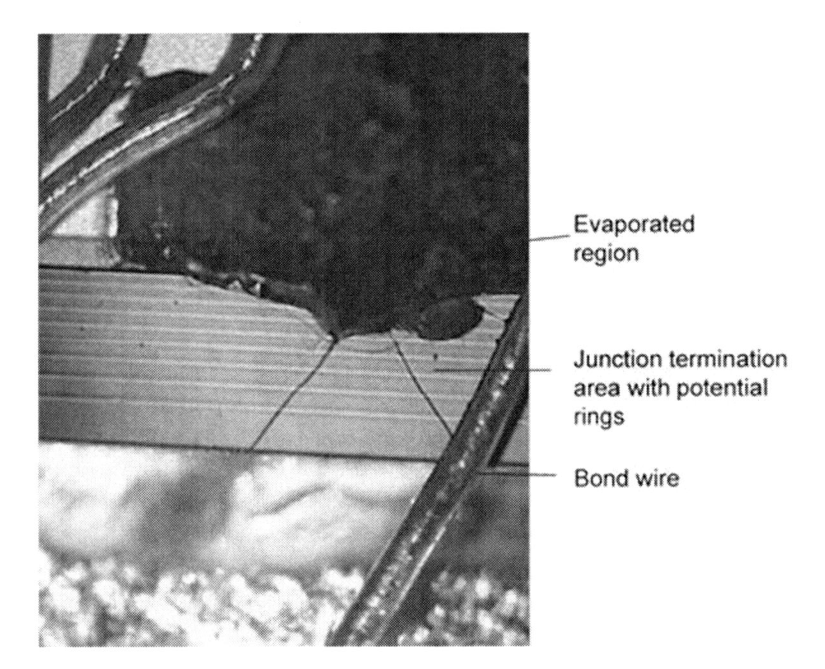

Fig. 12.10 3.3-kV diode possibly destroyed by overvoltage

A picture of a destroyed semiconductor, in which after failure a high current was flowing, is shown in Fig. 12.10. Part of the edge and a big part of the active area are evaporated. If such a picture occurs, one can presume that the destruction occurred at first at the edge and then it propagated toward the bond wires.

However, the cracks in the crystal lattice are not typical. These cracks indicate a local hot spot in a small, point-like position. Such failure pictures can also be found at strong dynamic avalanche (dynamic avalanche of the third degree, see Sect. 12.4.2). Therefore this failure picture is not unequivocal.

12.4 Dynamic Avalanche

12.4.1 Dynamic Avalanche in Bipolar Devices

During switching of all bipolar devices the increase of the voltage occurs at an instant, at which a large part of the stored carriers, which have conducted the forward current before, is still present in the device. This stored charge is partially removed during the voltage increase, and it flows as hole current through the space charge region.

Figure 12.11 shows the process in a simplified way. The pn-junction at the position $x = 0$ represents the blocking pn-junction of a bipolar device. Between the junction and w_{SC} the space charge region has extended, for supporting the applied voltage. Between w_{SC} and the end of the lowly doped layer exists a plasma zone, in which $n \approx p$ holds. The effects at the right side shall be neglected in this first approximation. At this position there will be either an nn^+-junction of a diode, or the collector layer of an IGBT, or the anode layer of a GTO thyristor, etc. As long as very hard switching conditions are not applied, no space charge is build up at this position.

Through the space charge the current flows as hole current, $j = j_p$. The density of holes p can be calculated from the current density at this instant:

$$p = \frac{j}{q \cdot v_{\mathrm{sat(p)}}} \tag{12.10}$$

In this equation $v_{\mathrm{sat(p)}}$ is the saturation drift velocity of holes under the condition of high fields, it amounts in silicon to approximately 1×10^7 cm/s and is close to the saturation drift velocity of electrons $v_{\mathrm{sat(n)}}$. A current density j of 100 A/cm^2 leads to $p = 8.2 \times 10^{13}$ cm^{-3}, which is already in the order of the background doping of a bipolar 1200 V device. The hole density p can no longer be neglected.

Holes have the same polarity as the positively charged ionized donors, hence their density now adds to the background doping to an effective doping N_{eff}:

$$N_{\mathrm{eff}} = N_{\mathrm{D}} + p \tag{12.11}$$

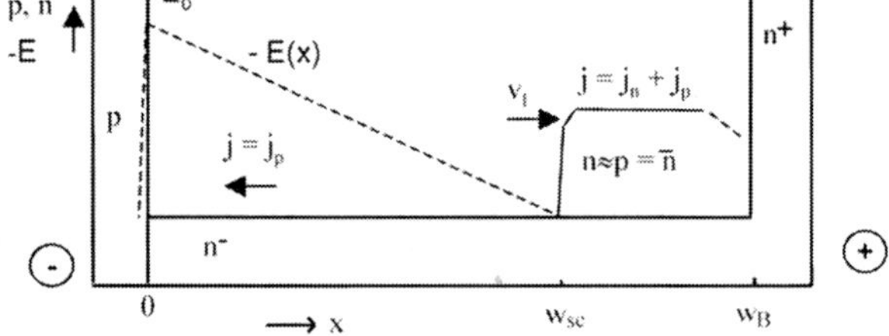

Fig. 12.11 Bipolar semiconductor device during the turn-off process

With the Poisson equation, N_{eff} determines the gradient of the electric field:

$$\frac{\mathrm{d}E}{\mathrm{d}x} = \frac{q}{\varepsilon}(N_{\mathrm{D}} + p) \tag{12.12}$$

so that $\mathrm{d}E/\mathrm{d}x$ is increased. With this, the field shape is steeper, E_0 is increased and the voltage, which drops across the space charge of width w_{SC}, is increased in the first instance. However, E_0 can rise only up to the avalanche field strength E_c. E_c will now be reached at an applied voltage far below the specified rated blocking voltage of the device and avalanche will set in. This process, which is now dominated by free carriers, is called dynamic avalanche.

This process occurs during turn-off of diodes, GTOs, and IGBTs, for further details, however, the specific peculiarities of the respective device and their physics must be considered.

12.4.2 Dynamic Avalanche in Fast Diodes

In the state which is drawn in Fig. 12.11, the shape of the electric field is approximately triangular, since no space charge can penetrate into the plasma layer. For a triangular field we can use the relation (3.84) between the avalanche breakdown voltage V_{BD} and doping:

$$V_{\mathrm{BD}} = \frac{1}{2} \cdot \left(\frac{8}{B}\right)^{\frac{1}{4}} \cdot \left(\frac{q \cdot N_{\mathrm{eff}}}{\varepsilon}\right)^{-\frac{3}{4}} \tag{12.13}$$

where for the ionization rates the proposal of Shields and Fulop is used with $B = 2.1 \times 10^{-35}\ \mathrm{cm}^6/\mathrm{V}^7$ and $n = 7$ at room temperature [Shi59, Ful67]. The derivation of this equation can be found in Chap. 3, Eqs. (3.75) to (3.84). This equation can be used since dynamic avalanche in this case is governed by holes.

If now Eqs. (12.10) and (12.11) are inserted in Eq. (12.13), we obtain a relation between the onset of dynamic avalanche and the current density in the space charge region, as shown in Fig. 12.12. For the background doping N_{D} the typical values for a 1700 V device and for a 3300 V device are used, $N_{\mathrm{D}} = 4.3 \times 10^{13}$ and $1.7 \times 10^{13}\ \mathrm{cm}^{-3}$, respectively.

From Fig. 12.12 one can recognize that for a device with high static blocking capability the avalanche onset voltage decreases strongly. In a 3.3 kV diode, dynamic avalanche sets in already at a reverse current density of 30 A/cm^2. For a reverse current density of 200 A/cm^2 the limit for the onset of dynamic avalanche has decreased down to 1050 V, and it is not much above the limit of a 1700 V device. There are only small differences between the two diodes voltage ratings, because the second term in Eq. (12.11) representing the free charge carriers is dominating.

Dynamic Avalanche of the First Degree

Close to the onset of dynamic avalanche, failures have been detected and assumed to be an unavoidable phenomenon of semiconductor physics [Por94]. Even a "silicon

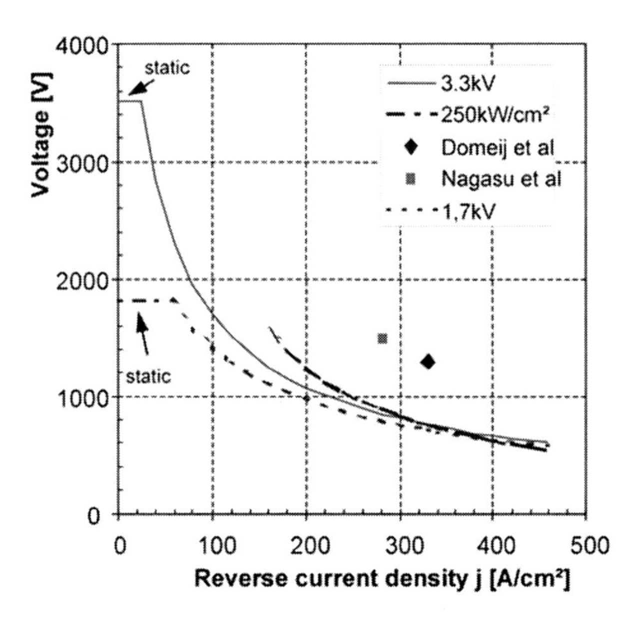

Fig. 12.12 Avalanche onset voltage depending on current density in reverse direction according to Eq. (12.13), former assumed limit of 250 kW/cm^2, reported operation points of 3.3-kV power diodes at max. power density of former work Domeij et al. [Dom99] and Nagasu et al. [Nag98] Reprinted from [Lut03] with permission from Elsevier

limit" at a power density $V{\cdot}I/A = 250\,\text{kW/cm}^2$ was sometimes assumed [Sit02]. The line for 250 kW/cm^2 shown in Fig. 12.12 is close to the calculations with Eq. (12.13), the onset of dynamic avalanche at the pn-junction.

However, Schlangenotto [Sco89b] published another point of view. Moderate dynamic avalanche, dynamic avalanche of the first degree, should not be critical. Dynamic avalanche generates electrons in the space charge which counteracts the positive charge of holes and then holds

$$N_{\text{eff}} = N_{\text{D}} + p - n_{\text{av}} \tag{12.14}$$

The increased hole density is partially compensated and this mechanism should be self-stabilizing. This was confirmed by diode designs bearing a significant amount of dynamic avalanche current [Lut97]. For 3.3 kV diodes Nagasu et al. [Nag98] and Domeij et al. [Dom99] reported measurements with 1.7 times higher voltages than the alleged limit. These results are also shown in Fig. 12.12.

The power loss density gives an estimation of the avalanche intensity, but it gives no physical explanation of a failure limit, since the high-power density occurs only during some 10 ns, while the dissipated energy is low and the temperature increase, if it is a one-time effect, is negligible.

However, for this to hold true it is a precondition that the device has no weak points in its design. In [Nag98] and [Tom96] it was shown that the edge of the active area is of great importance. In diodes with the common planar field limiting structures, the anode area is smaller than the cathode area. At the edge of the anode, additional current contributions from the n$^+$-region are occurring during forward conduction, as shown schematically in Fig. 12.13. The current density at the edge of

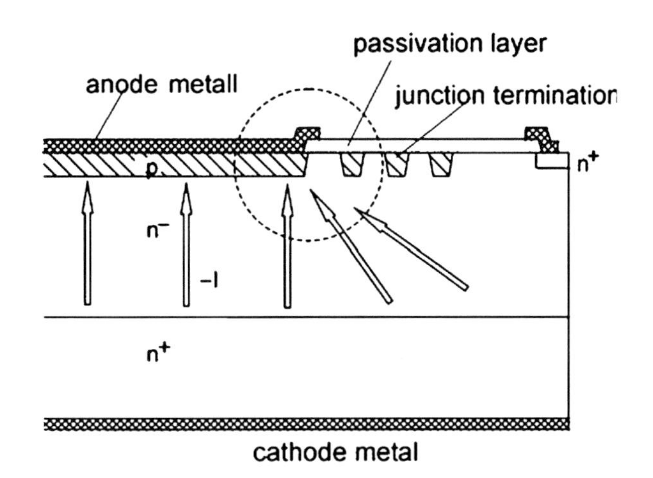

Fig. 12.13 Planar junction termination of a diode with increased current density at the edge of the active area. Reprinted from [Lut03] with permission from Elsevier

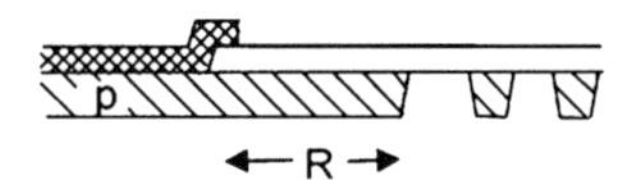

Fig. 12.14 Resistive layer for reduction of the current density at the edge of the anode. Reprinted from [Lut03] with permission from Elsevier

the anode is increasing and in device simulation one can observe a current filament already at the beginning of turn-off period. The implementation of a resistive zone at the edge, as shown in Fig. 12.14, is reducing this weak point. The p-layer is extended by a length R below the passivation layer. The anode side p-layers are only moderately doped in soft recovery diodes, therefore the region R acts as a pre-resistor for a current at the edge of the p-layer and the current density at the edge will be reduced. Such a structure at the edge was called "HiRC" structure (high reverse recovery capability) in [Mor00].

Dynamic Avalanche of the Second Degree

With an increasing current density, dynamic avalanche leads to filamentation of the current due to space charge effects as reported by Oetjen et al. [Oet00]. Figure 12.15 shows a static simulation of the electric field distribution and I–V characteristic of a p^+np^+ structure. This figure represents the process of Fig. 12.11; however, for an increased current density. The plasma layer of Fig. 12.11 is replaced by a p^+-layer which injects holes. Figure 12.15a displays the electric field at the current densities 500 and 1500 A/cm^2. The field has reached the avalanche field strength level at the pn-junction, in this figure at $x = 8\,\mu$m. The field shape for $j = 500$ A/cm^2 is approximately triangular. Avalanche generates electron–hole pairs in the region with high field strength. However, the generation is not local, since it happens within some length in the x-direction which is necessary for carrier acceleration. The holes are flowing to the left-hand side; the electrons are flowing to the right-hand side.

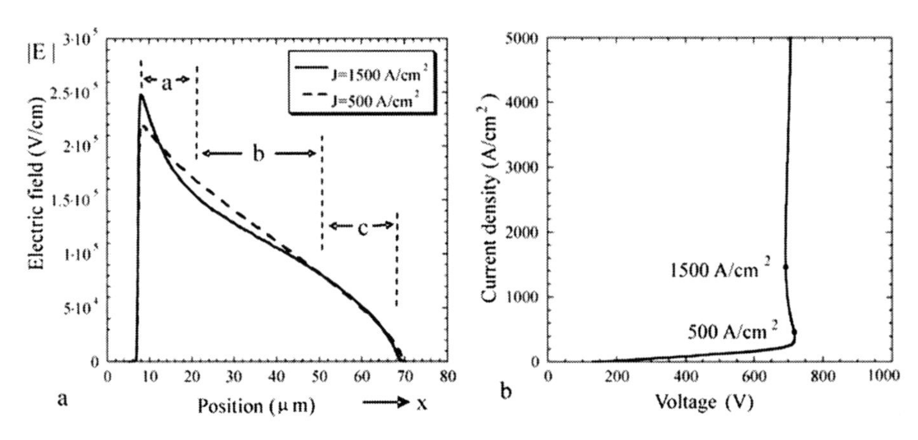

Fig. 12.15 Illustration for dynamic avalanche at an increased current density. Field shape (left-hand side), *I–V* characteristic (right-hand side). Reprinted from [Lut03] with permission from Elsevier

Close to the pn-junction – region a – the hole density is increased by the holes generated by avalanche. At the pn-junction holds

$$\frac{\mathrm{d}E}{\mathrm{d}x} = \frac{q}{\varepsilon}\left(N_{\mathrm{D}} + p + p_{\mathrm{av}}\right) \tag{12.15}$$

With this, $\mathrm{d}E/\mathrm{d}x$ at the pn-junction has become very high. This is drawn for $j = 1500$ A/cm^2. Further away from the pn-junction, in region b, the electrons n_{av} generated by dynamic avalanche are flowing to the right-hand side. There are also some generated holes p_{av}, but their density is decreasing with distance to the pn-junction. It can be summarized for region b:

$$\frac{\mathrm{d}E}{\mathrm{d}x} = \frac{q}{\varepsilon}\left(N_{\mathrm{D}} + p + p_{\mathrm{av}} - n_{\mathrm{av}}\right) \tag{12.16}$$

In region b a partial compensation of generated electrons and the flowing holes occurs, p_{av} is decreasing to the right-hand side, while n_{av} is increasing. Hence, the gradient of the electric field $\mathrm{d}E/\mathrm{d}x$ becomes flat. At the border to the plasma layer $E = 0$ must hold. Therefore in region c the gradient $\mathrm{d}E/\mathrm{d}x$ must increase again. The field in regions a to c will form a typical bowed shape.

The voltage correlates to the area under $E(x)$ which is slightly lower for the higher current density (see Fig. 12.15b). The *I–V* characteristic has a region of weakly negative differential resistance. Such a characteristic splits a homogenously distributed current to areas with lower current density and in filaments with high current density. The type of characteristic in Fig. 12.15b is described as S shaped in [Wak95], which can lead to the formation of stable current filaments.

Device simulation of such events shows a current density of 1000–2000 A/cm^2 in the filaments. Nevertheless, there are counteracting mechanisms, and destruction should be avoidable for the following reasons:

- The differential resistance is weakly negative, so that a further increase of current increases the voltage again. Therefore, the current density in the filaments is limited.
- The temperature inside the filament will increase leading to a reduction of impact ionization in the area of the filament, this counteracts the filamentation.
- The high local current density in a filament quickly removes the stored carriers locally from the plasma layer, counteracting the driving force for dynamic avalanche.

Therefore a state of rising, expiring, jumping, or moving filaments is to be expected, but a diode should still be capable to withstand this effect. Moving and jumping filaments are to be seen in two-dimensional device simulations [Nie05, Hei06]. It must be mentioned that two-dimensional simulations are limited, because filaments have a three-dimensional effect. However, such simulations are enormously time consuming and have not been possible up to now. In real devices, the edge of the active area is a strong inhomogenity, and it triggers the first filaments [Hei07]. For such a geometry including the edge of the active area, a two-dimensional simulation is capable of showing the main effects.

Most simulations in this field have been done under isothermal conditions that means without consideration of any temperature increase in filaments. However, the avalanche coefficients are strongly temperature dependent. The first temperature-dependent simulations of such problems showed an influence of temperature, filaments building up, and decreasing at different speeds, compared to isothermal simulations. However, the qualitative behavior was similar [Hei07].

Dynamic Avalanche of the Third Degree

Further increase of the stress by dynamic avalanche leads to the situation that a field at the nn^+-junction occurs while there is still dynamic avalanche at the pn-junction. The effects are shown in Fig. 12.16 in a simplified way. There is still remaining plasma, and the electric field between the plasma, and the pn-junction is bowed as in dynamic avalanche of the second degree. Between the plasma and the nn^+-junction a second electric field builds up, whose peak is at the nn^+-junction and whose gradient is inverted compared to the left space charge zone.

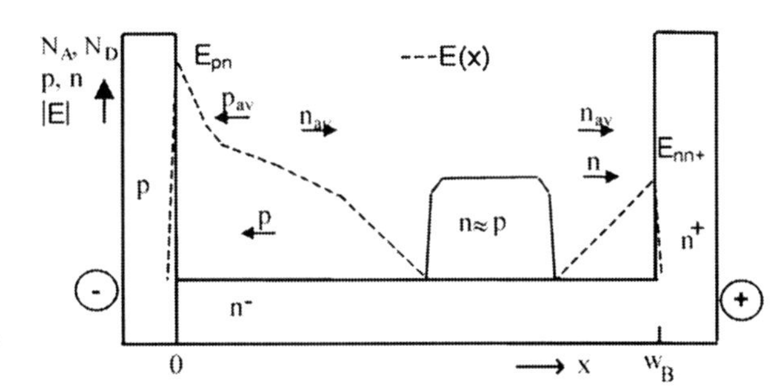

Fig. 12.16 Effects at strong dynamic avalanche. Fig adapted from [Lut09] © 2009 IEEE

This inverted gradient of the electric field can only occur, if the density of free electrons is so high that it overcompensates the positively charged donors of the background doping:

$$\left|\frac{\mathrm{d}E}{\mathrm{d}x}\right| = \frac{q}{\varepsilon}\left(n + n_{\mathrm{av}} - N_{\mathrm{D}}\right) \tag{12.17}$$

If n_{av} is increasing, the absolute value of $\mathrm{d}E/\mathrm{d}x$ will increase, and finally impact ionization sets on at the nn$^+$-junction also. This is usually initiated locally in a region where a high current density filament at the pn-junction has formed. A double-sided dynamic avalanche is given at this position.

Figure 12.17 shows results of a simulation of a diode under such conditions [Hei08]. In Fig. 12.17a the current density is shown, between $y = 3600$ and $4400\,\mu\mathrm{m}$ a wide filament at the pn-junction has formed, at the nn$^+$-junction occurs a narrow high-current filament at $y = 4000\,\mu\mathrm{m}$. In Fig. 12.17b the electric field is shown for

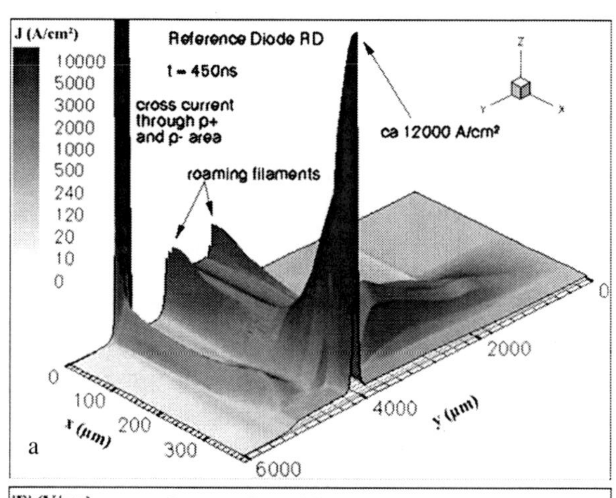

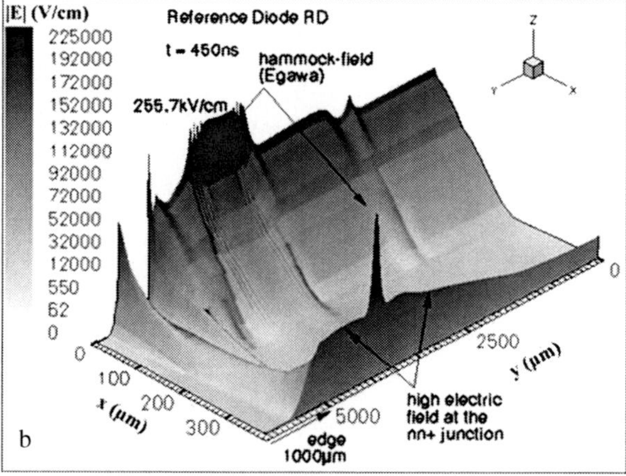

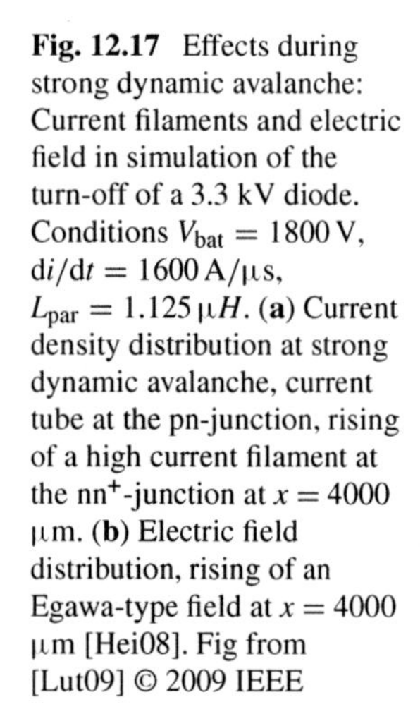
Fig. 12.17 Effects during strong dynamic avalanche: Current filaments and electric field in simulation of the turn-off of a 3.3 kV diode. Conditions $V_{\mathrm{bat}} = 1800\,\mathrm{V}$, $\mathrm{d}i/\mathrm{d}t = 1600\,\mathrm{A}/\mu\mathrm{s}$, $L_{\mathrm{par}} = 1.125\,\mu H$. (**a**) Current density distribution at strong dynamic avalanche, current tube at the pn-junction, rising of a high current filament at the nn$^+$-junction at $x = 4000$ µm. (**b**) Electric field distribution, rising of an Egawa-type field at $x = 4000$ µm [Hei08]. Fig from [Lut09] © 2009 IEEE

the same instant. At the nn^+-junction a second electric field appears with a high field peak at $y = 4000\,\mu$m.

This hammock-shaped electric field has already occurred as an Egawa-type field [Ega66] in Sect. 12.3. There are also similarities to the field at the nn^+-junction at second breakdown in bipolar transistors. Applying Wachutka's model for the destruction limit of GTO thyristors [Wac91] to fast recovery diodes using the boundary conditions in [Ben67] leads to the conclusion, that dynamic impact ionization at the pn-junction should be stable, whereas impact ionization at the nn^+-junction is highly unstable. It is sufficient, that the ionization integral $\int_0^w \alpha\,(E(x))\,\mathrm{d}x$ (see Eq. (3.67)) reaches a value of 0.3 to get to an instability mode. The current density in a filament will increase with a time constant in the range of some nanosecond [Dom03]. Impact ionization at the nn^+-junction is triggered by electrons and will occur at lower electric field strength because of the higher ionization rates of electrons. The rapid increase of the local current density is assumed here to be the cause of destruction.

In recent work, it has been shown by numerical simulation that the plasma-layer expands to the nn^+-junction in the vicinity of a cathode-side filament. This behavior differs from anode side filaments and results from the velocity saturation of the electrons and holes in the high-field region [Bab09]. The reduced depletion layer in the vicinity of the filament inhibits the lateral movement of the filament and, therefore, leads to strong local heating. This generates a destructive thermal filament.

The assumption that an Egawa-type field above a certain limit of 100 kV/cm at the nn^+-junction (Fig. 12.16, right-hand side) leads to the destruction of the diode could explain an experimentally found destruction limit. Figure 12.18 shows the

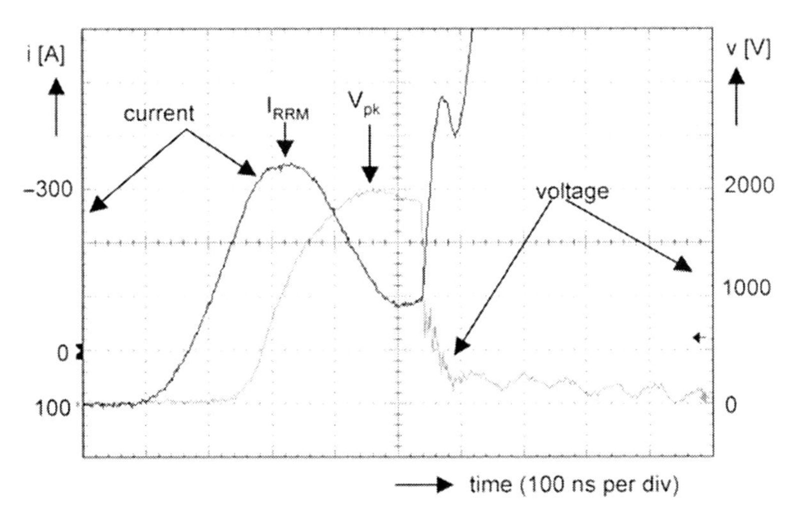

Fig. 12.18 Failure of a 3.3 kV diode under extremely strong dynamic avalanche conditions. Reprinted from [Lut03] with permission from Elsevier

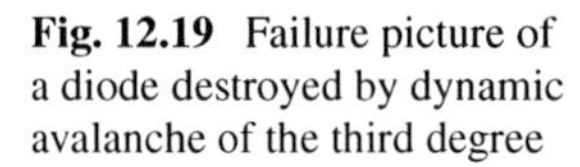

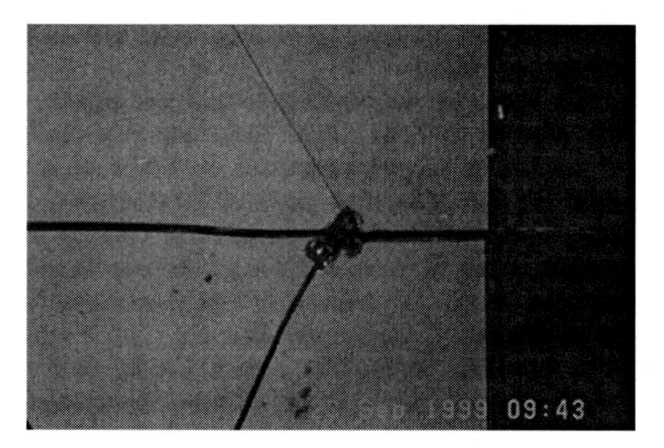

Fig. 12.19 Failure picture of a diode destroyed by dynamic avalanche of the third degree

failure of a 3.3 kV-rated diode under the condition of very fast commutation d*i*/d*t*. The reverse recovery current peak I_{RRM} of 360 A represents a reverse current density of 400 A/cm^2. The voltage across the diode is rising very fast, since in this experimental setup an additional capacitor of 22 nF was built-in between the gate and emitter of every switching IGBT. Already 200 ns after I_{RRM} the voltage has climbed up to 2000 V. The diode is destroyed shortly after the voltage peak.

If the employed IGBT withstands the short-circuit stress, which occurs after the destruction of the diode, and if it turns off the current successfully one can find a characteristic failure picture as shown in Fig. 12.19. At one point of the active area a small molten channel is found. In the case of dynamic avalanche of the third degree, cracks in angles of 60° were observed. The failure picture corresponds to the destruction of a 111-oriented silicon wafer by a point-shaped stress. This is a hint for a current filament of very high current density and very high temperature in a small area.

Such a failure limit was reproduced for devices of the same type of different production lots. Figure 12.20 shows the measurement results (black dots). The *x*-axis in

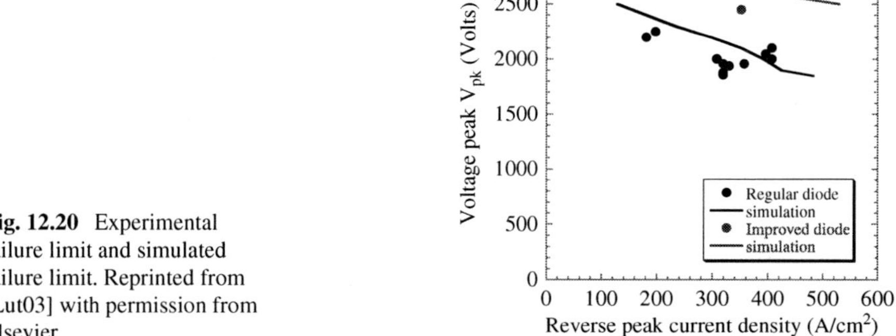

Fig. 12.20 Experimental failure limit and simulated failure limit. Reprinted from [Lut03] with permission from Elsevier

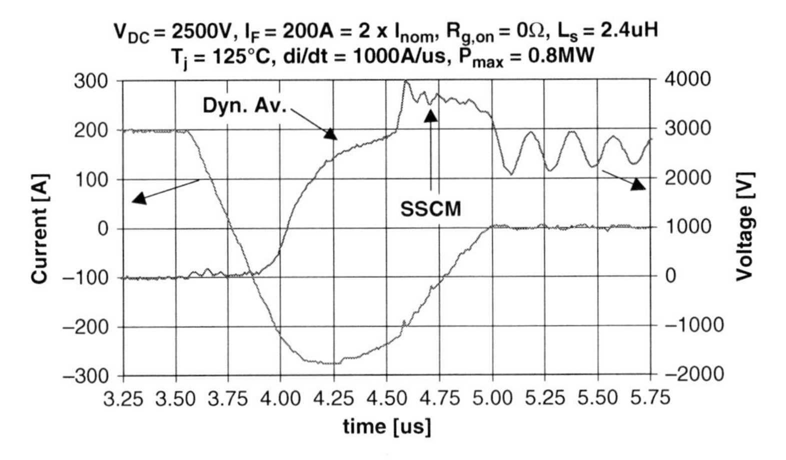

Fig. 12.21 3.3 kV diode with strong dynamic avalanche and SSCM mode [Rah04] © 2004 IEEE

Fig. 12.20 corresponds to the current density at reverse recovery current peak I_{RRM}, which can be adjusted by variation of d*i*/d*t*. The *y*-axis shows the value of the voltage peak V_{pk} (see Fig. 12.18) which occurred before failure. It can be recognized that there is a limit for the voltage and this limit depends only weakly on the current density in the measured interval.

A device simulation with AVANT Medici resulted in the strait line shown in Fig. 12.20. As condition for failure it was taken that a second electric field builds up at the nn$^+$-junction and that a failure occurs if this field grows up to 100 kV/cm.

The simulation predicts that an Egawa-type field occurs at higher voltages if the width of the n$^-$-layer of the diode is increased (dotted line in Fig. 12.20). This was confirmed by an experiment for a diode with a 50 μm thicker middle layer. However, the experiments yield that the limits of failure are lower than the simulated limits in tendency. Nevertheless, Fig. 12.20 shows that with the selection of the necessary thickness, the ruggedness can be adjusted.

More recent results showed that freewheeling diodes even up to the range of 3300 V could achieve a very high capability to withstand dynamic avalanche [Rah04]. Figure 12.21 shows a measurement with very high stress on the diode by the high battery voltage $V_{bat} = V_{DC}$, high d*i*/d*t*, and especially a high inductance of 2.4 μH. Already shortly after the voltage increase, strong dynamic avalanche occurs, and above a voltage of 1500 V at $t = 4.1$ μs a flattening in the increasing voltage is observed, the reverse recovery current peak is widened and the voltage increases very slowly. After strong dynamic avalanche, at $t = 4.55$ μs the internal plasma is suddenly exhausted. The hole current, which before this instant is fed by the internal plasma, suddenly disappears, thus the reason for dynamic avalanche is no longer present. Now the voltage climbs up steeply, but it is limited by the diode itself. This limit at approx. 3700 V is close to the static avalanche breakdown voltage of the diode.

However, in the current waveform in Fig. 12.21, at this point in time occurs no snap off, but only a small dip. Thereafter follows a current generated in static avalanche. This mode was called "switching self-clamping mode" (SSCM) [Rah04], where the diode clamps the voltage peak.

In the SSCM mode holds [Hei05]:

$$\frac{di}{dt} = \frac{V_{SSCM} - V_{bat}}{L_{par}} \tag{12.18}$$

where V_{SSCM} is a voltage close below the static breakdown voltage V_{BD}. The capability of the diode to transit from the mode of dynamic avalanche into static avalanche is a high progress in ruggedness. According to [Rah04] the design of this diode observed the necessary measures to achieve a high ruggedness: The p^+-anode layer was highly doped. The edge of the active area is designed in a way that local current crowding is avoided. Especially, the nn^+-junction of this diode has a very shallow gradient. If at the nn^+-junction the doping increases slightly, N_D is increased, according to Eq. (12.17), and more electrons would be necessary for inversion of the gradient of the electric field, and the danger that an Egawa-type field can arise is lessened.

12.4.3 Diode Structures with High Dynamic Avalanche Capability

Increased doping at the n^+-side by a buffer reduces Egawa-type fields, since the increased density of positively charged donor ions compensates part of the arriving electrons. However, most efficient are structures which inject holes for compensation of avalanche-generated electrons.

One of these structures is the "Field Charge Extraction" (FCE) structure [Kop05]. A part of the cathode area contains a p^+-layer, as shown in Fig. 12.22, at the cathode side. At the anode side, a He^{++}-implantation was applied to reduce locally the lifetime and to adjust for soft recovery behavior, see Chap. 5. Soft recovery can be achieved with different methods. The effect of the FCE structure is at the cathode side: If a space charge builds up at the cathode side, the p^+-layer will inject holes. The injected holes compensate the electrons generated by dynamic avalanche.

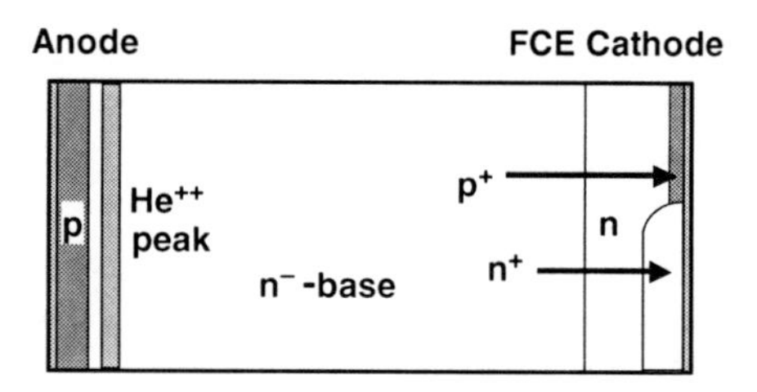

Fig. 12.22 Field Charge Extraction (FCE) diode, drawn after [Kop05] © 2005 IEEE

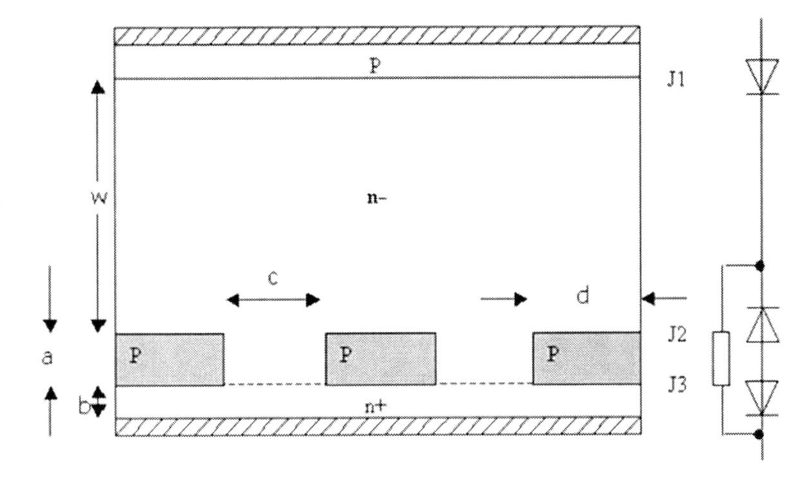

Fig. 12.23 Controlled injection of backside holes" (CIBH) diode [Chm06] © 2006 IEEE

The disadvantage of this structure is that part of the cathode area is lost for injection of carriers into the internal plasma. Therefore, the forward voltage drop of the diode will be increased.

The "Controlled Injection of Backside Holes" (CIBH) structure avoids this disadvantage. It contains floating p-layers in front of the cathode n^+-layer [Chm06]. The structure is shown in Fig. 12.23.

A continuous p-layer at the backside would realize a four-layer diode, this structure was discussed in [Mou88]. Such a four-layer diode acts in a similar way as a thyristor. It conducts in the forward direction after the cathode side pn-junction is overcome by breakover triggering. This leads to an additional voltage peak at turn-on of the diode and makes the structure unusable. Therefore in the CIBH diode this p-layer is interrupted by areas of the distance c, which form a resistor parallel to the additional junctions J_2 and J_3. The thickness of the p-layer between J_2 and J_3 is very low. Both sides of the junction J_3 are highly doped zones, which results in the onset of the avalanche breakdown at a small reverse bias like in an avalanche diode. Correspondingly the CIBH diode can be treated as a pin-diode with an integrated avalanche diode, which is connected with a parallel resistor. The distance c must be wide enough to avoid a deterioration of the turn-on behavior and it must be small enough to avoid high fields at the cathode side.

The effect of the suppression of electric fields is shown in Fig. 12.24. For comparison, Fig. 12.24a shows the electric field distribution during reverse recovery of a reference diode. The formation of the electric field peak at the nn^+ junction starts at $t = 400$ ns and leads to a critical Egawa-type field distribution after $t = 400$ ns, as described before.

Under the same switching condition the CIBH diode presents a completely different transient electric field distribution in the diode when compared to the reference diode, as illustrated in Fig. 12.24b. The situation at the backside is obviously improved. After the onset of the controlled avalanche at junction J_3, the voltage drop

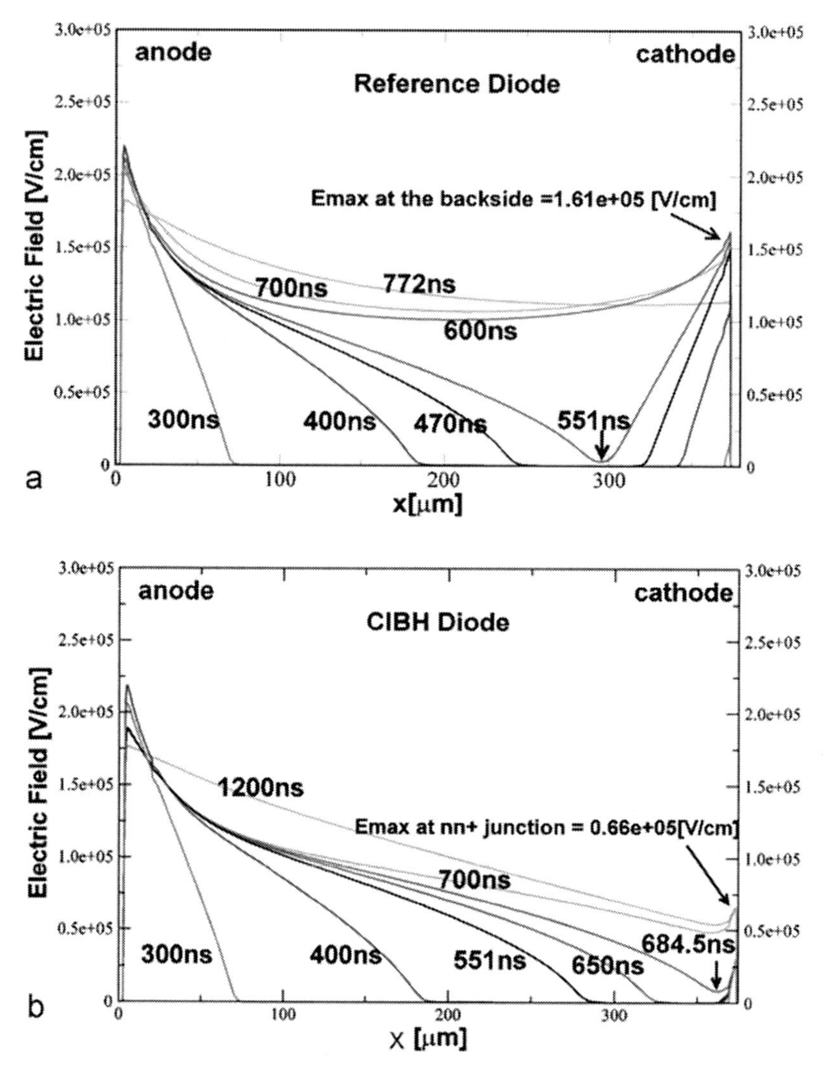

Fig. 12.24 Simulated electric field distribution of a reference diode (**a**) and of the CIBH diode (**b**) at very high stress in dynamic avalanche $T = 300\,\text{K}$, $J_F = 100\,\text{A/cm}^2$, $V_{bat} = 2500\,\text{V}$, $di/dt = 2000\,\text{A}/\mu\text{s}\,\text{cm}^2$, $L_{par} = 1.25\,\mu H$. Figure from [Chm06] © 2005 IEEE

at the backside is clamped at the breakdown voltage of the junction J_3. Avalanche at J_3 injects as much holes as necessary to compensate the electrons. The evolution of a second peak of electric field strength, which is correlated to a space charge region at the cathode side, is successfully suppressed.

Since Egawa-type fields can be effectively avoided, the CIBH diode has an extremely high ruggedness in dynamic avalanche. Figure 12.25 shows the turn-off of two 3.3.kV CIBH diodes under extreme stress. The maximal power density at turn-off is 2.5 MW/cm^2 – a factor of 10 above formerly assumed limits!

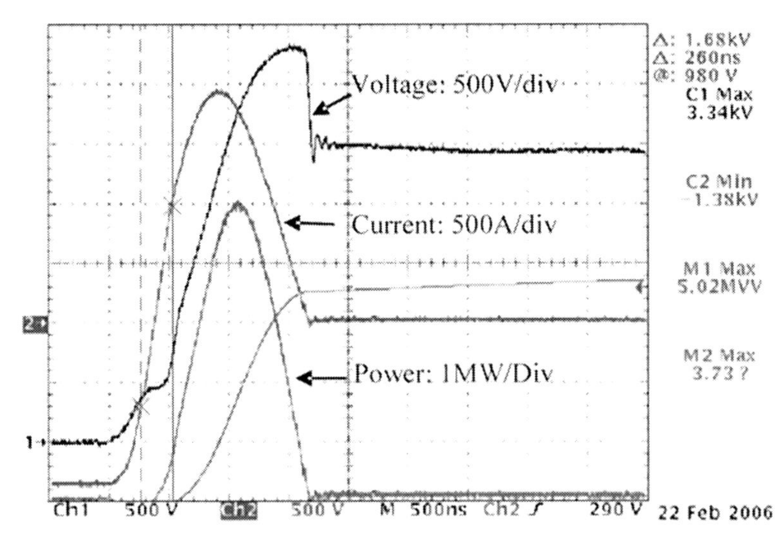

Fig. 12.25 Experimental ruggedness measurements of two CIBH diodes in a module at 5.5× nominal current and $T = 400\,\text{K}$, $V_{\text{bat}} = 2500\,\text{V}$, $\text{d}i/\text{d}t = 6500\,\text{A}/\mu\text{s}$, $L_{\text{par}} = 0.75\,\mu H$, voltage (CH1), current (CH2), dissipated power (Math1). Figure from [Chm06] © 2005 IEEE

The CIBH diode has additional advantages in the turn-off behavior: during plasma removal, the plasma does not detach the cathode zone, but remains connected to the cathode, see Fig. 12.24b. In contrast to the usual process at reverse recovery as described in Fig. 5.26, the plasma in the CIBH diode is removed only from the anode side.

This effect occurs, since early during reverse recovery, junction J_3 goes into the avalanche mode and injects a current $j_{\text{p,ava}}$. Then for the movement of the cathode-side plasma layer front holds [Bab08]

$$v_{\text{R}} = \frac{j_{\text{p}} - j_{\text{p,ava}}}{q \cdot n} \tag{12.19}$$

The absolute value of v_{R} is reduced and if $j_{\text{p,ava}} = j_{\text{p}}$, v_{R} becomes 0. The plasma stays connected to the nn$^+$-junction. In Eq. (5.94) v_{R} is equal to 0 and we obtain $w_{\text{x}} = w_{\text{B}}$. Therefore the whole base width w_{B} is supporting the voltage V_{sn} and in Eq. (5.95) $w_x = w_{\text{B}}$ holds.

If now the space charge region reaches w_{B} – this is the case at turn-off under conditions of low forward current, high parasitic inductance, and high voltage, where usually freewheeling diodes show snappy recovery – the p-layers inject additional holes. If the applied voltage is further increased, the reverse recovery is further improved. This was introduced as DSDM (dynamic self-damping mode) [Fel08].

The forward voltage drop of the CIBH diode is not distinctly increased compared to another diode of comparable thickness and stored charge, since during forward conduction we have a parallel connection of a triggered thyristor and a diode. A possible drawback could be a loss in the static breakdown voltage V_{BD} because of the hole injection into the electric field. By adjusting the process parameters even at high p-doses and high p area ratios the loss in the static breakdown voltage can be avoided while the softness is improved. The loss of breakdown voltage is in maximum 7% for diodes with a high implantation dose [Fel08].

12.4.4 Dynamic Avalanche: Further Tasks

The CIBH diode shows that power devices can be designed in a way that they can withstand high stress in dynamic avalanche. A review on dynamic avalanche in high-voltage devices is given in [Lut09]. However, manufacturers still restrict the allowed reverse current I_{RRM} with safe-operation area (SOA) diagrams. This SOA diagrams are usual for devices with blocking voltages of 3.3 kV and above and have a shape similar to Fig. 12.12. Such diagrams forbid the application of devices in dynamic avalanche.

We also conclude from Fig. 12.12: the higher the rated voltage of the device, the more the allowed reverse current density must be restricted if one wants to avoid dynamic avalanche. On the other hand, the stored charge Q_{RR} in a device increases with the base width of a device at a power of two – see Chap. 5, Eq. (5.64). With increased Q_{RR} also I_{RRM} is increasing. To keep I_{RRM} low, the current slope di/dt must be kept small. However, with a small di/dt during turn-on of a transistor, the turn-on losses in the transistor increase. Therefore, the required restriction of I_{RRM} restricts the possibility to reduce the switching losses in the application.

It was found that the higher the rated voltage, the lower the current densities at which formation of filaments and current tubes occur [Nie04]. Therefore, in spite of the progress, further research is necessary to understand the behavior in power devices at the border of the safe operation area. Similar effects of negative differential resistance and moving filaments have been found in ESD protection devices [Pog03]. Meanwhile one can understand some experimental results and attempts for improved designs have been successful, but there still exist open questions.

12.5 Exceeding the Maximum Turn-Off Current of GTOs

During turn-off of a GTO thyristor, the current below the emitter finger is extracted from the edge toward the middle. This was shown in the paragraph on GTO thyristors, see Fig. 8.16. Finally a narrow current-conducting area remains in the middle of the emitter finger, representing a current filament before the anode current decays. Even if high accuracy of the fabrication process technology is given, not all emitter fingers of the GTO thyristor are ideally identical, and even in a single emitter finger, there will be a position which will be the last for current conduction. If the maximum turn-off current is exceeded, a molten zone is found at this point. After chemical removal of the metallization and etching in concentrated potash KOH, which solves polysilicon faster than mono-crystalline silicon, a narrow hole can be found as shown in Fig. 12.26.

Fig. 12.26 GTO thyristor failed by exceeding the maximum turn-off current. Molten channel, created by a current filament in the middle of the cathode finger

The dark area in Fig. 12.26 shows the position of the final filament or current tube, while the molten channel reaches down to the backside of the device.

A failure picture as in Fig. 12.26 can be caused by exceeding the maximum turn-off current capability of the device. However, it can also occur because of a failure of an element in the RCD snubber (see Fig. 8.18), since during turn-off of a GTO thyristor the voltage slope dv/dt must be limited.

12.6 Short-Circuit and Over-Current in IGBTs

12.6.1 Short-Circuit Types I, II, and III

Three types of short circuit have to be distinguished for the IGBT [Eck94, Eck95, Let95]:

Short circuit I is a direct turn-on of the IGBT to a short circuit. The course of the voltage V_C, the collector current I_C, and the gate voltage V_G is shown in Fig. 12.27. Before turn-on, the voltage V_C is high, while at the gate a negative voltage is applied. After turn-on into the short circuit, the current increases to more than 6 kA, this is the value of the saturation current. This value is specified in some data sheets as I_{SC} and correspondents to the saturation current at $V_G = 15$ V. The IGBT is able to withstand the simultaneous load of a high current and a high voltage during the short-circuit pulse for some time. It must be turned off within a defined time, typically specified to 10 µs or lower, to ensure safe operation and to avoid failures by overheating.

In Fig. 12.27 I_{SC} decreases with time due to self-heating of the device. During turn-off of the short-circuit current, an inductive voltage peak is generated. In the figure this voltage peak amounts to approx. 2000 V. The condition for surviving of a short-circuit pulse is that the peak voltage must remain lower than the rated voltage within the specified safe operation area. To ensure this, short circuit must be turned off with a limited di/dt. Usually, for turn-off of the short circuit the driver uses a higher gate resistor to limit the di/dt. This soft turn-off is visible in Fig. 12.27.

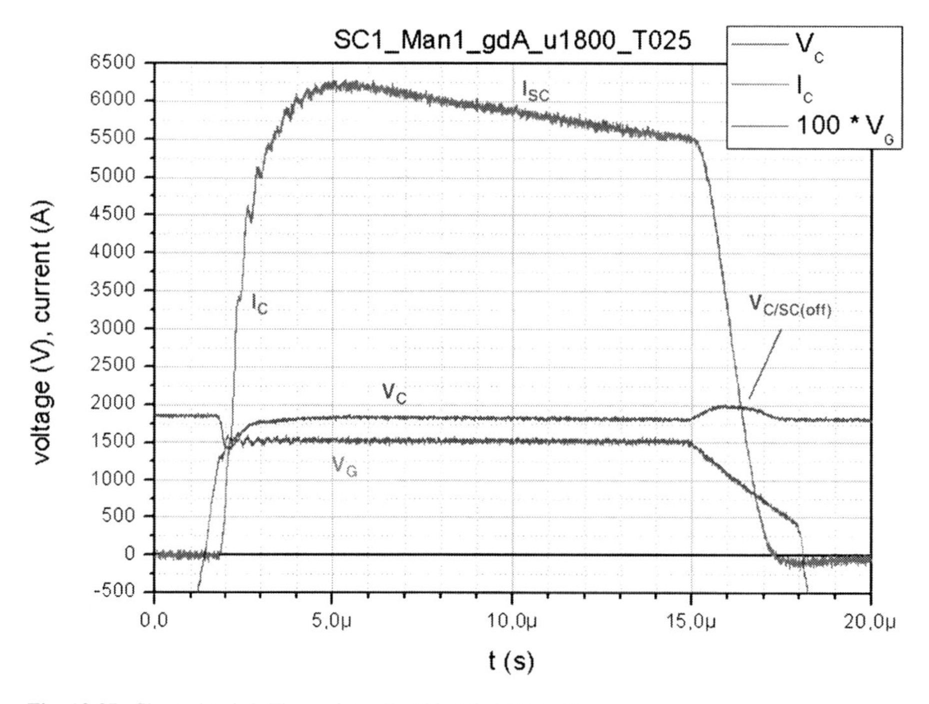

Fig. 12.27 Short circuit I. Figure from [Lut09b] © 2009 EPE

Short circuit II is the occurrence of a short circuit during the conducting mode of the IGBT [Eck94, Let95, Ohi02]. The occurrence is shown in Fig. 12.28 for the same IGBT as in Fig. 12.27. The IGBT is carrying the load current I_{Load} and the voltage drop is V_{CEsat}. As soon as the short circuit has occurred, the collector current will increase very steeply. The di/dt is determined by the DC-link voltage V_{bat} and the inductance of the short-circuit loop. During the time interval I, the IGBT is desaturated. The consequently high dv/dt of the collector–emitter voltage will induce a displacement current through the gate–collector capacitance, which increases the gate–emitter voltage. As can be seen in Fig. 12.28, V_{G} is dynamically increased up to 20 V due to this effect. This increase of V_{G} in turn causes a high short-circuit current peak $I_{\mathrm{C/SC(on)}}$, which in this case reaches a value of 14 kA.

Since the gate–collector capacitance is high at low V_{C}, the effect of the displacement current will be most significant at low voltage. The increased gate voltage during desaturation leads to a negative gate current back to the driver. The onset of a negative gate current together with the interconnection parasitics and IGBT input capacitances are supposed to cause the oscillations in V_{G} showed in Figs. 12.28 and 12.30.

After having completed the desaturation phase, the short-circuit current will drop to its static value I_{SC} (time interval II). Due to the current fall with negative di/dt, a voltage $V_{\mathrm{C/SC(on)}}$ will be induced over the parasitic inductances, which becomes visible as an voltage peak on the IGBT:

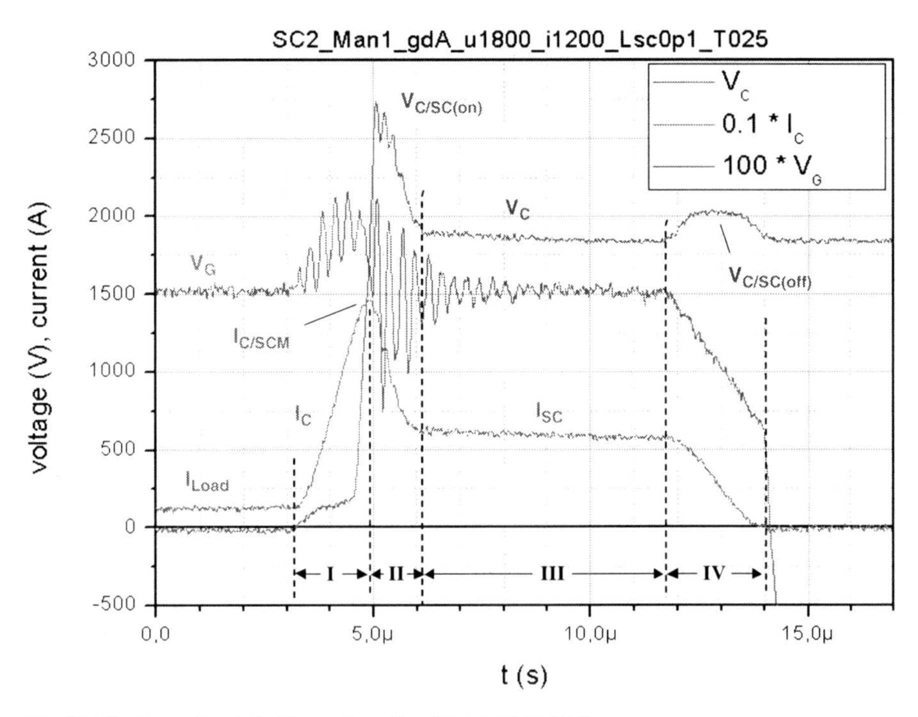

Fig. 12.28 Short circuit II. Figure from [Lut09b] © 2009 EPE

$$V_{C/SC(on)} = V_{bat} + L_{SC} \cdot \left| \frac{di_C}{dt} \right|_{max} \tag{12.20}$$

The stationary short-circuit phase (time interval III) is followed by the turn-off of the short-circuit current. Due to the negative di/dt, the commutation circuit inductance will again induce an overvoltage $V_{C/SC(off)}$ across the IGBT (time interval IV). Once again it must be ensured that this voltage peak stays below the rated voltage, respectively, within the specified safe operation area. To limit $I_{C/SCM}$ and to keep the gate–emitter voltage within the permissible limits, V_G has to be clamped.

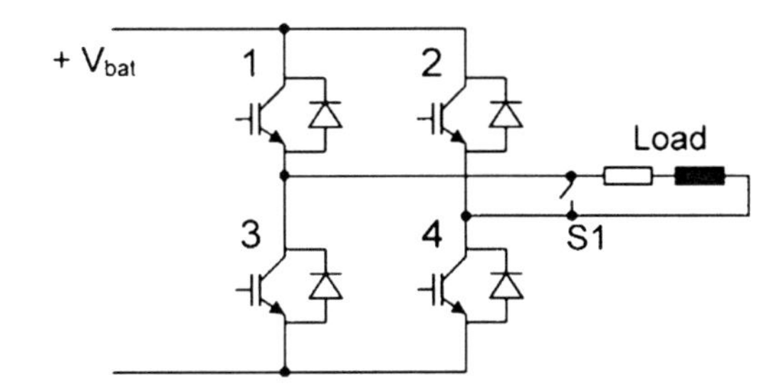

Fig. 12.29 Explanation of an SC III on example of a single phase inverter

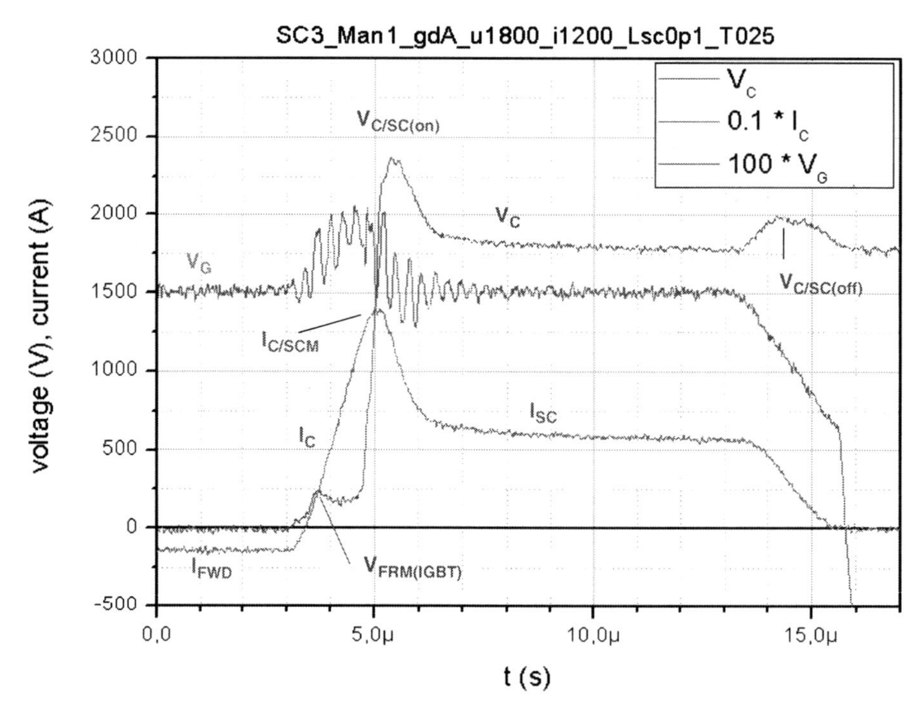

Fig. 12.30 Short circuit III. Figure from [Lut09b] © 2009 EPE

It should be noted that short circuit II is a harder condition than short circuit I. While $V_{C/SC(off)}$ can be limited by a soft turn-off function of the driver, this does not hold for $V_{C/SC(on)}$, which can easily overshoot the rated voltage of the device.

Short circuit III is the occurrence of a short circuit across the load during the conducting mode of the freewheeling diode. SC III can occur in all typical IGBT applications, since there is always an interval where the freewheeling diode is conducting. In motor drive applications, the duty cycle for the IGBT is higher than for the freewheeling diode, and SC II is more probable. However, if, e.g. a train drives down from a mountain, the motor is used as generator and energy is transferred back from the train to the grid. The duty cycle for the freewheeling diode is higher for the diode as for the IGBT in this operation mode and therefore SC III is more probable to occur than SC II.

The occurrence of a SC III event is explained using Fig. 12.29, in which a single-phase inverter is shown.

Inverters with pulse width modulation usually are driven with complementary signals. As starting point we assume that IGBT 1 and IGBT 4 are on and the current flows through IGBT 1, the load, and IGBT 4. As next, the IGBTs 1 and 4 are turned off, the gate signal of IGBTs 2 and 3 is set to "on" after a short dead time. Because the inductance of the load determines the current direction, the current will in this case flow through diodes 2 and 3 back to the voltage source. If now a short circuit occurs across the load – symbolized by closing the switch S1 – diode 2 and 3 will be commutated with a high di/dt and IGBTs 2 and 3, which at this time interval have a gate signal which is already set to "on," will be exposed to a short circuit. The current is thereby rapidly commutated from the diode to the IGBT, and the IGBT is turned on passively into the short circuit. During the short-circuit event, a reverse recovery process occurs at the freewheeling diode; however, the voltage course is determined by the IGBT and the occurring high dv/dt while the IGBT transits to its desaturation mode.

Before the short circuit, a positive $V_G = 15$ V is already applied at the IGBT; however, its current is zero because the inverse diode is conducting at this stage in the PWM pattern. When the short circuit occurs, the current changes its sign and the IGBT is turned on in a passive, diode-like manner. At usual turn-on, the voltage across the IGBT is high before turn-on. In the now given case, the voltage is low. Therefore a forward recovery peak V_{FRM} at the IGBT occurs. This is related to the forward recovery of a diode. It was observed in [Pen98] in a specially designed zero-voltage switching circuit with relatively low-voltage peaks. This forward recovery peak V_{FRM} can get very high, depending on di/dt. Several hundred volts can occur with wide-base high-voltage IGBTs. Forward recovery voltage peaks at IGBTs may be higher than at diodes [Bab09]. However, the measurement in Fig. 12.30 is at the outer terminals and the parasitic module inductance contributes significantly. Additionally, the diode parallel to the IGBT is in the reverse recovery process, the measured current at this instant contributes to diode and IGBT current and cannot be resolved in the used setup.

The dynamic short-circuit peak current $I_{C/SCM}$ amounts to 14 kA, this is almost the same value as in SC II. $I_{C/SCM}$ in SC III was found to be similar to $I_{C/SCM}$ in SC II for IGBTs from different manufacturers [Lut09b].

The occurrence of V_{FRM} is not supposed to be an extraordinary stress for the IGBT, since it occurs in the IGBT forward direction, where a high blocking capability is specified. The main additional stress in SC III for the module is the reverse recovery of the freewheeling diode during short circuit. The two-step voltage slope at reverse recovery of the diode, in which the second step may occur with an extraordinary high dv/dt, is a special stress event for the FWD. Usually, IGBTs fail in short circuit. However, in one of the SC III experiments, a diode failure was observed while the IGBT withstood [Lut09b].

A typical picture of a 3.3-kV IGBT, destroyed by short circuit is shown in Fig. 12.31. The large-area burned emitter regions are typical. A picture of a destroyed IGBT similar to Fig. 12.31 gives the specialist a hint that short circuit is probably the failure reason. The burned-off emitter regions are typically found for SC I as well as for SCII and SC III failures.

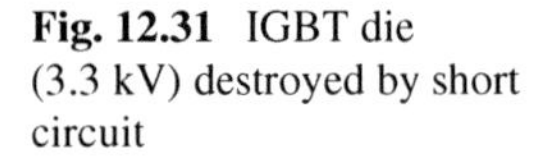

Fig. 12.31 IGBT die (3.3 kV) destroyed by short circuit

12.6.2 Thermal and Electrical Stress in Short Circuit

From its basic function, the IGBT has a short-circuit capability that limits the current. I_{SC} is the current in the active region of the IGBT forward *I–V* characteristic and can be approximated similar to Eq. (10.3):

$$I_{SC} = \frac{1}{1-\alpha_{pnp}} \cdot \frac{\kappa}{2} \cdot (V_G - V_T)^2 \left|_{V=V_{bat}}\right. \tag{12.21}$$

An electric field exists in the device (SC I) or is built up at desaturation (SC II, SC III). In short circuit II, there might be remaining plasma from the saturation mode before short circuit. The electric field is built-up quickly and the remaining plasma is extracted in a short time. Figure 12.32 shows the process in an NPT-IGBT in the short-circuit mode. An electric field has built up at the blocking pn-junction, which takes over the applied voltage V_{bat}. On the emitter side, the left-hand side in Fig. 12.32, the n-channel is conducting. Electrons are flowing into the space charge region, holes are injected from the p-emitter. In the electric field, the carriers flow with their drift velocities given in Chap. 2 with Eq. (2.38). The current density j_{SC}

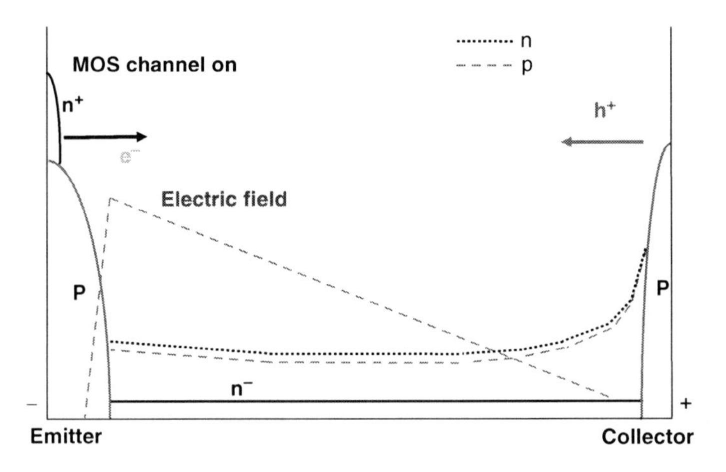

Fig. 12.32 Process in an NPT-IGBT during short circuit

is composed of $j_\mathrm{n} + j_\mathrm{p}$, where

$$j_\mathrm{n} = q \cdot n \cdot v_\mathrm{n} \\ j_\mathrm{p} = q \cdot p \cdot v_\mathrm{p} \tag{12.22}$$

Since v_n, v_p are now much higher than in a plasma layer, the total amount of electrons and holes is much lower than at forward conduction where *n,p* are in the range of 10^{16} cm^{-3}. Sometimes $v_{\mathrm{sat(n,p)}}$ is used for calculation of the carrier densities; however, especially for holes the drift velocity is often not saturated at the given fields. *n,p* are typically in the range of several 10^{14} cm^{-3} and clearly above the background doping N_D. The mobile carriers therefore lead to a feedback to the electric field and it holds

$$\frac{\mathrm{dE}}{\mathrm{d}x} = \frac{q}{\varepsilon}\left(N_\mathrm{D} - \frac{j_\mathrm{n}}{q \cdot v_\mathrm{n}} + \frac{j_\mathrm{p}}{q \cdot v_\mathrm{p}}\right) \tag{12.23}$$

where the terms in the bracket form an effective doping $N_\mathrm{eff} = N_\mathrm{D} - n + p$. The detailed feedback depends on the used IGBT technology and the conditions. In PT-IGBTs with the applied high p-emitter efficiency, typically the term caused by the hole current is dominating, and N_eff will be increased. In NPT-IGBTs, the electron density n is above the hole density and the reduced d*E*/dx leads to a wider extension of the space charge into the n$^-$-layer, reducing the electric field at the pn-junction [Las92]. In IGBTs with buffer layer, even $n > N_\mathrm{D} + p$ may occur, the gradient of the field will be inverted and the field peak shifts to the collector side.

The active region is a stable condition for a transistor; the deposited energy in the case of a SC I is

$$E = V_\mathrm{bat} \cdot I_\mathrm{SC} \cdot t_\mathrm{SC}$$

and due to the simultaneous high V_{bat} and I_{SC}, the temperature increases fast. If the short circuit is turned off within the requested time interval – 10 μs for former IGBT generations, 7 μs for new generations – the IGBT can in most cases withstand the thermal stress occurring during the short-circuit mode.

The holes are flowing through the p-well close to the emitter layer; this was shown in Fig. 10.14. The voltage drop in this path must be well below the built-in voltage V_{bi} of the junction. On the one hand, the density of carriers in short circuit is much lower than at rated current, where the IGBT is flooded with plasma. On the other hand, during short circuit very high temperatures occur. A high temperature decreases V_{bi}, therefore the danger of latch-up increases with increasing temperature. Nevertheless, there was much progress in manufacturing highly doped p^+-wells, therefore latch-up is usually no more a limit for the short-circuit capability of latest IGBT generations.

Equation (12.21) has given the parameters that determine I_{SC}. α_{pnp} is adjusted to be low in modern IGBTs by a low emitter efficiency at the collector side. It is typically in the range of 0.33 – 0.4. The other decisive factor in Eq. (12.21) is the channel parameter κ which gives the conductivity of the MOS channel. Since a comparatively high cell distance is of advantage in modern IGBTs – see Sect. 10.6 – the channel parameter κ is kept moderate. Therefore, in modern IGBTs a low I_{SC} can be achieved despite a high plasma density at the emitter side during usual forward conduction. As an example, Fig. 12.33 shows a comparison of a modern and conventional IGBT [Mor07]. For the modern IGBT (HiGT), the saturation current is not increased, despite the lower V_C at the typical operation condition (denoted V_{CE}(sat) in Fig. 12.33). The I_{SC} of the 50-A rated IGBT is approximately 175 A. Since the thermal short-circuit capability depends on the deposited energy in short

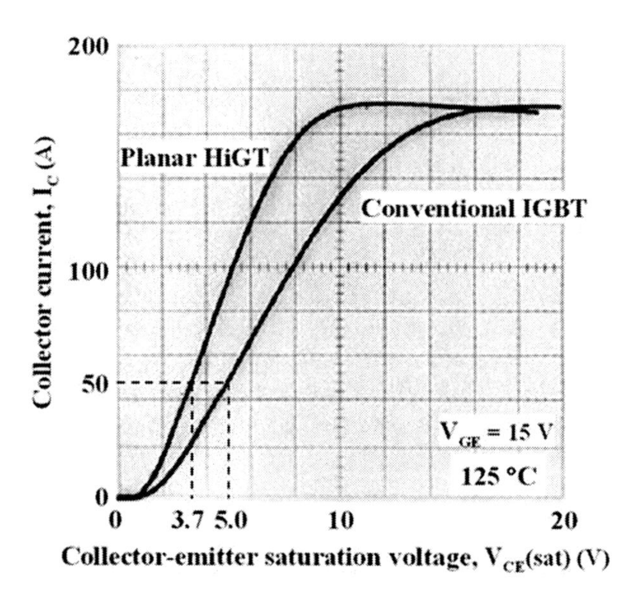

Fig. 12.33 Measured forward characteristic of a 3.3-kV conventional IGBT and planar HiGT (modern type). Figure from [Mor07] © 2007 IEEE

circuit, which is given by $I_{SC} \cdot V_{bat} \cdot t_{sc}$, the modern IGBT has the comparable short-circuit capability as the conventional one.

In several data sheets of 1200-V IGBTs, I_{SC} is even lower for the new IGBT generations. I_{SC} is specified to approx. 4 times the rated current, compared to 5–6 times the rated current for the older generations. In combination with a very high doping of the p-well (Fig. 10.14), leading to a very low resistance R_S (see Fig. 10.2b), very high short circuit ruggedness is achieved in modern IGBTs. However, in future IGBT generations with further reduced device area and device volume, the allowed time t_{sc} before turn-off of the short-circuit pulse will be limited from 10 μs to a lower value of 5 – 7 μs. This is possible in the application with meanwhile available fast-reacting improved gate drive units.

Thermal Limits for Medium-Voltage IGBTs

In medium-voltage IGBTs, the short-circuit capability is limited by temperature. The destruction of a 600-V IGBT by short circuit I is shown in Fig. 12.34. The applied stress in this case is beyond the short-circuit capability of the device. In Fig. 12.34a, the short circuit occurs at a battery voltage of 540 V during a time of approx. 60 μs, until the device is destroyed. In Fig. 12.34b the short-circuit current is turned off after approx. 40 μs. During the short-circuit pulse the device has been heated to a level that a high leakage current is flowing after the turn-off event. At such high temperatures additional charge carriers are generated thermally, see Eq. (2.6) and the leakage current increases further. After approx. 100 μs or more, the device is destroyed by overheat caused by the high leakage current.

Long-time tests with repetition of short-circuit events led to the conclusion that short circuit can be repeated up to 10,000 times without destruction of the device [Sai04]. This holds true as long as the deposited energy stays smaller than a certain critical energy E_C. A summary of the repeated short-circuit tests of a 600-V IGBT is shown in Fig. 12.35. A defined limit for the dissipated energy was found

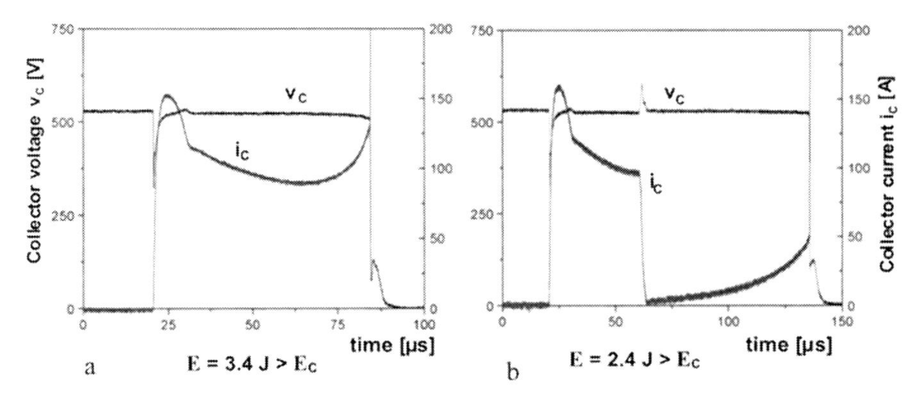

Fig. 12.34 600 V IGBTs in short circuit at $V_{bat} = 540\,\text{V}$, $T = 125°\text{C}$. Course of current and voltage at destruction of IGBTs at a deposited energy beyond the critical energy E_C. Figure from [Sai04]

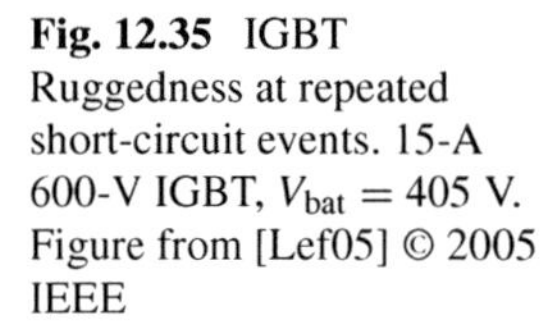

Fig. 12.35 IGBT Ruggedness at repeated short-circuit events. 15-A 600-V IGBT, $V_{bat} = 405$ V. Figure from [Lef05] © 2005 IEEE

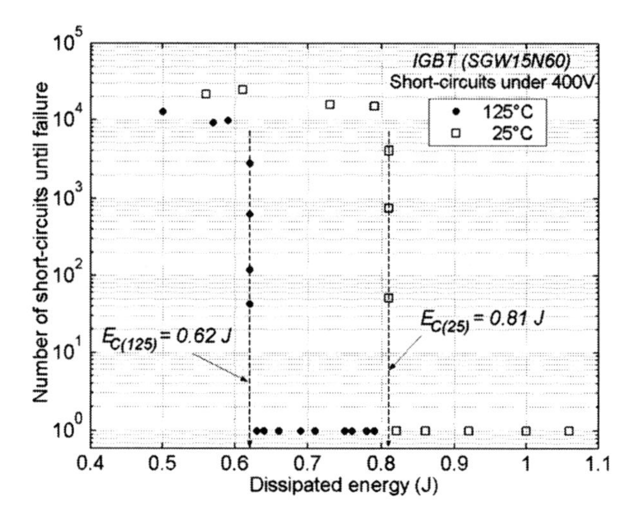

and this critical energy E_C is lower for the $T = 125°C$ compared to $T = 25°C$. Above the limit E_C, IGBTs are destroyed after one event by overheating. Since heat transport out of the device is small for the requested short time of the short-circuit load, the temperature increase can be calculated according to the thermal capacity of the device, compare Eq. (11.8), treating the deposited energy E as thermal energy Q_{th} by

$$\Delta T_{SC} = \frac{E}{C_{th}} = \frac{V_{bat} \cdot I_{SC} \cdot t_{SC}}{c \cdot \rho \cdot d \cdot A} \tag{12.24}$$

The evaluation of different IGBTs and the following calculation of the temperature increase for the volume of the semiconductor in which the electric field occurs resulted in temperatures in the range of 600°C [Sai04]. For IGBTs with different thicknesses, a similar final temperature was also found.

The results in Fig. 12.35 show that the failures in short circuit are purely thermal for the investigated 600-V IGBTs. Additionally, special attention was taken to find the ageing mechanisms of IGBTs which fail below E_C, but after a large number of short-circuit pulses. No trend in leakage current and threshold voltage was found. However, it was observed that the forward voltage drop V_C increases and the short-circuit current I_{SC} decreases with the number of cycles. Failure analyses showed the increase of the resistivity of the Al metallization layer, a strong degradation of the die metallization by Al reconstruction after approx. 10,000 cycles (Fig 12.36), and also a strong degradation of the bond wire attach [Lef08].

The degradation of metallization may lead to inhomogeneous current distributions and finally local overheating. The observed aging mechanisms have similarities to power cycling, where also the effects of aging of the metallization layer and bond wire lift off are observed [Lut08], see Sect. 11.6.

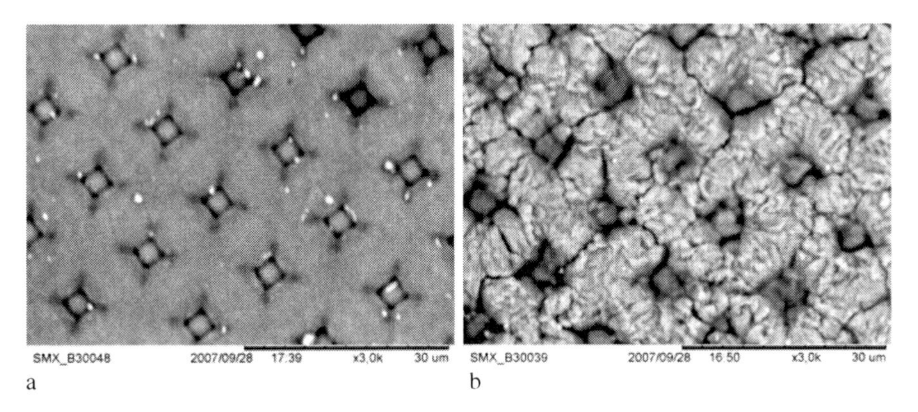

Fig. 12.36 Reconstruction of Al metallization at repeated short circuits: (a) before test (b) after 24,600 short-circuit cycles. Figure from [Lef08]

Since the main destructive effects are due to temperature, the critical energy E_C, which the device can take up as heat, depends on its thermal capacity. Modern IGBTs are designed with narrow widths w_B for the low-doped n-base layers to reduce the overall losses. Additionally the reduced voltage drop V_C during forward conduction allows higher rated currents for a given device area. Therefore the area as well as the thickness of the IGBT die is reduced and the thermal capacity is decreased correspondingly.

An example of the manufacturer Infineon for the die area of different 1200-V 75-A IGBT generations is shown in Fig. 12.37. The thickness of the dies for different IGBT generations and three different voltage ratings is shown in Fig. 12.38. For

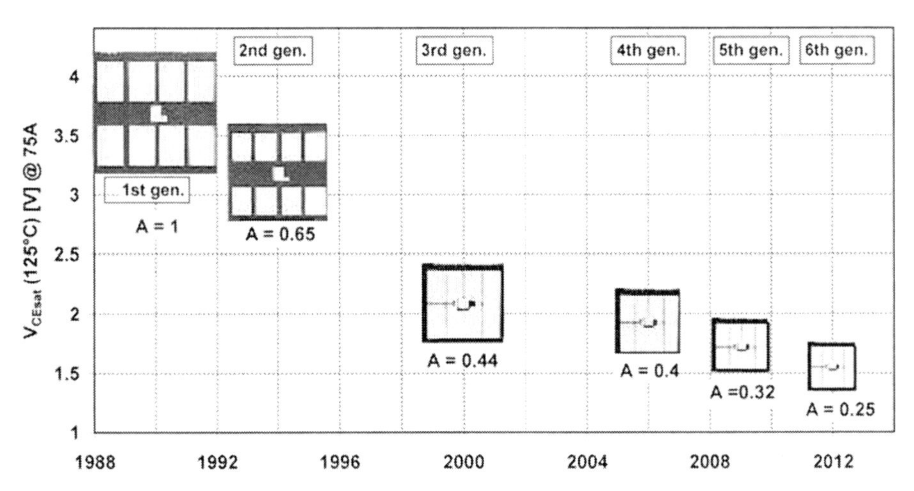

Fig. 12.37 Die area and forward voltage drop for different IGBT generations, on example of the manufacturer Infineon. Figure according to T. Laska, Infineon

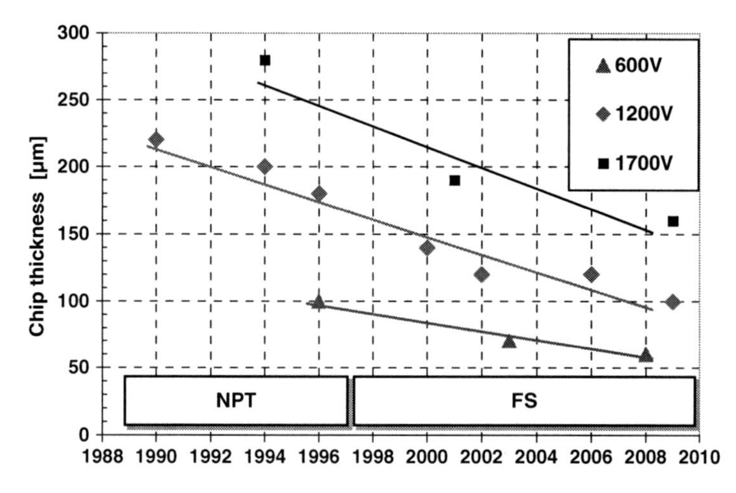

Fig. 12.38 Device thickness for different IGBT generations of the manufacturer Infineon. Figure from Infineon

the 1200-V 75-A IGBT chip, the area is reduced down to 44% and the thickness to 55% compared to the first IGBT generation from 1990. According to Eq. (11.8) this is a reduction of the thermal capacity down to 24%.

With this drastic reduction of the thermal capacity, also the thermal energy, which can be deposited in a device, is reduced in the same way. To maintain ruggedness during short circuit, I_{SC} has to be kept low.

Current Filamentation as Limit for the Short-Circuit Capability of High-Voltage IGBTs

While for 600-V and 1200-V IGBTs the short-circuit failure limit is mainly due to thermal reasons and aging [Lef08], this does not apply for high-voltage IGBTs. Due to the larger thickness of the device and due to the reduced current density, the temperature increase calculated with Eq. (11.24) is typically smaller. Especially, when short-circuit limits are determined, failures are not observed after a long t_{sc}, or at short-circuit turn-off, or after the turn-off have been completed, as shown in Fig. 12.34. They occur typically during the stationary phase in SC I. An example for such a measurement is shown in Fig. 12.39 [Kop09].

Figure 12.40 shows a summary of short-circuit failures for the 6.5 V IGBT as function of the applied DC-link voltage V_{bat}. At around $V_{bat} = 2000$ V the capability reaches a minimum and starts increasing again for higher V_{bat}. Note, the ruggedness is higher at 4500 V than at 1000 V!

A minimum for the short-circuit capability is found between 1500 and 2500 V, similar minima were found for a 4.5 kV rated IGBT between 1200 and 1800 V, and at about 1000 V for a 3.3 kV rated IGBT. This special voltage dependency needs new explanations.

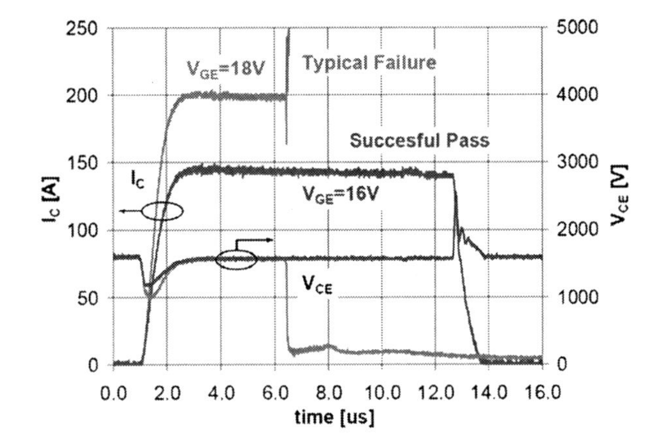

Fig. 12.39 Typical short-circuit pass and fail waveforms, SC I measurement of a 6.5 kV IGBT. Figure from [Kop09] © 2009 IEEE

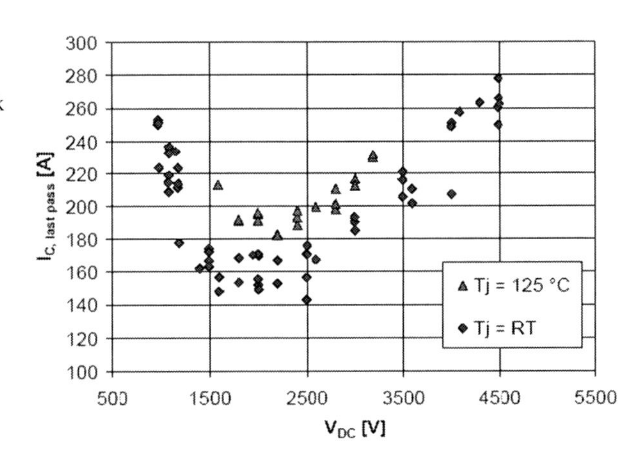

Fig. 12.40 Short-circuit capability for a 6.5 kV IGBT in dependency on the DC-link voltage V_{bat} and starting junction temperature. Figure from [Kop09] © 2009 IEEE

In these high-voltage IGBTs with a buffer layer at the collector side, a redistribution of the electric field to a field peak at the collector side under short-circuit stress was found [Kop08, Kop09], which results from a high electron density $n > N_D + p$, see Eq. (12.23). However, these field peaks at the nn$^+$-junction are moderate and the increasing ruggedness with increasing voltage excludes second breakdown and Egawa-type failure mechanisms. Further, it was found that an increased emitter efficiency of the collector-side p-emitter improves the ruggedness [Kop08, Kop09]. The increased α_{pnp} is, for the given low α_{pnp}, of low effect to I_{SC}. This experimental result also excludes latch-up as failure root cause. In [Kop09], the formation of filaments is suggested as failure mechanism. These filaments, found in numerical simulation, result from the different drift velocities of electrons and holes. Especially at medium voltage the electric fields are moderate and the drift velocity for the holes is far below that of electrons, see Fig. 2.14. Filaments with high local current density lead to IGBT destruction.

12.6.3 Turn-Off of Over-Current and Dynamic Avalanche

The turn-off of over-current is a very critical operation point of an IGBT. In difference to short circuit, the turn-off process is executed while the device is highly flooded with free carriers. The first IGBT generations for 3.3 kV restricted the turn-off capability of over-current to two times the rated current [Hie97]. A higher current was not allowed to be turned off. If a higher current occurs in these IGBTs, the driver has to wait until the IGBT is in the mode of saturation current, while the voltage increases and the IGBT then is turned off in the short-circuit mode. Note that in the short-circuit mode 5 – 6 times of the rated current are switched off without problems.

During turn-off of an over-current, the channel is first turned off. Hence, the electron current flowing via the channel becomes extinct. The total current must flow for a short instant completely as hole current. Figure 12.41 shows the process, for example, of an NPT-IGBT.

The hole current, fed by the remaining plasma, flows across the n^--layer, in which an electric field has build up. The density of free holes adds to the background doping. The gradient of the electric field gets steeper, as already treated with Eqs. (12.10) to (12.12). The blocking capability is therefore reduced as given in Eq. (12.13). Dynamic avalanche now generates electron–hole pairs in the region close to the blocking pn-junction. The holes are flowing to the left side in Fig. 12.41, while the electrons flow to the right side. The hole current, increased by dynamic avalanche, must flow through the p-well and via the resistor R_S (Fig. 10.2b). In this operation mode, the hole density is highest and the danger of turn-on of the parasitic npn-transistor and latching of the IGBT is greatest. If R_S is low enough, the IGBT will successfully withstand such conditions.

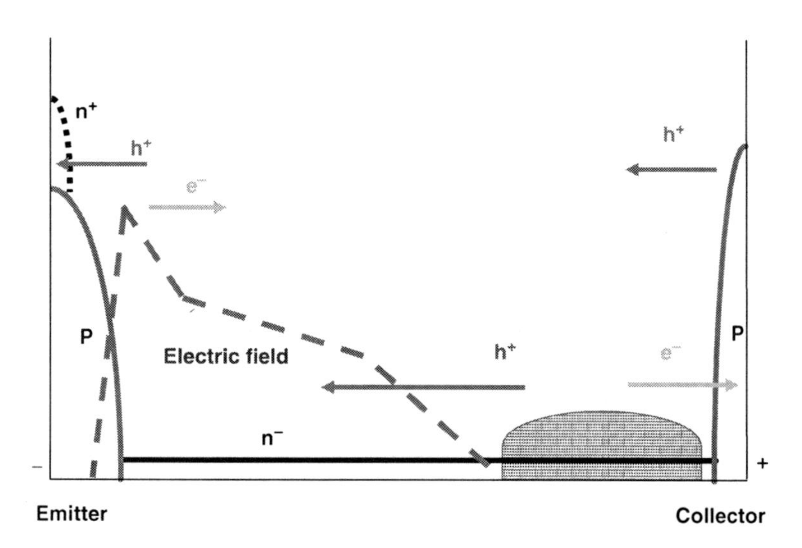

Fig. 12.41 NPT-IGBT at turn-off of an over-current and occurrence of dynamic avalanche

The electrons generated close to the pn-junction are flowing to the right-hand side and they compensate the hole current. An electric field will rise as drawn in Fig. 12.41 for a strong dynamic avalanche mode. This field has close similarity to the S-shaped field in dynamic avalanche of the second degree and the arguments given in context of Eqs. (12.15) and (12.16) hold, respectively.

An electric field shaped as in Fig. 12.41 leads to a weakly negative differential resistance in the *I–V* characteristics and an investigation of the effect using device simulation [Ros02] showed the occurrence of current tubes, which arise in certain regions and which jump to neighboring cells. This process also is similar to dynamic avalanche of the second degree.

However, a fundamental difference to a diode is that on the right side of the plasma a p-layer exists and the given collector-side pn-junction is forward biased. This layer injects holes, which compensate the electrons arriving from exhaustion of the remaining plasma and from dynamic avalanche. This opposes to a removal of the remaining plasma from the collector side. Due to the same polarity of negatively charged acceptor ions in the p-collector and the negative charge of electrons arriving at the collector side, no space charge region can build up at this point even with a high electron density. Therefore, dynamic avalanche of the third degree is not possible in an NPT-IGBT.

Another situation is given in a modern IGBT with a field stop layer or an n-buffer layer in front of the collector layer. This is shown in Fig. 12.42. The electron current flows to the right-hand side. If the hole current coming from the collector is not high enough to compensate the electron current, then at the $n^{-}n$-junction a space charge

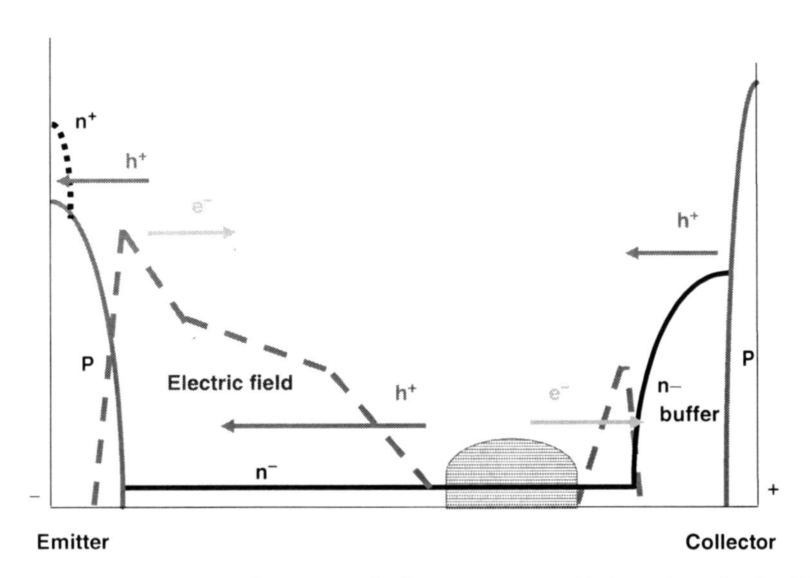

Fig. 12.42 IGBT with an n-buffer at turn-off of an over-current with dynamic avalanche. Under worse conditions a second electric field can rise in front of the n-buffer

can build up between the negative charged free electrons and the positively charged donor ions in the n buffer, and an electric field can rise.

In [Rah05] it is explained that this process is dangerous especially at the final plasma removal and the transition of the IGBT into the "switching self-camping mode" (SSCM, see Sect. 12.4). SSCM occurs in an IGBT during turn-off of an over-current under the condition of a high parasitic circuit inductance. In SSCM a field distribution with a second field peak at the nn$^+$-junction can occur, similar to second breakdown in a bipolar transistor. This effect is unstable. However, if a sufficiently high hole current is delivered from the p-collector layer and it compensates the electron current, then the SSCM event is stabilized in the IGBT. To achieve this, the emitter efficiency of the p-collector and the corresponding current gain α_{pnp} must not be too small.

Figure 12.43 shows the turn-off of a 3.3 kV-IGBT at a high over-current level. After the voltage has climbed up to approx. 2000 V, a flattening of the voltage course is to be seen. This is a sign of strong dynamic avalanche, where the holes flowing through the space charge limit the blocking capability in this time interval.

After the voltage has reached the applied DC link voltage of 2600 V, the current starts to decay, and still strong dynamic avalanche exists in the device. At a voltage of 3500 V the device transits into the SSCM. The voltage ramps up to 4000 V, which is close to the static avalanche breakdown voltage of the device. The course of the current and voltage waveforms seems to be more stable compared to the SSCM process in a diode (see Fig. 12.21). The hole current, which comes from the p-collector, adds a stabilizing effect during the SSCM [Rah05]. Figure 12.43 shows that very high stress in dynamic avalanche is also possible for IGBTs.

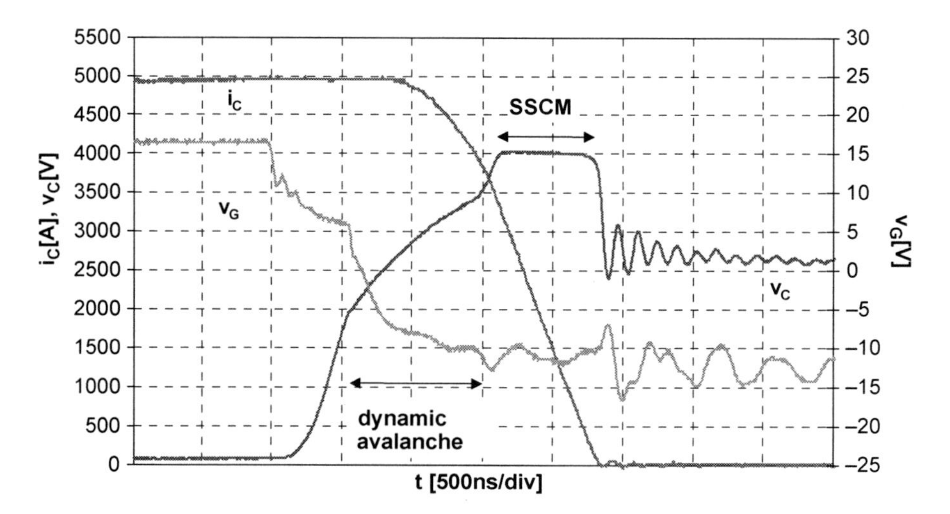

Fig. 12.43 Turn-off of a 3.3 kV 1200 A IGBT module at four times the rated current against a batters voltage $V_{bat} = 2600\,V$. After an interval of strong dynamic avalanche, the SSCM event follows. Figure from [Rah05] © 2007 IEEE

12.7 Cosmic Ray Failures

With the introduction of high-voltage semiconductors with turn-off capability in converters for electric traction in the beginning of the 1990s, failures in the application were observed which could not be explained with the available knowledge at the time. The failures occurred during the blocking mode of the devices. The application conditions were employed in the lab and long-term tests with high DC voltage in blocking direction were carried out. The tests confirmed the occurrence of spontaneous failures [Kab94]. The spontaneous character of the failures was strange, since no prior indications in device behavior, e.g., an increase of the leakage current, were found.

Figure 12.44 shows the results of the salt mine experiment. First, six failures occurred in the lab within 700 device hours. The test was interrupted and continued in a salt mine, with 140 m of solid rock above the test position. Under these conditions, no failure occurred. The test in the salt mine was interrupted again and continued in the lab, and now failures occurred once more at a comparable rate as before in the lab. The test position was moved again to a place in the cellar of a multi-storey building, at this position there was in sum 2.5 m of concrete above the test position. The failure rate was reduced again.

With these results, it was proven that cosmic ray was the main reason for failures of power semiconductor devices under such conditions.

The cosmic radiation consists of high-energy particles, which are created in supernova explosions or in the core of distant active galaxies; they hit the earth from all directions. A high-energy primary cosmic ray particle usually does not reach the

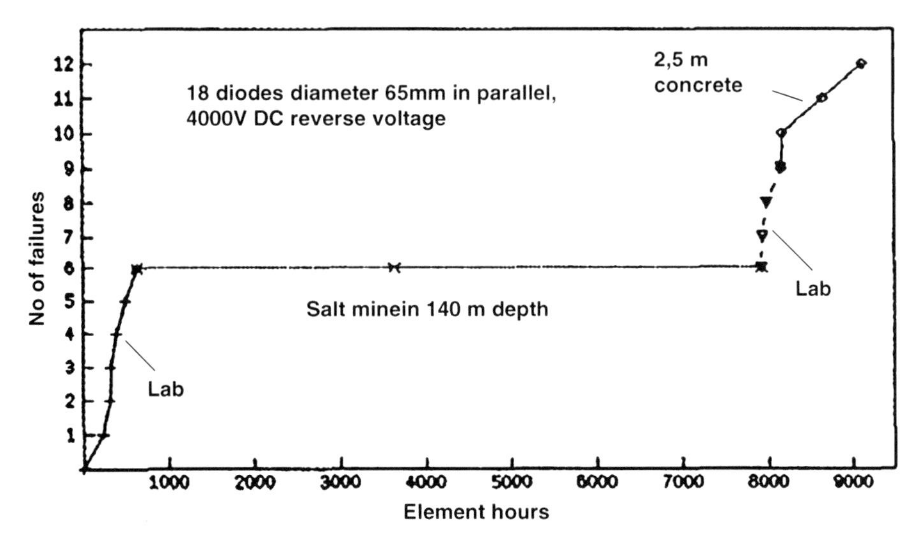

Fig. 12.44 Results of the salt mine experiment. Number of failures over the summarized time of DC voltage stress. Figure from [Kab94] © 1994 IEEE

surface of the earth directly, but collides with atmospheric particles. There it generates a variety of secondary high-energy particles; a cosmic-ray shower arrives at the earth surface as terrestrial cosmic radiation. What is relevant to device failures are high-energy neutrons and protons; however, also the effect of pions and myons is not excluded. A small part of the neutrons, traveling through a semiconductor device, collide with the cores of silicon atoms. Back scattered ions are created, which generate again locally a dense plasma of charge carriers. In the space charge region of a semiconductor device in the blocking mode, these carriers are separated and a current pulse occurs. If the electric field is higher than a certain threshold value – which depends on the initial plasma generation – then impact ionization creates more carriers than the carriers which flow out of the plasma region by the mechanism of diffusion. The discharge runs as a so-called streamer with high velocity through the device, in analogy to a discharge in a gas. The device is flooded locally with free carriers within some hundred picoseconds; hence, a local current tube occurs. Finally, the very high local current density destroys the semiconductor device.

The failed devices exhibit a pinhole-size molten channel from cathode to anode anywhere in the bulk. Failure pictures from device failures caused by cosmic ray are shown in Fig. 12.45. On the left-hand side in Fig. 12.45 a pinhole is to be seen. On the right-hand side bubbles in the metallization are to be seen and below the metallization a pinhole is hidden. The failure pictures show clearly an effect occurring in a very narrow region.

Figure 12.46 shows the cosmic ray failure picture of a 3.3-kV IGBT die. Again a pinhole is found, with a size in the order of the cell pitch of the IGBT die.

Several tests of different device designs were carried out, for acceleration of the failure rate tests stations at high altitudes were arranged (Zugspitze 2964 m, Jungfraujoch 3580 m). Terrestrial cosmic ray increases with altitude above sea level. The acceleration factor amounts to 10 in 3000 m, in 5000 m to approx. 45 [Kai05]. In parallel experiments, tests with particle accelerators were carried out, devices at

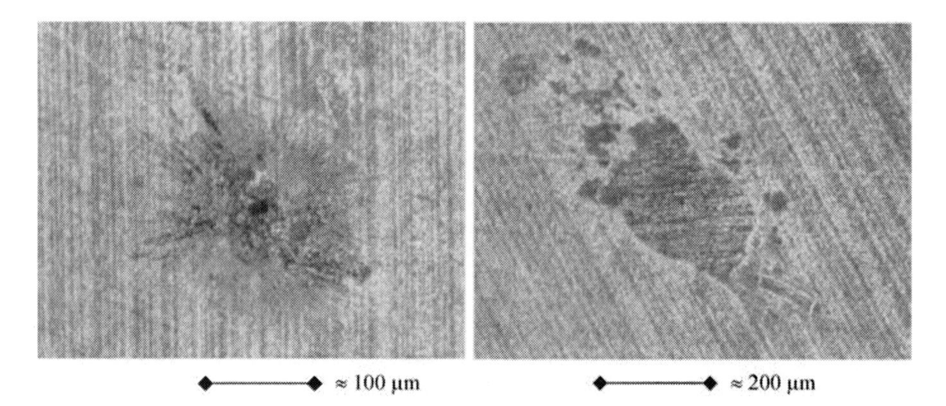

Fig. 12.45 Failure pictures after a cosmic ray destruction of 4.5-kV diodes with a diameter of <50 μm. The photos are from the cathode side. Left-hand side: small pinhole. Right-hand side: Molten area in the metallization with bubbles. Pictures from Jean-Francois Serviere, Alstom

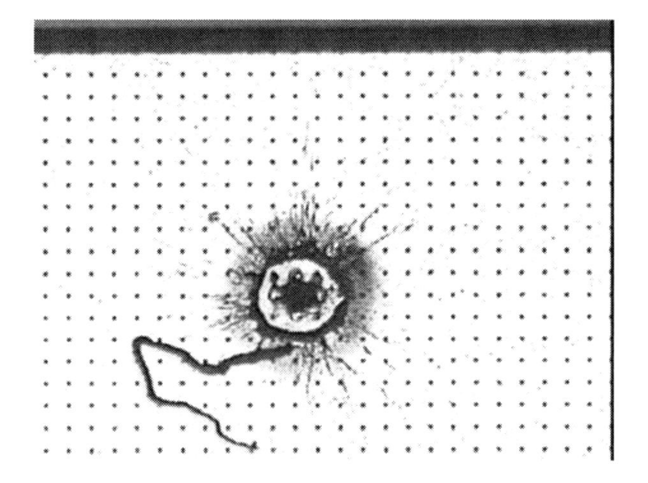

Fig. 12.46 Cosmic ray failure of a 3.3 kV IGBT in the cell area. Cell pitch 15 μm. Picture from G. Sölkner, Infineon

high-reverse voltages were irradiated with high-energy neutrons, protons, and other ions accelerated with high energy.

Irradiation with ions, especially the types of ions and energies which are generated at a neutron–silicon collision, is suited to study the detailed failure mechanism. A great amount of research work was done in this field, see, e.g., [Soe00].

An example of the results is shown in Fig. 12.47, taken from [Kai04]. Diodes rated at 3.3 kV have been irradiated with ^{12}C-ons and the reverse voltage applied during irradiation was increased. Two diode designs were compared having a triangular electric field shapes (NPT, see Fig. 5.3a) and another diode with a trapezoidal shape of the electric field (PT in Fig. 5.3b), which is assigned with FS (field stop). While at a small voltage the charge generated by a single ^{12}C ion is small for all

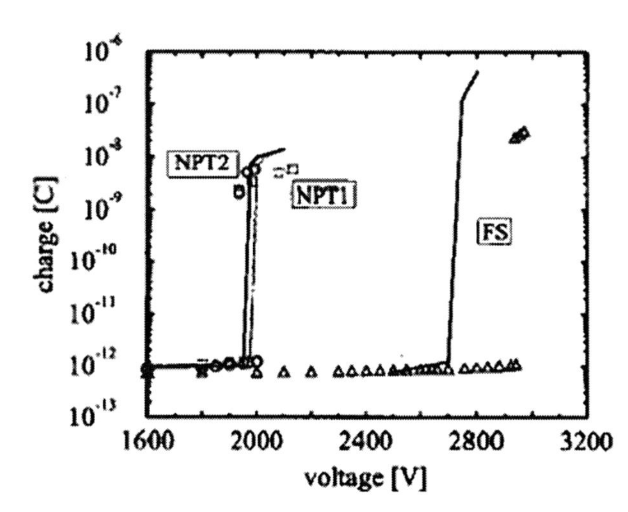

Fig. 12.47 Charge created at irradiation with single ^{12}C-ions of a kinetic energy of 1.7 MeV as a function of the applied reverse DC voltage to a 3.3 kV diode. Simulation (*straight line*) and experiment (*symbols*) for a triangular-filed shape NPT1 (□), NPT2 (O), and a trapezoidal field shape FS (Δ). Figure from [Kai04] © 2004 IEEE

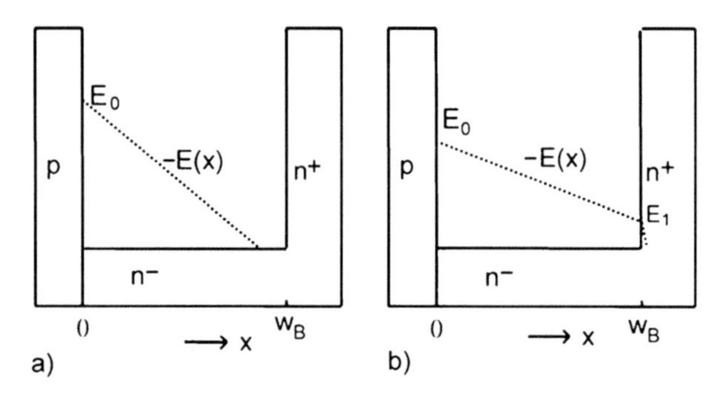

Fig. 12.48 Schematic drawing of the field shape for an NPT design (**a**) and PT design (**b**) at the same thickness and same applied voltage in the blocking direction

samples, above a defined threshold voltage a strong charge carrier multiplication sets on and the created charge increases suddenly more than three decades. For the FS diode, which has a PT design, this threshold voltage is more than 700 V higher when compared to the NPT diode.

Even though PT and NPT dimensioning in Fig. 12.47 are designed for the same rated voltage, the maximal electric field is much lower for the PT design. This is illustrated in Fig. 12.48. The shape of the electric field is drawn for two simplified diode structures with the same thickness at same applied blocking voltage. The area below the line $E(x)$, which corresponds to the reverse voltage, is the same for both devices. However, the value E_0 is much lower for the PT design. At the same voltage, impact ionization is still negligible for the device with PT design. For the occurrence of impact ionization in the PT diode, the applied voltage must be increased. Then the line of $-E(x)$ in Fig. 12.48b is shifted upward, until E_0 becomes close to a value as in Fig. 12.48a. Then impact ionization sets on for the diode with PT dimensioning too; however, at a much higher voltage.

The design of a device for increased cosmic ray stability is in substance done in the way that the maximal occurring DC voltage V_{bat} in the application is used and the value of E_0 is kept as low as possible. For this, the base doping N_D must be made low, so that a PT dimensioning is given. Additionally E_0 is lower, if the device is designed with a wider w_B. However, this is of disadvantage regarding the conduction losses and switching losses.

With these measures, the probability of cosmic ray-induced failures is decreased. However, the occurrence of such failures cannot be avoided completely, since the type and especially the energy of the striking particle cannot be predicted, and also the initial plasma discharge is dependent on some stochastic effects. The cosmic ray stability is specified in "Failure in Time" (FIT), where it holds

$$1\,\text{FIT} = 1 \text{ failure in } 10^9\,\text{h}$$

The requirement for a power electronic module for traction application is 100 FIT: One failure is to be expected statistically in 10^7 h, these are 1141 years. One must consider that a typical power module for such an application consists of 24 IGBTs

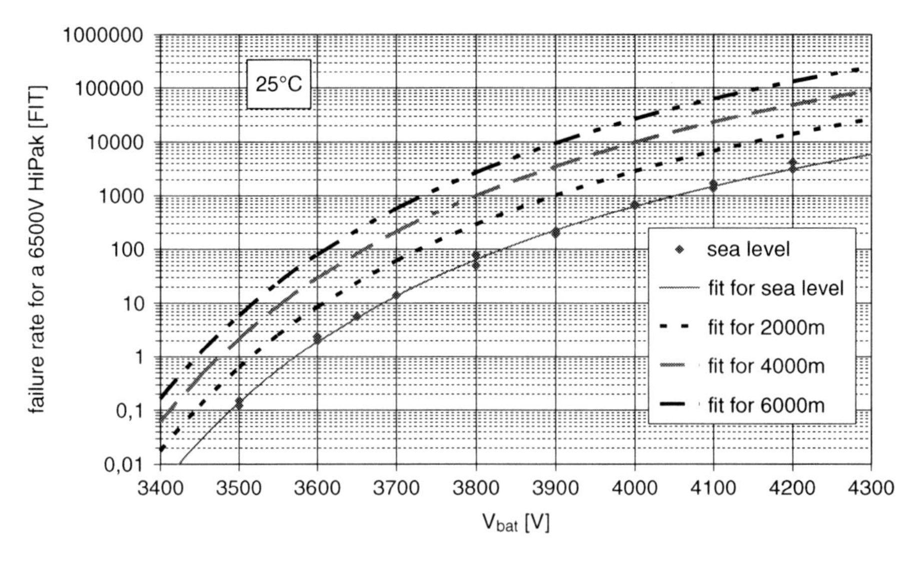

Fig. 12.49 Cosmic ray failure rate at $T = 25°\mathrm{C}$ for the 6.5 kV IGBT module 5SNA0600G650100 from ABB. Figure from [Kam04]

and 12 freewheeling diodes. For a single device holds a lower FIT rate, respectively (Fig. 12.49).

The failure rate is highest at low temperatures and is decreasing with temperature due to the decreasing avalanche ionization rates at increased temperature, see Chap. 2.

With the described measures, the given requirements can be fulfilled at specified conditions; however, the detailed final failure mechanism of these "single event burnout" [Alb05] is still a subject of research. In a MOSFET, the failure mechanism is explained to be finally the activation of the parasitic bipolar transistor and second breakdown of the bipolar transistor [Was86]. In diodes, no parasitic component of any transistor type is present. Even strong local avalanche breakdown should be stable, see Sect. 12.4. Some work [Alb05, Kai04, Soe00] explains the final failure mode by very high local heating caused by very strong dynamic avalanche. Field redistribution effects in analogy to dynamic avalanche of the third degree are considered in deep submicron CMOS devices for improved ruggedness against radiation-caused single event pulses [Das07].

For high-voltage devices, the dimensioning rules for cosmic ray stability contradict to other rules for optimizing the device characteristic; e.g. for diodes it contradicts to the requirement for soft-recovery behavior, which is hard to achieve if a strong PT dimensioning is used. A trade-off between different requirements must be made. To meet said demands, most of the nowadays high-voltage devices use designs with a middle layer thickness w_B much higher than necessary for the required blocking capability. However, this leads to increased losses in the forward conduction mode and/or increased turn-off losses.

12.8 Failure Analysis

Some failure mechanisms occurring in power circuits with IGBTs are discussed here in summary and interrelation. Table 12.1 gives different failure mechanisms, which are divided to failures caused by current respectively temperature, by voltage, and by dynamic effects. The failure reasons are printed in cursive letters.

Table 12.1 Some failure mechanisms in IGBT modules. *Cursive: failure reason.* Normal: failure picture

Current temperature	Voltage	Dynamic effects
– *Too high average current* The die shows molten areas with a diameter of several millimeter. Failures located in the active area	– *Production fault* Failure location starts at the edge	Applied voltage below the rated voltage – *Lack of dynamic ruggedness of the freewheeling diode* Diode and transistor in the associated commutation loop are destroyed
– *Surge current exceeded* Local molten area, size approx. 1 mm, sometimes cracks in the crystal, failure located in the active area	– *Voltage peaks above rated voltage* Failure location starts at the edge	– *Lack of dynamic ruggedness of the freewheeling diode* Only the diode is destroyed, pinhole with diameter <100 μm
– *Short-circuit capability of IGBT exceeded* Only IGBTs destroyed, large part of the emitter area burned off.	– *Lack of long term stability of the passivation layer* Failure location starts at the edge	– *Dynamic avalanche third degree* Pinhole with a diameter <100 μm, cracks in the crystal lattice originating
		– *Dynamic latch-up* eventually only 1 IGBT is destroyed

For failures caused by current, a molten zone within the active area of the device is typical. At very high average currents, one finds a destroyed area with a diameter of several millimeters.

If a surge current failure of a diode occurs, the molten area is usually smaller, in the range of 1 mm. For bonded diodes, often a melting of metallization beside the bond feet is observed. Surge current occurs in an application, for example, when a non-loaded DC link capacitor is connected by a diode rectifier circuit to the grid and a very high current pulse is generated in the first instant. In this case, an application fault is given. A loading circuit for the DC-link capacitor will be of help in this case. Manufacturers of power semiconductors know very well the typical surge current failure pictures of their devices, as shown for example in Fig. 12.6, and they can identify such failures.

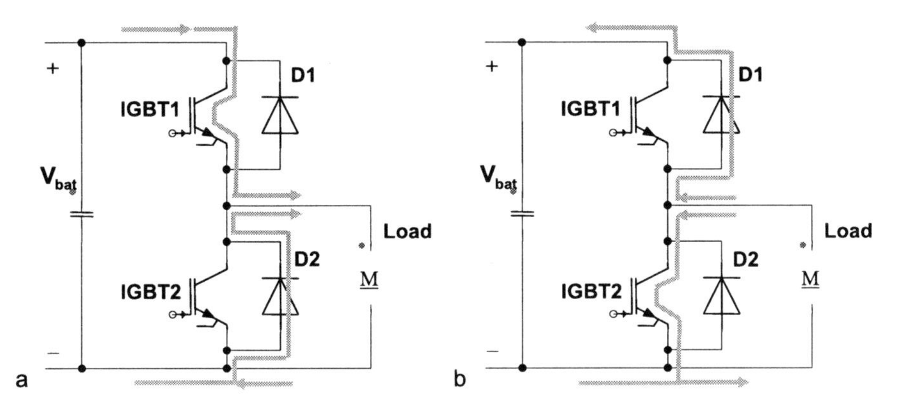

Fig. 12.50 Commutation loops in a half-bridge. (**a**) Power transmission from the DC link to the load. (**b**) Reverse power flux

If failures are caused by voltage, it is mostly observed that the failure location is at the edge of the device, such as in the junction termination structure. At these positions, the highest electric fields occur at the surface. The junction termination is very important in power device manufacturing and is most sensitive to contamination in the production line, to faults in photolithography due to dust particles, etc. If a device has a weak point due to a fault in production, this occurs usually at the edge of the device. While current-induced failures are mostly due to faults in the application, this must not be the case for voltage-induced failures. Application faults – voltage peaks higher than the rated voltage – as well as production faults must be taken into account.

Failures by dynamic effects are mainly related to switching events. The voltage stays below the rated voltage of the device. At switching events the transistors interact with their freewheeling diode. Figure 12.50 shows the corresponding commutation loops. At power transmission to the load (Fig. 12.50a) IGBT1 commutates with diode D2. If diode D2 fails during its turn-off, the associated transistor in the commutation loop turns on to a short circuit; hence, a short circuit within the bridge with very low inductance is given. Therefore, the IGBT may be destroyed by a short circuit. The same holds for the reverse power flux, where the commutation loop for this is shown in Fig. 12.50b. There IGBT2 commutates with diode D1.

If diodes as well as their associated transistors in the commutation loop are destroyed, the failure reason is usually due to the diode. If the diode fails, the IGBT may be destroyed. On the other hand, if the IGBT fails, this event causes usually no stress for the diode and the diode has no reason to fail. An exception might be, if a spark burns inside the module, which destroys further devices.

With such considerations, sometimes conclusions for failure reasons are still possible even if a module is heavily destroyed.

If the freewheeling diode fails and the IGBT turns-off successfully following short circuit, one finds afterward a typical pinhole in the diode. For freewheeling diodes rated at 1200–1700 V, these pinholes are a sign of lack of dynamic ruggedness. Freewheeling diodes with higher rated voltages, however, may be destroyed by a very high current density in the reverse direction and a contemporaneous high voltage. For such failures by dynamic avalanche of the third degree, one also finds cracks in the crystal lattice, originating at the pinhole. These cracks are a sign of very high local temperature spots. An example was given in Fig. 12.19.

If only the transistor is destroyed, the failure reason must be searched in the transistor. Short-circuit failure is one of the possible mechanisms and is always worth considering. For short-circuit failures, burned off emitter regions with large areas are typical. Note, however, at short circuit III (Sect. 12.6) even a diode failure may occur due to the extreme high dv/dt and voltage spike at diode commutation, while the IGBT may survive [Lut09b].

Another IGBT failure mode is dynamic latch-up, which is caused by single weak cells in the IGBT device. Such failures cannot be found during tests of static parameters at the device manufacturer. For modules with parallel connection of many IGBT dies, the manufacturers execute final tests in an application conform circuit including dynamic turn-off under very high stress conditions in order to find single weak devices and to avoid failures in the targeted application.

Failure analysis is a complex process and a lot of experience is necessary. There are other failure modes, e.g. cosmic ray failures. Also for such failures pinholes are typical. In most cases one cannot conclude a failure reason directly from a failure picture since there are different failure modes which may lead to similar failure pictures. The power circuit and the application condition must be investigated additionally.

With devices connected in parallel, a non-symmetric assembly may lead to oscillations and subsequently to overstress of devices at special positions. Often new questions come forth during failure analysis. In conclusion, failure research is very complex; however, the results are often extremely valuable.

References

[Alb05] Albadri AM, Schrimpf RD, Walker DG, Mahajan SV: "Coupled Electro-Thermal Simulations of Single Event Burnout in Power Diodes", IEEE Trans. Nucl. Sci., vol. 52, No. 6 (2005)

[Bab08] Baburske R, Heinze B, Lutz J, Niedernostheide FJ: "Charge-carrier Plasma Dynamics during the Reverse-recovery Period in p^+-n^--n^+ diodes, IEEE Trans. Electron Devices, Vol. 55 No 8, pp. 2164–2172 (2008)

[Bab09] Baburske R, Domes D, Lutz J, Hofmann W: "Passive turn-on process of IGBTs in Matrix converter applications", Proceedings EPE, Barcelona (2009)

[Ben67] Benda HJ, Spenke E: "Reverse Recovery Process in Silicon Power Rectifiers", Proceedings of the IEEE, vol. 55 No 8 (1967)

[Chm06] Chen M, Lutz J, Domeij M, Felsl H.P, Schulze HJ: "A novel diode structure with Controlled Injection of Backside Holes (CIBH)", Proceedings of the ISPSD, Neaples, pp 9–12 (2006)

[Das07] DasGupta S et al: "Effect of Well and Substrate Potential Modulation on Single Event Pulse Shape in Deep Submicron CMOS" IEEE Trans. Nucl. Sci., vol. 54, No. 6, pp. 2407–2412 (2007)

[Dom99] Domeij M, Breitholtz B, Östling M, Lutz J: "Stable dynamic avalanche in Si power diodes", Apllied Physics Letters vol. 74, No. 21, 3170 (1999)

[Dom03] Domeij M, Lutz J, Silber D: "On the Destruction Limit of Si Power Diodes During Reverse Recovery with Dynamic Avalanche", IEEE Trans. Electron Devices, vol. 50, No. 2, pp. 486–493 (2003)

[Eck94] Eckel HG, Sack L: "Experimental Investigation on the Behaviour of IGBT at Short-Circuit du. ring the On-State. In: 20th International Conference on Industrial Electronics, Control and Instrumentation, IECON'94, vol. 1, pp. 118–123 (1994)

[Eck95] Eckel HG, Sack L: "Optimization of the Short-Circuit Behaviour of NPT-IGBT by the Gate Drive", In: EPE 1995. Sevilla, pp. 213–218 (1995)

[Ega66] Egawa H: "Avalanche Characteristics and Failure Mechanism of High Voltage Diodes", IEEE Trans. Electron Devices, vol. ED-13, No. 11, pp. 754–758 (1966)

[Fel06] Felsl HP, Heinze B, Lutz J: "Effects of different buffer structures on the avalanche behaviour of high voltage diodes under high reverse current conditions" IEE Proceedings Circuits, Devices and Systems vol. 153 , Issue 1, pp. 11–15 (2006)

[Fel08] Felsl HP, Pfaffenlehner M, Schulze H, Biermann J, Gutt T, Schulze HJ, Chen M, Lutz J: "The CIBH Diode – Great Improvement for Ruggedness and Softness of High Voltage Diodes" ISPSD 2008, Orlando, Florida, USA (2008)

[Ful67]. Fulop W: "Calculation of Avalanche Breakdown Voltages of Silicon pn-Junctions", Solid-St. Electron., vol. 10, pp. 39–43 (1967)

[Gha77] Ghandhi SK, Semiconductor Power Devices, John Wiley and Sons, New York 1977

[Hei05] Heinze B, Felsl HP, Mauder A, Schulze HJ, Lutz J: "Influence of Buffer Structures on Static and Dynamic Ruggedness of High Voltage FWDs", Proceedings of the ISPSD, Santa Barbara (2005)

[Hei06] Heinze B, Lutz J, Felsl HP, Schulze HJ: "Influence of Edge Termination and Buffer Structures on the Ruggedness of 3.3 kV Silicon Free-Wheeling Diodes", Proceedings of the 8th ISPS, Prague (2006)

[Hei07] Heinze B, Lutz J, Felsl HP, Schulze HJ: "Ruggedness of High Voltage Diodes under very hard Commutation Conditions" Proceedings EPE 2007, Aalborg, Denmark (2007)

[Hei08] Heinze B, Lutz J, Felsl HP, Schulze HJ: "Ruggedness analysis of 3.3 kV high voltage diodes considering various buffer structures and edge terminations", Microelectronics Journal, vol. 39, Issue 6, pp. 868–877 (2008)

[Hei08b] Heinze B, Baburske R, Lutz J, Schulze HJ: "Effects of Metallisation and Bondfeets in 3.3 kV Free-Wheeling Diodes at Surge Current Conditions", Proceedings of the ISPS, Prague (2008)

[Hei97] Hierholzer M, Bayerer R, Porst A, Brunner H: "Improved characteristics of 3.3 kV IGBT modules", Proceedings PCIM Nuremberg (1997)

[How70] Hower PL, Reddi K: "Avalanche injection and second breakdown in transistors" IEEE Trans. Electron Devices, vol. 17, p. 320 (1970)

[How74] Hower PL, Pradeep KG, "Comparision of One- and Two-Dimensional Models of Transistor Thermal Instability", IEEE Trans. Electron Devices 21, No. 10, pp. 617–623 (1974)

[Kab94] Kabza H, Schulze H-J, Gerstenmaier Y, Voss P, Wilhelmi J, Schmid W, Pfirsch F, and Platzöder K: Cosmic radiation as a possible cause for power device failure and possible countermeasures", Proceedings of the 6th International Symposium on Power Semiconductor Devices & IC's, Davos, Switzerland (1994)

[Kai04] Kaindl W, Soelkner G, Becker HW, Meijer J, Schulze HJ, Wachutka G: "Physically Based Simulation of Strong Charge Multiplication Events in Power Devices Triggered by Incident Ions", Proceedings of the 16th International Symposium on Power Semiconductor Devices & IC's, Kitakyushu, Japan (2004)

[Kai05] Kaindl W, Modellierung höhenstrahlungsinduzierter Ausfälle in Halbleiterleistungsbauelementen, Dissertation, München 2005

[Kam04] Kaminski N: "Failure Rates of HiPak Modules Due to Cosmic Rays", ABB Application Note 5SYA 2042-02 (2004)

[Kin05] Kinzer D: "Advances in Power Switch Technology for 40 V - 300 V Applications", Proceedings of the EPE, Dresden (2005)

[Kop05] Kopta A, Rahimo M: "The Field Charge Extraction (FCE) Diode – A Novel Technology for Soft Recovery High Voltage Diodes" Proc. ISPSD Santa Barbara (2005), pp. 83–86.

[Kop08] Kopta A, Rahimo M, Schlapbach U, Kaminski N, Silber D: "Neue Erkenntnisse zur Kurzschlussfestigkeit von hochsperrenden IGBTs", Kolloquium Halbleiter-Leistungsbauelemente, Freiburg (2008)

[Kop09] Kopta A, Rahimo M, Schlapbach U, Kaminski N, Silber D: Limitation of the Short-Circuit Ruggedness of High-Voltage IGBTs, Proc. ISPSD Barcelona, pp. 33–37 (2009)

[Las92] Laska T, Miller G, Niedermeyr J: "A 2000 V Non-Punchtrough IGBT with High Ruggedness", Solid-St. Electron., Vol. 35, No. 5, pp. 681–685, 1992

[Las03] Laska T. et al: "Short Circuit Properties of Trench/Field Stop IGBTs Design Aspects for a Superior Robustness", Proc. 15th ISPSD, pp. 152–155, Cambridge (2003)

[Lef05] Lefebvre S, Khatir Z, Saint-Eve F: "Experimental behavior of single chip IGBT and COOLMOS^{TM} devices under repetitive Short-Circuit conditions, IEEE Trans. Electron Devices, vol. 52 Number 2, February 2005, pp. 276–283

[Lef08] Lefebvre S, Arab M, Khatir Z, Bontemps S: "Investigations on ageing of IGBT transistors under repetitive short-circuits operations", Proceedings of the PCIM (2008)

[Let95] Letor RR.; Aniceto GC.: "Short Circuit Behavior of IGBT′s Correlated to the Intrinsic Device Structure and on the Application Circuit", IEEE Transactions on Industry Applications, vol. 31, No. 2, March/April 1995

[Lin08] Linder S: "Potentials, Limitations, and Trends in High Voltage Silicon Power Semiconductor Devices", Proc. of the 9th ISPS, Prague, pp. 11–20 (2008)

[Lut97] Lutz J: "Axial recombination centre technology for freewheeling diodes" Proceedings of the 7th EPE, Trondheim, pp. 1502ff 1.502 (1997)

[Lut03] Lutz J, Domeij M: "Dynamic avalanche and reliability of high voltage diodes", Microelectronics Reliability 43, pp. 529–536 (2003)

[Lut08] Lutz J, Herrmann T, Feller M, Bayerer R, Licht T, Amro R: "Power cycling induced failure mechanisms in the viewpoint of rough temperature environment", Proceedings of the 5th International Conference on Integrated Power Electronic Systems, pp. 55–58 (2008)

[Lut09] Lutz J, Baburske R, Chen M, Heinze B, Felsl HP, Schulze HJ: "The nn^{+}-junction as the key to improved ruggedness and soft recovery of power diodes", IEEE Trans. Electron Devices, vol. 56, No. 11, pp. 2825–2832 (2009)

[Lut09b] Lutz J, Döbler R, Mari J, Menzel M: "Short Circuit III in High Power IGBTs" Proceedings EPE, Barcelona (2009)

[Mor00] Mori M, Kobayashi H, Yasuda Y: "6.5 kV Ultra Soft & Fast Recovery Diode (U-SFD) with High Reverse Recovery Capability", ISPSD 2000, 115–118 Toulouse (2000)

[Mor07] Mori M et al: "A Planar-Gate High-Conductivity IGBT (HiGT) With Hole-Barrier Layer" IEEE Trans. Electron Devices, vol. 54, NO. 6, JUNE 2007

[Mou88] Mourick P, Das Abschaltverhalten von Leistungsdioden, Dissertation, Berlin 1988

[Nag98] Nagasu M et al: "3.3 kV IGBT Modules having Soft Recovery Diodes with high Reverse Recovery di/dt Capability", Proceedings of the PCIM 98 Japan, 175 (1998)

[Nic00] Nicolai U, Reimann T, Petzoldt J, Lutz J: Application Manual Power modules, ISLE Verlag 2000. (http://www.semikron.com/skcomplue/en/application_manual-193.htm)

[Nie04] Niedernostheide FJ, Falck E, Schulze HJ, Kellner-Werdehausen U: "Avalanche injection and current filaments in high-voltage diodes during turn-off", Proceedings of the 7th ISPS′04, Prague (2004)

[Nie05] Niedernostheide FJ, Falck E, Schulze HJ, Kellner-Werdehausen U: "Periodic and traveling current-density distributions in high-voltage diodes caused by avalanche injection", Proceedings of the EPE (2005)

[Oet00] Oetjen J et al: "Current filamentation in bipolar devices during dynamic avalanche breakdown", Solid-St. Electron., Vol. 44, pp. 117–123 (2000)

[Ohi02] Ohi T, Iwata A, Arai K: Investigation of Gate Voltage Oscillations in an IGBT Module under Short Circuit Conditions, Proceedings of the Power Electronics Specialists Conference, 2002 IEEE 33rd Annual vol. 4, pp. 1758–1763

[Pen98] Pendharkar S, Shenai K: "Zero Voltage Switching Behavior of Punchthrough and Nonpunchthrough Insulated Gate Bipolar Transistors (IGBT's)", IEEE Trans. Electron Devices, vol. 45, Issue 8, pp. 1826–1835 (1998)

[Pog03] Pogany D, Bychikhin S, Gornik E, Denison M, Jensen N., Groos G, Stecher M: "Moving current filaments in ESD protection devices and their relation to electrical characteristics", Annual Proceedings - Reliability Physics (Symposium), pp. 241–248 (2003)

[Por94] Porst A: "Ultimate Limits of an IGBT (MCT) for High Voltage Applications in Conjunction with a Diode", Proceedings of the 6th ISPSD (1994)

[Rah04] Rahimo M, Kopta A et al: "Switching-Self-Clamping-Mode "SSCM", a breakthrough in SOA Performance for high voltage IGBTs and diodes", Proceedings of the ISPSD pp. 437–440 (2004)

[Rah05] Rahimo M et al: "A Study of Switching-Self-Clamping-Mode "SSCM" as an Over-voltage Protection Feature in High Voltage IGBTs" Proc. ISPSD, Santa Barbara (2005)

[Ron97] Ronan HR: "One Equation Quantifies a Power MOSFETs UIS Rating", PCIM Magazine August 1997, pp. 26–35 (1997)

[Ros02] Rose P, Silber D, Porst A, Pfirsch F: "Investigations on the Stability of Dynamic Avalanche in IGBTs", Proceedings of the ISPSD (2002)

[Sai04] Saint-Eve F, Lefebvre S, Khatir Z: "Study on IGBT lifetime under repetitive short-circuits conditions", Proceedings of the PCIM, Nürnberg (2004)

[Sco89b] Schlangenotto H, Neubrand H: "Dynamischer Avalanche beim Abschalten von GTO-Thyristoren und IGBTs", Archiv der Elektrotechnik 72, pp. 113–123 (1989)

[Shi59] Shields J: "Breakdown in Silicon pn-Junctions", Journ. Electron. Control No 6 pp 132ff (1959)

[Sil73] Silber D, Robertson M.J: "Thermal effects on the forward characteristics of silicon pin-diodes at high pulse currents", Solid-St. Electron., Vol. 16, pp. 1337–1346 (1973)

[Sit02] Sittig R: "Siliziumbauelemente nahe den Grenzen der Materialeigenschaften", ETG Fachtagung, Bad Nauheim, ETG Fachbericht 88 pp. 9 ff (2002)

[Soe00] Soelkner G, Voss P, Kaindl W, Wachutka G, Maier KH, Becker HW, "Charge Carrier Avalanche Multiplication in High-Voltage Diodes Triggered by Ionizing Radiation", IEEE Trans. Nucl. Sci., vol. 47, No. 6, pp. 2365–2372 (2000)

[Syn07] Advanced tcad manual. Synopsys Inc. Mountain View, CA. Available: http://www.synopsys.com (2007)

[Tom96] Tomomatsu Y et al: "An analysis and improvement of destruction immunity during reverse recovery for high voltage planar diodes under high dI_{rr}/dt condition", Proceedings of the ISPSD, pp. 353–356 (1996)

[Wac91] Wachutka G: "Analytical model for the destruction mechanism of GTO-like devices by avalanche injection", IEEE Trans. Electron Devices, vol. 38, p. 1516 (1991)

[Wak95] Wacker A., Schöll E., Criteria for stability in bistable electrical devices with S- or Z-shaped current voltage characteristic, J. Appl. Phys vol. 78 (12), pp. 7352–7357 (1995)

[Was86] Waskiewicz AE, Groninger JW, Strahan VH, and Long DM: "Burnout of power MOS transistors with heavy ions of 252-Cf," IEEE Trans. Nucl. Sci., vol. NS-33, no. 6, pp. 1710–1713, (1986)

Chapter 13
Power Device-Induced Oscillations and Electromagnetic Disturbances

13.1 Frequency Range of Electromagnetic Disturbances

Every power electronic switching action results in a deviation from the ideal sinusoidal AC current or the ideal homogeneous DC current. Switching events are usually done periodically in time. Every periodic event can be separated into a row of sinus and cosinus terms by means of Fourier transformation. With this tool, the generated frequencies, the harmonics, and their intensity can be calculated.

Figure 13.1 shows a rough overview of the disturbances and oscillations caused by power electronics. It is distinguished between disturbances created by switching

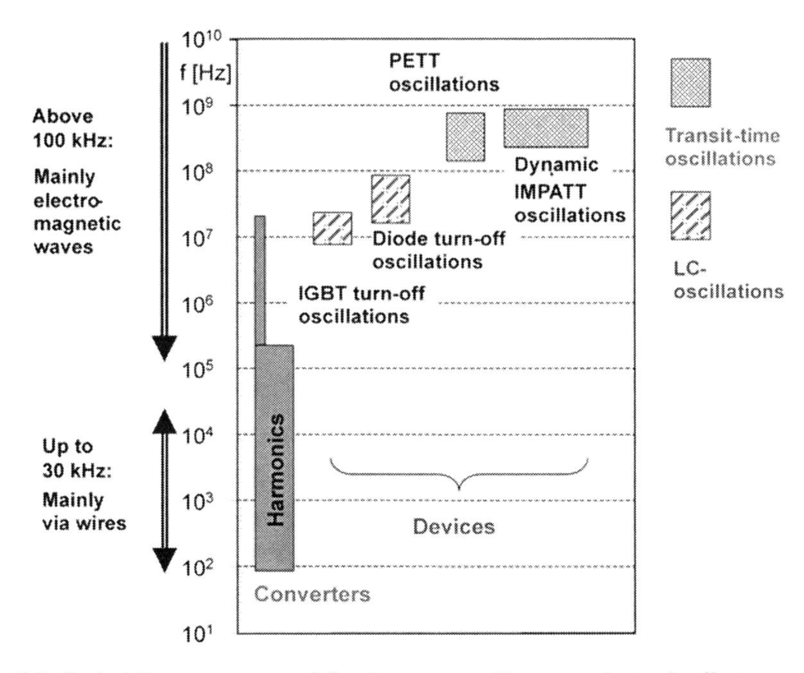

Fig. 13.1 Typical frequency ranges of disturbances caused by power electronic effects

J. Lutz et al., *Semiconductor Power Devices*, DOI 10.1007/978-3-642-11125-9_13,

events in power electronic converters, the harmonics of the switching frequency on the low- and medium-frequency range, and the device-induced high-frequency disturbances.

At low frequencies, e.g. in phase-commutated converters, the disturbances caused by the input rectifier occur as multiples of the grid frequency of 50 or 60 Hz, their intensity declines proportionally to $1/n$. In self-commutated converters with modern power devices, the typical switching frequencies for converters using IGBTs are in the range of 1–20 kHz; also in this case, harmonics of the respective switching frequency are to be expected. Higher switching frequencies can be applied with MOSFETs as power devices. In switch-mode power supplies, nowadays frequencies up to 1 MHz and above have become possible.

Device-induced oscillations, on the other hand, result from switching events. Since the switching times of the devices are much smaller than the period of the switching frequency, the electromagnetic disturbances caused thereby occur in a frequency range significantly higher than the switching frequency.

Electromagnetic disturbances of a frequency <30 kHz spread mainly via wiring and cables. They disturb the electric grid in the form of grid feedback. Electromagnetic disturbances of frequencies >100 kHz are radiated off to a large extent as electromagnetic waves, and this can cause incompatibility with other electronic and power electronic equipment.

Harmonics

Figure 13.2 shows two simplified examples of electric signals, either depicting the current or the voltage curve.

For a rectangular course with point symmetry to π and the amplitude a, as shown in Fig. 13.2a, the harmonics can be calculated using the Fourier transformation

$$y = \frac{4a}{\pi}\left(\sin\omega t + \frac{1}{3}\sin 3\omega t + \frac{1}{5}\sin 5\omega t + \cdots\right) \tag{13.1}$$

Multiples of the switching frequency $f = \omega/2\pi$ are generated, i.e. the 3rd, 5th, 7th, ... harmonics. With increasing ordinal number n, their intensity decreases proportionally to $1/n$.

However, for a trapezoidal course with point symmetry to π and the amplitude a as shown in Fig. 13.2b, the Fourier transformation results in

$$y = \frac{4}{\pi}\frac{a}{\mu}\left(\sin\mu\,\sin\omega t + \frac{1}{3^2}\sin 3\mu\,\sin 3\omega t + \frac{1}{5^2}\sin 5\mu\,\sin 5\omega t + \cdots\right) \tag{13.2}$$

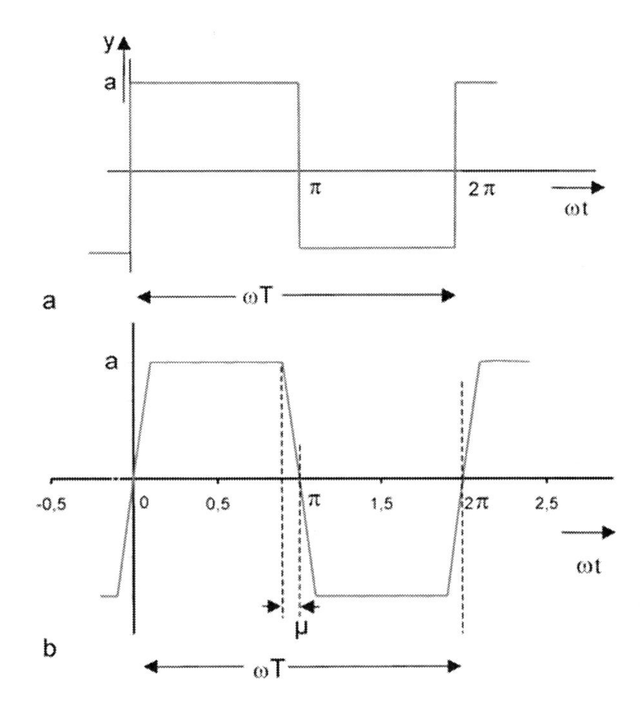

Fig. 13.2 (**a**) Rectangular course of a signal, (**b**) trapezoidal course of a signal

The amplitudes of the harmonics decrease proportionally to $1/n^2$. For a non-symmetric shape, additional terms will occur. Nevertheless, the faster decline of the harmonics and therefore a trapezoidal course is much more suitable. The slopes $\mathrm{d}i/\mathrm{d}t$ – represented in Fig. 13.2 with μ – can be adjusted by gate resistors if MOSFETs and IGBTs are used. To reduce harmonics, the switching times are reduced by increased gate resistors. However, this will increase switching losses. A trade-off must be made in many applications between switching losses on the one hand and electromagnetic emissions on the other hand.

Suitable filters are implemented as a countermeasure to improve the electromagnetic compatibility. This shall not be explained further in this chapter; the attention is riveted instead on oscillations created by power devices themselves.

13.2 LC Oscillations

13.2.1 Turn-Off Oscillations with IGBTs Connected in Parallel

In power modules, often a lot of single dies are connected in parallel. It is very difficult to give all single dies identical symmetrical conditions in respect to the

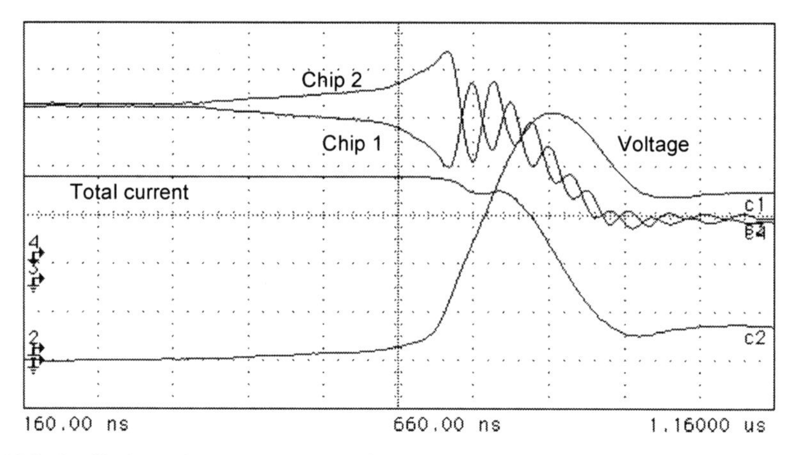

Fig. 13.3 Oscillations of the current in two IGBTs connected in parallel. Gate resistors 6.02 Ω for chip 1 and 6.45 Ω for chip 2. Current 10 A/div, voltage 50 V/div. Figure according to [Pal99] © 1999 EPE

length of the current-conducting path to the main terminals as well as in respect to the length of the wiring of the drive signals. Thermal conditions must be considered, too. Often trade-offs must be made. Figure 11.31 shows an example of parallel connection of five IGBT dies, in which asymmetric tracks for the main current to the respective die are given. The wires for the drive signals, for which also a symmetrical setup is of importance, are not drawn in the schematic circuit diagram of Figure 11.31b.

Figure 13.3 shows the measurement of a parallel connection of two IGBTs. To create differences, the gate resistors are chosen slightly deviating: for chip 1, a resistor of 6.02 Ω and for chip 2 of 6.45 Ω [Pal99]. The IGBT chips rated to 100 A are exposed to a forward current of 20 A in each chip. According to the lower gate resistor, chip 1 starts with the turn-off process. This has the consequence that the current in chip 2 increases in the first instance, while the total current is still at the same level. Then, during the current decline, an oscillation between the two chips builds up. A period of 50 ns can be read, which corresponds to a frequency of 20 MHz. The oscillations cannot be seen in the course of the total current in this case. Only the measurement of the single currents shows the oscillations occurring between the chips.

Figure 13.4a shows turn-off oscillations in a first version of a press-pack IGBT housing, see also Fig. 11.4. The current in a single chip of the many chips connected in parallel is measured. The current shows a high-frequency oscillation in the range of 10 MHz.

If the control terminals are realized with a printed circuit board (PCB) which leads to the same conditions for each of the 42 chips connected in parallel, the turn-off oscillations can be eliminated, as can be seen in Fig. 13.4b. Such a PCB, shown in Fig. 13.5, has copper layers on both sides; one side is the potential of the gate

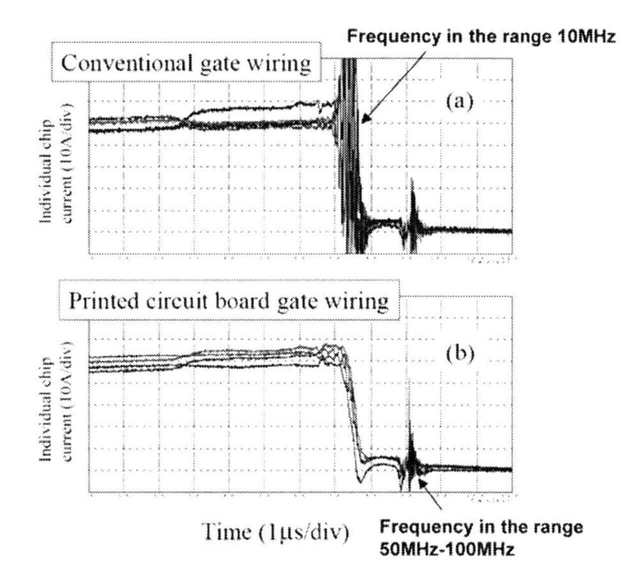

Fig. 13.4 Turn-off oscillations in a press-pack IGBT (**a**) and their elimination by a symmetrical arrangement of the gate signal connections with an integrated PCB (**b**) Figure from [Omu03] © 2003 IEEE

signal, the other side is the potential of the control emitter terminal. Very close to every single chip, gate resistors are arranged on the PCB.

The frequency range in which turn-off oscillations are found is between 10 and 20 MHz. This is clearly above the values which are to be expected for the harmonics of switching events (compare Fig. 13.1). Turn-off oscillations must be avoided not

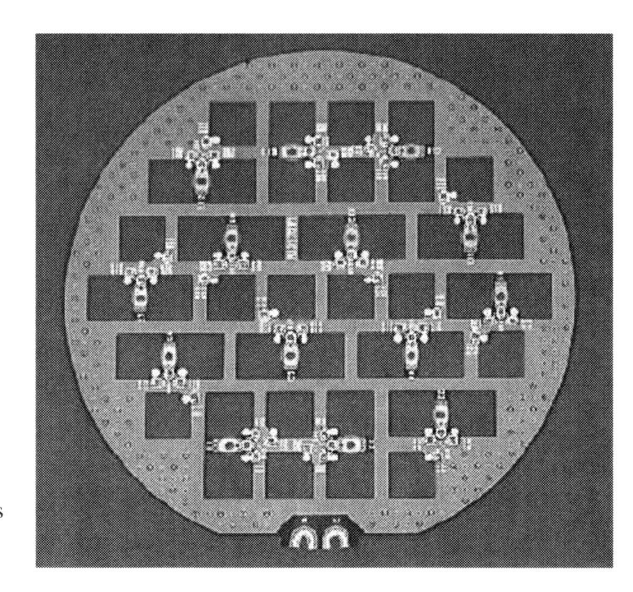

Fig. 13.5 PCB to ensure symmetrical control terminals in a press-pack IGBT. Figure from [Omu03] © 2003 IEEE

only because of electromagnetic emissions but also because they can additionally increase the turn-off losses of chips and can lead to thermal failures.

Possible countermeasures against turn-off oscillations are

- The setup of an arrangement as symmetrical as possible. This is also to be considered for parallel connection of discrete devices as well as single modules.
- If this is not possible because of mechanical and other reasons, the gate resistors R_G of IGBTs and MOSFETs can be increased. This counteracts oscillations, but in the same way it increases the turn-off losses. For IGBTs, compare Eq. (10.6) and the discussion of the influence of gate resistors on example of MOSFETs in context with Fig. 9.17.

13.2.2 Turn-Off Oscillations with Snappy Diodes

Fast diodes with insufficient reverse recovery behavior more often cause oscillations in power circuits than asymmetries in IGBTs connected in parallel. For details on snappy recovery behavior see Chap. 5. Figure 13.6 shows the course of the voltage at turn-off of a snappy diode in a circuit according to Fig. 5.19. The application is a step-down converter in a battery-fed electric vehicle. Instead of the IGBT in Fig. 5.19, a 100-V rated MOSFET is used as power switch. The snap-off of the reverse current in the diode leads to a voltage peak of 100 V. The abrupt snap-off leads to a current overshoot in forward direction, the diode is turned off again, a second and a third voltage peaks are generated, and finally the effect ends up in a damped LC oscillation.

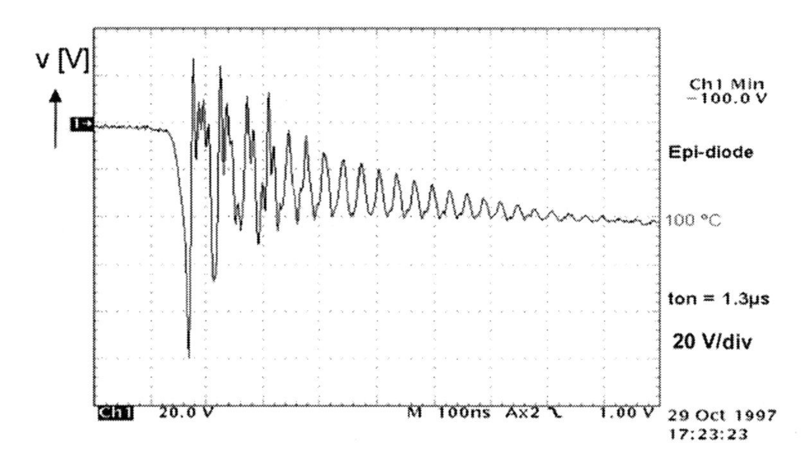

Fig. 13.6 Course of the voltage at turn-off of a snappy diode. Period 30 ns, frequency 33 MHz

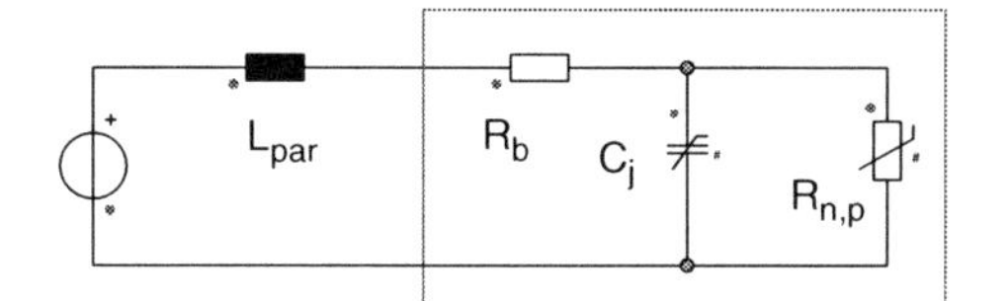

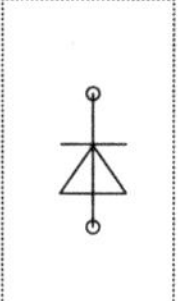

Fig. 13.7 Equivalent circuit for an oscillator consisting of diode and its parasitic components. Figure from [Kas97]

The frequency of the LC oscillation generated by snappy diodes is determined by the device capacity C_j and the parasitic inductance L_{par}

$$f = \frac{1}{2\pi}\sqrt{\frac{1}{L_{\text{par}} \cdot C_j}} \tag{13.3}$$

The equivalent electrical circuit for such an oscillator is shown in Fig. 13.7 [Kas97]. It must be noted that the capacity $C_j = c_j A$ is dependent on the voltage, see Eq. (3.109). The resistance R_{b} of the interconnections of the diode acts as damping component; the same holds for $R_{\text{n,p}}$ which stands for the base of the diode. During the turn-off process, electrons and holes are removed from the base and $R_{\text{n,p}}$ is not linear; in fact, it is strongly varying.

During the LC oscillation, C_j is not constant. For a diode rated to a blocking voltage of 1200 V with a steep doping profile of the p-layer at the pn-junction, C_j can be assumed to be 250 pF/cm^2 for the estimation of the expected frequency of oscillations [Kas97]. Considering the area of the diodes and some typical housings and their respective parasitic inductance, the values given in Table 13.1 are obtained.

For a diode rated to 100 A and 1200 V in an IGBT module of older construction type with a typical parasitic inductance of 100 nH, the frequency of LC oscillations caused by snappy diodes is expected to be in the range of 48 MHz. With a module

Table 13.1 Estimation of the frequency range of LC oscillations for 1200-V freewheeling diodes

	C_j	L_{par} (nH)	f(MHz)	$T = 1/f$(ns)
Bipolar 100-A diode, active area 0.44 cm^2	110 pF	20	107	9.3
		100	48	20.8
1200-A module, diodes	1.32 nF	100	13.9	72
		800	4.9	204

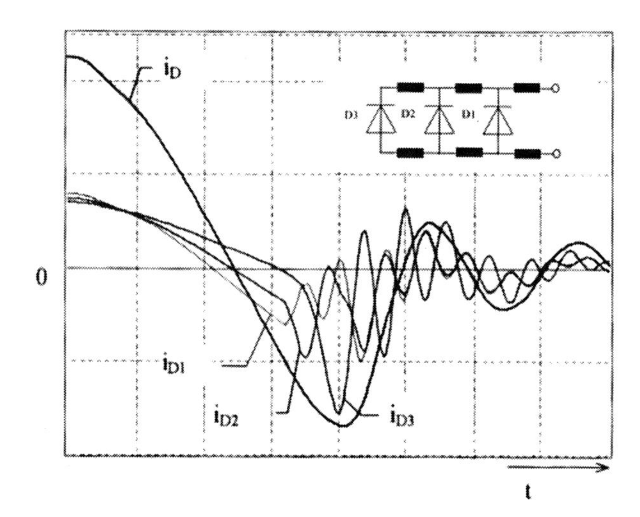

Fig. 13.8 Course of the current during reverse recovery in a parallel connection of diodes with different wiring inductances: 50 ns/div, 50 A/div, 25°C, V_{bat} approx. 300 V. Figure from [Eld98]

of modern architecture, approximately 20 nH are given as parasitic inductance; the frequency moves to a range of 100 MHz. In a high-power module rated to 1200 V, 12 diodes rated 100 A are connected in parallel and a much larger capacity is given. Because of the larger volume of the module, longer interconnections are necessary. Despite of this, the parasitic inductance in such modules has been reduced significantly. Depending on the inductance, frequencies in the range of 5 – 15 MHz are to be expected for the discussed oscillations.

If soft recovery diodes are used, the described oscillations can be avoided. Soft recovery diodes are producible today and they are indispensable for the operation of a modern power electronic converter.

However, it must be noted that not every oscillation in conjunction with diodes is originated by snappy turn-off behavior. If a non-suited unsymmetrical parallel connection is done, oscillations may occur even if soft-recovery diodes are used. An example is given in Fig. 13.8 [Eld98]. In this example, three diodes are connected in parallel; the diode D1 is located close to the main terminals, D2 and D3 with additional wiring are connected in parallel. The diode D1 with the lowest inductance is commutated with the highest di/dt. The reverse-recovery current maximum is first reached in D1. The process continues in D2, and then in D3, with an increased commutation velocity di/dt. It is maximal for D3, because the reverse current is already declining in D1 and D2. During the reverse-recovery process, the current oscillates between the devices. Finally, at the end of commutation, these internal oscillations superimpose to an oscillation of the total current i_D, too.

If a parallel connection is given and oscillations in the reverse recovery of freewheeling diodes are found, it also has to be investigated whether this is caused by asymmetrical arrangement of the wirings and interconnections.

Oscillations have even been found with a single soft-recovery diode, if the condition is fulfilled that the switching time of the diode t_{rr} agrees with half the period of the resonance of an LC oscillator circuit. In applications with MOSFETs and

IGBTs, this can be verified easily: the modification of the gate resistor R_G of the used IGBT or MOSFET can vary the turn-on slope of the transistor and with it the commutation velocity di/dt. So the switching time t_{rr} of the diode is modified and in this case the oscillations should vanish.

13.3 Transit-Time Oscillations

A power device has a lowly doped middle layer with a thickness w_B. At turn-off of a bipolar device, the existing charge carriers are removed; a part of them at an instant at which a space charge has already built up. The charge carriers flow through the space charge with at a drift velocity v_{sat}. This results in a transit time for which a first approximation can be given by

$$t_T = \frac{w}{v_{sat}} \tag{13.4}$$

The drift velocities for electrons and holes $v_{sat(n)}$ and $v_{sat(p)}$ are given in Eq. (2.38), the width of the space charge w during switching processes is smaller or equal to the base with w_B. The carrier transit time corresponds to a frequency of $1/t_T$. Dependent on the base width, transit-time oscillations occur with a frequency in this range; it is the range between 100 MHz and 1 GHz or even higher, see Fig. 13.1. The occurring frequency depends on the type of effect and on its phase relations. This is shown in the following.

Transit-time oscillations are used for the fabrication of microwave devices which are used as microwave oscillators [Sze81]. With power devices, such oscillations must be avoided since they are a danger for the power device itself, and since their electromagnetic emission can cause unwanted effects in driver circuits and other electronic components of the adjacent environment. The occurrence of transit-time oscillations was observed for two effects in power devices. They are described in the following paragraphs and measures to avoid them are discussed.

13.3.1 Plasma-Extraction Transit-Time (PETT) Oscillations

PETT oscillations can occur in a bipolar device at turn-off during the tail current interval. They have been observed with IGBTs as well as with soft-recovery freewheeling diodes [Gut01, Gut02]. Figure 13.9 shows an example with IGBTs.

The oscillations are observed in the course of the gate voltage, but they occur primarily as oscillations of the collector current I_C and collector voltage V_C where they cannot be resolved easily because of their low amplitude. Therefore, the stray effect into the gate signal is mainly observed. The oscillations occur after the device has taken up the voltage and the device is in the interval in which a tail current flows.

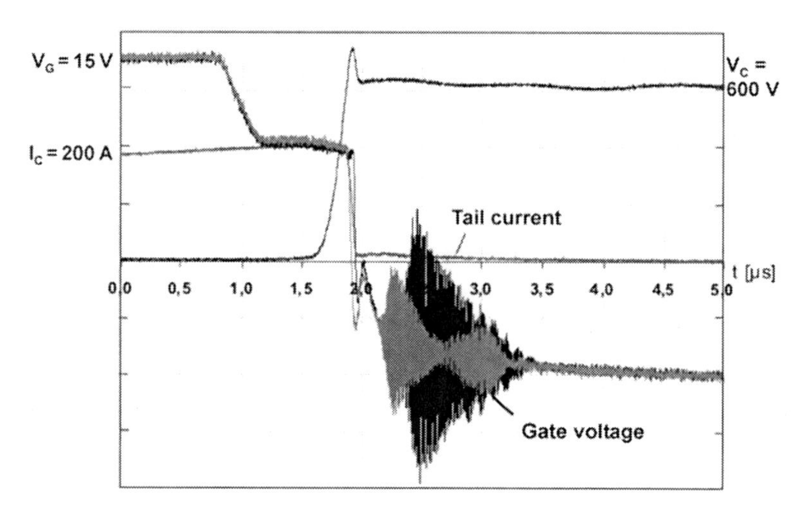

Fig. 13.9 PETT oscillations at turn-off of an IGBT. Figure from [Gut01] © 2001 IEEE

An example of the occurrence of PETT oscillations in a soft-recovery freewheeling diode is shown in Fig. 13.10 [Sie03]. The measurement was executed in a test setup of an IGBT module rated to 600 A and 1200 V. At turn-on of the IGBT, the freewheeling diode is turned off, the PETT oscillations occur in the tail current interval of the freewheeling diode. Only soft-recovery diodes can lead to PETT oscillations. However, soft recovery is essential for fast diodes in such applications. Since the resolution of the current measurement of the anode current of the diode is too low,

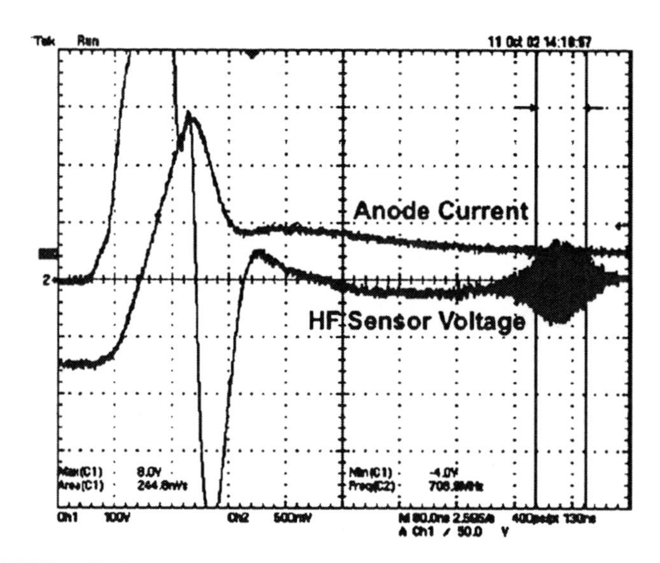

Fig. 13.10 PETT oscillation in the tail current of a soft-recovery freewheeling diode at turn-on of an IGBT. $V_{\mathrm{bat}} = 600$, $I_{\mathrm{F}} = 200\,\mathrm{A}$, $\mathrm{d}i/\mathrm{d}t = 400\,\mathrm{A}/\mu s$, $T = 300\,\mathrm{K}$. Figure from [Sie03] © 2003 EPE

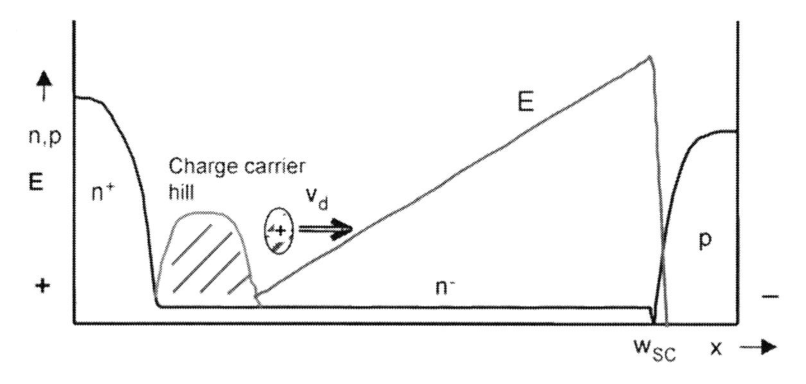

Fig. 13.11 Process in a freewheeling diode at occurrence of a PETT oscillation

the PETT oscillation is detected in this example with a wire loop which works as antenna and which is placed close to the diode.

The mechanism of oscillation is related to the mechanism of the Barrier Injection Transit Time (BARITT) diode which is intended as microwave oscillator. The BARITT diode has a metal–semiconductor–metal structure or a pn$^-$p-structure. At a voltage applied in blocking direction, the electric field reaches the opposite metal–semiconductor junction or the opposite p-layer and it releases an injection of carriers there [Sze81]. In contrast to the BARITT effect, with the PETT effect the space charge reaches the remaining charge carrier hill or remaining plasma zone which is still a reservoir of free carriers and feeds the tail current.

Figure 13.11 shows the process in the device at a PETT oscillation as example of a freewheeling diode. At the anode side (Fig. 13.11, right-hand side), an electric field has built up during the turn-off event and it takes up the voltage.

The charge carrier hill which feeds the tail current is still located at the cathode side (Fig. 13.11, left-hand side), compare Fig. 5.25 or Fig. 5.33. The tail current flows through the space charge as hole current. The shape of the electric field is triangular in this interval.

The occurrence of oscillations under such conditions is discussed in detail in [Eis98]. For a PETT oscillation, this is shown in Fig. 13.12. A high-frequency AC voltage $V_{\mathrm{RF}} \cdot \sin\omega t$ superimposed to the DC link voltage V_{DC} is assumed (Fig. 13.12a). In the same way as in the BARITT effect, an injected j_{inj} current is generated as the AC voltage has its maximum at $\omega t = \pi/2$ (Fig. 13.12b). This injected current flows through the space charge region with the velocity v_{d}. The corresponding current density at the terminals of the device j_{inf} is expressed by the Ramo–Shockley theorem [Eis98] and shown in Fig. 13.12c. The current flow at the device terminals starts at $\omega t = \pi/2$ and is found in the time interval ωt_{T}, needed by the carriers to transit the space charge region. The generated RF power is given by

$$P_{\mathrm{RF}} = \frac{A}{2\pi} \int_0^{2\pi} j_{\mathrm{inf}}(\omega t) \cdot V_{\mathrm{RF}} \sin \omega t \, \mathrm{d}\omega t \tag{13.5}$$

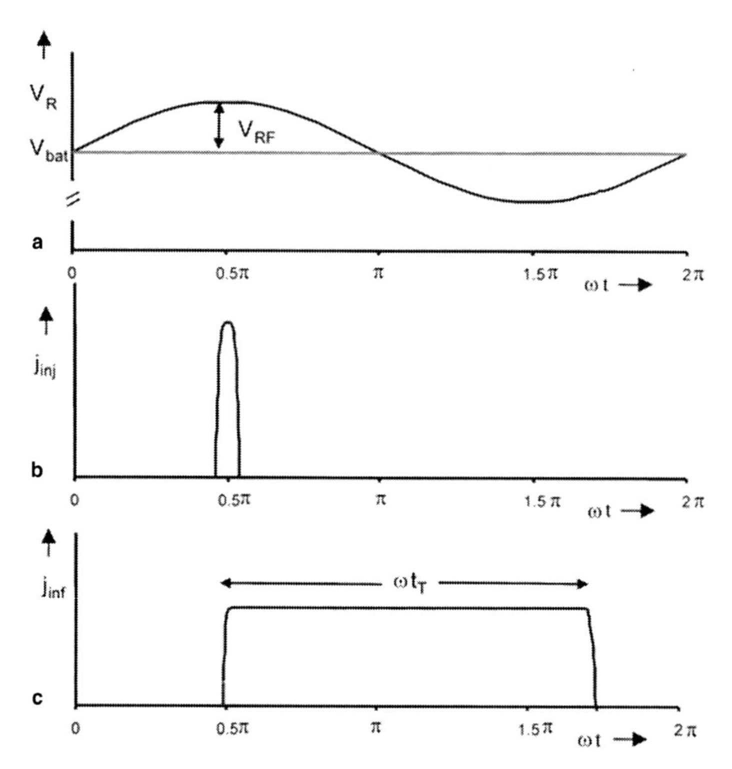

Fig. 13.12 Origin of PETT oscillations: (**a**) High-frequency AC voltage superimposed to the applied DC voltage, (**b**) Injection current at $w = w_P$ at the time $\omega t = \pi/2$. (**c**) Current at device terminals Fig. from [Sie08]. © 2006 IEEE

In Fig. 13.12a and c one can see that P_{RF} is positive for $\omega t_T < \pi$, zero at $\omega t_T = \pi$, and negative at $\omega t_T > \pi$. A negative value of P_{RF} means that RF power is generated. The created RF power is maximal for $\omega t_T = 3\pi/2$ and decreases again for $\omega t_T = 3\pi/2$. As long as P_{RF} is of negative sign, the device acts as current source and emits RF power.

The transit time t_T is given by [Gut02]

$$t_T = \int_{w_p}^{w_{SC}} \frac{1}{v_d(w)} dw \tag{13.6}$$

The velocity v_d in the space charge region depends on the strength of the electric field, compare Eq. (2.38) and Fig. 2.14. The electric field is of a triangular shape in the given case [Sie06b] (see Fig 13.11).

For the BARITT diode, the transit time is given in first-order approximation [Sze81] by Eq. (13.4) as $t_T = w_{SC}/v_{sat}$, where v_{sat} is the saturation drift velocity of holes under high-field conditions (approximately 10^7 cm/s in silicon). Note that the drift velocity v_d is lower than the saturation velocity v_{sat} for a significant part of the space charge region. Continuing with this simplification and taking into account the point of maximum RF power generation at $\omega_T = 3\pi/2$, the frequency of the PETT oscillations is approximated by

$$f_T = \frac{3 \cdot v_{sat}}{4 \cdot w_{sc}} \tag{13.7}$$

as it is given also in [Sze81].

From Fig. 13.12 it follows that excitation of the superimposed AC power is possible in a specific frequency range. It can be concluded that PETT oscillations occur only when the "negative-resistance" behavior found during one period is greater than all other resistive components in the complete circuit. Moreover, the phase shift between the oscillation voltage and the AC current at the device terminals is essential for an occurrence of this kind of oscillation. Furthermore, it can be concluded from Fig. 13.12 that the efficiency of the RF generation is low, since power is always dissipated during the interval $\pi/2$. Low efficiency is also a characteristic of BARITT diodes [Sze81]. Therefore, PETT oscillations only occur if there is a resonance circuit formed by the junction capacitance and the inductance of the bond wires close to the device, whose resonance frequency has to be close to the frequency f_T according to Eq. (13.7).

Additionally, PETT oscillations cannot occur as long as there is a high amount of remaining plasma in the device and the corresponding reverse current is high, since this stored charge acts as damping which hinders the occurrence of the oscillations. Further important parameters are the applied voltage V_{bat}, since this voltage determines w_{SC}, and the temperature, since the drift velocity is temperature dependent. It is typical for PETT oscillations that they only emerge at very special conditions. If one deviates from these conditions, no more PETT oscillations can be found. Therefore, the possible occurrence of PETT oscillations is easily overlooked in the procedure of the qualification of a power module.

PETT oscillations are radiated off as electromagnetic waves. This radiation can lead to the effect that the power electronic equipment violates the requirements of electromagnetic compatibility (EMC). The EMC requirements are fixed in different standards, e.g., by the European Standard EN55011 (International Standard IEC CISPR 11) [DIN00]. For details see the special literature in this field.

Figure 13.13a and b shows an experiment of measuring the electromagnetic radiation of the 600-A 1200-V module, from which the oscillations are shown in Fig. 13.10 [Sie03]. Since no chamber shielded against external electromagnetic radiation was available, first a measurement of existing electromagnetic disturbance of the environment was necessary. Figure 13.13a shows the result of the environmental EMC measurement which lasted 2 h, followed immediately by the subsequent measurements. The different gray-colored bars shown in Fig. 13.13a mark frequency ranges used by broadcasting and telecommunication. Obviously, the largest interfering signals were caused by mobile communication equipment. This measurement

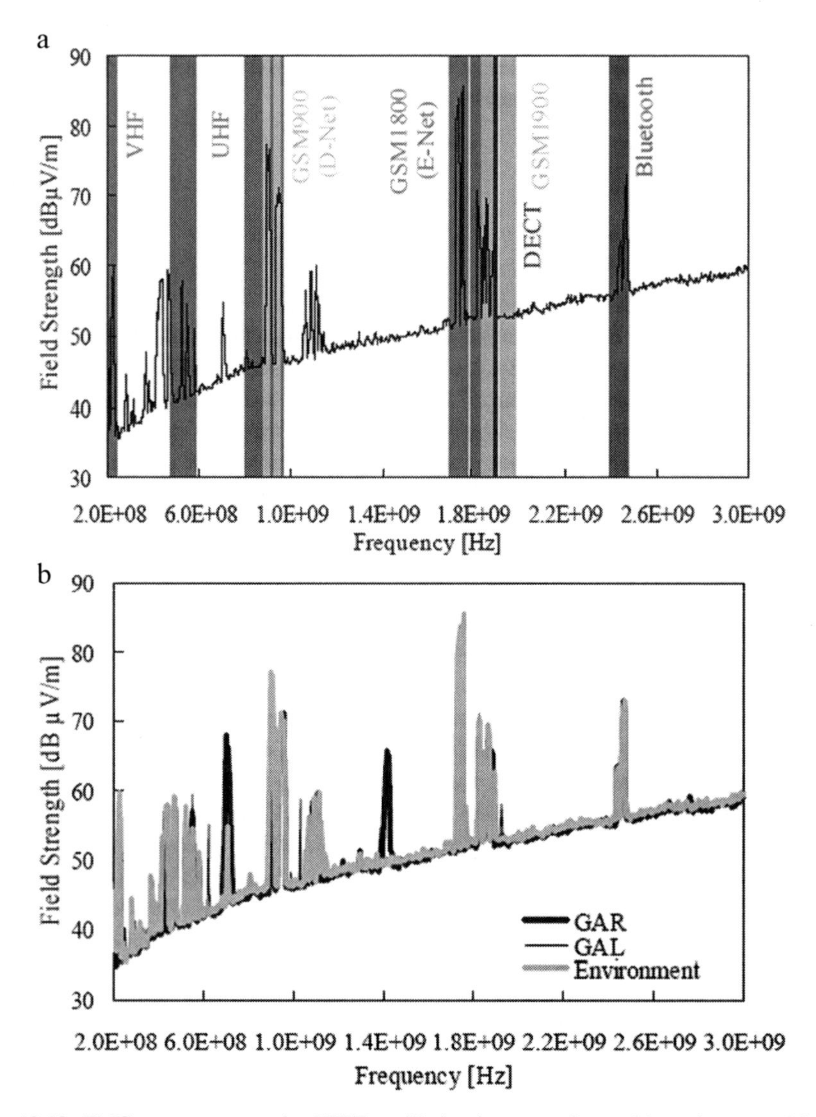

Fig. 13.13 EMC measurement of a PETT oscillation in comparison with environmental background radiation (**a**) Environment without operation of the module, (**b**) with operation of the module: PETT oscillations at the module GAR at 700 MHz and 1.4 GHz. Figure from [Sie06b] © 2006 IEEE

series is included in Fig. 13.13b to give evidence of additionally generated signals that were caused by the PETT oscillations.

The PETT oscillation during the turn-off (marked as GAR) caused two sharp peaks in the frequency spectrum, appearing at 700 MHz and 1.4 GHz, which could be assigned to the fundamental frequency and the second harmonic.

Fig. 13.14 Assembly of the module GAR which showed the PETT oscillations in the measurement in Fig. 13.10 and its internal electrical circuit. Figure from [Sie06b] © 2006 IEEE

The width of the space charge of the diode used in the module GAR amounts to approx. 85 μm for the given conditions. If the value of $8 \cdot 10^6$ cm/s is used as drift velocity for holes, a frequency of approx. 700 MHz results from Eq. (13.7), which agrees with the measured value.

Although the spurious radiation caused by PETT oscillation was relatively low in strength, exceeding the EMC limits could easily occur. In particular, this would be expected if more than one power module is used, as is typical for power electronic equipment, and the radiations of the single modules are summed up.

In another application, PETT oscillations were found in a 1.8-MW high-frequency converter with an operating frequency of around 100 kHz. The setup of the equipment consisted of more than 100 power modules. The onset of the oscillation generated an error signal in the control unit [Sie06b]. Therefore, PETT oscillations must be avoided.

To prevent PETT oscillations, it is therefore not particularly helpful to modify the semiconductor device itself. Every power semiconductor has a space charge if voltage is applied, and it has therefore the capability for oscillations according to Eq. (13.7). However, it is essential to avoid an LC circuit that is in resonance with the transit frequency given in Eq. (13.7).

Figure 13.14 shows the setup of the module GAR, for which PETT oscillations in Figs. 13.10 and 13.13b were found. The oscillations occurred at the diode which was used as freewheeling diode. It is marked as FWD in Fig. 13.14, left-hand side; in the equivalent circuit in Fig. 13.14 it is the diode at the bottom.

Figure 13.15 shows the impedance of this module on the left-hand side [Sie06b]. The three-dimensional EMC simulator FLO/EMC [FLO04] was used for the calculation. It solves the complete Maxwell equations numerically. The geometry of the module in Fig. 13.14 is used; the space is divided into cells modeled as the intersection of orthogonal transmission lines. There is no possibility to introduce real semiconductors into FLO/EMC. Therefore, a simplified model was used which reproduces the correct junction capacitance or on-state resistance of the devices (IGBT and FWD). For the characterization of the power modules, the excitation in the form of a delta pulse was applied across a FWD. In this way, the electric and magnetic fields and the resulting impedance can be calculated.

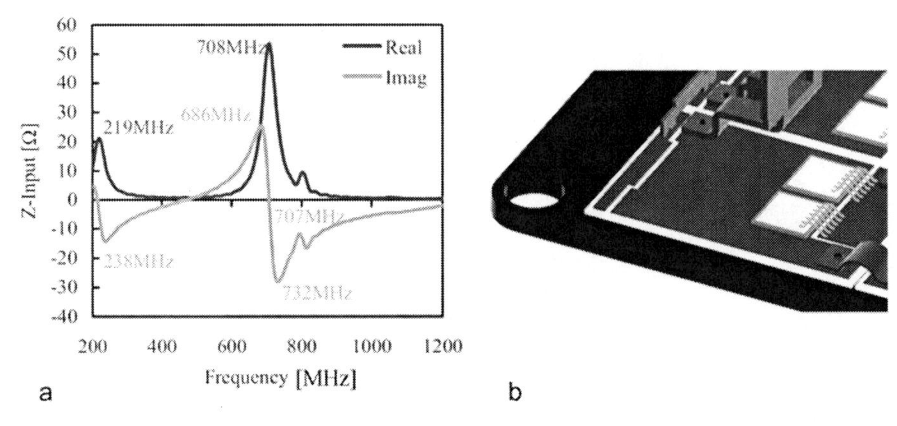

Fig. 13.15 Impedance of the module in Fig. 13.14 (**a**). Detail of the arrangement of the free-wheeling diodes at which the PETT oscillation occurs (**b**). Figure from [Sie06b] © 2006 IEEE

Figure 13.15 shows the simulation results for the impedance of the power module GAR (as seen from the FWD where the excitation was applied). A resonance point at a frequency of about 700 MHz resulted, which was in accordance with the oscillation frequency as given by the transit frequency f_T (Eq. (13.7)) of the FWD. This resonance point is a necessary condition for the appearance of PETT oscillation, and 3D EMC simulation can be used to predict resonance points of a complex mechanical construction.

However, it must be mentioned that already small modifications in the assembly may shift the occurrence of PETT oscillations to other conditions, or may weaken it or even eliminate it.

The arrangement of the bond wires in the module GAR is shown in Fig. 13.15, right-hand side. Considering the given chip area and the resulting space charge capacity, only a small inductance is possible for an LC circuit with resonance in the range of 700 MHz. It can be formed only by components in the immediate neighborhood of the chip, e.g. the next conductor paths on the DCB substrate, the bond wires, and their arrangement.

An obvious and efficient way for lowering inductance would be to provide additional shorts between the anode contact areas, as published in an older patent application of 1995 [Zim95] and shown in Fig. 13.16b. This results in a clear suppression of the module resonance in the transit-frequency range, as shown in the FLO/EMC simulation of this assembly in Fig. 13.16a [Sie06b]. No more resonance point is found in the range of 700 MHz.

The suggestion of [Zim95] is found to be an effective measure against PETT oscillations, in spite of the fact that no knowledge of details of the processes leading to PETT oscillations was available at that time.

To prevent PETT oscillations, the resonance point of the power module must be different from the transit-time frequency f_T governed by the power semiconductor structure. Three-dimensional EMC calculations are useful for this. Ideally, a future simulation system for such tasks should solve the full set of Maxwell equations for the complete construction and calculate the behavior of the semiconductor

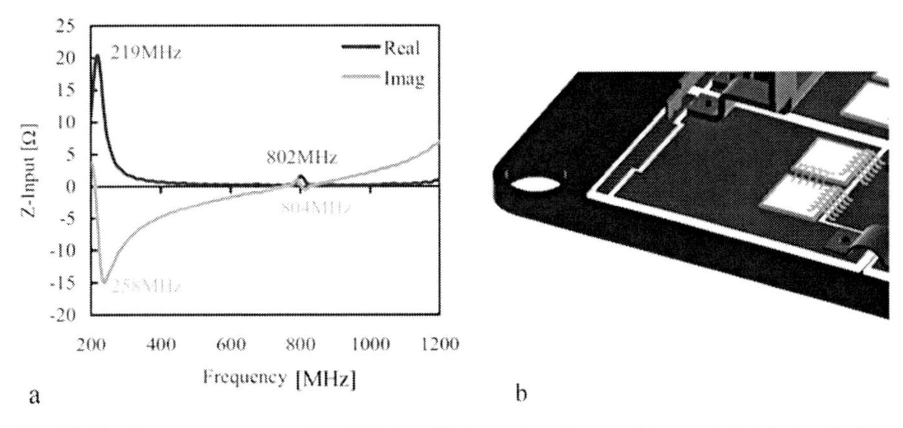

Fig. 13.16 Countermeasure against PETT oscillations: Impedance of the module (**a**), detail of the arrangement of the freewheeling diodes with shorting bond wires (**b**). Figure from [Sie06b] © 2006 IEEE

devices, e.g., solve the basic semiconductor equations. Although the capabilities of computers steadily increase, this task still seems to be too complex at the moment.

13.3.2 *Dynamic Impact-Ionization Transit-Time (IMPATT) Oscillations*

IMPATT oscillations have been found as a dynamic oscillatory effect at turn-off of soft-recovery freewheeling diodes [Lut98]. The attribute "dynamic" is used to indicate that these oscillations occur in connection with a switching event. The dynamic IMPATT oscillation is of high energy, it radiates off disturbances of high intensity, and leads to malfunctions of analogue and digital electronic circuits which are

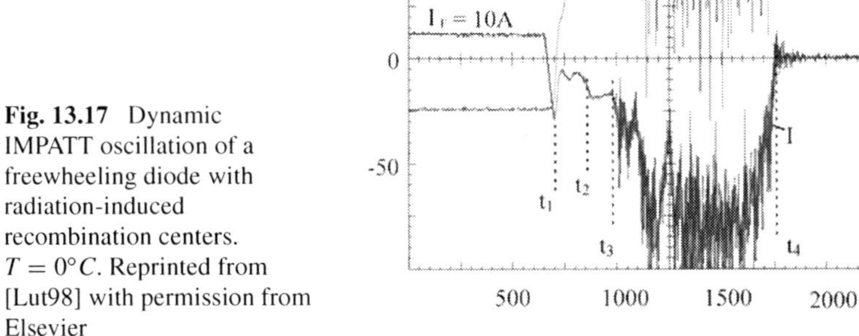

Fig. 13.17 Dynamic IMPATT oscillation of a freewheeling diode with radiation-induced recombination centers. $T = 0°C$. Reprinted from [Lut98] with permission from Elsevier

present, for example, in drive circuits. A measurement of such an effect is possible in a laboratory setup; such a measurement is shown in Fig. 13.17.

The measurement in Fig. 13.17 was executed in an application in the form of a step-down circuit according to Fig. 5.19. The temperature of the diode was 0°C. The course of the voltage is plotted inversely. The rated static blocking capability of this diode is 1200 V, its avalanche breakdown voltage is > 1300 V. After the reverse recovery current peak (time point t_1) the reverse current decreases. Between t_1 and t_2, the voltage climbs up close to the value of the applied battery voltage, while a tail current flows in the diode. After t_2, a current hump grows from the tail current. This hump occurs only if the battery voltage is above 910 V. A further increase in the battery voltage to 930 V effects that a high reverse current suddenly shoots up from the current hump whose amplitude is a multiple of the reverse recovery current peak. A high-frequency oscillation is superimposed. After some 100 ns (t_4), the oscillation is finished.

A reduction of the voltage by only 1 or 2 V, or an increase of the temperature by 1 or 2°C, removes the effect.

The mechanism of the dynamic IMPATT oscillation is related to that of the IMPATT diode, a device which is likewise intended as oscillator for microwaves [Sze81]. In IMPATT diodes, the static reverse-blocking capability is exceeded and the device is operated in the avalanche mode. In contrast, the dynamic IMPATT oscillation occurs at a voltage significantly lower than the avalanche breakdown voltage. The dynamic IMPATT oscillation is caused by the K-center, which is created at irradiation of the semiconductor with high-energy particles [Lut98]. The energy levels of the most important centers are shown in Fig. 4.25. The energy level of the K-center is located below the middle of the bandgap. Recent investigations have found its nature as C_iO_i, a defect consisting of an interstitial carbon and oxygen atom [Niw08]. Its contribution to recombination is low, but its density is higher than that of the density of the OV centers which determine the recombination in irradiated devices if typical annealing processes are used [Sie06].

The K-center has the characteristics of a temporary donor. At forward conduction, if the device is flooded with free carriers, it is occupied with a hole and it is positively charged. The effective doping then becomes

$$N_{\text{eff}} = N_{\text{D}} + N_{\text{T}}^{+} \tag{13.8}$$

wherein $N_{\text{T}}{}^{+}$ is the density of positively charged K-centers. After the voltage has changed its polarity, the center is discharged:

$$N_{\text{T}}^{+}(t) = N_{\text{T}}^{+} \cdot e^{-t/\tau_{\text{ep}}}. \tag{13.9}$$

The time constant τ_{ep} of this discharge process is temperature dependent; it is short (ca. 100 ns at 400 K) at high operation temperatures; it is in the order of some microseconds at temperatures below 300 K. This results in a temporarily increased doping of the device. The doping determines the onset voltage of avalanche breakdown V_{BD}; Eq. (3.84) can be used for the given situation. The increased doping N_{eff}

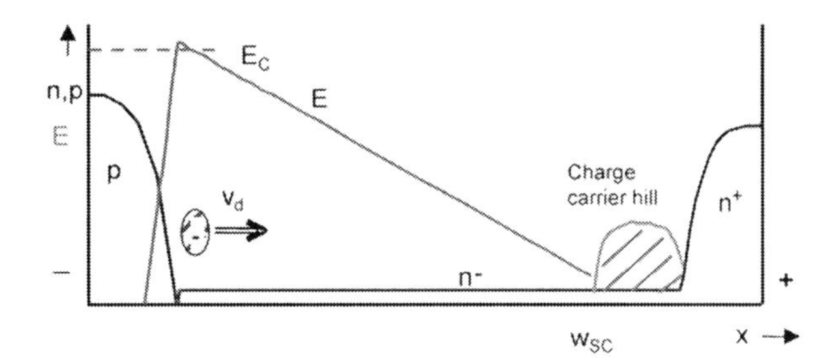

Fig. 13.18 Process in the device at a dynamic IMPATT oscillation

leads to a strongly reduced value of V_{BD}. If now a fast-switching transistor such as an IGBT applies the battery voltage to the diode in a very short time after the current zero-crossing point of the diode, it finds a device with reduced avalanche voltage, and impact ionization sets on. In Fig. 13.17, the temporarily reduced voltage V_{BD} is reached at the time point t_2 and a current created by avalanche occurs as a hump in the current curve. If the device is driven into the avalanche mode even more, the dynamic IMPATT oscillation starts.

The situation in the device during this process is shown in Fig. 13.18. Between t_1 and t_2 a small tail current flows (see Fig. 13.17); in this state, a charge carrier hill from the remaining plasma still exists close to the cathode and the shape of the electric field is triangular. Electron packets generated by impact ionization run through the electric field to the right-hand side.

A high-frequency AC voltage $V_{RF} \cdot \sin\omega t$ shall be assumed as superimposed to the DC-link voltage V_{DC}, see Fig. 13.19a. It generates a current pulse j_{inj}. Since the process of impact ionization needs time, the created current pulse is shifted by $\omega t = \pi/2$ to the peak value of the voltage and appears at $\omega t = \pi$, where the AC component of the voltage has its zero-crossing point (Fig. 13.19b). The injected current runs through the space charge at a velocity v_d. Figure 13.19c shows the corresponding current j_{inf} at the terminals of the device according to the Ramo–Shockley theorem [Eis98]. The current appears at the terminals during the time interval $\omega t = \pi$.

For the generated RF power P_{RF}, Eq. (13.5) holds. P_{RF} is of negative sign and is maximal for $\omega t_T = \pi$ it decreases again for $\omega t_T > \pi$. The diode acts as current source. It emits RF power. For the maximum of generated RF power at $\omega t_T = \pi$, one yields for the frequency of transit-time oscillation at an IMPATT effect [Sze81]

$$f_T = \frac{v_{sat}}{2 \cdot w_{SC}} \tag{13.10}$$

In contrast to oscillations according to the BARITT effect, there is no phase interval in which the emitted power is damped for the IMPATT effect. The IMPATT effect leads to a high-frequency oscillation of high energy. The RF efficiency of IMPATT diodes can amount up to 30%, as reported in the literature [Eis98]. Additionally, the signal of the IMPATT effect is not in a very narrow frequency

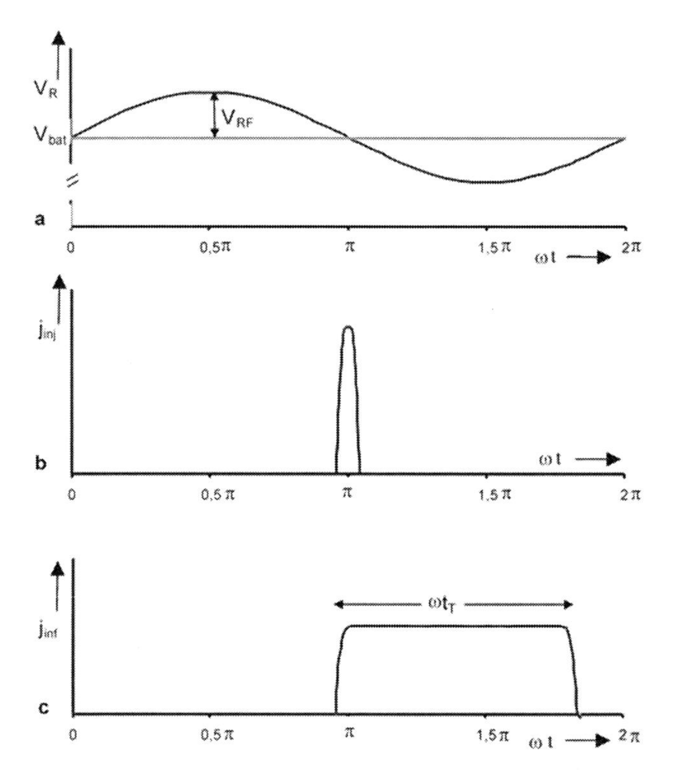

Fig. 13.19 Origin of IMPATT oscillations: (**a**) High-frequency AC voltage superimposed to the applied DC voltage. (**b**) Injected current at $w = 0$ at the time point $\omega t = \pi$. (**c**) Current at device terminals

band, as depicted in Fig. 13.15 for PETT oscillations, but it contains a lot of noise. This shows obviously that dynamic IMPATT oscillations must be avoided in any case.

IMPATT oscillations can occur if a power semiconductor contains too much K-centers. This can be due to too high electron irradiation doses [Lut98] or high doses of He irradiation projected to an unsuited position in the device [Sie04, Niw08]. For the case of electron irradiation which creates a homogeneous lifetime, a dimensioning rule is given in [Lut98]: Only so many K-centers are allowed that the device can sustain the maximal occurring DC-link voltage in the application (commonly 75% of the rated voltage) at its lowest operation temperature (typ. –40°C) without occurrence of avalanche. The generation rates of K-centers at irradiation with electrons are given in the literature [Sie06]. If such a rule is observed, the dynamic IMPATT oscillation can surely be avoided.

Sometimes IMPATT oscillations are found in device simulations, e.g., if a diode induces a high voltage peak by a reverse current snap-off, and the diode is driven by this in the static avalanche mode. Experimental reports on such oscillations are not known at the moment. However, such diodes are ruled-out in most power electronic applications because of dangerous voltage peaks and because of oscillations generated by snappy behavior, see Sect. 13.2.2. Soft recovery behavior is necessary.

References

[DIN00] DIN EN 55011 – Industrielle, wissenschaftliche und medizinische Hochfrequenzgeräte; Funkstörungen – Grenzwerte und Messverfahren, VDE-Verlag GmbH, Berlin, 2000

[Eis98] Eisele H, Haddad G: "Active microwave diodes", in S.M. Sze, Modern semiconductor device physics, New York 1998

[Eld98] El-Dwaik F, Ein Beitrag zur Optimierung des Wirkungsgrades und der EMV von Wechselrichtern für batteriegespeiste Antriebssysteme, Dissertation, Chemnitz 1998

[FLO04] Flomerics Ltd.: FLO/EMC Reference Manual Release 1.3, 2004

[Gut01] Gutsmann B, Silber D, Mourick, P: "Explanation of IGBT Tail Current Oscillations by a Novel "Plasma Extraction Transit Time" Mechanism", Proceeding of the 31st European Solid-State Device Research Conference, pp. 255–258 (2001)

[Gut02] Gutsmann B, Mourick P, Silber D: "Plasma Extraction Transit Time Oscillations in Bipolar Power Devices", Solid-St. Electron. 46 (5), 133–138 (2002)

[Kas97] Kaschani K, Untersuchung und Optimierung von Leistungsdioden, Dissertation, Braunschweig 1997

[Lut98] Lutz J, Südkamp W, Gerlach W: "IMPATT Oscillations in Fast Recovery Diodes due to Temporarily Charged Radiation Induced Deep Levels" Solid-St. Electron 42 No. 6, 931–938 (1998)

[Niw08] Niwa F, Misumi T, Yamazaki S, Sugiyama T, Kanata T, Nishiwaki K:"A Study of Correlation between CiOi Defects and Dynamic Avalanche Phenomenon of PiN Diode Using He Ion Irradiation", Proceedings of the PESC, Rhodos (2008)

[Omu03] Omura I et al: "Electrical and Mechanical Package Design for 4.5 kV Ultra High Power IEGT with 6kA Turn-off Capability" Proceedings of the ISPSD, Cambridge (2003)

[Pal99] Palmer PR, Joyce JC: "Causes of Parasitic Current Oscillation in IGBT Modules During Turn-Off", Proceedings of the EPE, Lausanne (1999)

[Sie03] Siemieniec R, Lutz J, Netzel M, Mourick P: "Transit Time Oscillations as a Source of EMC Problems in Bipolar Power Devices", Proceedings of the EPE, Toulouse (2003)

[Sie04] Siemieniec R, Lutz J, Herzer R: "Analysis of dynamic impatt oscillations caused by radiation induced deep centres with local and homogenous vertical distribution IEE Proceedings Circuits, Devices and Systems, Volume 151, Issue 3, pp. 219–224 (2004)

[Sie06] Siemieniec R, Niedernostheide FJ, Schulze HJ, Südkamp W, Kellner-Werdehausen U, Lutz J: "Irradiation-Induced Deep Levels in Silicon for Power Device Tailoring" Journal of the Electrochemical Society, 153 (2) G108–G118 (2006)

[Sie06b] Siemieniec R, Mourick P, Netzel M, Lutz J: "The Plasma Extraction Transit-Time Oscillation in Bipolar Power Devices – Mechanism, EMC Effects and Prevention", IEEE Trans El.Dev Vol 53 No 2, 369–379 (2006)

[Sze81] Sze SM, Physics of Semiconductor Devices. John Wiley & Sons, New York 1981

[Zim95] Zimmermann W, Sommer KH, Patent DE 19549011C2, 1995

Chapter 14
Power Electronic Systems

14.1 Definition and Basic Features

The expression "power electronic system" is used in different contexts with different meanings. A monograph on fundamental power electronic circuit topologies might well be found under the search criterion 'systems.' It is therefore appropriate to start with a precise definition of the object of discussion.

The term power electronic system will be used only for a complete assembly of all elements, necessary to perform an energy transformation task. Figure 14.1 illustrates this definition with an example.

The power electronic function is the transformation of energy from a battery to an electric traction motor and possibly vice versa. The fundamental switching processes are of course taking place in the silicon power devices, not shown here. On this level of abstraction, the principal item is a three-phase converter module

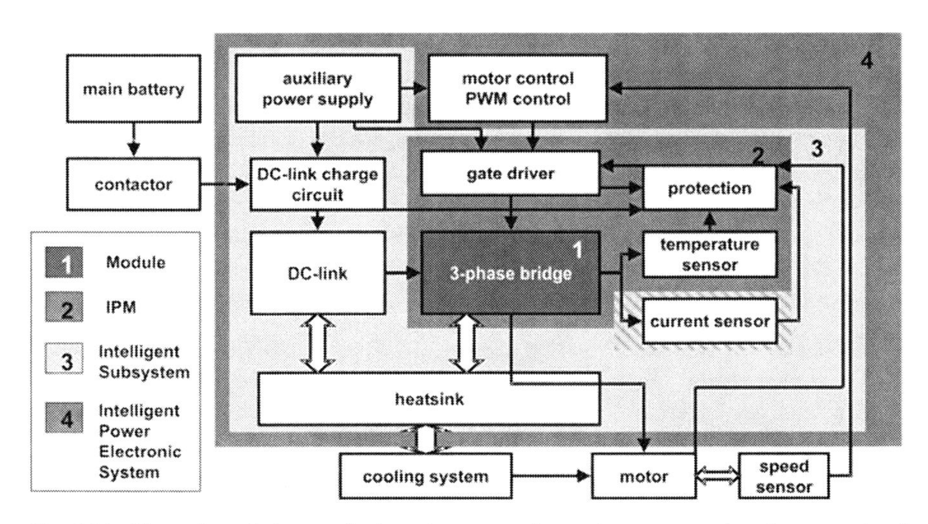

Fig. 14.1 Illustration of the terminology in power electronic systems using the example of automotive hybrid traction drives. Schematics from W. Tursky, Semikron

J. Lutz et al., *Semiconductor Power Devices*, DOI 10.1007/978-3-642-11125-9_14,

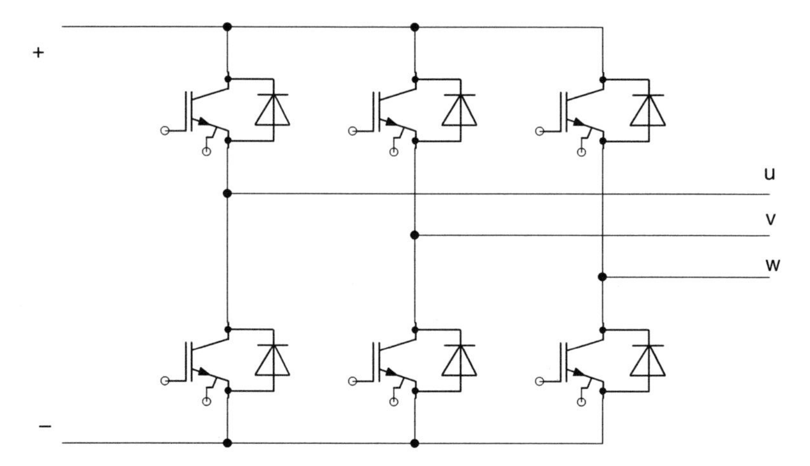

Fig. 14.2 Three-phase converter circuit diagram

with an internal circuit as shown in Fig. 14.2. This converter module contains the power electronic devices, six IGBTs, and six freewheeling diodes, as well as all the electrical contacts and the interface to a heat sink. Such converter topologies as shown in Fig. 14.2 are available in a single power module package.

The next level of functionality is reached by adding gate drives for the six IGBTs, sensors for temperature, and potentially current sensors and protection logic circuits. The integration of these functions in one package unit is named 'Intelligent Power Module' (IPM). IPM modules are commercially available for the small and medium power range.

Subjoining a DC-link energy storage, the DC-link charge circuit and the auxiliary power supply as well as the heat sink to the assembly leads to the next level of functionality: the intelligent sub-system. For specific applications, integrated custom-specific intelligent sub-systems are available on the market.

The only element missing to form a complete power electronic system is a 'digital signal processor' (DSP) or microprocessor, which generates the pulse width modulation (PWM) signals and is capable of controlling all aspects of the power electronic system by suitable software. The external heat exchanger as well as the source (battery with safety contactor) and the drain (motor with speed sensors) of the energy are not considered as part of the power electronic system.

A power electronic system must comprise the following features:

- *Full electrical functionality*: The power electronic circuit with the associated driver circuits, the sensor elements to monitor the operating status of the converter (output current, DC-link voltage, reference temperatures), and additionally start-up circuits, protection circuits, and auxiliary power supply circuits.
- *Thermal management components*: The extraction of heat from the silicon power devices is of highest priority for the operation of power electronic systems. The efficient function of these components should be permanently monitored to avoid thermal overload conditions.

- *Hardware and software for PWM and control*: The PWM and control algorithms are of fundamental relevance for the efficient and reliable performance of a power electronic system. They can enhance the energy efficiency of an application by selecting optimal operation conditions of the motor and they can protect the power electronic system from unexpected stress conditions and external failures like motor short circuits.

The system approach is an essential prerequisite for effective system optimization. Focusing on a single aspect of the system optimization, for example, the increase of the power density per unit volume, can lead to suboptimal system with respect to the reliability, if other functional aspects are not taken into account.

A general strategy to increase the power density without sacrificing the system reliability is the system integration. Integration of functions can eliminate connections and interfaces and can sometimes exploit synergy effects by combining more than one function in a single functional element. Examples for such synergy effects are frequently found in the integration of passive components, where, based on a layer technology derived from the PCB process, capacities and inductances are simultaneously integrated into a single building block. This concept has a high potential for improvements in power electronic systems in the future.

An effective concept is the monolithic integration. Today, components are available for small power, which have all power electronic, control, and logic functions integrated on a single chip. Hybrid integration, which assembles different components on a single substrate, is another way of system integration. Common to all strategies of integration is the reduction of footprint, the system volume, and the weight, while simultaneously reducing interconnections and interfaces.

A high level of integration reduces the effort necessary for the system assembly on the one hand, but increases the complexity of the components on the other hand. The increased complexity of highly integrated components requires a high quality of the manufacturing process and demands comprehensive test procedures to verify the component functionality. This enhanced effort in the manufacturing process is only commercially reasonable, if the production quantities are high enough.

The design of power electronic systems requires expert knowledge in several engineering disciplines, i.e. mechanical engineering, electrical engineering, material science, computer science. It requires a cooperation of engineers from all those fields for a successful design of a new power electronic system. A single engineer today can hardly cover all aspects of a system design by himself, so teamwork is mandatory. This has already changed the occupation requirements for engineers.

14.2 Monolithically Integrated Systems – Power ICs

Monolithic integration is the assembly of different functions on a single silicon chip. Sensor elements, analog and digital circuits, and power electronic devices are combined in a single integrated circuit.

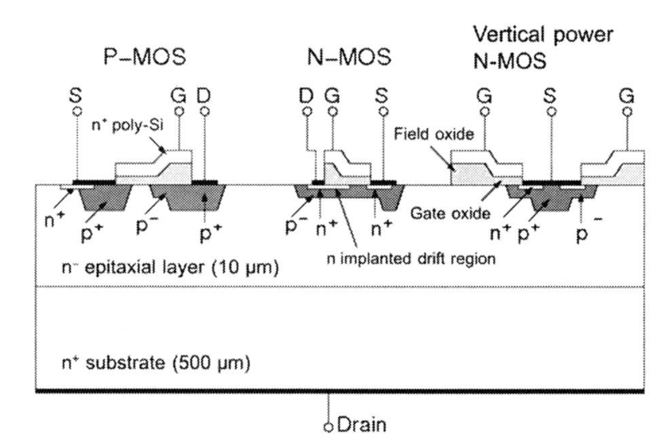

Fig. 14.3 Self-insulated vertical DMOS transistor integrated with CMOS logic elements. Drawing similar to [Thi88b]

The standard CMOS technology is the basis for the majority of integrated systems. The abbreviation CMOS stands for complementary MOSFET and indicates a technology adapted from n-channel and p-channel MOSFETs. Additional functional elements like sensors, storage cells, and power devices and the necessary interconnections can be added in the same technology by supplemental process steps and mask levels.

One of the first technologies that successfully implemented this process extension is the Smart SIPMOS technology [Pri96]. Figure 14.3 shows an example of use combining logic and power circuits on a single chip.

The vertical power transistor (power N-MOS) on the right-hand side in Fig. 14.3 has already been discussed in Chap. 9. The compatibility of its structure to the CMOS structures on the left-hand side is evident. Source structures, gate oxides and gate contacts, passivation layers, and contact metallization layers can extensively be inherited from the CMOS technology.

The mutual insulation of the different elements in the circuit shown in Fig. 14.3 is realized by pn-junctions; therefore, this concept is called 'junction insulation'. The blocking capability of devices based on this technology is with justifiable effort limited to 100–200 V [Gie02]. A closer look on Fig. 14.3 reveals a multitude of parasitic structures in the circuit: npn- and pnp-transistor structures and even pnpn-structures representing thyristor-type parasitic elements. In operation with significant current densities and particularly at elevated temperatures and high voltages, the interactions between different devices can provoke a latching of the parasitic thyristor structures. This effect is the dominant failure mode of integrated circuits; it limits the permitted current and voltage levels and primarily constrains the operation temperature range.

Adapted to the high voltage levels is the technology presented by ST Microelectronics as 'vertical intelligent power technology,' described in [And96]. Figure 14.4 illustrates an exemplary application. A vertical power MOSFET is shown on the left side, which serves as output stage. This device was discussed in Chap. 9. The elements of the logic circuit are insulated by several pn-junctions.

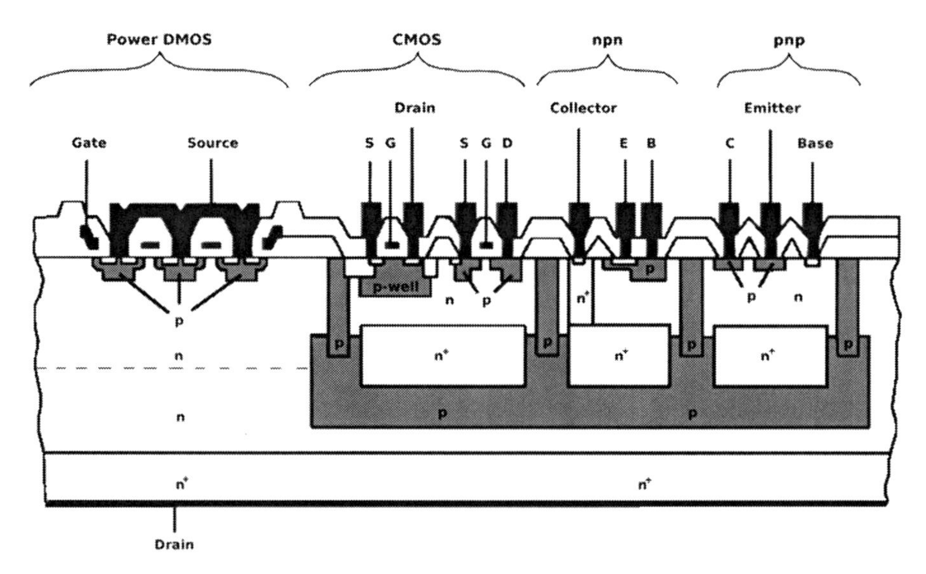

Fig. 14.4 Vertical intelligent power technology from ST Microelectronics. Figure prepared by R. Herzer based on ST Microelectronics datasheets

These pn-junctions are formed (beginning with the bottom n^+ substrate) by a first n^- epitaxy layer, local p and n^+ buried implantation layers, a secondary n-type epitaxy layer, and in lateral direction by deep diffusion p columns. The epitaxial growth process is interrupted for the implantation of the buried p and n^+ islands. During the resumed epitaxial growth, the elevated process temperatures initiate diffusion in the buried layers, so that they reach their projected dimensions at the end of the process. This elaborate multilayer insulation technique insulates the logic circuit elements for blocking voltages up to 600 V.

The collector of the npn bipolar transistor is connected to the buried n^+ collector island by a vertical n^+ column. This typical design structure for a high-gain npn transistor is therefore denoted as 'pseudo-vertical' device.

The technology depicted in Fig. 14.4 represents a platform for the integration of manifold structures and functions of analog and digital circuits. The fabrication of the deep p-zones for the electrical separation of the functional elements is rather elaborate. The deep diffusion of these columns from the surface is accompanied by a lateral diffusion with an aspect ratio of 0.8. For a 10 μm deep column, a lateral diffusion of 8 μm must be taken into account. This technique therefore requires relatively wide separation areas and thus provokes a loss of area available for active devices.

Another drawback is the generation of a leakage current with voltage applied to the pn-junctions, which is increasing for higher temperatures. The leakage currents limit the maximum blocking capability and restrict the operation temperature range as well. They comprise the potential of latching (the turn-on of parasitic thyristor structures), which results in the destruction of the circuit. pn-Insulated technologies are limited for high voltages to temperatures of maximum 150°C.

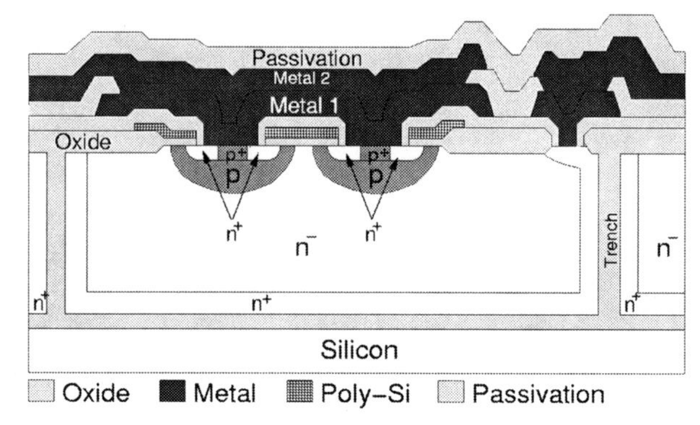

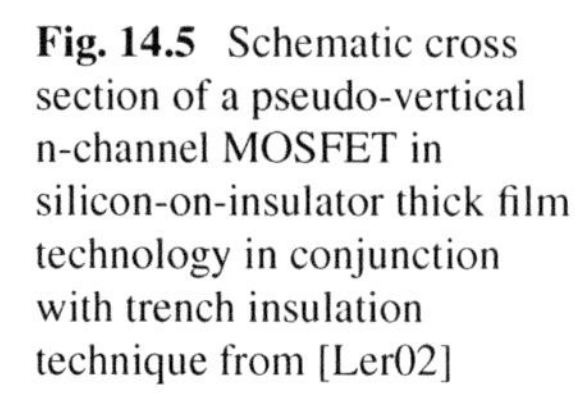

Fig. 14.5 Schematic cross section of a pseudo-vertical n-channel MOSFET in silicon-on-insulator thick film technology in conjunction with trench insulation technique from [Ler02]

These handicaps can be eliminated by a dielectric insulation technology, where oxide layers are implemented for the electrical separation of different devices. These oxide layers feature bidirectional insulation and generate much smaller leakage currents, so that a higher blocking capability can be realized with comparatively thin layers even at high operation temperatures.

An example of this so-called silicon-on-insulator (SOI) technology in conjunction with a trench technology is shown in Fig. 14.5. A wafer bonding process, in which two oxidized silicon wafers are merged together, produces the substrates for this technology. The top wafer was prepared with a diffused n^+ surface region prior to the bonding process. After the wafer bonding process, the top layer is grinded down to the desired thickness and afterward polished. A subsequent etching process forms the deep trenches. An implantation process followed by a diffusion process generates the n^+ layers on the trench sidewalls. Then the trenches are filled with SiO_2 and finally a plane surface is created by mechanical treatment. Now, the actual fabrication of the active devices can start.

Figure 14.5 shows a cross section of a pseudo-vertical n-channel MOSFET (see Chap. 9 for reference) in SOI technology for blocking voltages of 600 V, suitable for 230 V grid applications. The deep n^+ layer serves as the drain contact, which is routed via the vertical n^+ regions of the trench sidewalls to the surface contact on the right edge of the cell. In the adjacent dielectrically insulated cells, other independent power devices or arbitrary logic circuit structures (CMOS, bipolar elements) can be placed.

The production effort for SOI substrates is comparatively high. However, this technology effectively prevents any interaction between different elements of the circuit even at high current and voltage levels and under elevated temperatures. The cross talk between elements of the circuit – another limiting factor for an increasing integration density – is also reduced by the dielectric insulation. The SOI technology allows a high packing factor and exhibits a better exploitation of the wafer area due to the smaller insulation regions. SOI devices are immune against latch-up problems generated by parasitic pnpn-structures and the integrated devices remain functional up to operation temperatures of 200°C.

Monolithic integration has made tremendous progress during the last years with respect to packing density, blocking capability, and temperature stability. However, the conflicting requirements of high voltage and temperature stability and the immunity against cross talk effects on one the hand and the demand for increasing packing factors on the other are a challenge for further progress. Today, systems with a blocking capability up to 1200 V and current up to 10 A are integrated in smart power ICs. For higher voltages and currents, hybrid or discrete solutions are preferred.

14.3 System Integration on Printed Circuit Board

Discrete passive components in their conventional individual package consume a considerable part of the volume of grid-connected power electronic systems. Their main function is the preservation of the high quality of power grids. The progress of power electronic devices enables the transition to higher switching frequencies and therefore reduces the necessary capacitance and inductance values. This trend facilitates the integration of these passive components on printed circuit boards (PCB), which are a common platform for power electronic systems [Waf05]. The 'embedded passive integrated circuit' (emPIC) technology allows compact system designs and a high power density per unit volume [Pop05].

Layers with specific properties are required to design such a system, as can be seen in Fig. 14.6 [Waf05]. The application of printed electrical resistors on so-called prepregs (short form for pre-impregnated fibers, a semi-finished part in the PCB production process) is a state-of-the-art industrial process, although the alignment

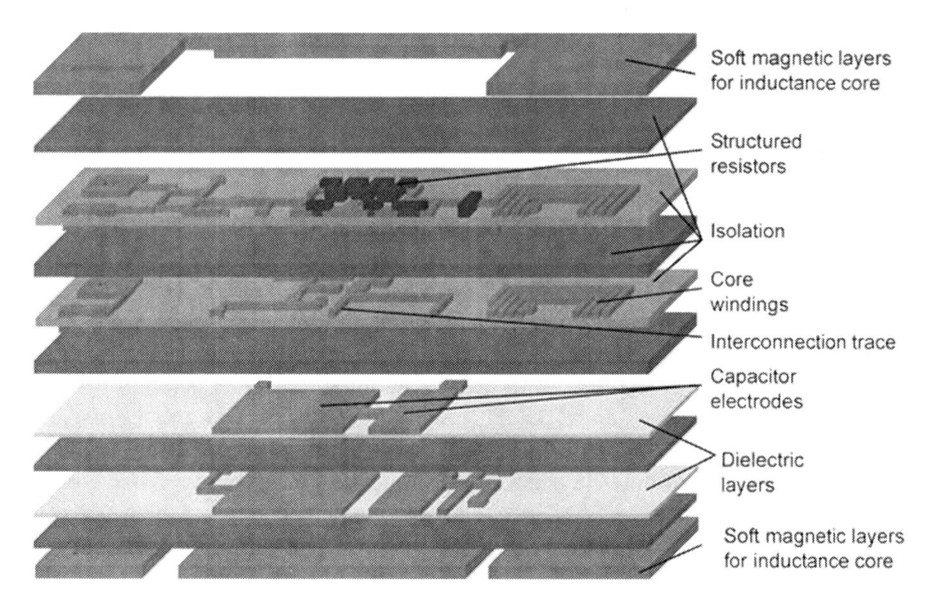

Fig. 14.6 Explosion view of the layers in a high- integration emPIC PCB

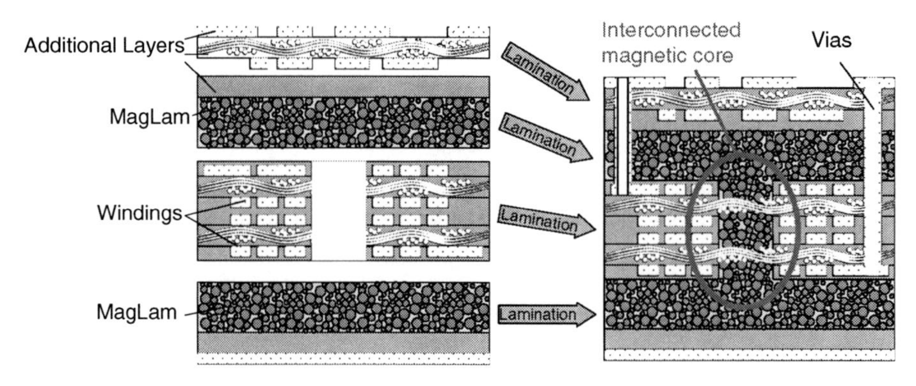

Fig. 14.7 Creation of a closed magnetic core with the ferrite polymer compound 'MagLam,' which is compatible with the PCB lamination process

tolerances requested by the emPIC technique is a requirement that is not easy to fulfill. The challenges for the research are to develop suitable layers with high dielectric constants and layers with high magnetic permeability.

Special PCB prepregs loaded with particles of a high dielectric material can be used to form capacitors. A commercially available layer material named 'C-Lam' features a relative dielectric constant of $\varepsilon_r = 12$. Applied as a dielectric layer with 40 μm thickness between two layers of copper, the so formed component has a capacity of 0.26 nF/cm^2 with a dielectric loss factor of 0.02 and a frequency response up to 1 GHz.

Embedding ferrite particles in a polymer matrix allows the fabrication of magnetic layers. MagLam is a brand name for a commercially available layer material, which is compatible with the PCB production process based on prepregs. MagLam exhibits a relative permeability of $\mu_r = 17$; the saturation flux density is 300 mT and the frequency application range exceeds 10 MHz. Figure 14.7 illustrates that an integrated transformer with a closed magnetic core can be created by appropriate structuring of the layers. During the lamination process, in which the layers are pressed together under elevated temperatures, material from the magnetic layers is pressed in the vertical duct and forms the closed magnetic core.

Another possible solution for the implementation of magnetic layers is 'μ-metal', which can be integrated as thin foils of 50 μm thickness. A relative permeability μ_r of more than 10,000 can thus be achieved. However, this material is electrically conductive and is therefore less suitable for high-frequency applications. The μ-metal foils can be laminated and structured as conventional copper layers and applied to thin flexible substrates as 'flexfoil' (polyimide), elastic coils can be created [Waf05b].

An example of a complete system based on the emPIC technology is depicted in Fig. 14.8. It shows an AC/DC converter in resonant topology, which delivers an output power of 60 W from a 230 V supply grid with an efficiency of up to 82% [Waf05]. The power MOSFETs and the driver ICs are assembled in discrete packages on the surface. Most of the other passive components have been integrated in the printed circuit board. The footprint of the systems has the size of a credit card.

Fig. 14.8 Slim line 60 W AC/DC converter for 230 V supply voltage – transformers and the majority of the capacitors are integrated in the PCB

Design and optimization of passive integration require the consideration of all interactions between different layers and elements to prevent undesirable effects. A three-dimensional simulation of the electromagnetic fields inside the PCB by solving the Maxwell equations is a powerful tool to avoid these problems. Software tools for this task, which were already discussed in Sect. 13.3, become more and more available. The analytical approach is nonetheless essential for a fast conceptual design especially for integrated transformers [Waf05].

The integration of passive components represents a tremendous progress in power electronic system design. The PCB, formerly only used as assembly platform and wiring element, advances to a functional component of the system. The traditional handicaps of the PCB traces like parasitic inductance and parasitic capacity are transformed into functional elements. The number of solder joints is dramatically reduced and the compact design with less externally assembled components makes the system less sensitive to mechanical vibration and shock. Those factors increase the system reliability. The passive integration has a high potential for system improvement in the future.

14.4 Hybrid Integration

Integration of power electronic systems constitutes a particular challenge, because the extraction and dissipation of the heat, generated in components during operation by power losses, implies narrow constrains to miniaturization. While microelectronics has achieved a tremendous progress over a period of several decades – continuously doubling the number of transistors on a single chip according to 'Moore's law' every 2 years – by confining the possible elements to a small set of standard elements on the one hand and in scaling down these standard elements to smaller and smaller sizes on the other hand, power electronics can adapt this strategy only rudimentarily.

Three major paradigm shifts were responsible for the revolutionary progress in information technology. The first step was the reduction of information to binary elements – a sequence of 0 and 1 – and thus the standardization of data. The second

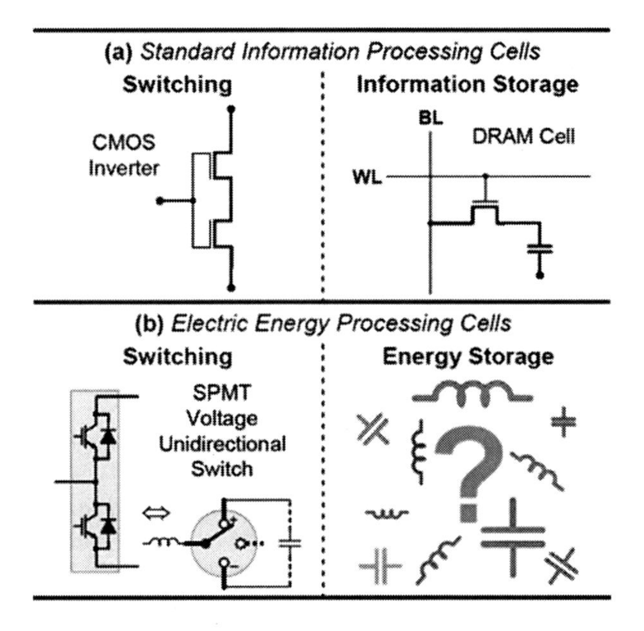

Fig. 14.9 Standardized modularization in (**a**) information technology and (**b**) power electronics according to [Bor05]

step was the introduction of the CMOS technology, consisting of CMOS inverter structures and DRAM storage cells as illustrated in Fig. 14.9a [Bor05]. Every microelectronic system was constructed by combining these basic elements to complex circuits and the progress was focused on the miniaturization of these elements. The third step was the 'very large scale integration' (VLSI) technology.

Can similar paradigm shifts in power electronics lead to a progress in integration comparable to the enormous success in microelectronics? The equivalence to the digital concept in information technology is the standard 'pulse width modulation' (PWM) technique in power electronics. The energy flow on the input side is chopped in single elements and is combined to the desired energy flow on the output side, which could be either a DC current with controlled amplitude in a DC regulator or else a sinusoidal current and voltage characteristic of selectable frequency in an AC inverter. There is also equivalence for the second step: The basic topology of two switching devices and two anti-parallel freewheeling diodes in half-bridge or phase-leg configuration as shown in Fig. 14.9b is a standard topology in power electronics. This configuration is used in the vast majority of power electronic applications.

The unsolved problem is step three in this analogy: the equivalence for a standardized storage cell like the DRAM in information technology, which would be a standard energy storage cell in power electronics (Fig. 14.9, right). A variety of single components in various technologies and package outlines are available without a noticeable trend for standardization. The final step in system assembly, the wiring of power devices and passive components, is elaborate and the passive components are to a great extend responsible for volume and weight of a system. A continuous progress by increasing integration density as in information technology is still not conceivable for power electronics.

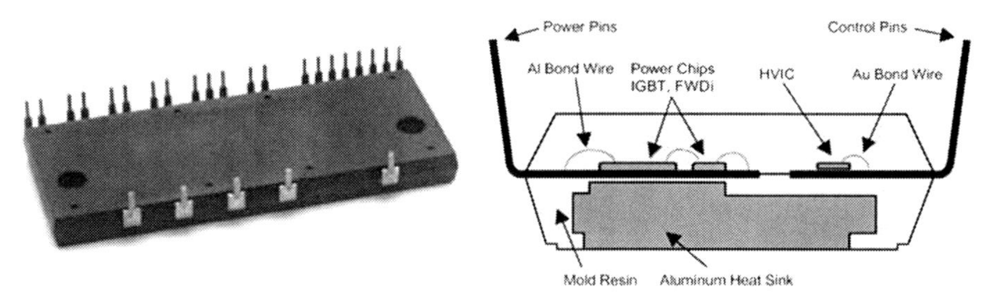

Fig. 14.10 Photographic image (*left*) and cross section (*right*) of a Mitsubishi DIP-IPM – the implemented heat sink structure increases the heat extraction [Mot99]

State-of-the-art power electronic devices can be operated today at high switching frequencies, especially power MOSFETs. This facilitates the reduction of the capacitive and inductive components. Materials with a high relative dielectric constant ε_r or a high relative permeability μ_r are available as discussed in Sect. 14.3. However, additional boundary conditions have to be accounted for besides the electrical requirements. The heat generated by power losses in passive components must be efficiently dissipated. Additional heat produced by the power devices in proximity can significantly increase the operation temperature of the passive component, so that high operation temperatures are mandatory. Finally, the stress provoked by differences in coefficient of thermal expansion in combination with high operation temperature is a challenge for the reliability of the component. These additional requirements impede the progress in integration.

While the integration of passive components is still in a state of infancy, manufacturers of power module have developed a different strategy: the integration of driver circuits into power modules. These 'intelligent power modules' (IPM) facilitate the system design and have reached a certain standard in the small power range.

As was already discussed in Sect. 11.2, the transfer mold technology based on copper lead frames is an ideal platform to integrate drivers in the low power range (see Fig. 11.11). With increasing power losses, the heat transfer through the contact leads is not sufficient. Integrated cooling structures enhance the heat dissipation of the package. Figure 14.10 illustrates this improvement for a DIP-IPM power module from Mitsubishi. This package is suitable for an output power of 1.5 kW with phase currents of 20 A maximum.

Higher output power levels require even more heat extraction capability. Implementing ceramic substrates allows designing IPM modules for output power of up to 22 kW. The example in Fig. 14.11 shows a power module without base plate, comprising a three-phase input converter, a three-phase inverter, a brake chopper, and a seven-channel SOI driver (refer to Fig. 14.5). The challenge in this design is the production of the small current tracks on the DBC substrate for the multiple SOI contacts. However, the excellent thermal contact of the SOI chip to the heat sink enables the device to dissipate more power losses, so that the output stages can produce higher gate currents and therefore are able to drive even larger chips.

Advanced level shifters are integrated in the SOI chip for the TOP and BOT gate drivers, which compensate shifts in the reference potential of both polarities. This

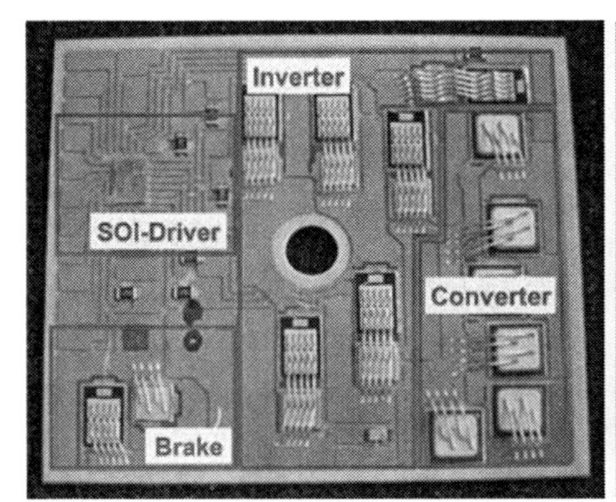

Fig. 14.11 IPM module without base plate from Semikron – an input converter, an inverter, a brake chopper, and a SOI driver assembled on a single substrate [Grs08]

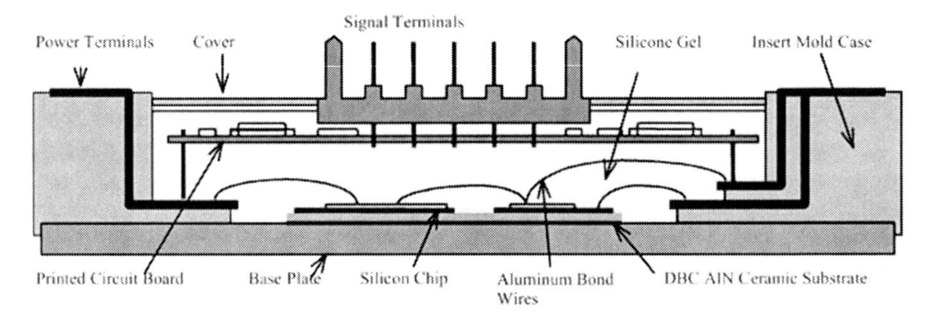

Fig. 14.12 Mitsubishi-IPM for engine ratings up to 30 kW [Mot93]

feature makes the driver immune to static and dynamic reference potential changes up to ±20 V.

Another approach for the IPM design is the integration of a common PCB driver into the package of a classical base plate module as shown in Fig. 14.12. The PCB is connected to the power circuit by soldered posts, which position the PCB below the top cover of the module housing.

These IPMs are typically equipped with a temperature sensor. The driver provides short-circuit detection and several additional protection functions for safe operation.

The next step in increasing complexity implements a micro-controller together with the control algorithms for pulse width modulation in a unique package. Thus, a 'motor control unit' evolves containing control circuit board including a computer chip with the appropriate control software (Fig. 14.13) [Ara05]. The application engineer receives a complete system as a black box, which can be adapted to the application demands simply by transmitting suitable control sequences.

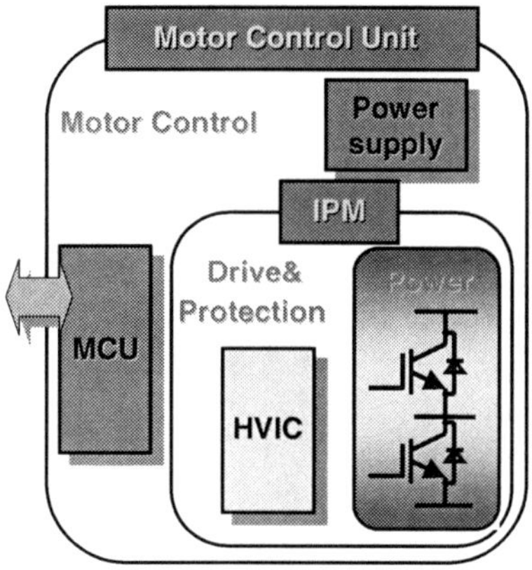

Fig. 14.13 Extension of an IPM to a complete motor control unit by adding a micro-controller and the power supply [Ara05]

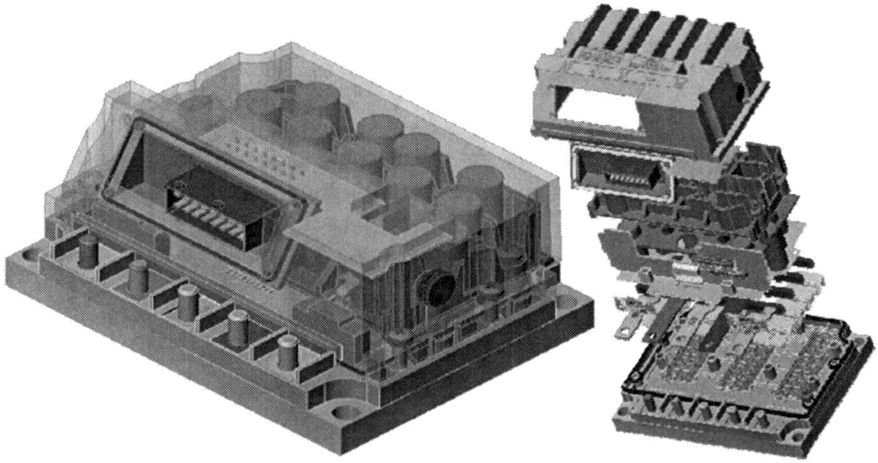

Fig. 14.14 Integrated power electronic system in one compact package with current and temperature sensors, DC-link capacitors, driver, and micro-controller – the power rating is 13 kW

The integration of the DC-link capacitors upgrades the package to a complete power electronic system, as shown in Fig. 14.14. The example shows a single housing equipped with a three-phase MOSFET inverter, current sensors, DC-link bus bars and DC-link capacitors, driver board, and a micro-controller. The sealed package design makes this compact architecture ideally suited for the application in industrial service vehicles.

While these concepts of integration aim to design a high-volume series product applicable to a multitude of different applications, the SKiiP platform introduced by Semikron focuses on a flexible design, which can be easily adapted to different requirements [Scn02]. Figure 14.15 exemplifies this concept with a module from the SKiiP family, comprising a three-phase inverter circuit with compensated current sensors for each output terminal. The load and control contacts between the DBC

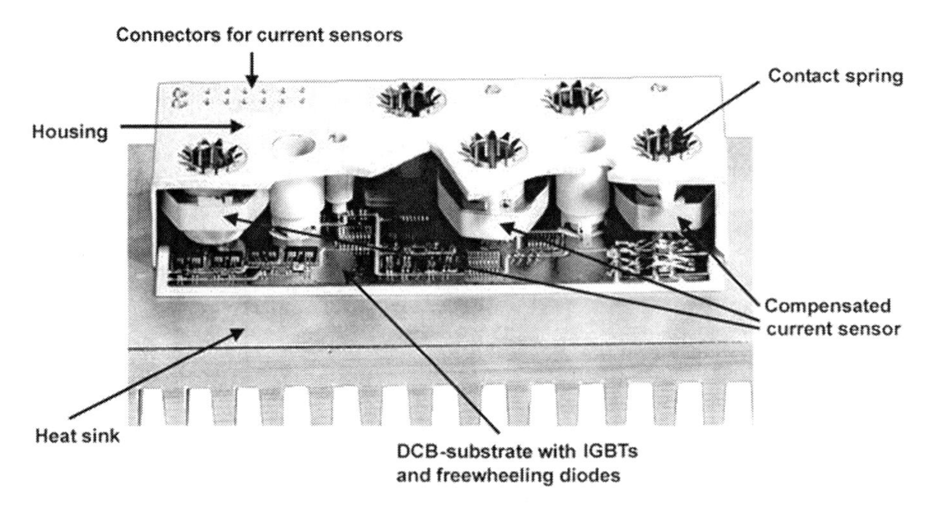

Fig. 14.15 Semikron SKiiP system with integrated compensated current sensors

substrate and the PCB (not shown) are accomplished by spring contacts. The control PCB and the DC-link are designed by application engineers and can be adapted to the specific requirements of the application. The module design can easily be customized by different DBC layouts and by implementation of suitable power devices. This flexibility allows for a commercially successful module production for moderate quantities.

Most of the above presented examples of hybrid integration have successfully penetrated the market. They simplify the system design for application engineers by solving the problems of supplying and controlling the load current and the extraction of heat. Nonetheless, essential aspects of integration remain unsolved:

- The package outline of passive components aggravates their efficient integration. Progressive foil capacitors and transformers integrated in PCBs are capable of opening new opportunities in the future.
- The requirement for an excellent thermal contact restricts the mounting plane of the power devices to the two-dimensional surface of the DBC substrate. If the load current is routed only by current tracks on the substrate, it is difficult to provide short current paths, which supply paralleled chips with identical path lengths in order to minimize parasitic effects and to prevent unbalance phenomena between parallel chips. Multi-contact internal bus bars in pressure contact systems could help to overcome this limitation in future architectures [Scn09].

A promising strategy is the exploitation of parasitic effects to implement new features by integration. An excellent example of this strategy is illustrated in Fig. 14.16, which shows a bus bar structure with an integrated high-frequency filter [Zha04].

For a power converter, bus bars are required to conduct the load current between the converter and the energy source. As state-of-the-art, the DC supply is

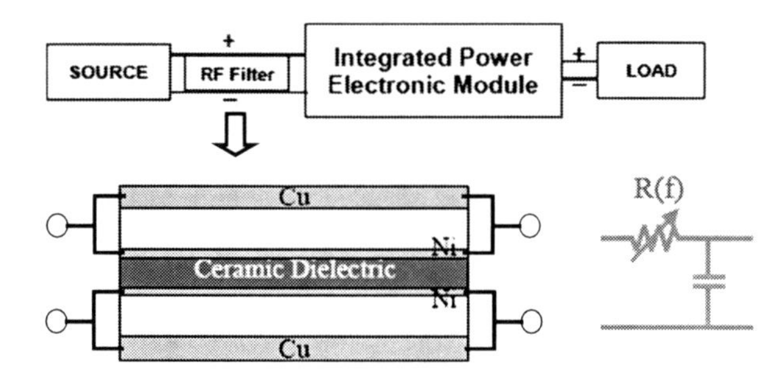

Fig. 14.16 Bus bar structure with integrated high-frequency filter [Zha04]

implemented by conductive sheets or bars, which are separated by an intermediate insulation layer, thus forming a parasitic capacity adjacent to the switching devices.

The structure in Fig. 14.16 implements a $BaTiO_3$ dielectric insulation layer between the +DC and –DC bus bars. The bus bars in this example are formed by a thin Ni layer on one side and a thicker Cu layer on the other side, separated by an Al_2O_3 ceramic insulator. They are assembled with the Ni layers toward the $BaTiO_3$ layer and the Cu layer to the outside to form a DC bus bar structure. Low-frequency currents are conducted primarily by the copper layer due to its lower electrical resistance. High-frequency currents on the other hand are squeezed into the Ni layer by the proximity effect. Therefore, the high-frequency content is damped stronger and the structure forms a low pass filter. The equivalent circuit of this bus bar structure is a capacitor with a frequency-dependent resistor, which is increasing with the frequency. This example demonstrates that a skilful arrangement of different elements can make use of physical effects – which are usually considered as handicap – to integrate a low-pass filter into a bus bar structure with little additional effort.

Combining multiple functions into single functional element is identified as synergy. Synergy effects represent a non-linear progress in the process of integration. They cannot be scheduled in a project plan or a road map, but when they are discovered, they generate an unexpected drive for further progress. Possible areas to search for synergy effects are as follows:

- *Synergy effects on the device level*: Electrical functions of different devices are integrated in a single chip. An example is the reverse conducting IGBT, which merges the traditional IGBT and the freewheeling diode into one single chip.
- *Synergy effects on the package level*: Different functions are merged into one functional element. An example is the integration of passive components into the PCB.

A survey of the existing opportunities in integration for a 12 V / 42 V DC–DC converter is presented in [Pop04]. Requirements with respect to cooling performance, electrical function and electro-magnetic interference are taken into account. This survey proposes a design in which all elements are assembled on a single thermally conductive rail, which allows to reduce the number of components from

38 to merely 19. This would diminish the required system volume to approximately one-eighth. The survey constitutes an indication of the conceivable miniaturization of power electronic systems by integration.

While the integration of driver electronics and sensor functions has reached a status of maturity in the successful concept of the IPM, the integration of passive components remains in a state of infancy. Details of research results in this area are currently being implemented in new products, but they remain incoherent. The thermal limit of power electronic systems impedes a fast progress as accomplished in the area of microelectronics.

The progress in increasing power density and volume reduction of power electronic systems will primarily be depending on the successful development of power devices with reduced on-state and switching losses. There is considerable potential for further progress in this direction, as has been indicated in many respects in this compendium. Improved devices will operate at higher switching frequencies, which will diminish the size of passive components. This will presumably be the major driving factor for future progress. It will continuously facilitate the boundary conditions for system integration, an indispensable requirement for a reduction of system cost and thus a prerequisite for realizing the potential of increasing energy efficiency provided by advanced power electronic systems.

References

[And96] Andreini A, Contiero C, Glabiati P: "BCD Technologies for Smart Power ICs". In: Murati B, Bertotti F, Vignola GA (eds) Smart Power ICs. Springer Berlin (1996)

[Ara05] Araki T: "Integration of Power Devices - Next Tasks", Proceedings of the EPE, Dresden, (2005)

[Bor05] Boroyevich D, van Wyk JD, Lee FC, Liang Z: "A View at the Future of Integration in Power Electronics Systems", Proceedings of the PCIM, pp. 11–20, Nürnberg, (2005)

[Gie02] Giebel T: Grundlagen der CMOS-Technologie, Teubner Verlag Stuttgart 2002

[Grs08] Grasshoff T, Reusser L: "Integration of a new SOI driver into a medium power IGBT module package", Proceedings of the PCIM Europe, Nuremberg (2008)

[Ler02] Lerner R, Eckoldt U, Knopke J: "High Voltage Smart Power Technology with Dielectric Insulation", Proc. of the 2nd International Conference on Integrated Power Systems (CIPS), pp. 83–88 (2002)

[Mot93] Motto ER: "New Intelligent Power Modules (IPMs) for Motor Drive Applications", Proc. IEEE IAS, Toronto (1993)

[Mot99] Motto ER, Donlon JF, Iwamoto H: "New Power Stage Building Blocks for Small Motor Drives", Proc. Powersystems World Conference '99, pp. 343–349, Chicago, (1999)

[Pop04] Popović J, Ferreira JA: "Concepts for High Packaging and Integration Efficiency", Proc. 35th Annual IEEE Power Electronics Specialists Conference PESC, pp. 4188–4194, Aachen (2004)

[Pop05] Popović J, Ferreira JA, Waffenschmidt E: "PCB Embedded DC/DC 42/14 V Converter for Automotive Applications", Proceedings of the EPE, Dresden (2005)

[Pri96] Pribyl W: "Integrated Smart Power Circuits Technology, Design and Application", Proceedings of the 22nd European Solid-State Circuits Conference, ESSCIRC (1996)

[Scn02] Scheuermann U, Tursky W: "IPMs zwischen Modul und intelligenten leistungselektronischen Antriebssystemen", Proc. Fachtagung Elektrische Energiewandlungssysteme, pp. 105–110, Magdeburg (2002)

[Sch09] Scheuermann U: "Power Module Design without Solder Interfaces – an Ideal Solution for Hybrid Vehicle Traction Applications", Proc. APEC 2009, pp. 472–478, Washington D.C.

[Tih88b] Tihanyi J: "Smart SIPMOS Technology", in: Siemens Forschungs- und Entwicklungsberichte Bd.17 Nr.1, Springer Berlin, pp. 35–42 (1988)

[Waf05] Waffenschmidt E, Ackermann B, Ferreira JA: "Design method and material technologies for passives in printed circuit board embedded circuits", Special Issue on Integrated Power Electronics of the IEEE PELS Transactions, Vol.20, No.3, pp. 576ff (2005)

[Waf05b] Waffenschmidt E, Ackermann B, Wille M: "Integrated ultra thin flexible inductors for low power converters", Proceedings of the Power Electronic Specialists Conference (PESC) 2005, Recife, Brazil (2005)

[Zha04] Zhao L, van Wyk JD: "A High Attenuation Integrated Differential Mode RF EMI Filter", Proc. 2004 CPES Power Electronics Seminar, pp. 74–77, Blacksburg (2004)

Appendix A
Modeling Parameters of Carrier Mobilities in Si and 4H-SiC

A.1 Mobilities in Silicon

They can be well described by the Caughey–Thomas formula [Cau67]:

$$\mu = \mu_\infty + \frac{\mu_0 - \mu_\infty}{1 + (N/N_{\mathrm{ref}})^\gamma}$$

see Fig. 2.12. The parameters μ_0, μ_∞, and N_{ref} at 300 K used for Fig. 2.12 have been determined by fitting the formula to the experimental carrier dependence at 300 K [Thu80, Mas83, Thu80b]. For the temperature dependence of parameters, the concentration dependence at temperatures between 250 and 450 K and the temperature dependence of resistivity at various doping densities have been used [Li77, Li78, Swi87] taking into account the incomplete ionization around 10^{18} cm^{-3}. The experimental results can be described well using the following temperature-dependent parameters:

Electrons:

$$\mu_0 = 1412 \cdot \left(\frac{300}{T}\right)^{2.28} \mathrm{cm^2/Vs}, \quad \mu_\infty = 66 \cdot \left(\frac{300}{T}\right)^{0.90} \mathrm{cm^2/Vs} \tag{A.1}$$

$$N_{\mathrm{ref}} = 9.7 \times 10^{16} \left(\frac{T}{300}\right)^{3.51} \mathrm{cm^{-3}} \quad \gamma = 0.725 \cdot \left(\frac{300}{T}\right)^{0.270} \tag{A.2}$$

Holes:

$$\mu_0 = 469 \cdot \left(\frac{300}{T}\right)^{2.10} \mathrm{cm^2/V\,s}, \quad \mu_\infty = 44 \cdot \left(\frac{300}{T}\right)^{0.80} \mathrm{cm^2/V\,s} \tag{A.3}$$

$$N_{\mathrm{ref}} = 2.4 \times 10^{17} \cdot \left(\frac{T}{300}\right)^{4.13} \mathrm{cm^{-3}} \quad \gamma = 0.70 \cdot \left(\frac{T}{300}\right)^{0.00} \tag{A.4}$$

The parameters at 300 K differ considerably from original values in [Cau67] because the early measurements [Irv62] have been noticeably corrected later.

J. Lutz et al., *Semiconductor Power Devices*, DOI 10.1007/978-3-642-11125-9,

A useful formula for the *field and concentration dependence of mobilities* in silicon has been proposed by Scharfetter and Gummel [Sch69], which reads as

$$\mu = \frac{\mu^{(0)}}{\left\{1 + \frac{N}{N/S + N_r} + \frac{(E/A)^2}{E/A + F} + \left(\frac{E}{B}\right)^2\right\}^{1/2}} \quad \text{(A1)}$$

N is again the concentration of donors or acceptors, $\mu^{(0)}$ are the respective mobilities from Eqs. (A.1), (A.2), and (A.3) for small $N < 1 \times 10^{14}$ cm^{-3}. The fitting parameters A, B, F, N_r, and S for electrons and holes are

	N_r	S	A	F	B
Electrons	3×10^{16}	350	3.5×10^{8}	8.8	7.4×10^{3}
Holes	4×10^{16}	81	6.1×10^{8}	1.6	2.5×10^{4}

A.2 Mobilities in 4H-SiC

The mobilities in 4H-SiC are weakly anisotropic as noted in Section 2.1, but in modeling the mobilities the anisotropy is mostly neglected. According to [Scr94] the dependence on the *total doping density N and temperature* can be described by

$$\mu = \mu_\infty + \frac{\mu_0 \cdot (T/300)^\alpha - \mu_\infty}{1 + (N/N_{ref})^\gamma}$$

with the following parameter values:

Electrons:

$$\mu_0 = 947\,\text{cm}^2/\text{Vs} \qquad \mu_\infty = 0,$$
$$N_{ref} = 1.94 \times 10^{17}\ \text{cm}^{-3}$$
$$\alpha = -2.15, \qquad \gamma = 0.61$$

Holes:

$$\mu_0 = 124\,\text{cm}^2/\text{Vs}, \qquad \mu_\infty = 15.9\,\text{cm}^2/\text{Vs}$$
$$N_{ref} = 1.76 \times 10^{19}\,\text{cm}^{-3}$$
$$\alpha = -2.15, \qquad \gamma = 0.34$$

Appendix B
Avalanche Multiplication Factors and Effective Ionization Rate

B.1 Multiplication Factors

The following derivation is based on a paper of McIntyre [McI66]. We consider a primary electron–hole pair at the point x in the depletion layer of a reverse-biased pn-junction and ask for the multiplication factor $M(x)$, which is the total number of pairs generated in the avalanche process initiated by the primary pair including the ionization by secondary carriers and so on. In the diode orientation of Fig. 3.14 the electrons will be swept to the left (the neutral n-region) and holes to the right (p-region). In traveling a path of length $\mathrm{d}x$ the probability that the electron will generate an electron–hole pair is $\alpha_\mathrm{n}\mathrm{d}x$. Similarly the hole will generate on average $\alpha_\mathrm{p} \cdot \mathrm{d}x$ pairs on a path of length $\mathrm{d}x$. Each of these secondary pairs generated at a point x' will itself generate carrier pairs on their path and experience a multiplication by a factor $M(x')$. Since this adds up to the multiplication of the primary pair, one has

$$M(x) = 1 + \int_0^x \alpha_\mathrm{n} M(x')\,\mathrm{d}x' + \int_x^w \alpha_p M(x')\,\mathrm{d}x' \tag{B.1}$$

where α_n and α_p depend over the field strength E on x'. Differentiating one obtains

$$\frac{\mathrm{d}M}{\mathrm{d}x} = (\alpha_\mathrm{n} - \alpha_\mathrm{p}) M(x) \tag{B.2}$$

This has the solution

$$M(x) = M(0)\,\exp\left(\int_0^x (\alpha_\mathrm{n} - \alpha_\mathrm{p})\,\mathrm{d}x'\right) \tag{B.3}$$

$$= M(w)\,\exp\left(-\int_x^w (\alpha_\mathrm{n} - \alpha_\mathrm{p})\,\mathrm{d}x'\right) \tag{B.4}$$

If (B.3) is inserted into the right-hand side of (B.1), one obtains for $x = 0$

$$M(0) = M_{\text{p}} = \frac{1}{1 - \int\limits_0^w \alpha_{\text{p}} \exp\left(\int\limits_0^x (\alpha_{\text{n}} - \alpha_{\text{p}})\,\mathrm{d}x'\right) \mathrm{d}x} \tag{B.5}$$

Since this is the multiplication factor of the primary carriers at the boundary $x = 0$ of the space charge layer, where holes from the neutral n-region enter the depletion layer, $M(0)$ is identical with the multiplication factor M_{p} of the saturation hole current density j_{ps}. Similarly, substituting (B.4) into the right-hand side of (B.1) one obtains for $x = w$

$$M(w) = M_{\text{n}} = \frac{1}{1 - \int\limits_0^w \alpha_{\text{n}} \exp\left(-\int\limits_x^w (\alpha_{\text{n}} - \alpha_{\text{p}})\,\mathrm{d}x'\right) \mathrm{d}x} \tag{B.6}$$

This is the multiplication factor for the electrons entering the depletion layer at $x = w$ from the neutral p-region. For arbitrary x it follows from (B.3) and (B.5) that

$$M(x) = \frac{\exp\left(\int\limits_0^x (\alpha_{\text{n}} - \alpha_{\text{p}})\,\mathrm{d}x'\right)}{1 - \int\limits_0^w \alpha_{\text{p}} \exp\left(\int\limits_0^{x'} (\alpha_{\text{n}} - \alpha_{\text{p}})\,\mathrm{d}x''\right) \mathrm{d}x'} \tag{B.7}$$

and a mathematically identical expression follows from (B.4) and (B.6). Assuming a homogeneous thermal generation G in the depletion layer, the mean value of $M(\mathrm{x})$ is the multiplication factor for the current j_{sc} generated in the space charge region:

$$\overline{M} = M_{\text{sc}} = \frac{\frac{1}{w}\int\limits_0^w \exp\left(\int\limits_0^x (\alpha_{\text{n}} - \alpha_{\text{p}})\mathrm{d}x'\right) \mathrm{d}x}{1 - \int\limits_0^w \alpha_{\text{p}} \exp\left(\int\limits_0^x (\alpha_{\text{n}} - \alpha_{\text{p}})\,\mathrm{d}x'\right) \mathrm{d}x} \tag{B.8}$$

Since according to (B.3) and (B.4)

$$\frac{M_{\text{n}}}{M_{\text{p}}} = \exp\left(\int\limits_0^w (\alpha_{\text{n}} - \alpha_{\text{p}})\,\mathrm{d}x\right) \tag{B.9}$$

it follows from (B.5) and (B.9) that M_{n}, M_{p}, and M_{sc} tend to infinity at the same field distribution and voltage. The breakdown condition can be written using the ionization integral in the denominator of (B.5)

$$I_{\mathrm{p}} \equiv \int_0^w \alpha_{\mathrm{p}} \exp\left(\int_0^x (\alpha_{\mathrm{n}} - \alpha_{\mathrm{p}})\,\mathrm{d}x'\right)\,\mathrm{d}x = 1 \tag{B.10}$$

If $\alpha_{\mathrm{n}} > \alpha_{\mathrm{p}}$, the multiplication factors obey the inequality $M_{\mathrm{n}} > M_{\mathrm{sc}} > M_{\mathrm{p}}$ for voltages below the breakdown voltage.

B.2 Effective Ionization Rate and Breakdown Condition

For a field-independent ratio $\alpha_{\mathrm{n}}/\alpha_{\mathrm{p}} = \gamma$, one obtains from the breakdown condition (B.10)

$$\begin{aligned} 1 &= \int_0^w \alpha_{\mathrm{p}} \exp\left(\int_0^x (\gamma - 1)\alpha_{\mathrm{p}}\,\mathrm{d}x'\right)\mathrm{d}x \\ &= \frac{1}{\gamma - 1}\left(\exp\left(\int_0^w (\gamma - 1)\alpha_{\mathrm{p}}\mathrm{d}x\right) - 1\right) \end{aligned} \tag{B.11}$$

since $\int_0^x f'(x')\exp\left(f(x')\right)\mathrm{d}x' = \left[\exp(f(x))\right]_0^x$. From (B.11) one has $\gamma = \exp\left[\int_0^w (\gamma - 1)\,\alpha_{\mathrm{p}}\,\mathrm{d}x\right]$ which can be written as

$$\int_0^w \frac{\alpha_{\mathrm{n}} - \alpha_{\mathrm{p}}}{\ln(\alpha_{\mathrm{n}}/\alpha_{\mathrm{p}})}\,\mathrm{d}x = \int_0^w \alpha_{\mathrm{eff}}\,\mathrm{d}x = 1 \tag{B.12}$$

where the effective ionization rate is defined as

$$\alpha_{\mathrm{eff}} = \frac{\alpha_{\mathrm{n}} - \alpha_{\mathrm{p}}}{\ln\,(\alpha_{\mathrm{n}}/\alpha_{\mathrm{p}})} \tag{B.13}$$

Hence this α_{eff} can be used together with the condition (B.12) to calculate the breakdown voltage of pn-junctions. In the relative small upper field range which contributes significantly to the integral, the preposition of constant $\alpha_{\mathrm{n}}/\alpha_{\mathrm{p}}$ is in most cases sufficiently fulfilled.

References for Appendices A and B

[Cau67] Caughey DM, Thomas RE, "Carrier Mobilities in Silicon Empirically Related to Doping and Field" Proceedings IEEE 23, pp. 2192–93 (1967)

[Irv62] Irvin JC, "Resistivity of Bulk Silicon and of Diffused Layers in Silicon", Bell Syst. Tech. J 41, pp. 387–410 (1962)

[Li78] Li SS, "The Dopant Density and Temperature Dependence of hole Mobility and Resistivity in Boron Doped Silicon" Solid-St. Electron., Vol.21, pp. 1109–1117 (1978)

[Li77] Li SS, Thurber WR, "The Dopant Density and Temperature Dependence of Electron Mobility and Resistivity in n-Type Silicon", Solid-St. Electron., Vol. 20, 609–616 (1977)

[Mas83] Masetti G, Severi M, Solmi S: "Modeling of carrier mobility against concentration in Arsenic-, Phosphorus-, and Boron-doped Silicon," IEEE Trans. Electron Devices ED-30, pp. 764–769 (1983)

[MCI66] McIntyre RJ, "Multiplication noise in uniform avalanche diodes", IEEE Trans. Electron Devices ED-13, pp. 164–168 (1966)

[Sch69] Scharfetter DL, Gummel HK "Large-signal analysis of a silicon Read Diode oscillator", IEEE Trans. Electron Devices ED-16, pp. 64–77 (1969)

[Sch87] Schaffer WJ, Negley GH, Irvine KG, Palmour JW, "Conductivity anisotropy in epitaxial 6H and 4H SiC" Materials Research Society Symposium Proceedings, Vol 339, pp. 595–600 (1994)

[Swi87] Swirhun SE: "Characterization of majority and minority carrier transport in heavily doped silicon", Ph.D. Dissertation, Stanford University, 1987

[Thu80] Thurber WR, Mattis RL, Liu YM, Filliben JJ: "Resistivity-Dopant Density Relationship for Phosporous-Doped Silicon" J. Electrochem. Soc., Vol 127, pp. 1807–1812 (1980)

[Thu80b] Thurber WR, Mattis RL, Liu YM, Filliben JJ: "Resistivity-Dopant Density Relationship for Boron-Doped Silicon", J. Electrochem. Soc., Vol 127, pp. 2291–2294 (1980)

Appendix C
Thermal Parameters of Important Materials in Packaging Technology

	Thermal conductivity (W mm^{-1} K^{-1})	Thermal capacity (J mm^{-3} K^{-1})	Thermal expansion (10^{-6}/K)	Source
Semiconductors				
Si	0.13	1.65×10^{-3}	2.6	[IoF01]
GaAs	0.055	1.86×10^{-3}	5.73	[IoF01]
SiC	0.37	2.33×10^{-3}	4.3	[IoF01]
Insulators				
SiO_2	0.0014	1.4×10^{-3}	0.55	[Sze81], Crystran
Al_2O_3	0.024	3.02×10^{-3}	6.8	Hoechst
AlN	0.17	2.44×10^{-3}	4.7	Hoechst
Si_3N_4	0.07	2.10×10^{-3}	2.7	Toshiba
BeO	0.251	2.98×10^{-3}	9	Brush-Wellman
Epoxyd	0.003		–	DENKA-TH1
Polyimide	3.85×10^{-4}		–	Kapton CR
Metals				
Al	0.237	2.43×10^{-3}	23.5	
Cu	0.394	3.45×10^{-3}	17.5	
Mo	0.138	2.55×10^{-3}	5.1	
Composite materials				
AlSiC	0.2	2.21×10^{-3}	7.5	
Solders				
Sn	0.063	1.65×10^{-3}	23	Demetron
SnAg(3.5)	0.083	1.67×10^{-3}	27.9	Demetron
SnPb(37)	0.07		24.3	Doduco 1/89
Interconnection layers				
Ag sinter layer	0.25	2.1×10^{-3}	18.9	[Thb06]
Thermal grease				
Wacker P 12	8.1×10^{-4}	2.24×10^{-3}	–	Wacker

Appendix D
Electric Parameters of Important Materials in Packaging Technology

	Electric resistivity (25°C) ($\mu\Omega$ cm)	Relative dielectric constant	Critical field strength (kV/cm)	Source
Semiconductors				
Si	*	11.7	150–300	
GaAs	*	12.9	400	
SiC	*	9.66	2000	
Insulators				
SiO_2	10^{20}–10^{22}	3.9	4000–10000	[Sze81]
Al_2O_3	10^{18}	9.8	150	Hoechst
AlN	10^{20}	9.0	200	Hoechst
Si_3N_4	10^{19}	8	150	Kyocera
BeO	10^{21}	6.7	100	Brush-Wellman
Epoxyd		7.1	600	DENKA-TH1
Polyimide		3.9	2910	Kapton CR
Metals				
Al	2.67	–	–	
Cu	1.69	–	–	
Mo	5.7	–	–	
Composite materials				
AlSiC	≈ 40	–	–	
Solders				
Sn	16.1	–	–	Demetron
SnAg(3.5)	13.3	–	–	Demetron
SnPb(37)	13.5	–	–	Doduco 1/89
Interconnection layers				
Ag-layer (NTV)	1.6	–	–	
Thermal grease				
Wacker P 12	5×10^{15}			Wacker

*doping dependent; – not defined

References for Appendices C and D

[IoF01] Ioffe Physical Technical Institute, St. Petersburg, Russia, 2001 http://www.ioffe.rssi.ru/SVA/NSM/Semicond/

[Thb06] Thoben M, Hong H, Hille F: "Hoch-zeitaufgelöste Zth-Messungen an IGBT-Modulen" Kolloquium Halbleiter-Leistungsbauelemente, Freiburg (2006)

[Sze81] Sze SM: Physics of Semiconductor Devices, John Wiley & Sons, NewYork (1981)

Appendix E
Often Used Symbols

Symbol	Unit	Meaning
A	cm^2	Area
B	cm^6/V	Fulop constant ($\approx 2.1 \times 10^{-35}$ in Si at 300 K)
$c_{n,p}$	cm^3/s	Capture coefficient of electrons/holes
$c_{An,p}$	cm^3/s	Auger capture coefficient of electrons/holes
C	As/V	Capacity
C_j	As/V	Junction capacity
D	cm^2/s	Diffusion constant
D_A	cm^2/s	Ambipolar diffusion constant
$D_{n,p}$	cm^2/s	Diffusion constant of electrons/holes
$e_{n,p}$	s^{-1}	Emission rate of electrons/holes
E	J, eV	Energy
E_c	eV	Lower edge of the conduction band
E_F	eV	Fermi level
E_g	V	Bandgap
E_V	eV	Upper edge of the valence band
E_{off}	J	Turn-off energy
E_{on}	J	Turn-on energy
E	V/cm	Electric field
E_c	V/cm	Electric field at avalanche breakdown
F	–	Statistic distribution function
$g_{n,p}$	$cm^{-3}s^{-1}$	Therm. generation rate of electrons/holes
$G_{n,p}$	$cm^{-3}s^{-1}$	Net generation rate of electrons/holes
G_{av}	$cm^{-3}s^{-1}$	Avalanche generation rate
$h_{n,p}$	cm^4/s	Emitter parameter of n/p emitter
i	A	$= I(t)$; current, time dependent
I	A	Current
I_C	A	Collector current
I_D	A	Drain current
I_E	A	Emitter current
I_F	A	Diode forward current
I_R	A	Current in blocking direction

I_{RRM}	A	Reverse recovery current maximum
j	A/cm^2	Current density
$j_{n,p}$	A/cm^2	Current density of electron/hole current
j_s	A/cm^2	Saturation current density
k	J/K	Boltzmann constant (1.38066×10^{-23})
L	H	Inductivity
L_{par}	H	Parasitic inductivity
L_A	cm	Ambipolar diffusion length
L_{DB}	cm	Debye length
$L_{n,p}$	cm	Diffusion length of electrons/holes
n,p	cm^{-3}	Density of free electrons/holes
n_0,p_0	cm^{-3}	Density in thermodynamic equilibrium
n^*,p^*	cm^{-3}	Density of minority carriers outside therm. equilibrium
n_i	cm^{-3}	Intrinsic carrier density
n_L,p_L	cm^{-3}	Density at the left edge of the flooded zone
n_R,p_R	cm^{-3}	Density at the right edge of the flooded zone
n_{av}, p_{av}	cm^{-3}	Density of electrons/holes generated by avalanche
N_A	cm^{-3}	Acceptor density
N_C	cm^{-3}	Density of states at the bottom of the conduction band
N_D	cm^{-3}	Donator density
N_{eff}	cm^{-3}	Effective density of positive charges
N_T	cm^{-3}	Density of deep centers
N_{T+}, N_{T-}	cm^{-3}	Density of positively/negatively charged deep centers
N_V	cm^{-3}	Density of states at the top of the valence band
q	As	Elementary charge (1.60218×10^{-19})
Q	As	Charge
Q_F	As	Charge carrying the forward current in a bip. device
Q_{RR}	As	Measured stored charge of a diode
$r_{n,p}$	$cm^{-3}s^{-1}$	Therm. recombination rates of electrons/holes
$R_{n,p}$	$cm^{-3}s^{-1}$	Net recombination rates of electrons/holes
R	Ohm	Resistor
R_{off}	Ohm	Gate resistor at turn-off
R_{on}	Ohm	Gate resistor at turn-on
R_{pr}	cm	Projected range
R_{th}	K/W	Thermal resistance
s	–	Soft-factor of a diode
S	cm^{-2}	Particles per area
t	s	Time
T	°C, K	Temperature
v	V	$= V(t)$; voltage, time dependent
V	V	Voltage
V_{bat}	V	Battery voltage/DC link voltage

V_B, V_{BD}	V	Avalanche breakdown voltage
V_C	V	Forward voltage of a transistor[1]
V_{drift}	V	Voltage drop across an n^--layer
V_{bi}	V	Built-in voltage of the pn-junction
V_F	V	Forward voltage (diode)
V_G	V	Gate voltage
V_{FRM}	V	Forward recovery voltage peak of a diode
V_M	V	Voltage peak
V_R	V	Voltage in blocking direction
V_s	V	Threshold voltage diode/thyristor/IGBT
V_T	V	Threshold voltage channel MOSFET, IGBT
$v_{n,p}$	cm/s	Velocity of electrons/holes
$v_{d(n,p)}$	cm/s	Drift velocity of electrons/holes
v_{sat}	cm/s	Saturation drift velocity at high electric field
w_B	cm	Width of the n^--layer
w, w_{SC}	cm	Width of the space charge
x	cm	Coordinate
x_j	cm	Depth of the pn-junction
α	–	Current gain in common-base circuit
α_T	–	Transport factor
$\alpha_{n,p}$	cm^{-1}	Ionization rates v of electrons/holes
α_{eff}	cm^{-1}	Effective ionization rate
β	–	Current gain in common-emitter circuit
γ	–	Emitter efficiency
ε_0	F/cm	Dielectric constant in vacuum (8.85418×10^{-14})
ε_r	–	Relative dielectric constant (Si: 11.9)
$\mu_{n,p}$	cm^2/V s	Mobility of free electrons/holes
ρ	As/cm^3	Space charge
σ	A/cm V	Electric conductivity
$\tau_{n,p}$	s	Carrier lifetime of electrons/holes
$\tau_{n0,p0}$	s	Minority carrier lifetime of electrons/holes
$\tau_{A,n}$, $\tau_{A,p}$	s	Auger carrier lifetime of electrons/holes
τ_{eff}	s	Effective carrier lifetime
τ_{HL}	s	Carrier lifetime at high injection
τ_{rel}	s	Relaxation time
τ_{sc}	s	Generation carrier lifetime
Φ	–	Ionization integral

Remark

[1]In data sheets of manufacturers usually instead of V_C the symbol V_{CE} (collector–emitter voltage), for V_G the acronym V_{GE} (IGBT) or V_{GS} (MOSFET) is used, for V_T the symbol $V_{GS(th)}$. Similar symbols are used for the current. The shorter symbols have been chosen in this work.

Index

U

V

W

Z

Breinigsville, PA USA
25 January 2011
254130BV00003B/18/P

9 783642 111242